Lecture Notes in Mathematics

A collection of informal reports and seminars
Edited by A. Dold, Heidelberg and B. Eckmann, Zürich

T0234377

292

W. D. Wallis
Anne Penfold Street
Jennifer Seberry Wallis

# Combinatorics:
## Room Squares, Sum-Free Sets, Hadamard Matrices

Springer-Verlag
Berlin · Heidelberg · New York 1972

**W. D. Wallis**
University of Newcastle, N. S. W., 2308/Australia

**Anne Penfold Street**
University of Queensland, St. Lucia, Queensland, 4067/Australia

**Jennifer Seberry Wallis**
University of Newcastle, N. S. W., 2308/Australia

AMS Subject Classifications (1970): 05-02, 05 B 05, 05 B 10, 05 B 15, 05 B 20, 05 B 30, 05 C 15, 05 C 20, 05 C 25, 05 C 99, 10 L 05, 10 L 10, 12 C 20, 15 A 21, 20 B 25, 22 D 99, 20 K 99, 62 K 10

ISBN 3-540-06035-9 Springer-Verlag Berlin · Heidelberg · New York
ISBN 0-387-06035-9 Springer-Verlag New York · Heidelberg · Berlin



Offsetdruck: Julius Beltz, Hemsbach/Bergstr.

# ACKNOWLEDGEMENTS

We wish to thank those authors who have sent us preprints of their results, notes on their unpublished work and whose correspondence has stimulated and improved our work.

We especially wish to thank the late Professor Leo Moser, Professor H.L. Abbott, Dr. E.G. Whitehead, Mr. K.R. Matthews, Dr. L.D. Baumert, Professor G. Szekeres and Professor A.L. Whiteman for their help and encouragement. We also wish to thank Dr. Sheila Oates Macdonald for reading and criticising much of Part 3 and Dr. Baumert for reading and improving some of Part 4. The remaining errors are our own.

We thank Academic Press, Inc., Professor J.G. Kalbfleisch and Professor R.G. Stanton for permission to use the results of chapter VIII of Part 3.

We pay tribute to Mrs. Karen Abraham for her splendid typing effort and finding and correcting many inconsistencies. We also thank Mrs. Roslyn Mills and Miss Anne Nicholls for coming to our rescue when Mrs. Abraham was ill.

# C O N T E N T S

PART 1. PRELIMINARIES . . . . . . . . . . . . 1

PART 2. ROOM SQUARES . . . . . . . . . . . . 29
W. D. Wallis

PART 3. SUM-FREE SETS . . . . . . . . . . . 123
Anne Penfold Street

PART 4. HADAMARD MATRICES . . . . . . . . . . 273
Jennifer Seberry Wallis

PART 5. AFTERMATH . . . . . . . . . . . . . . 491

# PART 1

## PRELIMINARIES

# CONTENTS

CHAPTER I. INTRODUCTION 5

CHAPTER II. BASIC DEFINITIONS 7

2.1 ARITHMETIC. Galois Fields, Quadratic Numbers, Legendre Symbol, Fermat Numbers, Cyclotomic Numbers 7

2.2 BALANCED INCOMPLETE BLOCK DESIGNS. BIBD, $(b,v,r,k,\lambda)$-configurations, SBIBD, $(v,k,\lambda)$-configurations 12

2.3 MATRICES. Incidence Matrices, Hadamard Matrices, Kronecker Products 13

2.4 DIFFERENCE SETS. Cyclic Difference Sets, Group Difference Sets 19

2.5 GRAPHS. Subgraphs, One-factorizations, Edge Colourings 21

2.6 GROUPS AND GENERALIZATIONS. Groupoid, Semigroup, Quasigroup, Loop, Latin Squares 23

2.7 PARTITIONS. $(k,A_i)$-subsets, Maximal Sum-free Sets, $\lambda(G)$, Schur Function, n-fold Regular 26

# CHAPTER I. INTRODUCTION

Marshall Hall (*Combinatorial Theory*, Blaisdell, Waltham, 1967, p.v) says that

> the central problem [of Combinatorial Theory] may be considered that of arranging objects according to specified rules and finding out in how many ways this may be done.

This is probably as close as we can come to a working definition. One then sees three basic areas:

(i) the theory of enumeration, wherein the "specified rules" are quite simple and it is required to find the number of essentially different ways in which the arrangement can be carried out;

(ii) discrete optimization such as linear programming; again the rules are simple but the problem is to find the "best" way of carrying them out according to some criterion;

(iii) the theory of designs, patterns and configurations, in which the rules are more demanding, where the basic problems are whether or not the objects can be arranged in a certain configuration and whether or not the configurations contain certain special features.

These three types of question are distinguished by Hall. Generally one adds another area:

(iv) graph theory, the study of the properties of a selection of subsets of size 2 from a given set.

Although graph theory is covered in areas (i) to (iii), special methods have been devised for the study of graphs.

This classification is not very precise, but it is fair to say that until recent years most research in combinatorial theory fell into one or another of these categories, and that very few papers dealt with more than one area or applied the techniques of one area to a problem of another area. The most notable exceptions have dealt with enumeration of graphs; but even here most papers have been almost entirely enumerative or almost entirely graph-theoretical.

It is by now trite to point out that we are in the midst of a great upsurge of combinatorial research. In an essay entitled: "Combinatorial Analysis" (in *The Mathematical Sciences: A Collection of Essays*, M.I.T. Press, 1969), Gian-Carlo Rota writes

> the next few years will probably witness an explosion of combinatorial activity, and the mathematics of the discrete will come to occupy a position at least equal to that of the applied mathematics of continua, in university curricula as well as in the importance of research.

An equally important point which has not been made so often is that the interrelation between the areas of combinatorial theory is becoming stronger. More and more research involves two or three of the areas we have listed.

This volume is part of the "explosion". It consists of three monographs on topics which have been the subject of attention recently. Each of the topics is currently "alive" and research on them is still going on; as witness to this, each monograph contains a selection of unsolved problems. The three monographs are separate, and at first sight they may appear to have little connection; however, it will be found that they have many concepts in common. Moreover, they share the common outlook which is found in many of the "constructive" parts of combinatorial theory.

The monographs also illustrate the ways in which the branches of combinatorial theory are coming together. For example, none of the three is about graph theory, but the concepts of graph theory are used in all three. In the chapter on Equivalence in *Hadamard Matrices*, generating functions and other enumerative techniques are applied. In *Room Squares* there is a close relationship to Latin squares; and the statistical application of a Room square depends on the interpretation of an Hadamard matrix as a block design. Various other examples exist. There are also applications of other parts of Mathematics: the elementary ideas of group theory are used throughout; many constructions in *Hadamard Matrices* depend on number theory and in particular the theory of cyclotomy, and this area is also used in the other monographs; both *Room Squares* and *Sum-Free Sets* use the idea of a quasigroup.

It was thought desirable to make the volume self-contained, so a chapter of basic definitions has been added to this introduction. The chapter might possibly have been sub-titled: "What every young combinatorial theorist should know". It is expected that the experienced reader will skim through this chapter and refer to it later as necessary.

Since most of the elementary ideas common to the monographs have been included here, we have been able to make the separate monographs independent of each other and still avoid repetition of material.

We have assumed only a familiarity with elementary group theory and linear algebra; there are no specific combinatorial prerequisites to the monographs.

## CHAPTER II. BASIC DEFINITIONS

2.1 ARITHMETIC. Galois Fields, Quadratic Numbers, Legendre Symbol, Fermat Numbers, Cyclotomic Numbers.

We denote the set of residue classes modulo the prime p by GF(p).

The concept of a polynomial in ordinary algebra can be extended to any field. Thus if $a_0, a_1, \ldots, b_0, b_1, \ldots$ are elements of any field F, then expressions of the type

$$a_0 + a_1x + a_2x^2 + \ldots$$

are the elements of a commutative ring F[x], when addition and multiplication are defined in the ordinary way by

$$\sum_{i=0}^{n} a_ix^i + \sum_{i=0}^{n} b_ix^i = \sum_{i=0}^{n} (a_i + b_i)x^i$$

$$\sum_{i=0}^{n} a_ix^i \times \sum_{i=0}^{n} b_ix^i = a_0b_0 + (a_1b_0 + a_0b_1)x + (a_2b_0 + a_1b_1 + a_0b_2)x^2 + \ldots$$

If in particular F is the field GF(p), $a_0, a_1, \ldots$ are residue classes (mod p), and the polynomial ring is denoted by $GF_p[x]$.

It is known that the number of elements contained in a finite field must be of the form $p^n$, where p is a prime integer and n any positive integer. Conversely, given a number of the form $p^n$, there always exists a field with $p^n$ elements in it, and any two finite fields with the same number of elements are isomorphic. The field with $p^n$ elements is called a *Galois field* and is denoted $GF(p^n)$.

Every element of $GF(p^n)$ can be expressed in the standard form

$$a_0 + a_1x + a_2x^2 + \ldots + a_{n-1}x^{n-1},$$

where the $a_i$ are integers ranging from 0 to p-1 and x is a root of an irreducible polynomial of degree n over GF(p). That is, $GF(p^n)$ is a quotient field of $GF_p[x]$. $GF(p^n)$ has characteristic p, and any non-zero element $\alpha$ of $GF(p^n)$ satisfies

$$\alpha^{p^n-1} = 1;$$

in fact, the multiplicative group of non-zero elements of $GF(p^n)$ is cyclic.

DEFINITION 1. A generator, x, of the cyclic multiplicative group, of order $p^n-1$, with elements of $GF(p^n)$ will be called a *primitive root* of $GF(p^n)$.

EXAMPLE. The elements of $GF(3^2)$ are

$$0, 1, 2, x, x+1, x+2, 2x, 2x+1, 2x+2$$

an additive elementary abelian group where x is a root of an irreducible equation. If we let x be a root of the irreducible equation $x^2 = x+1$ then we find

$$x, x^2 = x+1, x^3 = 2x+1, x^4 = 2, x^5 = 2x, x^6 = 2x+2, x^7 = x+2, x^8 = 1$$

are the elements of the cyclic multiplicative group of order $3^2-1$.

DEFINITION 2. Suppose p is an odd prime. If b is a non-zero residue modulo p, then b is called a *quadratic residue* if

$$x^2 \equiv b \pmod p$$

has solutions; otherwise b is called a quadratic non-residue. ($b = 0$ is specifically excluded from consideration.)

The following properties are standard (see, for example, I. Vinogradov, *Elements of Number Theory*, reprinted by Dover, New York, 1954):

(a) if b is a quadratic residue then $x^2 \equiv b$ has two solutions (of the form $x \equiv c$ and $x \equiv -c$);

(b) there are precisely $\frac{1}{2}(p-1)$ quadratic residues and $\frac{1}{2}(p-1)$ quadratic non-residues modulo p;

(c) if b is a quadratic residue modulo p, then

$$b^{\frac{1}{2}(p-1)} \equiv 1 \pmod p$$

and if b is a quadratic non-residue modulo p, then

$$b^{\frac{1}{2}(p-1)} \equiv -1 \pmod p;$$

(d) if x is a primitive root of GF(p) then the set of quadratic residues is precisely the set of even powers of x.

EXAMPLE. If we square all the numbers 1,2,3,4,5,6 and reduce them modulo 7 we get

$$1^2, 2^2, 3^2, 4^2, 5^2, 6^2$$

which reduce to

$$1, 4, 2, 2, 4, 1.$$

Thus the quadratic residues of 7 are 1,2,4 and the quadratic non-residues are 3,5,6. We notice that for $7 \equiv 3 \pmod 4$ that if x is a quadratic residue then $-x$ is a quadratic non-residue.

Similarly, squaring the non-zero residues of 13 we get

$$1^2, 2^2, 3^2, 4^2, 5^2, 6^2, 7^2, 8^2, 9^2, 10^2, 11^2, 12^2$$

which reduce, modulo 13, to

$$1, 4, 9, 3, 12, 10, 10, 12, 3, 9, 4, 1.$$

The quadratic residues of 13 are 1,3,4,9,10,12 and the quadratic non-residues are 2,5,6,7,8,11. Now 13 ≡ 1 (mod 4) and we note that if x is a quadratic residue then -x is also a quadratic residue.

DEFINITION 3. If b is any residue (mod p), then the *Legendre symbol* $\chi(b) = \left(\frac{b}{p}\right)$ is defined by

$$\chi(b) = \begin{cases} 0 & \text{if } b = 0, \\ 1 & \text{if } b \text{ is a quadratic residue}, \\ -1 & \text{otherwise.} \end{cases}$$

Then (see Vinogradov)

(e) $\chi(b) \equiv b^{\frac{1}{2}(p-1)} \pmod p$, see property (c),

(f) $\chi(1) = 1$,

(g) $\chi(-1) = \begin{cases} +1 & \text{if } p \equiv 1 \pmod 4, \\ -1 & \text{if } p \equiv 3 \pmod 4, \end{cases}$

(h) $\chi(-c) = \begin{cases} \chi(c) & \text{if } p \equiv 1 \pmod 4, \\ -\chi(c) & \text{if } p \equiv 3 \pmod 4, \end{cases}$

(i) $\chi(ab\ldots k) = \chi(a)\chi(b)\ldots\chi(k)$, so $\chi(ab^2) = \chi(a)$.

DEFINITION 4. The *quadratic character* $\chi$ is defined on the elements of a Galois field $GF(p^n)$ by

$$\chi(x) = \begin{cases} 0 & x = 0, \\ 1 & x = y^2, \text{ for some } y \in GF(p^n), \\ -1 & \text{otherwise.} \end{cases}$$

LEMMA 5. $\displaystyle\sum_{y \in GF(p^n)} \chi(y)\chi(y+c) = -1, \quad \text{if } c \neq 0.$

PROOF. $\chi(0)\chi(0+c) = 0$. For $y \neq 0$, there is a unique $s \neq 0$ such that $y + c = ys$. As y ranges of the non-zero elements of $GF(p^n)$, s ranges over all elements of $GF(p^n)$ except 1. (For $y = -c$, $s = 0$.) Hence

$$\begin{aligned} \sum_y \chi(y)\chi(y+c) &= \sum_{y \neq 0} \chi(y)\chi(y+c) \\ &= \sum_{y \neq 0} \chi(y^2)\chi(s) \\ &= \sum_{s \neq 1} \chi(s) \\ &= \sum_s \chi(s) - \chi(1) \\ &= -1. \end{aligned}$$

***

DEFINITION 6. The nth *Fermat number* $f_n$ is

$$f_n = 2^{2^n} + 1.$$

The first few Fermat numbers are $f_0 = 3$, $f_1 = 5$, $f_2 = 17$, $f_3 = 257$ and $f_4 = 65537$. Fermat conjectured that all Fermat numbers are prime, and the five examples given are all primes; however, $f_5$, $f_6$, $f_7$ and $f_8$ are all composite.

We use the following definitions and results from T. Storer, *Cyclotomy and Difference Sets*, Markham Publishing Company, Chicago, 1967, pp.24-25.

We suppose $G = GF(q)$ where $q = p^r = ef + 1$ and $x$ is a primitive root of $G$. Write $G^*$ for the set $G - \{0\}$ (or $G\setminus\{0\}$) of all members of $G$ other than 0; then

$$G^* = \{x^s: s = 0,1,\ldots,q-2\}.$$

The cyclotomic classes $C_i$ in $G$ are:

$$C_i = \{x^{es+i}: s = 0,1,\ldots,f-1\}.$$

These are a subgroup of $G^*$ and its cosets, so the $C_i$ are disjoint and their union is $G^*$. If $n$ is any integer then $C_{i+ne} = C_i$.

DEFINITION 7. For fixed $i$ and $j$, the *cyclotomic number* $(i,j)$ is the number of solutions of the equation

$$z_i + 1 = z_j, \qquad z_i \in C_i, \quad z_j \in C_j, \tag{1}$$

where $1 = x^0$ is the multiplicative identity element of $G$. That is, $(i,j)$ is the number of ordered pairs $s,t$ such that

$$x^{es+i} + 1 = x^{et+j}, \qquad 0 \leq s,t \leq f-1.$$

Thus for $e = 8$, $(i,j)$ is the number of ordered pairs $s,t$ such that

$$x^{8s+i} + 1 = x^{8t+j}, \qquad 0 \leq s,t \leq f-1.$$

For $f$ odd the numbers $(i,j)$ satisfy the relations (see Storer p.25, lemma 3),

$$(i,j) = (j+4,i+4) = (8-i,j-i).$$

These lead to the following array (Storer, p.29) in which the 64 constants $(i,j)$ $(i,j) = 0,1,\ldots,7 \pmod 8$ are expressed in terms of 15 of their number. $(i,j)$ is in row $i$ and column $j$.

| | 0 | 1 | 2 | 3 | 4 | 5 | 6 | 7 |
|---|---|---|---|---|---|---|---|---|
| 0 | A | B | C | D | E | F | G | H |
| 1 | I | J | K | L | F | D | L | M |
| 2 | N | O | N | M | G | L | C | K |
| 3 | J | O | O | I | H | M | K | B |
| 4 | A | I | N | J | A | I | N | J |
| 5 | I | H | M | K | B | J | O | O |
| 6 | N | M | G | L | C | K | N | O |
| 7 | J | K | L | F | D | L | M | I |

(2)

Explicit formulae have been found for these cyclotomic numbers. We quote from Storer [p.79].

LEMMA 8. *The cyclotomic numbers for $e = 8$, $f$ odd, are given by the array (2), and the relations:*

| *I. If 2 is a fourth power in G* | *II. If 2 is not a fourth power in G* |
|---|---|
| $64A = q - 15 - 2x$ | $64A = q - 15 - 10x - 8a$ |
| $64B = q + 1 + 2x - 4a + 16y$ | $64B = q + 1 + 2x - 4a$ |
| $64C = q + 1 + 6x + 8a - 16y$ | $64C = q + 1 - 2x$ |
| $64D = q + 1 + 2x - 4a - 16y$ | $64D = q + 1 + 2x - 4a$ |
| $64E = q + 1 - 18x$ | $64E = q + 1 + 6x + 24a$ |
| $64F = q + 1 + 2x - 4a + 16y$ | $64F = q + 1 + 2x - 4a$ |
| $64G = q + 1 + 6x + 8a + 16y$ | $64G = q + 1 - 2x$ |
| $64H = q + 1 + 2x - 4a - 16y$ | $64H = q + 1 + 2x - 4a$ |
| $64I = q - 7 + 2x + 4a$ | $64I = q - 7 + 2x + 4a + 16y$ |
| $64J = q - 7 + 2x + 4a$ | $64J = q - 7 + 2x + 4a - 16y$ |
| $64K = q + 1 - 6x + 4a$ | $64K = q + 1 + 2x - 4a$ |
| $64L = q + 1 + 2x - 4a$ | $64L = q + 1 - 6x + 4a$ |
| $64M = q + 1 - 6x + 4a$ | $64M = q + 1 + 2x - 4a$ |
| $64N = q - 7 - 2x - 8a$ | $64N = q - 7 + 6x$ |
| $64O = q + 1 + 2x - 4a$ | $64O = q + 1 - 6x + 4a$ |

*where $x, y, a$ and $b$ are specified by:*

*I.* $q = x^2 + 4y^2$, $x \equiv 1 \pmod 4$ *is the unique proper representation of* $q = p^\alpha$ *if* $p \equiv 1 \pmod 4$; *otherwise,*

$$q = (\pm p^{\alpha/2})^2 + 4.0^2; \quad i.e., \quad x = \pm p^{\alpha/2},\ y = 0.$$

*II.* $q = a^2 + 2b^2$, $a \equiv 1 \pmod 4$ *is the unique proper representation of* $q = p^\alpha$ *if* $p \equiv 1$ *or* $3 \pmod 8$; *otherwise,*

$$q = (\pm p^{\alpha/2})^2 + 2.0^2; \quad i.e., \quad a = \pm p^{\alpha/2},\ b = 0.$$

*The signs of $y$ and $b$ are ambiguously determined.*

In view of (1) the cyclotomic number $(i,j)$ may be expressed as the number of solutions of the equation

$$y - x = 1 \qquad y \in C_j, \quad x \in C_i. \tag{3}$$

If $d \in C_k$, then each solution of (3) yields a solution of

$$y_1 - x_1 = d \qquad y_1 \in C_{j+k}, \quad x_1 \in C_{i+k}.$$

It follows that if $d \in C_k$ then there are $(i-k, j-k)$ solutions of the equation

$$y - x = d \qquad y \in C_j, \quad x \in C_i.$$

This enables us to determine how often each difference arises from sets composed of given cyclotomic classes.

## 2.2 BALANCED INCOMPLETE BLOCK DESIGNS. BIBD, $(b,v,r,k,\lambda)$-configurations, SBIBD, $(v,k,\lambda)$-configurations.

DEFINITION 9. A *balanced incomplete block design* is a way of selecting b subsets called *blocks* from a set of v distinct objects so that each block contains exactly k distinct objects, each object occurs in exactly r different blocks, and every pair of distinct objects $a_i$, $a_j$ occurs together in exactly $\lambda$ blocks. This is called a BIBD or a $(b,v,r,k,\lambda)$-*configuration*.

For example, $b = 12$, $v = 9$, $r = 4$, $k = 3$, $\lambda = 1$ gives the following blocks:

| | | | |
|---|---|---|---|
| 1 2 3 | 1 4 7 | 1 5 9 | 1 6 8 |
| 4 5 6 | 2 5 8 | 2 6 7 | 3 5 7 |
| 7 8 9 | 3 6 9 | 3 4 8 | 3 4 9 . |

Counting the number of objects in two different ways we see

$$r(k-1) = \lambda(v-1) \tag{4}$$

and counting the number of replications of an element, we see each element is replicated r times but since each block has k elements we must have

$$bk = vr. \tag{5}$$

DEFINITION 10. The *complement* D' of a BIBD with parameters $(b,v,r,k,\lambda)$ is the configuration chosen by the rule: block j of D' = {i: i ∉ block j of D as i runs through the v possible objects}.

Clearly if $b,v,r',k',\lambda'$ are the parameters of the complementary design of a $(b,v,r,k,\lambda)$-configuration then if $0 < \lambda$ and $k < v - 1$ we have

$$\left.\begin{aligned} r' &= b - r \\ k' &= v - k \\ \lambda' &= b - 2r + \lambda \end{aligned}\right\} \tag{6}$$

and since $\lambda'(v-1) = r'(k'-1)$ we have $\lambda'(v-1) = (b-r)(v-k-1)$, so $0 < \lambda'$ and $k' < v - 1$.

DEFINITION 11. A symmetric $(b,v,r,k,\lambda)$-configuration has $b = v$ and $r = k$. This is called a $(v,k,\lambda)$-*configuration* or *SBIBD*.

For SBIBD's (4) becomes

$$k(k-1) = \lambda(v-1). \tag{7}$$

We state without proof some theorems from Marshall Hall Jr., *Combinatorial Theory*, Blaisdell [Ginn & Co.], Waltham, Mass., 1967, and H.J. Ryser, *Combinatorial Mathematics* (Carus Monograph No. 14), Wiley, New York, 1963.

THEOREM 12. *THE BRUCK-RYSER-CHOWLA THEOREM (Hall, pp.107-112).*

*If $v,k,\lambda$ are integers for which there is a $(v,k,\lambda)$-configuration, then:*

*(i) if $v$ is even $(k-\lambda)$ is a square;*

*(ii) if $v$ is odd, then the Diophantine equation*

$$x^2 = (k-\lambda)y^2 + (-1)^{\frac{1}{2}(v-1)} \tag{8}$$

*has a solution in the integers not all zero.*

Equation (7) and theorem 12 give necessary conditions for the existence of a $(v,k,\lambda)$-configuration but it is not known whether or not they are sufficient. No case has yet been found where these two conditions are both satisfied and the corresponding configuration has been proved not to exist.

Let $(k-\lambda)'$ and $\lambda'$ respectively, denote the square-free parts of $k-\lambda$ and $\lambda$ and let $d = ((k-\lambda)',\lambda')$; then (8) may be written as

$$dx^2 = \frac{(k-\lambda)'}{d} y^2 + (-1)^{\frac{1}{2}(v-1)} \frac{\lambda' z^2}{d}$$

or

$$ax^2 + by^2 + cz^2 = 0,$$

where $a,b,c$ are square-free and $(a,b,c) = 1$. This latter is called the *Legendre Equation* and

THEOREM 13. *(Ryser p.114) The Legendre equation has solutions for $x,y,z$ in the integers not all zero if and only if $-bc$, $-ac$ and $-ab$ are quadratic residues of $a,b$ and $c$ respectively.*

2.3 MATRICES. Incidence Matrices, Hadamard Matrices, Kronecker Products.

We use $I$ for the identity matrix and $J$ for the square matrix with every element $+1$.

DEFINITION 14. The *incidence matrix* of a BIBD is a (0,1)-matrix of size $v\times b$ with $a_{ij} = 1$ if object $i$ is in block $j$ and $a_{ij} = 0$ otherwise.

If $A = (a_{ij})$ is the incidence matrix of a BIBD with parameters $(b,v,r,k,\lambda)$, $A$ satisfies

$$J_{v\times v}A = kJ_{v\times b} \tag{9}$$

$$AA^T = (r-\lambda)I_{v\times v} + \lambda J_{v\times v}. \tag{10}$$

Conversely, if $A$ is a (0,1)-matrix satisfying (9) and (10) then this ensures there is a BIBD with parameters $(b,v,r,k,\lambda)$.

THEOREM 15. *(Marshall Hall Jr., p.104) If A is the incidence matrix of a $(v,k,\lambda)$-configuration then A satisfies*

$$AA^T = B = (k-\lambda)I + \lambda J \quad (11)$$

$$A^T A = B = (k-\lambda)I + \lambda J \quad (12)$$

$$AJ = kJ \quad (13)$$

$$JA = kJ. \quad (14)$$

THEOREM 16. *(Ryser's Theorem quoted from Hall, p.283) Suppose $A = (a_{ij})$ is a $v \times v$ matrix of integers and $k(k-1) = \lambda(v-1)$ and $AA^T = A^T A = (k-\lambda)I + \lambda J$. Then A or -A is composed entirely of zeros and ones and is the incidence matrix of a SBIBD.*

THEOREM 17. *If A is the matrix obtained from the incidence matrix of a $(v,k,\lambda)$-configuration by every element 0 being replaced by -1, then A satisfies*

$$AA^T = 4(k-\lambda)I_v + (v-4(k-\lambda))J_v.$$

PROOF. This follows by noting that if $B$ is the incidence matrix of the $(v,k,\lambda)$-configuration then $A = B-2J$. ***

DEFINITION 18. A *circulant matrix* $A = (a_{ij})$ of order n is one in which $a_{ij} = a_{1,j-i+1}$ where j-i+1 is reduced modulo n. For example:

$$\begin{bmatrix} 1 & 2 & 3 & 4 \\ 4 & 1 & 2 & 3 \\ 3 & 4 & 1 & 2 \\ 2 & 3 & 4 & 1 \end{bmatrix}.$$

DEFINITION 19. A (1,-1)-matrix is a matrix whose only elements are +1 and -1.

DEFINITION 20. A set $D = \{x_1, x_2, \ldots, x_k\}$ will be said to generate a circulant (1,-1)-matrix if

$$a_{1x} = \begin{cases} +1 & x \in D \\ -1 & x \notin D \end{cases}.$$

DEFINITION 21. A matrix $A = (a_{ij})$ of order n will be called *back circulant* if $a_{ij} = a_{1,i+j-1}$ where i+j-1 is reduced modulo n. For example:

$$\begin{bmatrix} 1 & 2 & 3 & 4 \\ 2 & 3 & 4 & 1 \\ 3 & 4 & 1 & 2 \\ 4 & 1 & 2 & 3 \end{bmatrix}.$$

LEMMA 22. *A back circulant matrix is symmetric.*

LEMMA 23. *The product of a back circulant matrix with a circulant matrix of the same order is symmetric. In particular, if A is back circulant and B is circulant*

$$AB^T = BA^T.$$

PROOF. Let $A = (a_{ij})$ where $a_{ij} = a_{1,j-i+1}$ and $B = (b_{ij})$ where $b_{ij} = b_{1,i+j-1}$. Then

$$BA^T = (\sum_k b_{ik}a'_{kj}) = (\sum_k b_{ik}a_{jk}) = (\sum_k a_{jk}b_{ik}) = (\sum_k a_{1,k-j+1}b_{1,i+k-1})$$

$$= (\sum_k a_{i,k-j+i}b_{j,i+k-j}) = (\sum_{\ell=i+k-j} a_{i\ell}b'_{\ell j}) = (\sum_\ell a_{i\ell}b'_{\ell j})$$

$$= AB^T.$$

***

LEMMA 24. *Any two circulant matrices of the same order commute.*

PROOF. With $A = (a_{ij})$ and $B = (b_{ij})$ both circulant

$$AB = (\sum_k a_{ik}b_{kj}) = (\sum_k b_{i,j-k+i}a_{j-k+i,j}) = (\sum_{\ell=j-k+i} b_{i\ell}a_{\ell j}) = BA.$$

***

DEFINITION 25. An *Hadamard matrix* is a (1,-1)-matrix whose row vectors are orthogonal.

EXAMPLE.

$$\begin{bmatrix} 1 & 1 \\ 1 & -1 \end{bmatrix}, \qquad \begin{bmatrix} -1 & 1 & 1 & 1 \\ 1 & -1 & 1 & 1 \\ 1 & 1 & -1 & 1 \\ 1 & 1 & 1 & -1 \end{bmatrix}.$$

These matrices were first considered as Hadamard determinants. They were so named because the determinant of an Hadamard matrix satisfies the equality of Hadamard's determinant theorem, which states that if $X = (x_{ij})$ is a matrix of order n then

$$|\det X|^2 \leq \prod_{i=1}^{n} \prod_{j=1}^{n} |x_{ij}|^2.$$

LEMMA 26. *Let H be an Hadamard matrix of order h. Then:*

*(a)* $HH^T = hI_h$;

*(b)* $|\det H| = h^{\frac{1}{2}h}$;

*(c)* $HH^T = H^TH$;

*(d) Hadamard matrices may be changed into other Hadamard matrices by permuting rows and columns and by multiplying rows and columns by -1. We call matrices which can be obtained from one another by these methods*

*H-equivalent; (we will show later that not all Hadamard matrices of the same order are H-equivalent);*

(e) *every Hadamard matrix is H-equivalent to an Hadamard matrix which has every element of its first row and column +1 - matrices of this latter form are called normalized;*

(f) *if H is a normalized Hadamard matrix of order 4n, then every row (column) except the first has 2n minus ones and 2n plus ones in each row (column), further n minus ones in any row (column) overlap with n minus ones in each other row (column);*

(g) *the order of an Hadamard matrix is 1,2 or 4n, n integer.*

PROOF. (a) Let $H = (h_{ij})$. Then the ij element of $HH^T = \sum_k h_{ik}h_{jk}$ = inner product of row i and row j = (h if i = j) and (0 if i ≠ j). Thus, $HH^T = hI$.

(b) $HH^T = hI_h$; $h^h = \det(HH^T) = (\det(H))^2$; so $|\det H| = h^{\frac{1}{2}h}$.

(c) $HH^T = hI$, so $H^T = hH^{-1}$; so $H^TH = hI = HH^T$.

(d) Clearly permuting the rows and columns neither changes the fact that every element is +1 or -1 nor the fact that the inner product of any two different rows is zero.

Suppose row i is multiplied throughout by -1. Then the inner product of the new row i with row j, j ≠ i, is

$$\sum_k - h_{ik}h_{jk} = - \sum_k h_{ik}h_{jk} = 0$$

and so the new matrix is still Hadamard.

Now suppose column i is multiplied throughout by -1. The inner product of the new column i with column j, j ≠ i, is

$$\sum_k -h_{ki}h_{kj} = - \sum_k h_{ki}h_{kj} = 0 \quad \text{(by part (c))}.$$

(e) We may use part (d) repeatedly, multiplying through each row and column which starts with -1 by -1 until the first row and column has every element +1.

(f) Normalize H as in (e). Now the inner product of row i (column i) with the first row (column) is given by

$$\sum_k h_{ik} = 0, \qquad (\sum_k h_{ki} = 0).$$

So exactly half the elements are +1 and half are -1. Rearrange the rows and columns using part (e) until

$$h_{mj} = +1 = h_{km}, \quad j,k = 1,2,\ldots,2n \quad \text{and} \quad h_{m,2n+j} = -1 = h_{2n+k,m}, \qquad m \neq 1.$$

Then if i ≠ 1 and i ≠ m the inner product of the ith row (ith column) with the mth is

$$\sum_{k=1}^{2n} h_{ik} - \sum_{j=2n+1}^{4n} h_{ij} = 0 \quad \left( \sum_{k=1}^{2n} h_{ki} - \sum_{j=2n+1}^{4n} h_{ji} = 0 \right) .$$

Let a be the number of minus elements in the ith row (ith column) which are in columns (rows) 2n + 1 to 4n, i.e., which overlap with minuses in the mth row. Then the number of minus elements in columns (rows) 1 to 2n of the ith row is 2n - a (since exactly 2n elements are plus and 2n are minus in every row (column) but the first). Thus

$$\sum_{k=1}^{2n} h_{ik} - \sum_{j=2n+1}^{4n} h_{ij}$$

= (sum of plus elements in rows 1 to 2n) - (sum of minus elements in rows 1 to 2n)
- (sum of plus elements in rows 2n + 1 to 4n) + (sum of minus elements in rows 2n + 1 to 4n)

= (a) - (2n -a) - (2n-a) + a

= 4a - 4n

= 0.

So a = n (and similarly for the columns).

(g) Hadamard matrices of order 1 and 2 are

$$[1] \qquad \text{and} \qquad \begin{bmatrix} 1 & 1 \\ 1 & -1 \end{bmatrix} .$$

Now suppose the order is h > 2. Normalize H and rearrange the first three rows to look like

$$\begin{array}{cccc} 1 \;\ldots\ldots\; 1 & 1 \;\ldots\ldots\; 1 & 1 \;\ldots\ldots\; 1 & 1 \;\ldots\ldots\; 1 \\ 1 \;\ldots\ldots\; 1 & 1 \;\ldots\ldots\; 1 & -1 \;\ldots\ldots\; -1 & -1 \;\ldots\ldots\; -1 \\ \underbrace{1 \;\ldots\ldots\; 1}_{x} & \underbrace{-1 \;\ldots\ldots\; -1}_{y} & \underbrace{1 \;\ldots\ldots\; 1}_{z} & \underbrace{-1 \;\ldots\ldots\; -1}_{w} \end{array} .$$

Where x, y, z and w are the number of columns of each type. Then because the order is h

$$x + y + z + w = h$$

and taking the inner products of rows 1 and 2, 1 and 3, and 2 and 3 respectively we get

$$x + y - z - w = 0$$

$$x - y + z - w = 0$$

$$x - y - z + w = 0 .$$

Solving we get

$$x = y = z = w = h/4$$

and so h ≡ 0 (mod 4).

***

DEFINITION 27. An Hadamard matrix H is said to be *regular* if the sum of all the elements in each row or column is a constant k. Hence $HJ = JH = kJ$.

DEFINITION 28. If $M = (m_{ij})$ is a $m \times p$ matrix and $N = (n_{ij})$ is an $n \times q$ matrix, then the *Kronecker product* $M \times N$ is the $mn \times pq$ matrix given by

$$M \times N = \begin{bmatrix} m_{11}N & m_{12}N & \cdots & m_{1p}N \\ m_{21}N & m_{22}N & \cdots & m_{2p}N \\ \vdots & & & \\ m_{m1}N & m_{m2}N & \cdots & m_{mp}N \end{bmatrix}$$

LEMMA 29. *The following properties of Kronecker product follow immediately from the definition:*

(a) $p(M \times N) = (pM) \times N = M \times (pN)$ $\qquad$ *p a scalar,*

(b) $(M_1 + M_2) \times N = (M_1 \times N) + (M_2 \times N)$

(c) $M \times (N_1 + N_2) = M \times N_1 + M \times N_2$

(d) $(M_1 \times N_1)(M_2 \times N_2) = M_1M_2 \times N_1N_2$

(e) $(M \times N)^T = M^T \times N^T$

(f) $(M \times N) \times P = M \times (N \times P)$ .

EXAMPLE. Let $M = \begin{bmatrix} 1 & 1 \\ 1 & -1 \end{bmatrix}$ and $N = \begin{bmatrix} -1 & 1 & 1 & 1 \\ 1 & -1 & 1 & 1 \\ 1 & 1 & -1 & 1 \\ 1 & 1 & 1 & -1 \end{bmatrix}$ then

$$M \times N = \begin{bmatrix} N & N \\ N & -N \end{bmatrix} = \begin{bmatrix} -1 & 1 & 1 & 1 & -1 & 1 & 1 & 1 \\ 1 & -1 & 1 & 1 & 1 & -1 & 1 & 1 \\ 1 & 1 & -1 & 1 & 1 & 1 & -1 & 1 \\ 1 & 1 & 1 & -1 & 1 & 1 & 1 & -1 \\ -1 & 1 & 1 & 1 & 1 & -1 & -1 & -1 \\ 1 & -1 & 1 & 1 & -1 & 1 & -1 & -1 \\ 1 & 1 & -1 & 1 & -1 & -1 & 1 & -1 \\ 1 & 1 & 1 & -1 & -1 & -1 & -1 & 1 \end{bmatrix}$$

LEMMA 30. *Let $H_1$ and $H_2$ be Hadamard matrices of orders $h_1$ and $h_2$. Then from (d) and (e) of lemma 29 $H = H_1 \times H_2$ is an Hadamard matrix of order $h_1h_2$.*

NOTATION. We will write $\Sigma' A \times B \times C \times \ldots \times D$, where $\times$ is the Kronecker product, to mean the sum obtained by circulating the letters formally. Thus

$$\Sigma' A \times B \times C = A \times B \times C + B \times C \times A + C \times A \times B.$$

The sum $\Sigma'$ over p letters has p terms in the sum.

2.4 DIFFERENCE SETS. Cyclic Difference Sets, Group Difference Sets

DEFINITION 31. A set of k residues $D = \{a_1,\dots,a_k\}$ modulo v is called a $(v,k,\lambda)$-*difference set* (or a *perfect difference set* or *cyclic difference set*, if for every $d \not\equiv 0 \pmod v$ there are exactly $\lambda$ ordered pairs $(a_i,a_j)$, $a_i$, $a_j \in D$ such that $a_i - a_j \equiv d \pmod v$.

For example the numbers 0, 1, 3, 8, 12, 18 modulo 31 form a (31,6,1)-difference set since

| | | |
|---|---|---|
| $1 - 0 \equiv 1$ | $12 - 1 \equiv 11$ | $8 - 18 \equiv 21$ |
| $3 - 1 \equiv 2$ | $12 - 0 \equiv 12$ | $3 - 12 \equiv 22$ |
| $3 - 0 \equiv 3$ | $0 - 18 \equiv 13$ | $0 - 8 \equiv 23$ |
| $12 - 8 \equiv 4$ | $1 - 18 \equiv 14$ | $1 - 8 \equiv 24$ |
| $8 - 3 \equiv 5$ | $18 - 3 \equiv 15$ | $12 - 18 \equiv 25$ |
| $18 - 12 \equiv 6$ | $3 - 18 \equiv 16$ | $3 - 8 \equiv 26$ |
| $8 - 1 \equiv 7$ | $18 - 1 \equiv 17$ | $8 - 12 \equiv 27$ |
| $8 - 0 \equiv 8$ | $18 - 0 \equiv 18$ | $0 - 3 \equiv 28$ |
| $12 - 3 \equiv 9$ | $0 - 12 \equiv 19$ | $1 - 3 \equiv 29$ |
| $18 - 8 \equiv 10$ | $1 - 12 \equiv 20$ | $0 - 1 \equiv 30$ |

An equivalent definition is

DEFINITION 32. A set of k residues $D = \{a_1,\dots,a_k\}$ modulo v is called a $(v,k,\lambda)$-difference set if among the collection of elements $[a_i - a_j : i \neq j,\ 1 \leq i,\ j \leq k]$ all the non-zero residues occur $\lambda$ times

DEFINITION 33. If D is a $(v,k,\lambda)$-difference set, define

$$D + x = \{a + x : a \in D\}$$

for each residue x(mod v). Then D + x is called a *shift* of D. Every shift of a $(v,k,\lambda)$-difference set is itself a $(v,k,\lambda)$-difference set.

DEFINITION 34. If D is a $(v,k,\lambda)$-difference set and x is any residue (mod v), define

$$xD = \{xa : a \in D\}.$$

If xD is a shift of D then x is a *multiplier* of D; in particular, if xD = D then x *fixes* D.

The trivial multiplier 1 always fixes a different set. Other examples: if D is the (31, 6, 1)-difference set

$$D = \{0,1,3,8,12,18\}$$

of the earlier example. Then

$$D + 28 = \{0,5,9,15,28,29\};$$

we observe that 5D = D + 28, so that 5 is a multiplier of D. If D is the (7,3,1)-difference set {1,2,4}, then 2D = D, so 2 is a multiplier which fixes D.

$(v,k,\lambda)$-difference sets can be generalized. These sets have been defined on the integers modulo v, an additive abelian group. We now define a group difference set on an additive abelian group and note that the properties we state for group difference sets also apply to $(v,k,\lambda)$-difference sets.

DEFINITION 35. A subset $S = \{x_i\}$ of elements of an additive abelian group G of order v is called a *$(v,k,\lambda)$-group difference set* if for every $y \neq 0 \in G$ there are exactly $\lambda$ ordered pairs $(x_i,x_j)$ such that

$$x_i - x_j = y;$$

or equivalently among the collection of elements $[x_i-x_j : i \neq j,\ 1 \leq i,\ j \leq k]$ the non-zero elements of G each occur $\lambda$ times.

EXAMPLE. In $GF(3^3)$ with $x^3 = x+2$

$\{x^2,\ x^2+2x,\ x^2+x+1,\ 2x^2+2,\ x^2+x,\ x^2+2,\ 2x,\ 2x+1,\ x^2+2x+1,\ 2x^2+x+1,\ 2x+2,\ 2x^2+2x+1\}$

is a (27,13,6)-group difference set.

DEFINITION 36. The incidence matrix $A = (a_{ij})$ of a $(v,k,\lambda)$-group difference set S is defined by ordering the elements of the group $G = \{g_i\}$ (or the integer modulo v) as $g_1,g_2,\ldots,g_v$ and defining $a_{ij} = \begin{cases} 1 & g_j-g_i \in S, \\ 0 & \text{otherwise.} \end{cases}$

If G is the integers modulo v and $g_1,g_2,\ldots,g_v$ have the usual ordering of the residues modulo v

$$a_{ij} = \begin{cases} 1 & g_j-g_i = j-i \in S \\ 0 & \text{otherwise} \end{cases}$$

$$= \begin{cases} 1 & j-i+1-1 \in S \\ 0 & \text{otherwise} \end{cases}$$

$$= a_{1,j-i+1}$$

and so A is circulant or cyclic in this case S is called a *cyclic difference set.*

LEMMA 37. *Suppose $S = \{x_i\}$ is a $(v,k,\lambda)$-group difference set, defined on an additive abelian group G. Then*

$$\lambda(v-1) = k(k-1).$$

PROOF. The set S is a k-set, so the collection $[x_i-x_j : i \neq j,\ 1 \leq i,\ j \leq k]$ has $k(k-1)$ elements. By definition this collection has each of the $v-1$ non-zero elements of G occurring $\lambda$ times and so has $\lambda(v-1)$ elements. Thus

$$\lambda(v-1) = k(k-1).$$

***

LEMMA 38. *The incidence matrix $A = (a_{ij})$ of a $(v,k,\lambda)$-group difference set $S = \{x_i\}$ defined on an additive abelian group G, satisfies*

$$AJ = kJ, \quad AA^T = (k-\lambda)I + \lambda J.$$

PROOF. The i,j element of AJ is $\sum_m a_{im}$ as J has every element plus one. Now

$$\begin{aligned}\sum_m a_{im} &= \text{number of times } g_m - g_i \in S \text{ as } g_m \text{ takes all the values of } G\\ &= \text{number of elements in } S\\ &= k.\end{aligned}$$

So AJ = kJ. The (i,j) element of $AA^T$ is

$$\begin{aligned}\sum_m a_{im}a_{jm} &= \text{number of times } \{(g_m - g_i \in S) \text{ and } (g_m - g_i \in S)\},\ m = 1,\ldots,v\\ &= \text{number of times } \{(d \in S) \text{ and } (d+g_i-g_j \in S)\} \text{ where } d = g_m - g_i,\\ &\qquad m = 1,\ldots,v.\end{aligned}$$

For i = j, $\sum_m a_{im}a_{jm}$ = number of times $d \in S$ as d takes all the values of G. For $i \neq j$ consider $T = \{s_\ell : s_\ell \in S,\ s_n - s_\ell = g_i - g_j,\ \text{some } s_n \in S\}$ which has $\lambda$ elements, then $T = \{s_\ell : s_\ell \in S,\ s_\ell + g_i - g_j \in S\}$ and so $\sum_m a_{ij}a_{jm} = \lambda$.

Thus $AA^T$ has k on the diagonal and $\lambda$ elsewhere; i.e.,

$$AA^T = (k-\lambda)I + \lambda J.$$

***

## 2.5 GRAPHS. Subgraphs, One-factorizations, Edge Colourings.

DEFINITION 39. A *graph* G is a set V of *vertices* (or *points*) together with a set E of *edges* (or *lines*), where E is a subset of the set of all unordered pairs of elements of V. As consequences of the definition, no edge joins a vertex to itself, and it cannot happen that two vertices are joined by two edges.

DEFINITION 40. Two vertices x and y are *adjacent* if and only if {x,y} is an edge; the number of vertices adjacent to x is the *valency* or *degree* of x; if every vertex has the same valency k then the graph is called *regular of valency* k.

It is convenient to represent a graph in a diagram. Vertices are represented by dots or small circles and edges by lines joining vertices.

Examples.

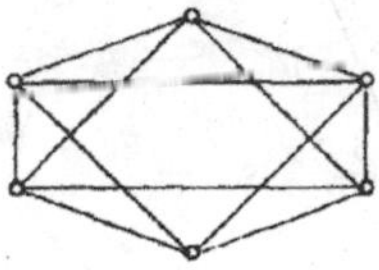

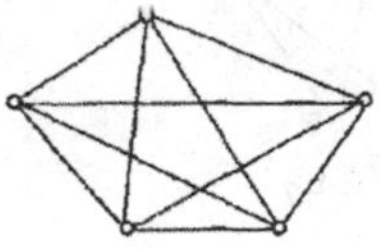

Both graphs are regular of valency four.

DEFINITION 41. The *complement* $\overline{G}$ of a graph is defined as follows: if G has vertex-set V and edge-set E, then $\overline{G}$ has vertex-set V and edge-set consisting of every unordered pair of members of V which is not in E.

Example. The following graphs are complements.

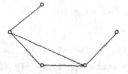

One particular graph which is of importance is the *complete graph* $K_n$, which consists of n vertices and all $\frac{1}{2}n(n-1)$ possible edges, and its complement the *null graph* $\overline{K}_n$, which has n vertices and no edges. The right-hand graph in the first example is $K_5$.

DEFINITION 42. If $G_1$ and $G_2$ are graphs with vertex-sets $V_1, V_2$ respectively and edge-sets $E_1, E_2$ respectively, then their *union* and *intersection* are defined as follows:

$G_1 \cup G_2$ has vertex-set $V_1 \cup V_2$, edge-set $E_1 \cup E_2$;

$G_1 \cap G_2$ has vertex-set $V_1 \cap V_2$, edge-set $E_1 \cap E_2$.

Example.

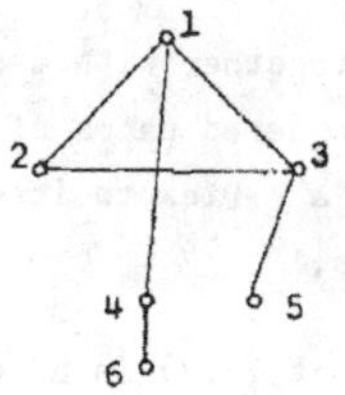


$G_1$

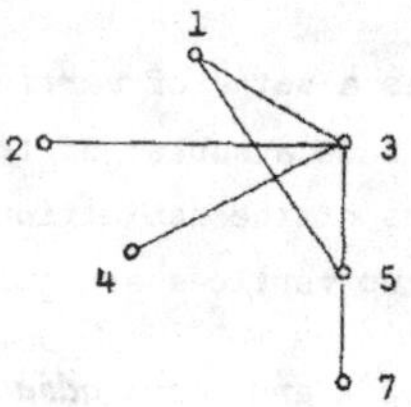


$G_2$

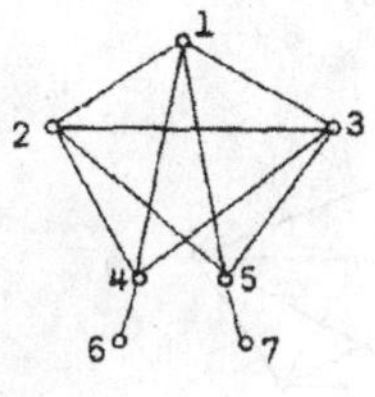


$G_1 \cup G_2$

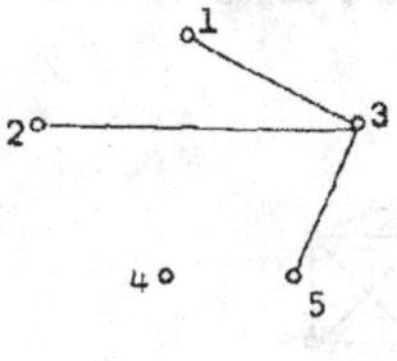


$G_1 \cap G_2$

DEFINITION 43. Two graphs are *edge-disjoint* if their intersection is a null graph.

DEFINITION 44. H is a *subgraph* of G if its vertex-set and edge-set are subsets of the vertex-set and edge-set of G respectively. H is a *spanning subgraph* of G if its vertex-set equals the vertex-set of G. A spanning subgraph is also called a *factor* of G.

DEFINITION 45. A one-factorization of G is a set of factors of G which are regular of valency 1 (called *one-factors*) which are edge-disjoint and have union G.

THEOREM 46. *$K_n$ has a one-factorization if and only if n is even.*

For example, the complete graph $K_6$ with vertex-set {0,1,2,3,4,5} has one-factorization

| | | |
|---|---|---|
| 01 | 23 | 45 |
| 02 | 14 | 35 |
| 03 | 15 | 24 |
| 04 | 13 | 25 |
| 05 | 12 | 34 |

where, for example, the second line means that one of the factors has edges {0,2}, {1,4} and {3,5}.

DEFINITION 47. If $G_1$ has vertex-set $V_1$ and edge-set $E_1$, and $G_2$ has vertex-set $V_2$ and edge-set $E_2$, then the map $\alpha: V_1 \to V_2$, denoted by $x \to x\alpha$, is an *isomorphism* from $G_1$ to $G_2$ provided:

(i) $\alpha$ is one-to-one and onto;

(ii) $\{x,y\} \in E_1$ if and only if $\{x\alpha, y\alpha\} \in E_2$.

The idea of an *edge-colouring* of a graph is an alternative view of factorization. Given a set of 4 colours, allot one of the set to each edge of the graph. If the set of all edges given a certain colour is interpreted as the set of edges of a factor, then the collection of factors corresponding to the various colours is a factorization of G. A one-factorization corresponds to a colouring in which every vertex is on one edge of each colour.

A suitable reference for all the ideas in this section is F. Harary, *Graph Theory*, Addison-Wesley, 1969.

2.6 GROUPS AND GENERALIZATIONS. Groupoid, Semigroup, Quasigroup, Loop, Latin Squares.

It has been assumed that the reader is familiar with the basic ideas of group theory. If not, a suitable reference is Marshall Hall Jr., *The Theory of Groups*, Macmillan, New York, 1959. One definition which is not in that book is:

DEFINITION 48. If G is a finite group, an *exponent* of G is a positive integer n such that $g^n = 1$ (the identity of the group) for every $g \in G$.

DEFINITION 49. A *groupoid* G consists of a non-empty set G together with a binary operation $\otimes$ which is closed on G; that is,

A1. $a \otimes b \in G$ for all a and b in G.

DEFINITION 50. A *semigroup* G is a groupoid which satisfies the associativity axiom

A2. $a \otimes (b \otimes c) = (a \otimes b) \otimes c$ for all $a,b,c \in G$.

DEFINITION 51. A *quasigroup* G is a groupoid which satisfies the rule:

A3. For every a and b in G there are uniquely defined elements x and y of G which satisfy

$$a \otimes x = b, \quad y \otimes a = b.$$

DEFINITION 52. A quasigroup G is called a *loop* if

A4. There is an element e of G, the *identity* element, satisfying

$$a \otimes e = e \otimes a = a \quad \text{for all a in G.}$$

DEFINITION 53. A quasigroup G is called an *idempotent quasigroup* if

A5. $a \otimes a = a$ for all a in G.

DEFINITION 54. Any of the above structures G is *abelian* or *commutative* if

$$a \otimes b = b \otimes a \quad \text{for all } a,b \in G.$$

Examples.

(i) The set G of non-negative real numbers, with

$$a \otimes b = |a-b|,$$

is a commutative groupoid and has an identity element 0. It is not a semigroup since

$$1 \otimes (2 \otimes 3) \neq (1 \otimes 2) \otimes 3,$$

and it is not a quasigroup since

$$2 \otimes 3 = 2 \otimes 1.$$

(ii) The set of all positive integers is a semigroup under multiplication, but it is not a quasigroup since

$$3 \otimes x = 2$$

has no solution.

(iii) The three objects a,b and c, with the rules that the product of any element with itself is itself and the product of two different elements is the third element, is a commutative idempotent quasigroup.

The various axioms are not independent. For example, A3, A4 and A5 cannot all hold if G has more than one element. If G is a finite set then the axioms A1, A2 and A3 are a set of axioms for a group and between them they imply A4.

DEFINITION 55. A *Latin square* of order r based on a set R of size r is an r×r array whose entries are chosen from R in such a way that every member of R appears precisely once in each row and once in each column.

THEOREM 56. *If G is a quasigroup of order r with elements $g_1, g_2, \ldots, g_r$, then the array with (i,j) entry $g_i \times g_j$ is a Latin square based on G. Conversely, given a Latin square based on a set G, the reversal of this construction defines a binary operation X under which G is a quasigroup.*

DEFINITION 57. The *join* of two Latin squares of side r is the array whose (i,j) entry is the ordered pair with the (i,j) entry of the first Latin square as its first member and the (i,j) entry of the second Latin square as its second member; two Latin squares based are *orthogonal* if the ordered pairs which occur once as entries in their join are all different.

THEOREM 58. *Suppose L and M are orthogonal Latin squares. If L' and M' are obtained from L and M by reordering the rows simultaneously; that is, by carrying out a permutation on the rows of L and the same permutation on the rows of M, then L' and M' are orthogonal; and similarly for columns. If M is formed from M' by relabelling the members of the set on which M is based, then L and M' are orthogonal.*

This theorem tells us, for example, that there is no loss of generality in assuming that two orthogonal Latin squares of side r are both based on {1,2,...,r} and both have first row (1,2,...,r).

If every pair of Latin squares in a set is an orthogonal pair, the squares are called *mutually* or *pairwise* orthogonal.

THEOREM 59. *If N(r) is the largest possible number of pairwise orthogonal Latin squares of side r, then*

*(i) $N(r) \leq r-1$; equality holds when r is a prime power;*

*(ii) $N(1) = N(2) = N(6) = 1$; in all other cases $N(r) \geq 2$.*

Example. The following arrays are pairwise orthogonal Latin squares of side 4 based on {1,2,3,4}:

| | | | | | | | | | | | | | |
|---|---|---|---|---|---|---|---|---|---|---|---|---|---|
| 1 | 3 | 4 | 2 | | 1 | 4 | 2 | 3 | | 1 | 3 | 4 | 2 |
| 3 | 1 | 2 | 4 | | 3 | 2 | 4 | 1 | | 4 | 2 | 1 | 3 |
| 4 | 2 | 1 | 3 | | 4 | 1 | 3 | 2 | | 2 | 4 | 3 | 1 |
| 2 | 4 | 3 | 1 | | 2 | 3 | 1 | 4 | | 3 | 1 | 2 | 4 |

DEFINITION 60. We say that two orthogonal Latin squares L and M of side r have a *common transversal* if there is a set of r positions, one in each row and each column, such that L has different entries in each of the positions and so does M.

If L and M are Latin squares with a transversal we can reorder rows simultaneously so that the transversal lies on the diagonal positions of L and M, and then relabel the entries of L and M so that the ith diagonal entry of each square becomes i. In other words, if two orthogonal Latin squares have a common transversal then there is no loss of generality in assuming that the diagonal entries of their join is

$$(1,1),\ (2,2),\ (3,3),\ \ldots\quad .$$

Example. The second and third Latin squares in the above example have a common transversal which lies in the diagonal positions.

THEOREM 61. *If $r \neq 1,2,3$ or 6 then there are orthogonal Latin squares of order r with a common transversal.*

For further material on the algebraic structures, refer to R.H. Bruck, "What is a loop?", in *Studies in Modern Algebra* (ed. A.A. Albert), M.A.A., Washington, 1963. A proof of Theorem 58 is given in Marshall Hall Jr., *Combinatorial Theory*, Blaisdell, Waltham, 1967. Theorem 61 is more recent; see A. Hedayat, E.T. Parker and W.T. Federer, "The existence and construction of two families of designs for two successive experiments", *Biometrika*, 57 (1970), 351-355.

2.7 PARTITIONS. $(k,A_i)$-subsets, Maximal Sum-free Sets, $\lambda(G)$, Schur Function, n-fold Regular.

DEFINITION 62. Let S be an s-set, and let $\Pi_r(S)$ denote the collection of all r-subsets, $A_i$, of S. Further, let

$$\Pi_r(S) = A_1 \cup A_2 \cup \ldots \cup A_n$$

be a partition of $\Pi_r(S)$ into n mutually disjoint subsets. Suppose that for some $k \geq r$, there exists a k-subset K of S, such that all the r-subsets of K belong to the same $A_i$, for some i. Then we call K a *$(k,A_i)$-subset of S.*

THEOREM 63. *(Ramsey) Let $n,k_1,k_2,\ldots,k_n,r$ be positive integers with $k_i \geq r$, $i = 1,\ldots,n$. Then there exists a least positive integer $R(k_1,\ldots,k_n,r)$ such that* *the following statement is true for any* $s \geq R(k_1,k_2,\ldots,k_n,r)$:

*For any s-set S and for any partition of* $\Pi_r(S)$ *into n classes* $A_1,\ldots,A_n$ *then there exists a subset* $K_i$ *of S which is a* $(k_i,A_i)$*-subset for some* $i = 1,2,\ldots,$ *or n.*

To illustrate these ideas, we consider the following old puzzle:

> There are s people at a party, where $s \geq 6$. Show that there are either three mutual acquaintances or three mutual strangers.

We write S for the s-set of all people at the party. If $r = 2$ then $\Pi_r(S)$ is the collection of unordered pairs of people. $\Pi_r(S)$ is partitioned into two classes, $A_1$ being the set of all pairs who know each other and $A_2$ the set of all pairs of strangers. We put $k_1 = k_2 = 3$. Then Ramsey's Theorem tells us that there is a number $R(3,3,2)$ such that when $s \geq R(3,3,2)$, S contains a subset K of size 3 in which either every pair of elements belongs to $A_1$ or every pair belongs to $A_2$; that is, there are three people of whom either every pair are acquaintances or every pair are strangers. The puzzle asks us to show that $R(3,3,2) \leq 6$.

This puzzle can be interpreted in graph-theoretic terms:

> The edges of the complete graph on s vertices, $s \geq 6$, are all coloured, some red and some blue. Show that the graph contains either a red triangle or a blue triangle.

The vertices, of course, correspond to the people at the party and red edges to pairs of acquaintances. Generalizing this we have the following analogue of Ramsey's Theorem in the case $r = 2$.

THEOREM 64. *Given positive integers* $n,k_1,k_2,\ldots,k_n$ *with each* $k_i \geq 2$, *there is a least positive integer* $R(k_1,k_2,\ldots,k_n,2)$ *such that the following statement is true for every* $s \geq R(k_1,k_2,\ldots,k_n,2)$:

*For any edge-colouring of* $K_s$ *into n colours there exist an i,* $1 \leq i \leq n$, *and a subset L of the vertices of size* $k_i$ *such that every edge which joins two vertices in L has colour i.*

DEFINITION 65. A set S of elements of a group G is said to be a *sum-free set* if the equation $x_1 + x_2 - x_3 = 0$ has no solution for $x_1,x_2,x_3 \in S$.

DEFINITION 66. Suppose S is a sum-free set in G and for every $T \leq G$ which is sum-free $|T| \leq |S|$. Then S is a *maximal sum-free set* in G and $|S| = \lambda(G)$.

EXAMPLE. In the integers modulo 12, {1,3,5,7,9,11} is a maximal sum-free set and $\lambda(G) = 6$.

The simplest example of the use of sum-free sets is in finding lower bounds for Ramsey numbers:

EXAMPLE (due to Greenwood and Gleason). Let G be the elementary abelian group of order 16 with generators a,b,c,d. Partition the non-zero elements of G into three disjoint sum-free sets:

$$S_1 = \{a,b,c,d,a+b+c+d\},$$
$$S_2 = \{a+b,\ b+c,\ c+d,\ a+b+c,\ b+c+d\},$$
$$S_3 = \{a+c,\ a+d,\ b+d,\ a+c+d,\ a+b+d\},$$

and assign to each of these sets one of three colours. In the complete graph on 16 vertices, label each vertex with a group element. The edge joining vertex x to vertex y, we label $x+y$. Now $x+y$ belongs to exactly one of our sum-free sets and we colour this edge with the colour assigned to this set.

Suppose $x+y$ and $y+z$ belong to the same set, so that we have a triangle with two edges the same colour. Then $(x+y)+(y+z) = x+z$ must belong to a different set since each is sum-free, so the third edge of the triangle is a different colour and no monochromatic triangle is possible. Hence $R_3(3,2) \geq 16$; in fact, Greenwood and Gleason showed that $R_3(3,2) = 17$. (Notation: $R_3(3,2) = R(3,3,3,2)$.)

DEFINITION 67. The *Schur function f(n)* is the largest positive integer such that it is possible to partition the integers $\{1,2,\ldots,f(n)\}$ into n sets, none of which contains a solution to the equation $x_1 + x_2 - x_3 = 0$; that is, into n sum-free sets.

More general functions are defined analogously:

DEFINITION 68. If m,n are positive integers, Turan defines *f(m,n)* to be the largest integer such that $\{m,m+1,\ldots,m+f(m,n)\}$ can be partitioned into n sum-free sets.

We could also consider other functions such that $\{1,\ldots,g(n)\}$, say, could be partitioned into n sets, none of which contain a solution to the equation $\sum_{i=1}^{m} a_i x_i = 0$ where the $a_i$'s are integers; or even more generally we could define a Schur function on a system of simultaneous linear equations.

DEFINITION 69. Rado called the equation $\sum_{i=1}^{m} a_i x_i = 0$ *n-fold regular* if there exists a least positive integer f(n) such that whenever $\{1,\ldots,f(n)+1\}$ is partitioned into n classes in any manner, there is at least one of the classes containing a solution to the given equation. The equation is said to be *regular* if it is n-fold regular for every n.

# PART 2

## ROOM SQUARES

W.D. WALLIS

# CONTENTS


| | | |
|---|---|---|
| CHAPTER I. | INTRODUCTION | 33 |
| 1.1 | First definitions | 33 |
| 1.2 | The existence problem | 33 |
| 1.3 | Isomorphism | 35 |
| 1.4 | Description of Room squares | 35 |
| 1.5 | Isomorphisms of standardized Room squares | 37 |
| 1.6 | Subsquares | 38 |
| 1.7 | Room squares and one-factorizations | 39 |
| CHAPTER II. | STARTERS AND THEIR APPLICATION | 42 |
| 2.1 | The starter-adder approach | 42 |
| 2.2 | Orthogonal starters | 45 |
| 2.3 | Strong starters in groups of prime power order | 48 |
| 2.4 | Strong starters from Steiner triple systems | 50 |
| 2.5 | Quintuplication of a strong starter | 53 |
| CHAPTER III. | THE MOORE-TYPE CONSTRUCTION AND GENERALIZATIONS | 56 |
| 3.1 | Introduction | 56 |
| 3.2 | The Moore-type construction | 56 |
| 3.3 | First extension | 60 |
| 3.4 | Second extension | 62 |
| 3.5 | A square of side $4r+1$ | 63 |
| CHAPTER IV. | THE FRAME CONSTRUCTION | 66 |
| 4.1 | Background | 66 |
| 4.2 | The array $\mathcal{A}_n$ | 67 |
| 4.3 | The Room square construction | 68 |
| CHAPTER V. | DOUBLING CONSTRUCTIONS | 71 |
| 5.1 | Introduction | 71 |
| 5.2 | First doubling theorem | 71 |
| 5.3 | Second doubling theorem | 72 |
| CHAPTER VI. | OTHER CONSTRUCTIONS | 76 |
| 6.1 | A construction using finite projective spaces | 76 |
| 6.2 | Construction of adders for patterned starters | 78 |
| 6.3 | Construction from pairwise balanced designs | 81 |
| 6.4 | Notes on the method of pairwise balanced designs | 83 |
| CHAPTER VII. | THE EXISTENCE PROBLEM | 85 |
| 7.1 | Fermat numbers | 85 |
| 7.2 | General existence results | 87 |

CHAPTER VIII. ROOM SQUARES OF SIDE 7 91

8.1 One-factorizations of $K_8$ 91
8.2 Room squares of side 7 94
8.3 Skew squares of side 7 97

CHAPTER IX. ROOM DESIGNS OF HIGHER DIMENSION 100

9.1 The number of orthogonal symmetric Latin squares 100
9.2 Construction from starters 101
9.3 Room designs of higher dimension 103
9.4 Higher designs of side 7 104

CHAPTER X. MISCELLANEA 106

10.1 Some isomorphism results 106
10.2 Howell rotations 109
10.3 A statistical application 110
10.4 Some unsolved problems 112

BIBLIOGRAPHIC NOTES 115

REFERENCES 118

# CHAPTER I. INTRODUCTION

1.1 FIRST DEFINITIONS. T.G. Room introduced the idea of a Room square to mathematics in a short note [33] in 1955. Unknown to him, equivalent objects had previously been used by the directors of duplicate bridge tournaments under the name of Howell rotations. Room considered the squares as a mathematical recreation only, but a statistical application was soon found by Archbold and Johnson [2]; moreover there are connections between Room squares and other mathematical objects - quasigroups, Latin squares, graph factorizations, Steiner triple systems - as we shall presently see.

A Room square $\mathcal{R}$ of side r, where r is an odd integer, is an r×r array whose cells may be empty or may contain an unordered pair of entries from a set R of size r + 1, satisfying the following conditions:

(i) every unordered pair of elements of R occurs precisely once in $\mathcal{R}$;

(ii) every member of R occurs precisely once in each row and precisely once in each column of $\mathcal{R}$.

We shall speak of $\mathcal{R}$ as being a Room square based on the set R.

Obviously we shall be dealing with unordered pairs of objects throughout the monograph, so it is important to distinguish between the ordered pair (x,y) and the unordered pair {x,y}. We sometimes write x,y or xy for an unordered pair when there is no chance of confusion, especially in the figures, and we regularly refer to an edge of a graph in the form xy.

1.2 THE EXISTENCE PROBLEM. The first question about Room squares is: when do they exist?

It is easy to investigate small values of r. For r = 1 the 1 × 1 array

| |
|---|
| 0, 1 |

is a Room square. The case r = 3 is solved almost as easily: suppose we had a Room square of side 3 based on {0,1,2,3}. By reordering rows and columns we can ensure that {0,1} lies in the (1,1) position. In order to complete the first row, {2,3} must lie in cell (1,2) or in cell (1,3); and to complete the first column, {2,3} must lie in cell (2,1) or in cell (3,1). This is a contradiction, and no square can exist.

In the case r = 5, we can assume the first row is

| | | | | |
|---|---|---|---|---|
| 0,1 | 2,3 | 4,5 | - | - |

.

It makes no matter whether the first column contains {2,4}, {3,5} or {2,5}, {3,4}, as the first case can be converted into the second by the element permutation $4 \leftrightarrow 5$, and this does not alter the (unordered) pairs in the first row. So there is no loss of generality in assuming a Room square of side 5 has the form

| | | | | |
|---|---|---|---|---|
| 0,1 | 2,3 | 4,5 | - | - |
| 2,4 | | | | |
| 3,5 | | | | |
| - | | | | |
| - | | | | |

where a dash indicates an empty cell.

The entries {2,5} and {3,4} must appear somewhere, but they cannot be in rows 1 to 3 or columns 1 to 3. Moreover, they cannot be in the same row or the same column, as this would entail {0,1} occurring again. Say {2,5} is in the (4,4) cell and {3,4} is in the (5,5). If there is an entry in cell (4,5) it must be {0,1}, which has already been used, so the (4,5) cell is empty, as is the (5,4) cell.

Cells (4,2), (4,3), (5,2) and (5,3) must be occupied in order that the fourth and fifth rows shall each contain three pairs. Cell (5,2) cannot contain 2, 3 or 4, so it may have {0,5} or {1,5} in it; if it were {1,5} we could interchange 0 and 1 throughout the square, so assume the entry is {0,5}. We can now fill in cells (5,3), (4,2) and (4,3), there being only one possibility in each case. So the square is a completion of

| | | | | |
|---|---|---|---|---|
| 0,1 | 2,3 | 4,5 | - | - |
| 2,4 | - | - | | |
| 3,5 | - | - | | |
| - | 1,4 | 0,3 | 2,5 | - |
| - | 0,5 | 1,2 | - | 3,4 |

.

The pairs {0,2} and {0,4} remain to be placed. They cannot both be in the third row, but neither one can be in row 2. So completion of the square is impossible.

So far the only Room square we have found is of side 1, and is trivial; in the next two cases there is no answer. At this point one begins to suspect that there might be no further Room squares, and that this monograph might turn out to be very short indeed! But the reality is very different. We shall prove that there is a Room square of every odd side except 3, 5 and 257; in view of the above remarks, only side 257 remains in doubt.

1.3 ISOMORPHISM. We shall define two Room squares to be *isomorphic* if it is possible to obtain one from the other by a sequence of the following operations:

(i) permute the rows;

(ii) permute the columns;

(iii) relabel the elements of the set on which it is based.

(Clearly these three operations commute.) If $\mathcal{S}$ is a square of side r based on S, relabel the elements of S according to any scheme which transforms S into $R = \{0,1,2,\ldots,r\}$. Then it is possible to permute the rows and columns in such a way as to make $\{0,i\}$ lie in the $(i,j)$ cell. We shall say the resulting square is *standardized*. There can be many standardized squares corresponding to the same Room square.

It is sometimes natural to consider two Room squares essentially the same when one is the transpose of the other; sometimes it is not. We shall call squares $\mathcal{R}$ and $\mathcal{S}$ *equivalent* when $\mathcal{R}$ is isomorphic to $\mathcal{S}$ or to the transpose of $\mathcal{S}$.

1.4 DESCRIPTION OF ROOM SQUARES. One useful descriptor of a Room square is its incidence matrix. If $\mathcal{R}$ is a Room square of side r then its incidence matrix is of size $r \times r$ and has entry 1 or 0 in its $(i,j)$ position according as the $(i,j)$ cell of $\mathcal{R}$ is occupied or empty. This incidence matrix is regular, as every row or column contains $\frac{1}{2}(r+1)$ entries 1; when $r+1$ is divisible by 4, say $r = 4m-1$, the matrix may possibly be the incidence matrix of a symmetric balanced incomplete block design with parameters $(4m-1, 2m, m)$, and in that case we call the square an *Hadamard Room square* (since the block design corresponds to an Hadamard matrix).

In addition to Hadamard Room squares there is another special class; namely, Room squares with complements. Two subclasses, skew Room squares and embedded Room squares, have been used in the literature.

Suppose $\mathcal{R}_1$ and $\mathcal{R}_2$ are standardized Room squares of side r, based on the set $R = \{0,1,2,\ldots,r\}$. We say that $\mathcal{R}_1$ and $\mathcal{R}_2$ are *complementary* if, in every case when $i \neq j$, one of $\mathcal{R}_1$ and $\mathcal{R}_2$ has an entry in position $(i,j)$ but not both. Putting it another way, if $A_i$ is the incidence matrix of $\mathcal{R}_i$, then

$$A_1 + A_2 = J + I.$$

The importance of complementary Room squares stems from the usefulness of a particular array which we shall call the *sum* of $\mathcal{R}_1$ and $\mathcal{R}_2$. This sum is formed by first deleting all diagonal entries from $\mathcal{R}_1$ and $\mathcal{R}_2$, then adding r to every member of every pair remaining in $\mathcal{R}_2$, and finally superimposing the two squares. By the complementary property we are not asked to put two entries in the same cell in this process, so the sum is an array of unordered pairs.

Suppose $\mathcal{R}$ is a standardized Room square of side r in which the entry {0,x} has been replaced by {x,x} for every x. It may be possible to order every entry of $\mathcal{R}$, replacing x,y either by (x,y) or by (y,x), and then to add further entries in such a way that the resulting array is the join of a pair of orthogonal Latin squares. If this can be done we call $\mathcal{R}$ an *embedded* Room square, and say it is embedded in the join of the two Latin squares.

A *skew* Room square $\mathcal{R}$ is a standardized Room square in which, whenever i and j are different, $\mathcal{R}$ has an entry in exactly one of its (i,j) and (j,i) cells and has the other one empty. Alternatively, if A is the incidence matrix of a standardized square $\mathcal{R}$, $\mathcal{R}$ is skew if

$$A + A^T = J + I.$$

A matrix A of zeros and ones which has every diagonal entry 1 and satisfies this equation $A + A^T = J + I$ will be called *skew-type*.

THEOREM 1.1. *If there is a skew Room square or an embedded Room square of side r then there are complementary Room squares of side r.*

PROOF. It is obvious that a skew Room square and its transpose are complements.

Suppose $\mathcal{R}$ is a standardized Room square of side r embedded in the join $\mathcal{L}$ of two orthogonal Latin squares. Delete from $\mathcal{L}$ every entry which is in the same position as an entry of $\mathcal{R}$, and interpret the remaining entries as unordered pairs. Then put in diagonal entries {0,1}, {0,2},..., {0,r}. The result will be Room, and will be complementary to $\mathcal{R}$.

Suppose L is a Latin square which is orthogonal to its transpose. (Such squares are called self-orthogonal.) The entry (x,x) occurs on the diagonal of the join $\mathcal{L}$ of L and $L^T$; and if (x,y) occurs in position (i,j) then (y,x) occurs in position (j,i). Exactly one of the entries (x,y) and (y,x) is retained (as an unordered pair) in a Room square embedded in $\mathcal{L}$, so that exactly one of the (i,j) and (j,i) positions would be occupied. This means that a Room square which is embedded in the join of a self-orthogonal Latin square with its transpose will necessarily be skew.

Although the definitions of complementary, skew and embedded Room squares are different, we know of no side for which one of these types of squares exists and another does not. The author feels that this probably indicates the paucity of knowledge on special Room squares. However, an investigation of this problem would be very interesting.

1.5 ISOMORPHISMS OF STANDARDIZED ROOM SQUARES. Suppose $\mathcal{R}$ is a standardized Room square based on $R = \{0,1,\ldots,r\}$. If $\mathcal{S}$ is a Room square isomorphic to $\mathcal{R}$ then $\mathcal{S}$ can be obtained from $\mathcal{R}$ by first permuting the elements of R, by a permutation $\phi$ say, and then permuting rows and columns. If $\mathcal{S}$ is to be standardized also, then $\phi$ determines the row and column permutations: if $x\phi = 0$, and $\{x,y\}$ lies in cell $(i_y,j_y)$ of $\mathcal{R}$, then the row and column permutations will be

$$i_y \mapsto y\phi \text{ and } j_y \mapsto y\phi$$

respectively. Given $\mathcal{R}$ and $\phi$, we shall write $\mathcal{R}\phi$ for the standardized square obtained from $\mathcal{R}$ by using the element permutation $\phi$; from what we have said, $\mathcal{R}\phi$ is uniquely determined.

LEMMA 1.2. *If $\mathcal{R}$ is standardized and $0\phi = 0$ then the row and column permutations induced by $\phi$ are each equal to $\phi$.*

THEOREM 1.3. *Suppose $\mathcal{R}$ is a standardized Room square based on R whose incidence matrix is A, and $\phi$ is a permutation of R under which $x \mapsto 0$. If there is an automorphism $\psi$ of $\mathcal{R}$ which maps $x$ to 0, then $\mathcal{R}\phi$ has incidence matrix $P^{-1}AP$ for some permutation matrix P and $\mathcal{R}\phi$ is skew if and only if $\mathcal{R}$ is skew.*

PROOF. Since $\psi$ is an automorphism of $\mathcal{R}$, the incidence matrix of $\mathcal{R}\psi$ is A. Now

$$\mathcal{R}\phi = \mathcal{R}\psi(\psi^{-1}\phi);$$

$\psi^{-1}\phi$ maps 0 to 0, so by the lemma the incidence matrix of $\mathcal{R}\phi$ is found by permuting both the rows and the columns of A by $\psi^{-1}\phi$. So it is

$$P^{-1}AP,$$

where P is the permutation matrix corresponding to $\psi^{-1}\phi$. Finally, it is clear that A is skew-type if and only if $P^{-1}AP$ is.

COROLLARY 1.4. *In order to find out whether there is a skew Room square isomorphic to the standardized Room square $\mathcal{R}$, it is sufficient to examine the squares $\mathcal{R}\psi_x$, where $x$ ranges through a set of representatives of the orbits of R under the action of the automorphism group of $\mathcal{R}$ and $\psi_x$ is the cycle $(0x)$.*

PROOF. Suppose $\mathcal{R}\phi$ is skew, and suppose $x\phi = 0$. Then

$$\mathcal{R}\phi = \mathcal{R}\psi_x(\psi_x\phi),$$

and $\psi_x\phi$ maps 0 to 0. There is an automorphism of $\mathcal{R}\psi_x$ which maps 0 to 0 ( for example, the identity map), so $\mathcal{R}\phi$ is skew if and only if $\mathcal{R}\psi_x$ is skew. So it is sufficient to examine the $\mathcal{R}\psi_x$.

Now suppose x and y belong to the same orbit of the automorphism group of $\mathcal{R}$. Let $\chi$ be an automorphism of $\mathcal{R}$ such that $x\chi = y$. Then $\psi_x\chi\psi_y$ is an automorphism of

$\mathcal{R}\psi_y$ since

$$\mathcal{R}\psi_y(\psi_y\chi\psi_y) = \mathcal{R}\chi\psi_y = \mathcal{R}\psi_y ;$$

$\psi_y\chi\psi_y$ maps x to 0. The permutation $\psi_y\psi_x$ is the cycle (0yx), which maps x to 0. Now

$$\mathcal{R}\psi_x = \mathcal{R}\psi_y(\psi_y\psi_x).$$

Replacing $\mathcal{R}$ by $\mathcal{R}\psi_y$, $\phi$ by $\psi_y\psi_x$ and $\psi$ by $\psi_y\chi\psi_y$ in theorem 1.3 we see that $\mathcal{R}\psi_y$ and $\mathcal{R}\psi_x$ are both skew or else neither is skew. So it is sufficient to consider a set of $\psi_x$ where only one x in each orbit occurs.

COROLLARY 1.5. *If $\mathcal{R}$ is a standardized Room square whose automorphism group is transitive then either every standardized square isomorphic to $\mathcal{R}$ is skew or none is.*

Very little is known about Room square isomorphism. One interesting problem which presents itself is the following:
Given a Room square of side r based on an arbitrary set R, it can be standardized in (r+1)! ways, according to the way in which R is mapped to {0,1,2,...,r}. Is it possible to do this in such a way that the resulting Room square is skew, and also to find another way which does not yield a skew square? We shall show in section 8.3 that the answer is 'yes' by finding an example of side 7. Expressed another way, the example proves that skewness is not an isomorphism-invariant.

1.6 SUBSQUARES. Suppose $\mathcal{R}$ is a Room square of side r based on R, and consider the result of selecting a set of s rows and a set of s columns of $\mathcal{R}$ and deleting all other rows and columns, so that the result is an s×s array $\mathcal{S}$. It may happen that all the entries in $\mathcal{S}$ are pairs of members of a subset S of R with s + 1 elements, and that $\mathcal{S}$ is in fact a Room square based on S. We then call $\mathcal{S}$ a *subsquare* of $\mathcal{R}$. According to this definition every Room square has itself as a subsquare, and every Room square has a subsquare of side 1.

THEOREM 1.6. *If there is a Room square of side r with a subsquare of side s, where $r > s$, then*

$$r \geq 3s + 2.$$

PROOF. Suppose $\mathcal{R}$ is a Room square of side r based on R, and $\mathcal{S}$ is a subsquare of $\mathcal{R}$ of side s based on S. Without loss of generality we may assume $\mathcal{S}$ occupies the first s rows and columns of $\mathcal{R}$, so that

$$\mathcal{R} = \begin{array}{|c|c|} \hline \mathcal{S} & \mathcal{A} \\ \hline \mathcal{B} & \mathcal{C} \\ \hline \end{array} .$$

Let x be a member of R\S. (This set is not empty, since r > s.) Then x occurs once in each of rows 1,2,...,s, and all those occurrences must be in the block $\mathcal{A}$. So, if the pair containing x in row i is $\{x,x_i\}$, none of $x_1,x_2,\ldots,x_s$ is in S. Similarly, the occurrence of x in column i, $1 \leq i \leq s$, must be in block $\mathcal{B}$; if the pair is $\{x,y_i\}$ then $y_i \notin S$. The objects $x_1,x_2,\ldots,x_s,y_1,y_2,\ldots,y_s$ must all be different, since they all occur with x in different positions, and they are all different from x. So we know at least 3s + 2 elements of R; namely, $x,x_1,x_2,\ldots, x_s,y_1,y_2,\ldots,y_s$, and the s + 1 members of S. So

$$r \geq 3s + 1.$$

But r and s are both odd, so r = 3s + 1 is impossible. So

$$r \geq 3s + 2.$$

In the case of a *skew* Room square $\mathcal{R}$ we define a special sort of subsquare called a *skew subsquare*. This is one which is situated symmetrically about the diagonal of $\mathcal{R}$. More formally, we say $\mathcal{S}$ is a skew subsquare of the skew Room square $\mathcal{R}$ if $\mathcal{S}$ is a subsquare of $\mathcal{R}$ and if $\mathcal{S}$ is situated at the intersection of rows $a_1,a_2,\ldots,a_s$ with columns $a_1,a_2,\ldots,a_s$ of $\mathcal{R}$, for some $a_1,a_2,\ldots,a_s$. A skew subsquare is a subsquare and is a skew Room square, but the converse need not hold. The reason for this definition will be seen in section 3.2.

1.7 ROOM SQUARES AND ONE-FACTORIZATIONS. Suppose $\mathcal{R}$ is a Room square of side r based on R; write K for the complete graph $K_{r+1}$ with R for its set of vertices. We construct a one-factorization of K, the *row factorization* $\mathcal{F}$, as follows: if $\{x_0,y_0\}$, $\{x_1,y_1\},\ldots,\{x_n,y_n\}$ are the pairs in one row of $\mathcal{R}$, then $x_0y_0,x_1y_1,\ldots,x_ny_n$ are the edges of one factor in $\mathcal{F}$. Analogously we define the *column factorization* $\mathcal{G}$ of $\mathcal{R}$. The two one-factorizations $\mathcal{F}$ and $\mathcal{G}$ have the property that if F is any factor in $\mathcal{F}$ and G is any factor in $\mathcal{G}$, then F and G have at most one edge in common. Two one-factorizations with this property will be called *orthogonal*. Given two orthogonal one-factorizations of $K_{r+1}$ it is easy to obtain a Room square by reversing the construction; the resulting Room square is unique only up to row and column permutation because there is no fixed ordering on the factors of a one-factorization.

If $\phi$ is a permutation of the vertices of a $K_{r+1}$, and $\mathcal{F}$ is a one factorization of the graph, we define $\mathcal{F}\phi$ to be the one-factorization whose factors are derived from the factors in $\mathcal{F}$ using $\phi$: if

$$x_0y_0\ x_1y_1\ \ldots$$

is a factor in $\mathcal{F}$ then

$$(x_0\phi)(y_0\phi)\ (x_1\phi)(y_1\phi)\ \ldots$$

is a factor in $\mathcal{F}\phi$. $\mathcal{F}$ and $\mathcal{G}$ are isomorphic one-factorizations when $\mathcal{F}\phi = \mathcal{G}$ for some $\phi$. It will be convenient to assume a standard ordering of the factors in a one-factorization, and accordingly, we shall always assume that one vertex (0 say) has been distinguished and that 0i always belongs to the ith factor. If we talk of a one-factorization $\mathcal{F}\phi$ it shall be assumed that the factors in $\mathcal{F}\phi$ have been reordered if necessary to attain this standardized form. Clearly we have

THEOREM 1.7. *Standardized Room squares $\mathcal{R}$ and $\mathcal{S}$ are isomorphic if and only if there is a permutation which is simultaneously an isomorphism from the row factorization of $\mathcal{R}$ to the row factorization of $\mathcal{S}$ and an isomorphism from the column factorization of $\mathcal{R}$ to the column factorization of $\mathcal{S}$.*

This approach to Room squares through one-factorizations is useful when discussing isomorphism of Room squares, for the following reason: there is a convenient set of isomorphism-invariants for a one-factorization. As in [46], we define a *k-division* in a one-factorization to be a set of k factors whose union is disconnected. Given an $\ell$-division, where $\ell > k$, any k factors in the $\ell$-division will be a k-division; we call a k-division *maximal* if it cannot be embedded in a (k+1)-division. Then, if $d_k$ is the number of maximal k-divisions in a factorization $\mathcal{F}$, the numbers $d_k$ are invariant under isomorphisms of $\mathcal{F}$. In a factorization of $K_{2s}$ it is clear that

$$d_1 = 2s-1$$

$$d_k = 0 \text{ if } \begin{cases} s \text{ is even and } k > s-1 \\ s \text{ is odd and } k > s-2 \end{cases}.$$

Suppose $\mathcal{F}$ is a one-factorization of $K_{r+1}$,

$$\mathcal{F} = \{F_1, F_2, \ldots, F_r\},$$

where r is odd; and suppose further that 0i belongs to $F_i$. We define a square array $F = (f_{ij})$ of side r according to the rules:

(i) $f_{ii} = i$ for all i;

(ii) if $ij \in F_k$ then $f_{ij} = k$.

Then

$$\{f_{ij}: 1 \le i \le r\} = \{i\} \cup \{k: ij \in F_k \text{ some } j > 0\}.$$

Of the edges ij, where j varies, exactly one will occur in each $F_k$; the edge 0i is in $F_i$. So

$$\{k: ij \in F_k \text{ some } j > 0\} = \{1,2,\ldots,i-1,i+1,\ldots,r\},$$

and the entries $f_{ij}$, $1 \le j \le r$, comprise $\{1,2,\ldots,r\}$. This means that every row contains every number from 1 to r once; and the corresponding remark applies to columns. In other words, F is a Latin square, and we shall call it the *Latin square associated with* $\mathcal{F}$. F will be symmetric.

Now suppose $\mathcal{R}$ is a standardized Room square of side r, whose row and column factorizations $\mathcal{F}$ and $\mathcal{G}$ have associated Latin squares F and G. Write $\mathcal{L}$ for the join of F and G.

If the entry (x,y) occurs in the (i,j) cell of $\mathcal{L}$, where $i \neq j$, then $\mathcal{R}$ must have had {i,j} in row x and column y. No other pair can occur in that position, so (x,y) does not appear anywhere in $\mathcal{L}$ outside positions (i,j) and (j,i). Since $\mathcal{R}$ is standardized, $x \neq y$. The entries (x,x) occur on the diagonal of $\mathcal{L}$, but not otherwise. This is as close as we can come to orthogonality in symmetric squares. Accordingly, we shall define F and G to be *orthogonal symmetric Latin squares** when both F and G are symmetric Latin squares and their join satisfies:

every ordered pair (x,x) occurs precisely once, on the diagonal;

no ordered pair (x,y) occurs twice above the diagonal.

It is clear that a Room square of side r is equivalent to a pair of orthogonal symmetric Latin squares of side r.

Suppose the set G forms one commutative idempotent quasigroup under the binary operation # and another under the operation Δ. To distinguish these quasigroups we denote them (G,#) and (G,Δ); we write L(#) and L(Δ) for the Latin squares equivalent to (G,#) and (G,Δ) respectively. If $\mathcal{L}$ is the join of L(#) and L(Δ), then the ordered pairs $(g_i,g_i)$ will occur once each on the diagonal of $\mathcal{L}$, since the quasigroups are idempotent. If, further,

(i) whenever x,y and $p \in G$ satisfy

$$x \# y = x \Delta y = p,$$

$x = y = p$;

(ii) whenever $p,q \in G$, $p \neq q$, there exists at most one unordered pair of elements x and y of G such that

$$x \# y = p, \; x \Delta y = q;$$

then L(#) and L(Δ) will be orthogonal symmetric Latin squares. In this situation (G,#) and (G,Δ) will be called a *Room pair of quasigroups*, and the study of Room squares could be conducted from this point of view. It is easy to see that if we start with a pair of orthogonal symmetric Latin squares, then the corresponding quasigroups form a Room pair.

---

* by analogy with "free Abelian group".

## CHAPTER II. STARTERS AND THEIR APPLICATION

2.1 THE STARTER-ADDER APPROACH. In this section we discuss an approach to Room squares first used by Stanton and Mullin in [39], and given in its more general form by Mullin and Nemeth [25].

Suppose $r = 2s+1$, where s is any positive integer, and suppose G is an abelian group of order r written in additive notation. By a starter in G we shall mean a set X of unordered pairs of elements of G,

$$X = \{\{x_1,y_1\}, \{x_2,y_2\},\ldots,\{x_s,y_s\}\}$$

with the properties that $\{x_1,x_2,\ldots,x_s,y_1,y_2,\ldots,y_s\}$ contains every element of G other than zero once each and that the differences $\pm(x_1-y_1)$, $\pm(x_2-y_2)$,..., $\pm(x_s-y_s)$ also comprise all the non-zero elements. By a counting argument we see that the elements $\pm(x_i-y_i)$ must be all different.

Given an ordering of the elements of X, an adder $A_X$ for the starter X will mean an ordered set of s distinct non-zero elements of G,

$$A_X = (a_1,a_2,\ldots,a_s),$$

such that $\{x_1+a_1, x_2+a_2,\ldots,x_s+a_s, y_1+a_1, y_2+a_2,\ldots,y_s+a_s\}$ again consists of all non-zero elements of G.

There are two special types of starter. On the one hand, it may be that $x_i+y_i = 0$ for every i; that is, $y_i = -x_i$. Then the starter is called patterned. On the other hand, if the $x_i+y_i$ are all non-zero and all different, the starter is called strong.

THEOREM 2.1. *Every abelian group G of odd order r has a patterned starter.*

PROOF. There is only one candidate for a patterned starter X, since the ordering of the members of X is irrelevant. X is formed by pairing every non-zero member of G with its negative. Since every element except 0 has odd order, this process always results in a set of unordered pairs. Moreover,

$$x_i-y_i = x_i+x_i$$
$$y_i-x_i = y_i+y_i$$

so that $\{\pm(x_i-y_i)\}$ consists of all doubles of non-zero members of G; since $|G|$ is odd,

$$\{x_i+x_i \mid x_i \in G\setminus\{0\}\} = G\setminus\{0\}$$

and X is a starter.

THEOREM 2.2. *If X is a strong starter in a group G of order r, then $A_X$ is an adder for X where*

$$A_X = (-x_1-y_1, -x_2-y_2, \ldots, -x_s-y_s).$$

PROOF. If we write $a_i$ for $-x_i-y_i$ then

$$x_i+a_i = -y_i$$

$$y_i+a_i = -x_i$$

so $\{x_i+a_i, y_i+a_i\} = \{-x \mid x \in G, x \neq 0\}$

$$= \{x \mid x \in G, x \neq 0\}.$$

So $A_X$ is an adder for X.

We shall refer to an adder $A_X$ as skew if $a \in A_X$ implies that $-a \notin A_X$.

THEOREM 2.3. *If there is an abelian group G of odd order r with starter and an adder, then there is a Room square of side r. If the adder is skew, then the square is skew Room.*

PROOF. Suppose the entries of G are $0 = g_1, g_2, \ldots, g_r$. It is convenient to construct an array $\mathcal{R}$ whose rows and columns are labelled $g_1, g_2, \ldots, g_r$. Denote the ith pair in a starter X as $\{x_i, y_i\}$ and the ith member of an adder $A_X$ for X as $a_i$.

We first construct the first row (row $g_1$) of $\mathcal{R}$. Cell $(g_1, g_1)$ contains $\{g_0, g_1\}$, where $g_0$ is a new object, not a member of G. When $k \neq 1$, cell $(g_1, g_k)$ is empty unless $-g_k$ is a member of $A_X$, but if $-g_k = a_i$ then cell $(g_1, g_k)$ contains $\{x_i, y_i\}$.

In row $g_j$ we leave cell $(g_j, g_k)$ empty when cell $(g_1, g_k-g_j)$ is empty; however, if $\{x_i, y_i\}$ lies in the $(g_1, g_k-g_j)$ position then we place $\{x_i+g_j, y_i+g_j\}$ in position $(g_j, g_k)$. In order to make this construction apply to the diagonal positions we define $g_0 + g = g_0$ for all g in G; then the jth diagonal entry is $\{g_0, g_j\}$. (The whole procedure is consistent with the case $j = 1$, since $g_1 = 0$.)

We shall show that $\mathcal{R}$ is a Room square based on $R = G \cup \{g_0\}$. There are precisely $\frac{1}{2}(r+1)r$ entries in $\mathcal{R}$, and they are all the pairs $\{x_i+g, y_i+g\}$ where g ranges through all entries of G and i goes from 1 to s (recall that $s = \frac{1}{2}(r-1)$), as well as the diagonal entries $\{g_0, g_i\}$. If $\{g_a, g_b\}$ is any unordered pair of members of G, then from the second property of a starter there will be a pair $(x_i, y_i)$ in X such that

$$\pm(x_i-y_i) = g_a-g_b. \tag{2.3}$$

If we write $g = g_b-y_i$ or $g_a-y_i$ according as the + or - sign occurs in (2.3), then

$$\{x_i+g, y_i+g\} = \{g_a, g_b\}.$$

Therefore, every unordered pair of elements of G is a member of the set $\{x_i+g, y_i+g\}$, so every unordered pair of elements of R occurs somewhere in $\mathcal{R}$. By counting we see that each pair must occur precisely once, so $\mathcal{R}$ satisfies property (i) of a Room square.

From the way in which $\mathcal{R}$ is constructed, it follows that cell $(g_j, g_k)$ contains the entry $\{x_i+g_j, y_i+g_j\}$ if and only if cell $(g_j+g_h, g_k+g_h)$ contains the entry $\{x_i+g_j+g_h, y_i+g_k+g_h\}$. So not only are the entries in row $g_j$ obtained by adding $g_j$ to each member of the entries of row $g_1$, but also the entries in column $g_k$ are obtained by adding $g_k$ to each member of the entries in column $g_1$. (Put $g_h = -g_k$ in the above working.) Now if we add $g_j$ to every member of G we again obtain every member of G, in some order, and $g_0 + g_j = g_0$; so in the obvious notation

$$R + g_j = R.$$

So we will know that every row of $\mathcal{R}$ contains every member of R precisely once if we show that the first row has this property, and similarly for columns. This will prove that $\mathcal{R}$ satisfies the final property of a Room square. Row $g_1$ contains all the pairs $\{x_i, y_i\}$ of X, which constitute all the non-zero members of G once each by the definition of a starter, and also $\{g_0, g_1\}$, so it contains every member of R precisely once. For column $g_1$ we observe that when $j > 1$ cell $(g_j, g_1)$ contains $\{x_i+g_j, y_i+g_j\}$ if and only if cell $(g_1, -g_j)$ contains $\{x_i, y_i\}$; that is, if and only if $g_j = a_i$. So the entries in the first column are

$$\{g_0, g_1\} \cup \{x_i+a_i,\ y_i+a_i \mid i = 1,2,\ldots,s\}$$

which is R by the definition of an adder.

Finally, assume that $A_X$ is skew: if $a \in A_X$ then $-a \notin A_X$. We can standardize $\mathcal{R}$ by replacing each $g_i$ by i throughout and relabelling row and column $g_i$ as row and column i. If position (j,k) of this square is occupied where $j \neq k$ then $g_j - g_k \in A_X$. Consequently, $g_k - g_j \notin A_X$, so the (k,j) square is unoccupied. So $\mathcal{R}$ is skew Room.

This construction is much easier to understand in the special case where G is the cyclic group of order r. We make the identification $g_1 = r$, $g_2 = 1, \ldots, g_r = r-1$, and $g_0 = 0$. After constructing the first row, we find the entry in cell (j,k) by adding 1 to each entry in cell (j-1,k-1), with the addition and the co-ordinates being taken modulo r except that $0+1 = 0$. For example, in the case $r = 9$, the starter

$$\{1,2\}, \quad \{3,7\}, \quad \{4,6\}, \quad \{5,8\}$$

and the corresponding adder

$$1, \quad 7, \quad 2, \quad 8$$

give rise to the square

| | | | | | | | | |
|---|---|---|---|---|---|---|---|---|
| 09 | 58 | 37 | -- | -- | -- | -- | 46 | 12 |
| 23 | 01 | 69 | 48 | -- | -- | -- | -- | 57 |
| 68 | 34 | 02 | 17 | 59 | -- | -- | -- | -- |
| -- | 79 | 45 | 03 | 28 | 16 | -- | -- | -- |
| -- | -- | 18 | 56 | 04 | 39 | 27 | -- | -- |
| -- | -- | -- | 29 | 67 | 05 | 14 | 38 | -- |
| -- | -- | -- | -- | 13 | 78 | 06 | 25 | 49 |
| 15 | -- | -- | -- | -- | 24 | 89 | 07 | 36 |
| 47 | 26 | -- | -- | -- | -- | 35 | 19 | 08 |

The starter-adder method was used in [39] to construct Room squares of side r for all odd integers r from 7 to 49. The results in that paper were found by computer. In each case the group involved was the cyclic group of order r. Starters and adders for squares of sides 51 and 55 have been found by computer and are given in [6]. In nearly all of these cases the patterned starter was used; however, it was found that there is no possible adder for the patterned starter when r = 9.

2.2 ORTHOGONAL STARTERS. In this section we examine a slightly different view of starters, and define orthogonal starters. We prove that the existence of two orthogonal starters is equivalent to the existence of a starter with an adder; however, the relationship between one-factorizations and starters can be made explicit in terms of orthogonal starters.

Suppose H is an abelian group of order $r = 2s+1$ and suppose X and Y are starters in H. If $\{x_i, x'_i\}$ is any member of X then there are unique non-zero elements $h_i$ and $-h_i$ of H such that

$$x_i - x'_i = h_i,\ x'_i - x_i = -h_i,$$

and there will also be a unique member of Y, $\{y_j, y'_j\}$ say, such that

$$y_j - y'_j = \pm h_i.$$

Assume $y_j - y'_j = h_i$ (if necessary, the labels $y_j$ and $y'_j$ can be interchanged, since a starter contains unordered pairs). Then

$$x_i - x'_i = y_j - y'_j$$

and the pair $\{y_j, y'_j\}$ is uniquely determined by $\{x_i, x'_i\}$. Write

$$d_i = y_j - x_i = y'_j - x'_i.$$

If the $d_i$ are all distinct and non-zero, we shall say that X and Y are orthogonal.

Suppose we have a starter X and another starter Y orthogonal to it. Write

$$A_X = \{d_i\}.$$

Then $\{x_i+d_i\} \cup \{x'_i+d_i\} = \{y_i\} \cup \{y'_i\} = H \setminus \{0\}$, so $A_X$ is an adder for X in H. Consequently, if there is a pair of orthogonal starters in a group of order r, there is a starter in that group with an adder. So there is a Room square of side r. On the other hand, if X is a starter in H and $A_X$ is an adder for X, we can construct a starter Y orthogonal to X. If

$$X = \big\{\{x_1,x'_1\}, \{x_2,x'_2\},\ldots, \{x_s,x'_s\}\big\},$$

$$A_X = \{\ d_1\ \ ,\ \ d_2\ ,\ldots,\ \ d_s\}\ ,$$

where $d_i$ is the element of $A_X$ corresponding to $\{x_i,x'_i\} \in X$, write

$$y_i = x_i+d,$$

$$y'_i = x'_i+d,$$

$$Y = \big\{\{y_1,y'_1\},\{y_2,y'_2\},\ldots,\{y_n,y'_n\}\big\}.$$

There is no trouble checking that Y is a starter and is orthogonal to X.

THEOREM 2.4. *A starter in an abelian group of order r implies the existence of a one-factorization of $K_{r+1}$. If there are two orthogonal starters in a group then the corresponding one-factorizations are orthogonal.*

PROOF. Suppose X is a starter in a group H of order r = 2s + 1. We shall write the group elements as $h_1,h_2,\ldots,h_r$ where $h_1 = 0$. From the properties of a starter, any unordered pair of members of H, $\{h_i,h_j\}$ say, can be expressed in the form

$$\{h_i,h_j\} = \{x+h_k,x'+h_k\},$$

where $\{x,x'\}$ is a member of X and $h_k$ is a member of H; moreover, $\{x,x'\}$ and $h_k$ are uniquely determined by $\{h_i,h_j\}$.

We construct a one-factorization $\mathcal{F} = \{F_1,F_2,\ldots,F_r\}$ of the $K_{r+1}$ based on $\{h_0\} \cup H$:

$$F_k = \{h_0h_k\} \cup \{h_ih_j:\ h_i = x+h_k, h_j = x'+h_k,\ \text{some } \{x,x'\} \in X\}.$$

$$S = \{h_0,h_k\} \cup \{x+h_k,x'+h_k:\ \{x,x'\} \in X\}\ \text{ is the set of all vertices of } F_k.$$

Since X is a starter, $\{x,x':\ \{x,x'\} \in X\} = H \setminus \{0\}$, so

$$S = \{h_0,h_k\} \cup \big((H \setminus \{0\}) + h_k\big)$$

$$= \{h_0,h_k\} \cup H \setminus \{h_k\},$$

so $F_k$ contains every vertex; as there are only s + 1 edges in $F_k$, no vertex can occur in more than one edge. So the $F_k$ are one-factors. From the preceding paragraph every edge of the graph belongs to precisely one of the factors $F_k$, so $\mathcal{F}$ is a one-factorization.

Now suppose X and Y are orthogonal starters which give rise to one-factorizations $\mathcal{F}$ and $\mathcal{G}$. Suppose $\mathcal{F}$ and $\mathcal{G}$ are not orthogonal. Then there will be two edges $h_ih_j$ and $h_ph_q$ which belong to the same factor in $\mathcal{F}$ and also belong to the same factor in $\mathcal{G}$. Provided none of i,j,p,q is 0, $h_ih_j$ and $h_ph_q$ belonging to $F_k$ mean that there exist pairs $\{x_1,x'_1\}$ and $\{x_2,x'_2\}$ in X such that

$$\left.\begin{aligned} h_i &= x_1+h_k,\ h_j = x'_1+h_k, \\ h_p &= x_2+h_k,\ h_q = x'_2+h_k. \end{aligned}\right\} \quad (2.5)$$

Consequently

$$h_i-h_p = x_1-x_2.$$

Since $h_ih_j$ and $h_ph_q$ belong to one factor in $\mathcal{G}$ we also obtain

$$h_i-h_p = y_1-y_2,$$

for some $\{y_1,y'_1\}$ and $\{y_2,y'_2\}$ in Y. We derive

$$y_1-x_1 = y_2-x_2,$$

whence $d_1 = d_2$. As the $d_i$ are all different, $\{x_1,x'_1\} = \{x_2,x'_2\}$. But from this and (2.5)

$$\{h_i,h_j\} = \{h_p,h_q\},$$

contradicting the fact that we started with two edges. Similarly, a contradiction is obtained when one of $\{i,j,p,q\}$ is 0. So the two one-factorizations are orthogonal.

THEOREM 2.6. *If there is a strong starter in an abelian group H of order $r = 2s+1$ then there are three pairwise orthogonal starters in H.*

PROOF. Consider the starters

$$X = \{\{x_i,y_i\}:\ 1 \le i \le s\}$$

$$Y = \{\{-x_i,-y_i\}:\ \{x_i,y_i\} \in X\}$$

$$Z = \{\{h_i,-h_i\}:\ h_i \in H,\ h_i \ne 0\}$$

where X is the strong starter which was given. Y and Z are easily checked to be starters. Since X is strong, all the $x_i+y_i$ are distinct and non-zero. It follows that all the $-x_i-y_i$ are distinct and non-zero, so Y is a strong starter. Moreover, if $y_i = -x_i$ then $x_i+y_i = 0$, so $X \ne Z$ (in fact, X and Z can have no pair in common), and similarly $Y \ne Z$. If $\{x_i,y_i\} = \{-x_j,-y_j\}$ for some $i \ne j$, then $x_i-y_i = \pm(x_j-y_j)$, so X would not be a starter; and if $\{x_i,y_i\} = \{-x_i,-y_i\}$ then either $x_i = -y_i$

(impossible as we noticed above), or $x_i = -x_i$ (impossible in a group of odd order); consequently, X and Y are not equal.

To prove that X and Z are orthogonal, we first observe that (since r is odd)

$$B = \{h+h: \; h \in H\setminus\{0\}\} = H\setminus\{0\},$$

and for any $b \in H\setminus\{0\}$ there is a non-zero element whose sum with itself equals b; the element will necessarily be unique, since otherwise B would have less than r-1 elements. Call this entry $\frac{1}{2}b$. Then, given $\{x_i,y_i\} \in X$, write $d_i = -\frac{1}{2}b_i$; the $d_i$ will all be distinct since the $b_i$ are distinct, and will be non-zero. Define $h_i$ by $h_i = d_i+x_i$. Then $d_i = h_i-x_i$ and $y_i+d_i = -h_i-y_i$, so the $d_i$ are the elements whose being distinct and non-zero is required in the definition of orthogonal starters. A similar proof shows that Y and Z are orthogonal.

The orthogonality of X to Y was essentially shown when we proved that a strong starter always has an adder. In this case we put $d_i = -x_i-y_i$. Then the $d_i$ are all distinct and non-zero, and

$$d_i = (-y_i)-x_i = (-x_i)-y_i.$$

So there are three pairwise orthogonal starters in the group.

2.3 STRONG STARTERS IN GROUPS OF PRIME POWER ORDER. The most important starter construction, by Mullin and Nemeth [25], finds strong starters for almost all prime powers:

THEOREM 2.7. *If p is an odd prime such that $p^n = 2^k t+1$ where t is an odd integer greater than 1 then there is a strong starter in $GF(p^n)$. The corresponding adder is skew.*

PROOF. Suppose x is a generator of the multiplicative group of non-zero elements of $GF(p^n)$. (It is well known that this group is cyclic.) For convenience write $\delta$ for $2^{k-1}$; then $x^{2\delta t} = 1$, but $x^\alpha \neq 1$ when $1 \leq \alpha < 2\delta t$. We shall verify that X,

$$X = \{X_{ij}: \; 0 \leq i \leq \delta-1, \; 0 \leq j \leq t-1\},$$

is a strong starter, where

$$X_{ij} = \{x^{i+2j\delta}, \; x^{i+(2j+1)\delta}\}.$$

It is easy to see that the totality of the entries in the $X_{ij}$ constitute all non-zero elements of $GF(p^n)$ - in fact, the left-hand members of $X_{00}, X_{10},\dots,X_{\delta-1,0}$ are $x^0,x^1,\dots,x^{\delta-1}$, and the right-hand members are $x^\delta,x^{\delta+1},\dots,x^{2\delta-1}$; similarly, $\{X_{i1}\}$ contain the members $x^{2\delta},x^{2\delta+1},\dots,x^{4\delta-1}$, and so on. The differences between the elements of $X_{ij}$ are $\pm x^{i+2j\delta}(x^\delta-1)$. Since $x^\delta-1$ is non-zero and $GF(p^n)$ is a field, the entries $x^{i+2j\delta}(x^\delta-1)$ and $x^{i_1+2j_1\delta}(x^\delta-1)$ will be equal if and only if $x^{i+2j\delta} = x^{i_1+2j_1\delta}$, and this can occur only when $i = i_1$ and $j = j_1$; so the $\delta t$

differences with a + sign are all different. Similarly the differences with a - sign are all different. Moreover, if

$$x^{i+2j\delta}(x^{\delta}-1) = -x^{i_1+2j_1\delta}(x^{\delta}-1)$$

then

$$x^{i+2j\delta-i_1-2j_1\delta} = -1$$

and squaring

$$x^{2(i-i_1)+4\delta(j-j_1)} = 1.$$

This must mean

$$2(i-i_1)+4\delta(j-j_1) \equiv 0 \pmod{2\delta t}$$

since $2\delta t$ is the order of x. In particular

$$2(i-i_1) \equiv 0 \pmod{2\delta}$$

and since $0 \leq i < \delta$ and $0 \leq i_1 < \delta$ we must have $i - i_1 = 0$. Therefore

$$4\delta(j-j_1) \equiv 0 \pmod{2\delta t}$$

from which

$$2(j-j_1) \equiv 0 \pmod{t};$$

but $-t < j-j_1 < t$, so $j = j_1$ or $2(j-j_1) = \pm t$. The second case contradicts the fact that t is odd, and if $j = j_1$ then we have

$$2x^{i+2j\delta}(x^{\delta}-1) = 0$$

which is impossible since $p^n$ is odd and consequently $2, x$ and $x^{\delta}-1$ are all non-zero in $GF(p^n)$. So no two of the $2\delta t$ differences can be equal. Therefore, the set of differences constitutes all non-zero elements of $GF(p^n)$.

This means that the set X satisfies the definition of a starter. To show that the starter is strong we have to show that the $\delta t$ elements formed by *adding* the two members of $X_{ij}$ are different. The sum resulting from $X_{ij}$ is

$$x^{i+2j\delta}(x^{\delta}+1),$$

and these objects are all different because $x^{\delta}+1$ is non-zero.

The adder corresponding to a strong starter has as its members the negatives of the sums of the two elements of a member of the starter. Therefore, the members of the adder are all the

$$-x^{i+2j\delta}(x^{\delta}+1).$$

To show that an adder is skew we must show that it never contains both an element and its negative; that is, to show that

$$-x^{i+2j}(x^{\delta}+1) = x^{i_1+2j_1}(x^{\delta}+1)$$

is impossible (in the present case). But if the equation held then

$$-x^{i+2j} = x^{i_1+2j_1},$$

and we have already shown that to be impossible. So the adder is always skew.

COROLLARY 2.8. *If* $p^n$ *is an odd prime power not of the form* $2^k+1$, *then there is a skew Room square of side* $p^n$.

The technique of the above theorem cannot be applied to the case of a prime power of the form $2^k+1$, since $1+x^\delta = 0$ in that case.

2.4 STRONG STARTERS FROM STEINER TRIPLE SYSTEMS. A Steiner triple system $\mathcal{S}$ of order v is simply a balanced incomplete block design on v treatments with parameters $k = 3$ and $\lambda = 1$. So $\mathcal{S}$ consists of $v(v-1)/6$ 3-sets (blocks) of the set V of v treatments with the property that for any treatments x and y there is precisely one 3-set containing both x and y.

Let $\mathcal{S}$ be a Steiner triple system of order v, and write K for the complete graph on $v+1$ points, v of them named after the v treatments of $\mathcal{S}$ and the other one labelled 0. For any point x other than 0 we define $F_x$ to be the collection of edges

$$0x \cup \{yz: \{x,y,z\} \in \mathcal{S}\}.$$

If y occurred in two of these edges, yz and yt say, then {x,y,z} and {x,y,t} are two 3-sets both containing {x,y}; this is impossible. So the edges in $F_x$ are disjoint. However, every point of K lies in an edge, since 0 and x are both represented and if $y \neq x$ and $y \neq 0$ then there will be a 3-set containing x and y. So $F_x$ is a one-factor of K. If yz is any edge of K and $y \neq 0$, $z \neq 0$, then there will be precisely one treatment x such that $\{x,y,z\} \in \mathcal{S}$, so yz belongs to precisely one factor, and clearly 0y belongs to precisely one factor. Therefore

$$\mathcal{F} = \{F_x: x \in V\}$$

is a one-factorization of K.

LEMMA 2.9. *In the one-factorization derived from a Steiner triple system* $\mathcal{S}$, *there is a 3-division corresponding to each 3-set. Each of these 3-divisions is such that one of the components of the union of the three factors is a* $K_4$.

PROOF. If $\{x,y,z\} \in \mathcal{S}$, then

$F_x$ contains 0x and yz,

$F_y$ contains 0y and xz,

$F_z$ contains 0z and xy,

so $\{F_x, F_y, F_z\}$ is the required 3-division and the $K_4$ is spanned by {0,x,y,z}.

Two Steiner triple systems $\mathcal{S}$ and $\mathcal{T}$, based on the same set of treatments, are called *orthogonal* if there is no 3-set common to $\mathcal{S}$ and $\mathcal{T}$ and if whenever {x,y,z} and {a,b,z} belong to $\mathcal{S}$ and {x,y,t} and {a,b,c} belong to $\mathcal{T}$, $t \neq c$. It is easy to see that the two one-factorizations derived from such an $\mathcal{S}$ and $\mathcal{T}$ are orthogonal; so orthogonal Steiner systems of order v imply a Room square of side v. However, the converse does not hold; it is shown in [26] that there are no orthogonal Steiner systems of side 9. In fact, it is not always possible to go back from a one-factorization to a Steiner triple system; the one-factorization derived from a triple system will always contain a 3-division, and not every one-factorization contains such divisions (as we shall see in section 8.1).

Suppose G is a finite abelian group. If {a,b,c} is any 3-set of elements of G, then we associate the six elements $\pm(a-b)$, $\pm(b-c)$ and $\pm(a-c)$ with {a,b,c}. A set $\mathcal{B}$ of t 3-sets from G will be called a *set of Steiner difference blocks* if the 6t elements associated with the t blocks are all different and comprise all the non-zero elements of G. (Necessarily, G will have order $6t+1$.) It is well known that the $t(6t+1)$ 3-sets

$$\{a+g,\ b+g,\ c+g\}:\quad \{a,b,c\}\in\mathcal{B},\quad g\in G$$

are all different and form a Steiner triple system with G as its set of treatments.

LEMMA 2.10. *If $\mathcal{B}$ is a set of Steiner difference blocks in G, and the 3-sets in the associated Steiner triple system which contain the identity element e of G are*

$$\{e,x_1,y_1\},\{e,x_2,y_2\},\ldots,\{e,x_{3t},y_{3t}\},$$

*then $\{x_1,y_1\},\{x_2,y_2\},\ldots,\{x_{3t},y_{3t}\}$ are a starter in G.*

PROOF. Since e occurs in precisely one 3-set with each other entry of G, $\{x_1,y_1,x_2,\ldots,y_{3t}\}$ will constitute $G\setminus\{e\}$ with each entry occurring once.

The 3t 3-sets containing e can be partitioned into t sets of three: if {a,b,c} is any member of $\mathcal{B}$ then we group together the three 3-sets

$$\{e,b-a,e-a\},\ \{a-b,e,c-b\},\ \{a-c,b-c,e\}.$$

If these three 3-sets are $\{e,x_1,y_1\}$, $\{e,x_2,y_2\}$ and $\{e,x_3,y_3\}$ respectively, then

$$\pm(x_1-y_1) = \pm(b-c)$$

$$\pm(x_2-y_2) = \pm(a-c)$$

$$\pm(x_3-y_3) = \pm(a-b),$$

so the six differences are the differences associated with {a,b,c}. As we go through the 3t pairs of differences $\pm(x_i-y_i)$ we obtain the 6t differences associated with the members of $\mathcal{B}$, so we have every non-zero group element once. So the given set is a starter.

We call a set of Steiner difference blocks *strong* if the 3t entries $x_i+y_i$ obtained from the 3-sets $\{e,x_i,y_i\}$ in the Steiner triple system are all different.

It follows that a strong set of Steiner difference blocks gives rise to a strong starter, and consequently to a Room square.

THEOREM 2.11. *If p is a prime and n is a positive integer such that $p^n$ is congruent to 1 modulo 6, then there is an abelian group G of order $p^n$ admitting of a strong set of Steiner difference blocks.*

PROOF. G is the additive group of $GF(p^n)$. We write $p^n = 6t+1$, and select a primitive root x in $GF(p^n)$. The 3-sets

$$\{x^i, x^{2t+i}, x^{4t+i}\}: \ i = 0,1,2,\ldots,t-1$$

are the required Steiner difference blocks.

We first check that they are Steiner difference blocks. The differences associated with $\{x^i, x^{2t+i}, x^{4t+i}\}$ are

$$\pm x^i(x^{2t}-1), \ \pm x^i(x^{4t}-1), \ \pm x^i(x^{4t}-x^{2t}),$$

where 1 here denotes the multiplicative identity in $GF(p^r)$. These differences could be written as

$$\pm x^i(x^{2t}-1), \ \pm x^{i+4t}(x^{2t}-1), \ \pm x^{i+2t}(x^{2t}-1)$$

since $x^{6t} = 1$; and, since $x^{3t} = -1$, they are

$$x^i y, \ x^{i+3t}y, \ x^{i+t}y, \ x^{i+4t}y, \ x^{i+2t}y, \ x^{i+5t}y,$$

where $y = x^{2t}-1$. As i ranges from 0 to t-1, the totality of differences are all the

$$x^{i+jt}y, \quad 0 \leq i \leq t-1, \quad 0 \leq j \leq 5.$$

As i and j go through the range, $x^{i+jt}$ ranges through $G\setminus\{0\}$. Since x is primitive, y is non-zero; and since $G\setminus\{0\}$ is a multiplicative group, $(G\setminus\{0\})y = G\setminus\{0\}$. That is, the totality of associated differences contains every non-zero member of G once. So the 3-sets form a set of Steiner difference blocks.

Now we show the set is strong. The three members of the starter defined in Lemma 2.10 which arise from $\{x^i, x^{2t+i}, x^{4t+i}\}$ are

$$\{x^{2t+i}-x^i, x^{4t+i}-x^i\}, \ \{x^i-x^{2t+i}, x^{4t+i}-x^{2t+i}\}, \ \{x^i-x^{4t+i}, x^{2t+i}-x^{4t+i}\}.$$

The sums of these pairs are

$$a_i = x^i(x^{2t}+x^{4t}-2) = x^i(x^{2t}-x^{4t})^2,$$
$$b_i = x^i(1+x^{4t}-2x^{2t}) = x^i(x^{2t}-1)^2,$$
$$c_i = x^i(1+x^{2t}-2x^{4t}) = x^i(x^{4t}-1)^2,$$

where 2 means 1+1, 1 being the multiplicative identity. Clearly $a_i$, $b_i$ and $c_i$ are always non-zero.

Suppose $a_i = a_j$ for some $i \neq j$. Since $x^{2t}-x^{4t} \neq 0$, we have $x^i = x^j$; this is impossible. Similarly, $b_i = b_j$ or $c_i = c_j$ can never occur when $i \neq j$.

Finally, say $a_i = b_j$ for some i and j. Then

$$x^i(x^{2t}-x^{4t})^2 = x^j(x^{2t}-1)^2,$$

that is

$$x^{i+4t}(1-x^{2t})^2 = x^j(x^{2t}-1)^2,$$

whence $x^{i+4t} = x^j$. This cannot happen, because $0 \leq i < t$ and $0 \leq j < t$ and x is of order 6t. Similarly, $a_i = c_j$ or $b_i = c_j$ can never occur. This completes the proof that all 3t sums are different, and the starter (and consequently the set of Steiner difference blocks) is strong.

COROLLARY 2.12. *There are orthogonal Steiner systems of every prime power order congruent to 1 modulo 6.*

PROOF. The strong starter found in theorem 2.11 gives rise to a Room square whose row factorization is the factorization derived from a Steiner triple system. If the column factorization also comes from a Steiner triple system then the two systems will be orthogonal. It is found that this happens; in fact, the column factorization may be derived from the set of Steiner difference blocks

$$\{-x^i, -x^{2t+i}, -x^{4t+i}\}: \quad 0 \leq i \leq t-1.$$

## 2.5 QUINTUPLICATION OF A STRONG STARTER.

THEOREM 2.13. *If G is a finite abelian group of order prime to 6, which admits of a strong starter, then there is a strong starter in the direct sum of G with the cyclic group of order 5.*

PROOF. Let us write $n = 2s+1$ for the order of G. Since G is finite and abelian, it is a direct sum of cyclic groups; we can interpret any cyclic group of order m as the ring of integers modulo m under addition, and consider G as the additive group of the direct sum of these rings. In this way we endow G with a multiplication. This multiplication will have an identity element, 1 say; we write $1+1$ as 2 and $1+1+1$ as 3, and $2^{-1}$ and $3^{-1}$ will exist since $(n,6) = 1$.

Let the strong starter in G be

$$X = \{\{x_1,y_1\},\{x_2,y_2\},\ldots,\{x_s,y_s\}\}.$$

Choose two non-zero elements a and b of G such that neither a nor b equals $x_i + y_i$ for any i. (This requires that $s+2 \leq n-1$; as no strong starter exists in a group of size less than 7, there is no problem.) Write $h = 2^{-2}(b-a)$ and $g = 2^{-1}a$.

Finally, partition the set of all non-zero elements of G into two classes P and N, in such a way that

$$h \in P$$

$$-3^{-1}h \in P$$

$$x \in P \text{ iff } -x \in N.$$

We now write down six sets, A,B,C,D,E,F, of unordered pairs of elements of $G \oplus Z_5$, where $Z_5$ is the cyclic group of order 5, written as

$$Z_5 = \{0,1,2,3,4,\}.$$

(We are using 1,2 and 3 in two senses, but no confusion arises.) The union of this set will be shown to be a strong starter.

$$A = \{\{(x,0),(y,0)\}:\{x,y\}\varepsilon X\}$$

$$B = \{\{(x+g,1),(2x+g,2)\}: \ x \in P,\ x \neq h\}$$

$$C = \{\{(x+g,4),(2x+g,3)\}: \ x \in P,\ x \neq h\}$$

$$D = \{\{(x+g,1),(2x+g,3)\}: \ x \in N\}$$

$$E = \{\{(x+g,4),(2x+g,2)\}: \ x \in N\}$$

$$F = \{\{(h+g,1),(g,2)\}; \ \{(h+g,4),(g,3)\}; \ \{(g,1),(g,4)\}; \ \{(2h+g,2),(2h+g,3)\}\}.$$

To prove that this is a starter, we must show that every non-zero element of $G \oplus Z_5$ occurs in one pair and that the set of all differences between members of a pair also consists of all non-zero elements.

Every element of the form (y,0), with x ≠ 0, occurs in a pair in A. The pairs of B and D will contain all elements of the form (x+g,1) except for (h+g,1) and (0+g,1); the latter elements arise in F. If x runs through G then so does x+g, so we have every element of the form (y,1). Similarly, all elements (y,2), (y,3) and (y,4) are included.

Since X is a starter, the elements x - y, where {x,y}ε X, comprise $G \setminus \{0\}$. So all elements of the form (x,0), x ≠ 0, come up as differences in A. The differences in B contain all the elements of the form (x,1) where x ε P but x ≠ h. From C we get all the (-x,1) with x in the same range; these will equal the (x,1) with x ε N but x ≠ -h, since N consists of the negatives of elements of P. The remaining elements (h,1), (-h,1) and (0,1) come from F. Similarly, all elements with second component 2,3 or 4 arise as differences.

In both cases, to see that 0 does not come up and that no element appears twice, it is sufficient to observe that there are only 5s + 2 pairs in the union, and consequently 5n - 1 elements in each of the sets we have been discussing.

We complete the proof by showing that the starter is strong, which means that the $5s+2$ elements formed by summing the two members of the pair are all distinct and non-zero. The sums from A have the form $(x+y,0)$ where $\{x,y\} \varepsilon X$; since X is strong, these are all different and non-zero. From B and C we get all things of the form

$$(3x+2g,3),\ (3x+2g,2),$$

where x ranges through $P \setminus \{h\}$. D and E give all the elements

$$(3x+2g,4),\ (3x+2g,1),$$

for $x \varepsilon N$. These are all non-zero, and are distinct unless $3x+2g = 3x_1+2g$ where $x \neq x_1$. But $3^{-1}$ exists, so

$$3x+2g = 3x_1+2g \Rightarrow 3x = 3x_1 \Rightarrow x = x_1.$$

Finally we have the pairs from F. These have sums $(h+2g,3)$, $(h+2g,2)$, $(2g,0)$ and $(4h+2g,0)$. If the first two have already arisen then

$$h+2g = 3x+2g$$

for some x in $P \setminus \{h\}$. But this would imply $3^{-1}h \varepsilon P$, and we have specified $3^{-1}h \varepsilon N$. We can rewrite $(2g,0)$ as $(a,0)$ and $(4h+2g,0)$ as $(b,0)$, and these are not yet in our list since neither a nor b is a sum of a pair in X.

Therefore we have a strong starter.

# CHAPTER III. THE MOORE-TYPE CONSTRUCTION AND GENERALIZATIONS

3.1 INTRODUCTION. In 1893, E.H. Moore [21] proved that if there are Steiner triple systems on $v_1, v_2$ and $v_3$ points, where the $v_3$-system is a subsystem of the $v_2$-system, then there is a Steiner triple system on

$$v_1(v_2-v_3)+v_3$$

points. Recently J.D. Horton [11] found ways to prove similar theorems about other combinatorial objects. For Room squares he proved that the existence of a Room square of side $v_1$ and of a square of side $v_2$ with a subsquare of side $v_3$ implies the existence of a Room square of side $v_1(v_2-v_3)+v_3$ *provided* $v_2-v_3 > 6$. Subsequently, the theorem has been generalized to show that a skew square results when the original squares were skew and the subsquare was a skew subsquare, and to include the case $v_2-v_3 = 6$ whenever complementary Room squares of side $v_1$ exist.

The construction remains valid when $v_3 = 0$, even though we cannot have a null subsquare of a Room square. So we obtain, as a corollary, the result that Room squares of sides $r_1$ and $r_2$ can be used to construct a Room square of side $r_1r_2$. For historical purposes we point out that this multiplication theorem was proved first. It was announced in [4], but the proof given there was erroneous (see [23]); a correct proof was given in [37] and a proof which shows that skewness is preserved may be found in [43].

We shall also prove that the existence of a Room square of side r implies the existence of a square of side $4r+1$ with a subsquare of side r. The proof is a generalization of one of the extensions of the Moore-type construction.

## 3.2 THE MOORE-TYPE CONSTRUCTION

THEOREM 3.1. *Suppose there exists a Room square $\mathcal{R}_1$ of side $v_1$, a Room square $\mathcal{R}_2$ of side $v_2$ with a subsquare $\mathcal{R}_3$ of side $v_3$, and a pair of orthogonal Latin squares L and M of side $n = v_2-v_3$. Then there is a Room square $\mathcal{R}$ of side*

$$v = v_1(v_2-v_3)+v_3$$

*with subsquares isomorphic to $\mathcal{R}_1$, $\mathcal{R}_2$ and $\mathcal{R}_3$. If $\mathcal{R}_1$ and $\mathcal{R}_2$ are skew squares and $\mathcal{R}_3$ is a skew subsquare of $\mathcal{R}_2$ then $\mathcal{R}$ will be skew and the three subsquares will be skew subsquares.*

PROOF. We shall construct a Room square of side v based on the set of $v+1$ symbols

$$X = \{0, 1_1, 1_2, \ldots, 1_{v_1}, 2_1, \ldots, n_{v_1}, n+1, \ldots, v_2\}.$$

We write $L_i$ for $L$ with each entry $x$ replaced by $x_i$, and define $M_j$ similarly; and we denote the join of $L_i$ and $M_j$ as $\mathcal{L}_{ij}$. The entries of each $\mathcal{L}_{ij}$ will be interpreted as *unordered* pairs.

We standardize $\mathcal{R}_2$ in a particular way. We relabel the objects so that $\mathcal{R}_3$ is based on $\{0, n+1, n+2, \ldots, v_2\}$, and then reorder rows and columns so that $\mathcal{R}_3$ occupies the last $v_3$ rows and columns of $\mathcal{R}_2$. Then standardize $\mathcal{R}_2$ without relabelling the elements $0, n+1, \ldots, v_2$ or reordering the last $r_3$ rows and columns. Write

$$\mathcal{R}_2 = \begin{array}{|c|c|}\hline \mathcal{A} & \mathcal{B} \\ \hline \mathcal{C} & \mathcal{R}_3 \\ \hline \end{array} \,.$$

We shall use $\mathcal{A}_i$, $\mathcal{B}_i$ and $\mathcal{C}_i$ to denote $\mathcal{A}$, $\mathcal{B}$ and $\mathcal{C}$ with each entry $\{x,y\}$ replaced by $\{x_i, y_i\}$, it being understood that $x_i = x$ when $x = 0$ or $x > n$.

We shall assume that $\mathcal{R}_1$ is standardized.

We first convert $\mathcal{R}_1$ into an $nv_1 \times nv_1$ array $\mathcal{S}$ by replacing each of its cells by an $n \times n$ array. An empty cell is replaced by an empty $n \times n$ array; the entry $\{0,i\}$ is replaced by $\mathcal{A}_i$; and the entry $\{i,j\}$, $i \neq 0$, $j \neq 0$, is replaced by $\mathcal{L}_{ij}$.

Now write

$$\mathcal{R} = \begin{array}{|c|c|}\hline & \mathcal{B}_1 \\ & \mathcal{B}_2 \\ \mathcal{S} & \vdots \\ & \mathcal{B}_{v_1} \\ \hline \mathcal{C}_1 \;\; \mathcal{C}_2 \; \ldots \; \mathcal{C}_{v_1} & \mathcal{R}_3 \\ \hline \end{array}$$

Row $i$ of $\mathcal{R}_1$ contains $\{0,i\}$ and a set of entries $\{j,k\}$ whose elements collectively exhaust $\{1, \ldots, i-1, i+1, \ldots, v_1\}$. So the ith row of blocks of $\mathcal{R}$, $1 \leq i \leq v_1$, contains $\mathcal{A}_i$, $\mathcal{B}_i$ and a set of $\mathcal{L}_{jk}$ such that the collection of all the $j$ and $k$ is $\{0, 1, \ldots, i-1, i+1, \ldots, v_1\}$. Each row of $\mathcal{L}_{jk}$ contains all the $x_j$ and $y_k$, $1 \leq x, y \leq n$, once each. Each row of

$$\begin{array}{|c|c|}\hline \mathcal{A}_i & \mathcal{B}_i \\ \hline \end{array}$$

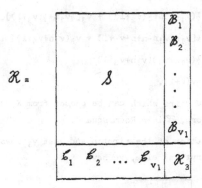

contains all the objects

$$\{0_i, 1_i, \ldots, (v_2)_i\},$$

that is

$$\{0, 1_i, \ldots, n_i, n+1, \ldots, v_2\},$$

once each. So every row in the ith row of blocks in $\mathcal{R}$ contains every element of X once, for $1 \leq i \leq v_1$. In the last row of blocks, since each row of $\mathcal{C}_i$ contains $\{1_i, 2_i, \ldots, n_i\}$ once each, $\{\mathcal{C}_1, \mathcal{C}_2, \ldots, \mathcal{C}_{v_1}\}$ will contain every element of X except for $0, n+1, n+2, \ldots,$ and $v_2$ precisely once, and the missing elements are contained once in each row of $\mathcal{R}_3$. Consequently, every row of $\mathcal{R}$ contains every member of X once. A similar proof applies to columns.

If we take the whole of $\mathcal{R}$, we have every entry of $\mathcal{L}_{ij}$ once, for $1 \leq i \leq v_1$ and $1 \leq j \leq v_1$. So every possible unordered pair of the form $\{x_i, y_j\}$, $1 \leq x, y \leq n$, $1 \leq i, j \leq v_1$, appears once. We also have all entries of $\mathcal{A}_i$, $\mathcal{B}_i$ and $\mathcal{C}_i$, $1 \leq i \leq v_1$; for a fixed i, these three arrays will contain every pair of the form $\{x_i, y_i\}$, $1 \leq x, y \leq n$, and every entry $\{x_i, z\}$, $1 \leq x \leq n$, $z = 0$ or $n < z \leq v_2$. $\mathcal{R}_3$ will contain all the pairs $\{z, t\}$, where both z and t belong to $\{0, n+1, n+2, \ldots, v_2\}$. So contains every unordered pair of members of X.

Finally, $\mathcal{A}_i$ and $\mathcal{B}_i$ together contain $\frac{1}{2}(v_2+1)$ pairs per row, and each row of each $\mathcal{L}_{ij}$ contains n pairs. $\mathcal{C}_i$ contains $\frac{1}{2}n$ pairs in each row and $\mathcal{R}_3$ has $\frac{1}{2}(v_3+1)$. So the number of pairs in $\mathcal{R}$ is

$$\begin{aligned} v_1 n\,[\tfrac{1}{2}(v_1-1)n+\tfrac{1}{2}(v_2+1)] &+ v_3[\tfrac{1}{2}v_1 n+\tfrac{1}{2}(v_3+1)] \\ &= \tfrac{1}{2}[v_1 n(v_1 n-n+n+v_3+1) + v_3(v_1 n+v_3+1)] \\ &= \tfrac{1}{2}(v_1 n+v_3)(v_1 n+v_3+1) \\ &= \tfrac{1}{2}v(v+1). \end{aligned}$$

This is the number of unordered pairs which can be chosen from X, so each pair must appear precisely once. Therefore, $\mathcal{R}$ is a Room square.

If we take the intersection of the first n and last $v_3$ rows with the corresponding columns and delete everything else, we have

| | |
|---|---|
| $\mathcal{A}_1$ | $\mathcal{B}_1$ |
| $\mathcal{C}_1$ | $\mathcal{R}_3$ |

which is isomorphic to $\mathcal{R}_2$. $\mathcal{R}_3$ is exhibited as a subsquare in the last $v_3$ rows and columns. To discover $\mathcal{R}_1$, take the intersection of rows $1, n+1, \ldots, n(v_1-1)+1$ and the corresponding columns. The array formed has entry $\{0, 1_i\}$ where $\mathcal{R}_1$ had $\{0, i\}$ and $\{1_i, 1_j\}$ where $\mathcal{R}_1$ had $\{i, j\}$, so it is isomorphic to $\mathcal{R}_1$.

We now assume that $\mathcal{R}_1$ and $\mathcal{R}_2$ were skew and that $\mathcal{R}_3$ was a skew subsquare of $\mathcal{R}_2$. The standardization of $\mathcal{R}_2$ can be carried out simply by permuting rows and columns to place $\mathcal{R}_3$ in the bottom right corner. Since $\mathcal{R}_3$ is placed symmetrically in $\mathcal{R}_2$, the same permutation is used on rows and columns, so that the skewness of $\mathcal{R}_2$ is retained.*

Write $R, R_1, R_2$ and $R_3$ for the incidence matrices of $\mathcal{R}, \mathcal{R}_1, \mathcal{R}_2$ and $\mathcal{R}_3$, and write

$$R_2 = \begin{pmatrix} A & B \\ C & R_3 \end{pmatrix}.$$

As the $\mathcal{R}_i$ are skew,

$$R_1 + R_1^T = I + J,$$
$$R_2 + R_2^T = I + J,$$
$$R_3 + R_3^T = I + J,$$

and since

$$R_2 + R_2^T = \begin{pmatrix} A+A^T & B+C^T \\ C+B^T & R+R^T \end{pmatrix}$$

we have

$$A + A^T = I + J$$
$$B + C^T = J.$$

Now the incidence matrix of $\mathcal{S}$ is S, where

$$S = I \times A + (R_1 - I) \times J$$

(where $\times$ denotes Kronecker product, I is $v_1 \times v_1$ and J is $n \times n$). So

$$\begin{aligned} S + S^T &= I \times A + I \times A^T + (R_1 - I) \times J + (R_1^T - I) \times J \\ &= I \times (A + A^T) + (R_1 + R_1^T - 2I) \times J \\ &= I \times (I + J) + (J - I) \times J \\ &= I \times I + J \times J \\ &= I \times J \qquad (nv_1 \times nv_1). \end{aligned}$$

---

* This explains the definition of "skew subsquare". If $\mathcal{R}_3$ were simply a subsquare of $\mathcal{R}_2$ which was itself a skew square, it might not be possible to place $\mathcal{R}_3$ in the bottom right corner of $\mathcal{R}_2$ without destroying the skewness of $\mathcal{R}_2$.

Thus

$$R+R^T = \left(\begin{array}{ccc|c} & & & B+C^T \\ & & & B+C^T \\ & S+S^T & & \cdot \\ & & & \cdot \\ & & & \cdot \\ & & & B+C^T \\ \hline C+B^T & C+B^T & C+B^T & R_3+R_3{}^T \end{array}\right)$$

$$= \begin{pmatrix} I+J & J \\ J & I+J \end{pmatrix}$$

$$= I+J.$$

So $\mathcal{R}$ is skew. The $\mathcal{R}_i$ are skew squares which are symmetrically placed subsquares, so they are skew subsquares.

THEOREM 3.2. *If there are Room squares of sides $v_1$ and $v_2$ then there is a Room square of side $v_1v_2$ with subsquares of sides $v_1$ and $v_2$ which are isomorphic to the original squares. If the original squares were skew then the resulting square is skew and the subsquares are skew subsquares.*

PROOF. The construction of $\mathcal{S}$ in the proof of theorem 3.1 is carried out with $n = v_2$ and with $\mathcal{R}_2$ replacing $\mathcal{A}$. By methods analogous to those in that proof it may be shown that $\mathcal{S}$ is a Room square with the required properties.

3.3 FIRST EXTENSION. In this section we give a construction for a Room square of side $v = 6r+1$ which assumes the existence of complementary Room squares of side $r$ and also of a suitable set of Latin squares of side $r$.

Observe that we do not seem to get a square with a subsquare of side $r$, and that we cannot be sure of obtaining a skew square.

Suppose $\mathcal{R}_1$ and $\mathcal{R}_2$ are complementary Room squares of side $r$. We shall define $r \times r$ arrays $\mathcal{R}_{ij}$ which are a sort of generalized sum of $\mathcal{R}_1$ and $\mathcal{R}_2$. We first delete all diagonal entries from $\mathcal{R}_1$ and $\mathcal{R}_2$. Then the entry $\{x,y\}$ of $\mathcal{R}_1$, where $x < y$, is replaced by $\{x_i,y_j\}$, and the same entry of $\mathcal{R}_2$ is replaced by $\{x_{i+3},y_{j+3}\}$. Finally, the two arrays formed are superimposed.

If we write $R_i = \{1_i,2_i,\ldots,r_i\}$, then $\mathcal{R}_{ij}$ and $\mathcal{R}_{ji}$ between them contain all the pairs with one entry from $R_i$ and the other from $R_j$, and also all pairs with one entry from $R_{i+3}$ and the other from $R_{j+3}$. If $\mathcal{R}_{ij}$ is placed beside $\mathcal{R}_{jk}$ then row $x$ of the combined array will contain in its entries all members of $R_j$ and $R_{j+3}$ except for

$x_j$ and $x_{j+3}$; a similar remark applies to columns when $\mathcal{R}_{ij}$ is placed above $\mathcal{R}_{jk}$.

Given a pair of Latin squares L and M based on $\{1,2,\ldots,r\}$ which have a common transversal, we shall denote by $\mathcal{L}_{ij}$ the $r\times r$ array whose $(x,y)$ entry is $(\ell_i,m_j)$, where $\ell$ is the $(x,y)$ entry of L and $m$ is the $(x,y)$ entry of M, except that the diagonal entries of $\mathcal{L}_{ij}$ shall be empty.

THEOREM 3.3. *If there are complementary Room squares $\mathcal{R}_1$ and $\mathcal{R}_2$ of side r, then there is a Room square of side $6r+1$ with a subsquare of side 7.*

PROOF. We shall construct a Room square based on the set

$$\{\infty,0,1_1,2_1,\ldots,r_1,1_2,\ldots,r_6\}.$$

The theorem is trivial when $r = 1$. If $r > 1$ then $r \geq 7$, so we can assume the existence of orthogonal Latin squares L and M of side r which have a common transversal.

We shall use the notations $\mathcal{L}_{ij}$ and $\mathcal{R}_{ij}$ as defined above. $\mathcal{D}_{ij}$, $\mathcal{E}_{ij}$ and $\mathcal{F}_{ij}$ respectively shall denote a square array of size r with entries only on the diagonal, a one-row array of length r and a one-column array of length r; in each case the xth entry shall be $\{x_i,x_j\}$. We make the special convention that $x_0$ shall equal 0 and $x_7$ shall equal $\infty$ for $1 \leq x \leq r$.

The required array is shown in Figure 3.1, and is easily shown to be Room. The entries common to rows $1, r+1, 2r+1, \ldots, 6r+1$ and the corresponding columns form a Room square of side 7 based on $\{0,1_1,2_1,3_1,4_1,5_1,6_1,\infty\}$.

| | | | | | | |
|---|---|---|---|---|---|---|
| $\mathcal{R}_{11}\cup\mathcal{D}_{01}$ | $\mathcal{R}_{22}$ | $\mathcal{R}_{33}\cup\mathcal{D}_{45}$ | $\mathcal{D}_{67}$ | | | $\mathcal{F}_{23}$ |
| $\mathcal{R}_{23}\cup\mathcal{D}_{57}$ | $\mathcal{R}_{31}\cup\mathcal{D}_{02}$ | $\mathcal{R}_{12}$ | | | $\mathcal{D}_{13}$ | $\mathcal{F}_{46}$ |
| $\mathcal{R}_{32}$ | $\mathcal{R}_{13}\cup\mathcal{D}_{56}$ | $\mathcal{R}_{21}\cup\mathcal{D}_{03}$ | $\mathcal{D}_{12}$ | | $\mathcal{D}_{47}$ | |
| | $\mathcal{D}_{37}$ | | $\mathcal{L}_{14}\cup\mathcal{D}_{04}$ | $\mathcal{L}_{25}\cup\mathcal{D}_{26}$ | $\mathcal{L}_{36}$ | $\mathcal{F}_{15}$ |
| $\mathcal{D}_{36}$ | $\mathcal{D}_{14}$ | $\mathcal{D}_{27}$ | $\mathcal{L}_{26}$ | $\mathcal{L}_{34}\cup\mathcal{D}_{05}$ | $\mathcal{L}_{15}$ | |
| $\mathcal{D}_{24}$ | | | $\mathcal{L}_{35}\cup\mathcal{D}_{35}$ | $\mathcal{L}_{16}\cup\mathcal{D}_{17}$ | $\mathcal{L}_{24}\cup\mathcal{D}_{06}$ | |
| | | $\mathcal{E}_{16}$ | | $\mathcal{E}_{34}$ | $\mathcal{E}_{25}$ | $\{0,\infty\}$ |

FIGURE 3.1

3.4 SECOND EXTENSION. We now present a second way of obtaining squares of side $6r+1$.

This construction demands that there should exist an embedded Room square of side r. An embedded square implies the existence of complementary Room squares, so the construction is theoretically no stronger than the one given in the preceding section, and may not be as strong. However, the square of side $6r+1$ which we construct contains a subsquare of side r.

Suppose $\mathcal{R}$ is a standardized Room square of side r which is embedded in the join $\mathcal{L}$ of a pair of orthogonal Latin squares. Let $\mathcal{S}$ be a Room square of side 7 based on $\{0,1,2,3,4,5,6,7\}$. We require $\{0,7\}$ to be in the (7,7) position, and we shall write

$$\mathcal{S} = \begin{array}{|c|c|} \hline \mathcal{A} & \mathcal{B} \\ \hline \mathcal{C} & \{0,7\} \\ \hline \end{array}\ ;$$

but we do not want $\mathcal{S}$ standardized, and in fact we shall assume that $\{0,3\}$ lies in the (6,1) cell.

Define arrays $\mathcal{A}_i$, $\mathcal{B}_i$ and $\mathcal{C}_i$, $1 \leq i \leq r$, to be formed from $\mathcal{A}$, $\mathcal{B}$ and $\mathcal{C}$ respectively by replacing x by $x_i$ when $1 \leq x \leq 6$. When $1 \leq i \leq 6$, $1 \leq j \leq 6$, $i \neq j$, 6×6 arrays $\mathcal{F}_{ij}$ and $\mathcal{G}_{ij}$ are constructed as follows: if the (i,j) cell of $\mathcal{L}$ contains $\{p,q\}$ then $\mathcal{F}_{ij}$ has (5,5) entry $\{6_p,6_q\}$, (6,6) entry $\{5_p,6_q\}$, and all other cells empty, and $\mathcal{G}_{ij}$ is formed from

| | | | | | |
|---|---|---|---|---|---|
| 51 | 62 | 35 | 46 | 13 | 24 |
| 26 | 15 | 61 | 52 | 34 | 43 |
| 64 | 53 | 16 | 25 | 41 | 32 |
| 45 | 36 | 54 | 63 | 22 | 11 |
| 12 | 44 | 23 | 31 | 55 | |
| 33 | 21 | 42 | 14 | | 56 |

by replacing xy by $\{x_p,y_q\}$ throughout.

The required Room square of side $6r+1$ is

| | | | | | |
|---|---|---|---|---|---|
| $\mathcal{A}_1$ | $\mathcal{D}_{12}$ | $\mathcal{D}_{13}$ | --- | $\mathcal{D}_{1r}$ | $\mathcal{B}_1$ |
| $\mathcal{D}_{21}$ | $\mathcal{A}_2$ | $\mathcal{D}_{23}$ | --- | $\mathcal{D}_{2r}$ | $\mathcal{B}_2$ |
| . . . | | | | | . . . |
| $\mathcal{D}_{r1}$ | $\mathcal{D}_{r2}$ | $\mathcal{D}_{r3}$ | --- | $\mathcal{A}_r$ | $\mathcal{B}_r$ |
| $\mathcal{C}_1$ | $\mathcal{C}_2$ | $\mathcal{C}_3$ | --- | $\mathcal{C}_r$ | $\{0,7\}$ |

where $\mathcal{D}_{ij} = \mathcal{F}_{ij}$ when the (i,j) cell of $\mathcal{R}$ is empty and $\mathcal{D}_{ij} = \mathcal{G}_{ij}$ when the (i,j) cell of $\mathcal{R}$ is occupied.

Suppose that cells (i,j) and (k,ℓ) of $\mathcal{L}$ contain the entries (p,q) and (q,p) respectively, and that cell (i,j) of $\mathcal{R}$ is occupied (the entry being, of course $\{p,q\}$, since $\mathcal{R}$ is embedded in $\mathcal{L}$). Then $\mathcal{D}_{ij}$ will contain every unordered pair with one member of the form $x_p$ and one of the form $y_q$, $1 \leq x,y \leq 6$, except for $\{6_p,6_q\}$ and $\{6_p,5_q\}$. Cell (k,ℓ) of $\mathcal{R}$ will be empty, by the properties of embedded Room squares, so $\mathcal{D}_{ji}$ contains the missing pairs $\{6_p,6_q\}$ and $\{6_p,5_q\}$. $\mathcal{A}_p$, $\mathcal{B}_p$ and $\mathcal{C}_p$ contain between them all unordered pairs of the form $\{x_p,y_p\}$, $\{0,x_p\}$ and $\{x_p,7\}$ where $1 \leq x,y \leq 6$. So the array contains all unordered pairs which can be chosen from

$$\{0,1_1,2_1,\ldots,6_1,2_2,\ldots,6_r,7\}.$$

It is now straightforward to check that it is a Room square based on this set.

The cells of the array which are common to rows and columns 1,2,3,4,5,6 and $6r+1$ form a Room square of side 7. To find a subsquare of side r, consider the intersection of rows 6,12,...,6r with columns 1,7,...,6r-5; it is a Room square based on

$$\{0,3_1,\ldots,3_r\}.$$

In summary, we have

THEOREM 3.4. *If there is an embedded Room square of side r, then there is a Room square of side 6r+1 with subsquares of sides r and 7.*

3.5 **A SQUARE OF SIDE 4r+1.** The construction of a square of side 6r+1 from a square of side r in section 3 depended on the existence of a Room square of side 7. Even though there is no Room square of side 5, we can find a similar construction for a square of side 4r+1. We use the fact that the array

| | | | | |
|---|---|---|---|---|
| A | | | | |
| | {0,4} | | {1,5} | {2,3} |
| | {3,5} | {0,2} | | {1,4} |
| | {1,2} | {4,5} | {0,3} | |
| | | {1,3} | {2,4} | {0,5} |

where A = $\{\{0,1\},\{2,5\},\{3,4\}\}$, is very nearly a Room square. The square of side $4r+1$ which we construct is particularly interesting because it has a subsquare of side r.

We shall construct a Room square based on

$$\{\infty,0,1_1,2_1,\ldots,r_1,1_2,\ldots,r_4\}.$$

The notations $\mathcal{D}_{ij}$, $\mathcal{E}_{ij}$ and $\mathcal{F}_{ij}$ of section 3 will be used, with the convention that $x_0 = 0$ and $x_5 = \infty$. If $\mathcal{R}$ is a standardized Room square of side r, then $\mathcal{R}_{ij}$ is the array derived from $\mathcal{R}$ by deleting the diagonal and then replacing every entry {x,y} by $\{x_i,y_j\}$ when $x<y$.

THEOREM 3.5. *If there is a Room square $\mathcal{R}$ of side r, $r > 1$, then there is a Room square of side $4r+1$ with a subsquare of side r isomorphic to $\mathcal{R}$.*

PROOF. Assume $\mathcal{R}$ to be standardized. Choose permutations $\phi$ and $\psi$ on $\{1,2,\ldots,r\}$ such that:

(i) the $(i,i\phi)$ and $(i,i\psi)$ cells of $\mathcal{R}$ are empty for every i;

(ii) $i\phi, i\psi$ and i are all different for every i.

(The existence of such permutations is guaranteed by theorem 5.3.3 of [34].) Now define $\mathcal{A}$ to be the $r\times r$ array with

$$\{\infty,x_2\} \text{ in position } (x,x)$$

$$\{0,x_1\} \text{ in position } (x,x\phi)$$

$$\{x_3,x_4\} \text{ in position } (x,x\psi)$$

for $x = 1,2,\ldots,r$, and every other cell empty. Then the desired Room square is shown in Figure 3.2, where $\mathcal{R}_{ij}\chi$ means the result of carrying out the column permutation $\chi$ on $\mathcal{R}_{ij}$.

| | | | | |
|---|---|---|---|---|
| $\mathcal{R}_{22} \cup \mathcal{A}$ | $\mathcal{R}_{31}$ | $\mathcal{R}_{13}$ | $\mathcal{R}_{44}$ | |
| $\mathcal{R}_{43}\ \psi$ | $\mathcal{R}_{14} \cup \mathcal{D}_{04}$ | $\mathcal{R}_{32}$ | $\mathcal{R}_{21} \cup \mathcal{D}_{15}$ | $\mathcal{F}_{23}$ |
| $\mathcal{R}_{34}\ \psi$ | $\mathcal{R}_{23} \cup \mathcal{D}_{35}$ | $\mathcal{R}_{41} \cup \mathcal{D}_{02}$ | $\mathcal{R}_{12}$ | $\mathcal{F}_{14}$ |
| $\mathcal{R}_{11}\ \phi$ | $\mathcal{R}_{42} \cup \mathcal{D}_{12}$ | $\mathcal{R}_{24} \cup \mathcal{D}_{45}$ | $\mathcal{R}_{33} \cup \mathcal{D}_{03}$ | |
| | | $\mathcal{E}_{13}$ | $\mathcal{E}_{24}$ | $\{0,\infty\}$ |

FIGURE 3.2

There is no difficulty in proving that the square is Room. The entries common to rows and columns $3r+1$ to $4r$ form a subsquare of side $r$ based on

$$\{0, 1_3, 2_3, \ldots, r_3\},$$

isomorphic to $\mathcal{R}$.

## CHAPTER IV. THE FRAME CONSTRUCTION

4.1 BACKGROUND. Some of the Room square constructions we present are complicated, and it is not easy to see how they were originally reached. It seems worthwhile to include, in at least one case, the background to a construction.

One natural way of proceeding is to try to build up a Room square from a smaller one - part of the structure is already there, so the task should be easier. (This argument is not necessarily valid, but it is appealing.)

Suppose $\mathcal{R}$ is a Room square of side r based on the set $R = \{0,1,\dots,r\}$. To construct a square whose side is about nr, it seems reasonable to start with n squares of side r. For convenience suppose they are based on the disjoint sets $R_1, R_2, \dots, R_n$, where $R_i = \{0_i, 1_i, \dots, r_i\}$.

The problem can be approached as follows: first, try to put all the unordered pairs of elements, one from each of $R_i$ and $R_j$, into a small array. Then try to arrange these small arrays into a large array.

If we denote by [i,j] an $r \times r$ array containing all pairs from $R_i$ and $R_j$, the problem is to arrange these objects [i,j] into an $n \times n$ array. Clearly [i,j] and [i,k] cannot both lie in the same row and same column. So the $n \times n$ array will be very like a Room square of side n with [i,j] replacing {i,j}. It seems probable that the $n \times n$ array will differ from a Room square in the placing of the objects [i,i], if at all.

I had reached this point, and could find very little advantage in going on. I could use Room squares of sides r and n to construct a Room square of side rn, but this result was already known - and my proof would be really the same as the known one. Modifications to find interesting new squares, for example, side $rn + 2$, did not work.

At the back of my mind, I was always aware of a very useful way of putting together *ordered* pairs - use the join of two Latin squares. But, if I did this, I would have the unordered pair {i,j} twice in the $n \times n$ array. The only answer was to find *two* $r \times r$ blocks, $\mathcal{R}_{ij}$ and $\mathcal{R}_{ji}$ say, which would have no common entry and which would contain most or all possible pairs from $R_i$ and $R_j$ between them.

Suppose I could do this. Then the $3r \times 3r$ array

$$\begin{matrix} \mathcal{R}_{11} & \mathcal{R}_{22} & \mathcal{R}_{33} \\ \mathcal{R}_{23} & \mathcal{R}_{31} & \mathcal{R}_{12} \\ \mathcal{R}_{32} & \mathcal{R}_{13} & \mathcal{R}_{21} \end{matrix}$$

might be something like a Room square of side 3r.

There are more problems. First of all, I was headed for a Room square of side 3r based on a set of 3r+ 3 elements. I mentally shelved this by deciding that I would somehow merge the elements $0_i$ with one another. In consequence I replaced each set $R_i$ by the set $R_i^o = R_i \setminus \{0_i\}$, and decided that $\mathcal{R}_{ij}$ and $\mathcal{R}_{ji}$ should contain all the pairs with one entry from each of $R_i^o$ and $R_j^o$. The second problem is to ensure that column k of $\mathcal{R}_{ij}$ and column k of $\mathcal{R}_{ji}$ between them contain all the members of $R_i^o$ and $R_j^o$ once each. This can be done by basing $\mathcal{R}_{ij}$ on a Room square $\mathcal{R}$ with $\{x,y\}$ being replaced by $\{x_i,y_j\}$ in $\mathcal{R}_{ij}$ and by $\{x_j,y_i\}$ in $\mathcal{R}_{ji}$. In order to have a straightforward rule, I decided to replace $\{x,y\}$ by $\{x_i,y_j\}$ in $\mathcal{R}_{ij}$ when $x<y$, and to delete the entry $\{0,x\}$ completely.

My remaining worry was the "diagonal" elements - pairs of the form $\{0,x_i\}$ and $\{x_i,x_j\}$. It is necessary to place these pairs somewhere. Further, it must be done in such a way that $0,x_1,x_2$ and $x_3$ are put into rows x,2x,3x and columns x,2x and 3x. What I wanted was to place $\{0,x_1\}$ and $\{x_2,x_3\}$ in row x and column x, and so on; but this would mean two pairs in the (x,x) cell. The only possibility was to reorder rows or columns in the blocks in some way so that $x_2$ and $x_3$ would be missing in column y, for some $y \neq x$; then, provided the (x,y) cell was originally empty in $\mathcal{R}$, all would be well. I needed a column permutation, $\theta$ say, which could be applied to $\mathcal{R}_{23}$ and $\mathcal{R}_{32}$, with the property that cell $(x,x\theta)$ was always empty in $\mathcal{R}$. A half-remembered theorem in Ryser's book [34, theorem 5.3.3] supplied the necessary permutation.

I now had a construction for a square of side 3r. The extension to nr may be trivial, but it took three months for me to realize this. I first had to see that the essential property of the block array is that $\mathcal{R}_{ij}$ and $\mathcal{R}_{ji}$ always occur in the same column. It was then necessary to disabuse myself of the idea that using more than one permutation would lead to trouble.

The next two sections contain a formal proof of the theorem whose genesis has been outlined above.

## 4.2 THE ARRAY $\mathcal{A}_n$.

Suppose n is an odd integer. We shall denote by $\mathcal{A}_n$ the $n \times n$ array whose entries are ordered pairs, and whose (i,j) entry is the pair

$$(j-i+1, i+j-1),$$

where the elements j-i+1 and i+j-1 are taken modulo n as members of the set $N = \{1,2,\ldots,n\}$.

If we write

$$x = j-i+1, \quad y = i+j-1,$$

we have $i = \frac{1}{2}(y-x)+1$ and $j = \frac{1}{2}(x+y)$ (reduced modulo n); $\frac{1}{2}$ means the inverse of 2 in the arithmetic modulo n; that is, $\frac{1}{2}(n+1)$. So any ordered pair of members of N appears in a specific position in $\mathcal{A}_n$. From the definition we can never be required to put two pairs in the same cell.

If you look along row i, the left-hand members of the entries are

$$\{2-i,3-i,\ldots,-i,1-i\}$$

and the right-hand members are

$$\{i,i+1,\ldots,i-2,i-1\}.$$

In column j, the left-hand members are

$$\{j,j-1,\ldots,j+2,j+1\},$$

while the right-hand positions contain

$$\{j,j+1,\ldots,j-2,j-1\}.$$

In every case the set is N.

Most importantly, suppose the pair (x,y) occurs in column j of $\mathcal{A}_n$. That is,

$$j = \tfrac{1}{2}(x+y).$$

Then the pair (y,x) will also lie in column j, since

$$\tfrac{1}{2}(y+x) = \tfrac{1}{2}(x+y).$$

We summarize in the following lemma.

LEMMA 4.1. *The entries of* $\mathcal{A}_n$ *consist of the ordered pairs of members of N taken once each. The entries in a given row or column of* $\mathcal{A}_n$ *contain between them every member of N once as a left member and once as a right member. If the pair (x,y) occurs in a given column then (y,x) also occurs in that column.*

4.3 THE ROOM SQUARE CONSTRUCTION. We now show how to construct a Room square of odd side nr based on $\{0,1,2,\ldots,nr\}$, given a square of side r where $r > n$. It is convenient to write

$$i_j = i+r(j-1),$$

so that every integer from 1 to nr has a unique representation. Given a standardized Room square $\mathcal{R}$ of order r we shall write $\mathcal{R}_{ij}$ for the array formed from $\mathcal{R}$ in the following way:

(i) delete all diagonal entries;

(ii) if $x < y$ replace the entry $\{x,y\}$ of $\mathcal{R}$ by $\{x_i,y_j\}$.

These are the same arrays as were used in section 3.5. The set of all arrays $\mathcal{R}_{ij}$ with $1 \leq i \leq n$ and $1 \leq j \leq n$ will contain between them all unordered pairs of integers between 1 and nr except for the pairs $\{x_i, x_j\}$. If $\mathcal{R}_{ij}$ and $\mathcal{R}_{jk}$ are placed next to one another then row x of the resulting array will contain each of $1_j, 2_j, \ldots, r_j$ except for $x_j$ precisely once; a similar remark applied to columns when $\mathcal{R}_{ij}$ is placed above $\mathcal{R}_{jk}$.

LEMMA 4.2. *Given a Room square $\mathcal{R}$ of side r, where $r = 2s+1$, there are s permutations $\phi_1, \phi_2, \ldots, \phi_s$ of $\{1,2,\ldots,r\}$ with the properties that $k\phi_i = k\phi_j$ never occurs unless $i = j$, and that cell $(k, k\phi_i)$ is empty for $1 \leq k \leq r$, $1 \leq i \leq s$.*

PROOF. Consider the $r \times r$ matrix M whose $(k,\ell)$ position contains 1 if $\mathcal{R}$ has its $(k,\ell)$ cell empty, but is 0 otherwise. M is a matrix of zeros and ones with every row and column sum equal to s. So [34, theorem 5.5.3] M is a sum of s permutation matrices, say

$$M = P_1 + P_2 + \ldots + P_s.$$

We define $\phi_i$ to be the permutation corresponding to $P_i$; if $P_i$ has its $(k,\ell)$ entry equal to 1 then $k\phi_i = \ell$. If $k\phi_i = k\phi_j$ occurred when $i \neq j$ then $P_i$ and $P_j$ would both have 1 in position $(k, k\phi_i)$, so M would have an entry equal to 2 or more. That the $(k, k\phi_i)$ cell of $\mathcal{R}$ is empty follows from the definition of M.

THEOREM 4.3. *If r and n are odd integers such that $r \geq n$, and if there is a Room square $\mathcal{R}$ of side r, then there is a Room square of side rn.*

PROOF. For convenience write $r = 2s+1$ and $n = 2t+1$. We construct the Room square by replacing every entry of $\mathcal{A}_n$ by an $r \times r$ block.

For a given j, select permutations $\phi_{j1}, \phi_{j2}, \ldots, \phi_{jn}$ satisfying the following conditions:

(i) $\phi_{jk} = \phi_{j\ell}$ if and only if $(k,\ell)$ and $(\ell,k)$ appear in column j of $\mathcal{A}_n$;

(ii) if the entry in row j and column j of $\mathcal{A}_n$ is $(x,y)$ then $\phi_{jx}$ and $\phi_{jy}$ equal the identity permutation;

(iii) all the $\phi_{jk}$ except for the identity permutation are selected from the set of s permutations associated with $\mathcal{R}$ according to lemma 4.2.

(This will be possible since $r \geq n$ and consequently $s \geq t$.) Now replace the entry $(k,\ell)$ in column j of $\mathcal{A}_n$ by the array $\mathcal{R}_{k\ell}\phi_{jk}$ which is obtained by performing the permutation $\phi_{jk}$ on the columns of $\mathcal{R}_{k\ell}$.

From lemma 4.1 and the properties of the arrays $\mathcal{R}_{k\ell}$ it follows that the resulting array will contain every number from 1 to rn in each row and each column, except that every $x_k$ is missing from every row $x_j$ and that when j is such that $(k,\ell)$ is an entry in column j of $\mathcal{A}_n$ then $x_k$ and $x_\ell$ are missing from column $(x\phi_{jk})_j$.

Moreover, the array contains every unordered pair of members of $\{1,2,\ldots,rn\}$ except for the pairs of the form $\{x_k,x_\ell\}$, and contains each precisely once.

We now insert some more entries into the jth diagonal block of the new array. For each k, if $(k,\ell)$ was an entry of column j of $\mathcal{A}_n$, we place $\{x_k,x_\ell\}$ in the $(x,x\phi_k)$ position of this block (that is, in the $\left(x_j,(x\phi_k)_j\right)$ position of the new square) except that $\{0,x_j\}$ is used instead of $\{x_j,x_j\}$ in the relevant position. Lemma 4.2 tells us that we shall not finish with two entries in the same cell, so this step is possible. The finished array now contains every entry from $\{0,1,2,\ldots,rn\}$ once per row and once per column, and contains every unordered pair from that set precisely once. So it is the required Room square.

## CHAPTER V. DOUBLING CONSTRUCTIONS

5.1 INTRODUCTION. In this chapter we give constructions for Room squares of sides $2r-1$ and $2r+1$ which depend on squares of side $r$.

The two constructions grow from the following basic idea. If you wish to construct a Room square based on a set $R$, partition $R$ into two subsets, $R_1$ and $R_2$ of equal size. Suppose you can find a square array $\mathcal{A}$ which contains every unordered pair of members of $R_1$ and every unordered pair of members of $R_2$, and a square array $\mathcal{B}$ which contains every unordered pair of objects, one from $R_1$ and the other from $R_2$. Then if

| | |
|---|---|
| $\mathcal{A}$ | |
| | $\mathcal{B}$ |

contained every member of $R$ one per row and once per column it would be Room.

This plan cannot be carried out: if $R$ has $2n$ elements then either $\mathcal{A}$ or $\mathcal{B}$ would have at most $n-1$ rows and columns, but would contain $n$ unordered pairs in each row. In the two constructions we show how to build arrays which have nearly all the entries required, and then show where additional entries must be inserted into the master array.

In each case the array $\mathcal{A}$ will be of size $r$ and will be closely related to the sum of a pair of complementary Room squares; in one case we make further restrictions on the complementary squares. $\mathcal{B}$ will be derived from the join of a pair of orthogonal Latin squares.

5.2 FIRST DOUBLING THEOREM. For our theorem we shall need complementary Room squares on $\{0,1,\ldots,r\}$ with the property that, if $\{i,r\}$ lies in cell $(x_1,y_1)$ in one square and in cell $(x_2,y_2)$ in the other, and $i \neq 0$, then

$$x_1 \neq x_2, \quad y_1 \neq y_2.$$

For convenience we will refer to such a pair as "suitable" for the duration of this section. Notice that a skew square and its transpose form a suitable pair.

LEMMA 5.1. *If there exists a suitable pair of Room squares of side $r$ and two orthogonal Latin squares of side $r-1$ with a common transversal, then there is a Room square of side $2r-1$.*

PROOF. Denote the two Latin squares by $L$ and $M$; write $N$ for the square obtained by adding r to every entry in M. Write $\mathcal{B}$ for the square array formed by deleting all the diagonal entries from the join of L and N. Delete every entry involving r or 2r from the sum of the suitable pair of Room squares of side r, then add in diagonal entries $\{1,r+1\},\{2,r+2\},\ldots,\{r,2r\}$; call the result $\mathcal{A}$. Then

$$\begin{array}{|c|c|}\hline \mathcal{A} & \\ \hline & \mathcal{B} \\ \hline \end{array}$$

contains every unordered pair of elements of $S = \{1,2,\ldots,2r\}$ except for those pairs which involve precisely one of r and 2r.

Suppose $\{i,r\}$ occupied cell $(x_i,y_i)$ of $\mathcal{R}_1$ and cell $(z_i,t_i)$ of $\mathcal{R}_2$. Then the array above contains every element of S once per row and once per column, except that r and 2r are missing from every row and column except row and column r, and that for $1 \leq i \leq r$

$i$ is missing from rows $i+r$ and $x_i$ and columns $i+r$ and $y_i$,

$i+r$ is missing from rows $i+r$ and $z_i$ and columns $i+r$ and $t_i$.

Insert the entries

$\{i,r\}$ in cell $(x_i,i+r)$

$\{i,2r\}$ in cell $(i+r,y_i)$

$\{i+r,r\}$ in cell $(i+r,t_i)$

$\{i+r,2r\}$ in cell $(z_i,i+r)$

for $i = 1,2,\ldots,r-1$. Since $x_i \neq z_i$ and $y_i \neq t_i$, it is never required that two entries be placed in the same cell, so the insertions can be carried out. It is easy to see that the resulting square is Room.

THEOREM 5.2. *If there are suitable Room squares of side r, and in particular if there is a skew Room square of side r, then there is a Room square of side 2r - 1.*

PROOF. In view of the lemma, we need only consider the cases where $r-1 = 2$ or 6. In the case $r-1 = 2$ there is no Room square of side r, so there is nothing to prove. When $r-1 = 6$, $2r-1 = 13$, and a Room square of side 13 is known.

5.3 SECOND DOUBLING THEOREM. For the second construction we need a pair of orthogonal Latin squares of side $n = r+1$ with the property that the pairs (n,i) and (i,n) do not appear in the same row or column in their join for $i = 1,2,\ldots,n-1$. We first prove that this is always possible when there are two orthogonal Latin squares of side n, provided $n > 3$; the proof is complicated and I am sure a simpler version must exist.

LEMMA 5.3. *Suppose L and M are orthogonal Latin squares of side n. Then there are orthogonal Latin squares $P = (p_{xy})$ and $Q = (q_{xy})$ of side n with the property that if*

$$p_{xy} = q_{zt} = n,$$

$$q_{xy} = p_{zt} = i,$$

*then*

$$x \neq z,\ y \neq t,$$

*for $i = 1,2,\ldots,n-1$. Moreover*

$$p_{nn} = q_{nn} = n.$$

PROOF. Write $\ell_{xy}$ and $m_{xy}$ for the (x,y) entries of L and M respectively. For convenience assume that the rows and columns of both squares and the entries of L have been permuted in such a way that

$$\ell_{xn} = \ell_{nx} = m_{xn} = x$$

for $1 \leq x \leq n$.

Consider the $2 \times (n-1)$ array

$$A = \begin{pmatrix} 1 & 2 & \ldots & n-1 \\ m_{n1} & m_{n2} & \ldots & m_{n,n-1} \end{pmatrix}.$$

Since the pair (x,x) appears in column n of the join of L and M, $m_{nx}$ will never equal x. So A is a Latin rectangle. Since $n-1 > 2$, we can add at least one more row to A such that the resulting rectangle is still Latin. (In fact, we could add $n-3$ rows; see [34], theorem 6.2.2.) Let

$$(b_1, b_2, \ldots, b_{n-1})$$

be such a row. For every x from 1 to $n-1$,

$$x \neq b_x \neq m_{nx}.$$

Reorder the columns of L and M simultaneously so that the old column x becomes column $b_x$, and denote the new squares obtained from L and M as $L_1 = (\lambda_{xy})$ and $M_1 = (\mu_{xy})$ respectively. $L_1$ and $M_1$ are orthogonal Latin squares.

We now construct P and Q as follows. If $\lambda_{ij} = x$ and $\mu_{ij} = y$ then $p_{xy} = i$ and $q_{xy} = j$. Since (x,y) appears precisely once in the join of $L_1$ and $M_1$, we have uniquely defined every entry in P and Q.

To prove that P is Latin, observe that row x of P contains the entries

$$\{i:\ \lambda_{ij} = x \text{ for some } j\};$$

since $L_1$ was a Latin square, x occurs once in every row, so every possible value of i appears once. Each column of P contains $\{1,2,\ldots,n\}$ by a similar argument depending on the Latin property of $M_1$. The proof that Q is Latin is analogous.

Given any ordered pair (i,j) of members of $\{1,2,\ldots,n\}$, there will be precisely one ordered pair (x,y) such that

$$\lambda_{ij} = x,\ \mu_{ij} = y.$$

So there will be an x and a y satisfying

$$p_{xy} = i,\ q_{xy} = j.$$

So P and Q are orthogonal.

Finally, we observe that when

$$p_{xy} = q_{zt} = n,$$
$$p_{zt} = q_{xy} = i,$$

then

$$\lambda_{ni} = x,\ \mu_{ni} = y,$$
$$\lambda_{in} = z,\ \mu_{in} = t.$$

If i = n we have x = y = z = t = n, so $p_{nn} = q_{nn} = n$. Otherwise, z = t = i, $b_x = i$ and $y = m_{nj}$ where $b_j = i$. If x = z then $i = b_i$, which is impossible. If y = t, then $b_j = m_{nj}$, which is also impossible. So

$$x \neq z,\ y \neq t.$$

THEOREM 5.4. *If there are complementary Room squares of side r, r > 1, then there is a Room square of side 2r + 1.*

PROOF. The Room square we construct shall be based on the set $S = \{0,1,2,\ldots,2r,2r+1\}$. We denote the complementary Room squares as $\mathcal{R}_1$ and $\mathcal{R}_2$, and write $\mathcal{A} = \mathcal{R}_1 + \mathcal{R}_2$.

Since r > 1 we have r > 6, so there are two orthogonal Latin squares of side r + 1. Let P and Q be orthogonal Latin squares of side r + 1 of the type whose existence is guaranteed by lemma 6, and write

L = P with each r + 1 replaced by 0,

M = Q with r added to every entry.

We denote by $\mathcal{B}$ the join of L and M with every entry which involves 0 or 2r + 1 deleted except the entry (0,2r+1); say (0,i) is deleted from position $(x_i,y_i)$ and (i,2r+1) is deleted from position $(z_i,t_i)$.

The array

| | |
|---|---|
| $\mathcal{A}$ | |
| | $\mathcal{B}$ |

would be a Room square except for the following omissions:

all pairs $\{x,y\}$, $1 \leq x \leq 2r$, $y = 0$ or $2r+1$, are missing;

elements 0 and $2r+1$ are missing from every row and column except the last;

$i$ is missing from rows $i$ and $x_i + r$ and columns $i$ and $y_i + r$, and $i+r$ is missing from rows $i$ and $z_i + r$ and columns $i$ and $t_i + r$, for $1 \leq i \leq r$.

We insert entries as follows:

$\{0,i\}$ in cell $(x_i+r,i)$

$\{i,2r+1\}$ in cell $(i,y_i+r)$

$\{0,i+r\}$ in cell $(i,t_i+r)$

$\{i+r,2r+1\}$ in cell $(z_i+r,i)$

for $1 \leq i \leq r$. We are never asked to put two entries in the same cell unless $x_i = z_i$ or $y_i = t_i$, but from lemma 5.3 this is impossible. So we have a Room square.

## CHAPTER VI. OTHER CONSTRUCTIONS

**6.1 A CONSTRUCTION USING FINITE PROJECTIVE SPACES.** The first two general construction of Room squares after Room's original note was given by Archbold and Johnson in 1958.

The construction has been superceded by later constructions; but it remains interesting not only for historical reasons but also because it relates Room squares to finite geometries. We have rewritten the construction so as to utilize the idea of a starter.

The construction takes place in the finite projective geometry PG(m,2), which can be realized as the set of $r = 2^{m+1}-1$ points which are non-zero vectors of the length $m+1$ with elements in the field of order 2 and the set of lines with points $\{x,y,x+y\}$. Each line has 3 points. These sets obey the definition of a projective space of dimension m.

An automorphism of a projective geometry is a one-to-one map from the points to themselves which preserves the lines; if $\phi$ is an automorphism of PG(m,2), then

$$P\phi + Q\phi = (P+Q)\phi.$$

By a theorem of Singer, there is an automorphism $\phi$ of PG(m,2) such that $\Phi = \{\phi,\phi^2,\ldots,\phi^r\}$ is isomorphic to the set of non-zero elements of the finite field $GF(2^{m+1})$: if $\rho$ is a primitive element of $GF(2^{m+1})$ then $\phi^i$ corresponds to $\rho^i$, and if $\rho^i+\rho^j = \rho^k$ then

$$P\phi^i+P\phi^j = P\phi^k.$$

Select such an automorphism $\phi$ of the PG(m,2). Select one point, $P_r$ say, and label the other points $P_1,P_2,\ldots,P_{r-1}$ according to $P_i = P_r\phi^i$. Then define a set X of pairs of elements of $\Phi$ by $\{\phi^i,\phi^j\}\in X$ whenever $P_iP_jP_r$ is a line of the PG(m-1,2).

We show that when $\Phi$ is considered as a cyclic group of order r, X is a starter for $\Phi$. We first observe that X contains every member of $\Phi$ except the identity $\phi^r$, since there is a unique line through $P_i$ and $P_r$ whenever $i \neq r$. Moreover, the set of "differences" is

$$S = \{\phi^{i-j},\phi^{j-i}:\ P_iP_jP_r \text{ is a line}\}.$$

$\phi^{i-j} = \phi^{j-i}$ is impossible since r is odd, and if $\phi^{i-j} = \phi^{k-\ell}$ then $P_iP_jP_r$ and $P_kP_\ell P_r$ are both lines where $\ell \equiv j-i+k \pmod r$. But, since $P_iP_jP_r$ is a line and $\phi$ is an automorphism,

$$(P_i\phi^{k-i})(P_j\phi^{k-i})(P_r{}^{k-i})$$
$$= P_kP_{j-i+k}P_{r-i+k}$$

is a line. If $k = j-i+k$ then $j = i$, which is impossible. Otherwise there is exactly one line through $P_k$ and $P_{j-i+k}$, so

$$P_r = P_{r-i+k}$$

and consequently $i = k$. So S contains no duplications. It contains $r-1$ elements, so it contains every element except $\phi^r$ once. So X is a starter.

We now show that X is strong provided that m is even. The necessary conditions are that for $\{\phi^i,\phi^j\}\varepsilon\ X$,

$$\phi^i\phi^j \neq \phi^r,$$

and that if $\{\phi^k,\phi^\ell\}$ is another pair in X,

$$\phi^k\phi^\ell \neq \phi^i\phi^j.$$

The fact that $\{\phi^i,\phi^j\}\varepsilon\ X$ means that

$$P_i+P_j = P_r,$$

so the necessary conditions can be restated by saying that there should be at most one unordered pair $\{i,j\}$ satisfying

$$\left.\begin{aligned} P_i+P_j &= P_r \\ \phi^i\phi^j &= \phi^x \end{aligned}\right\} \quad (6.1)$$

for any x, and that there should be no solution in the case $x = r$. Using the correspondence between $\Phi$ and $GF(2^{m+1})$, the equations (6.1) are equivalent to

$$\sigma+\tau = 1$$

$$\sigma\tau = \chi$$

where 1 is the multiplicative identity and $\sigma = \rho^i$, $\tau = \rho^j$, $\chi = \rho^x$. Eliminating $\tau$ we have

$$\sigma^2+\chi = \sigma,$$

which can have at most two solutions for $\sigma$; if $\phi = \rho^i$, $\tau = \rho^j$ is a solution, then $\sigma = \rho^j$, $\tau = \rho^i$ will be the other one, so there is at most one unordered pair $\{i,j\}$ satisfying (6.1). (The possibilities of $\sigma = 0$ or $\sigma = \tau$ can be ignored, since they can only *reduce* the number of solutions of (6.1).) When $\chi = 1$ we have

$$\sigma^2+1 = \sigma,$$

which has a solution if and only if there is an element $\sigma$ of $GF(2^{m+1})$ satisfying $\sigma \neq 1$, $\sigma^3 = 1$; since m is even, 3 does not divide $2^{m+1}-1$, so this is impossible.

The finite geometry PG(m,2) can be considered as a Steiner triple system on $r = 2^{m+1}-1$ objects, so the result of this construction is superceded by theorem 2.11.

6.2 CONSTRUCTION OF ADDERS FOR PATTERNED STARTERS. In 1960, Archbold [1] proved the following result.

THEOREM 6.2. *Suppose X is a (4m-1,2m,m) difference set in the additive group of integers modulo 4m - 1, and suppose there is an integer t modulo 4m - 1 which fixes X and satisfies*

(i) $t^{2m-1} \equiv 1 \pmod{4m-1}$ *but* $t^k \not\equiv 1$, $1 \leq k < 2m-1$,

(ii) *the permutation* $x \to tx$ *on X is a cycle of length* $2m-1$.

*Choose an element* $x_1$ *of X for which* $tx_1 \neq x$. *Suppose there is an integer u such that both u and* $x_1+u$ *are prime to* $4m-1$ *and such that there is an integer v for which the unordered pair* $\{v,x_1+v\}$ *does not belong to the set of unordered pairs*

$$\Big\{\{0,x_1\},\ \{u,x_1+u\},\dots,\{t^{2m-2}u,x+t^{2m-2}u\},\ \{-x_1,0\},\ \{-x_1+t(x_1+u),\ t(x_1+u)\},\dots,\{-x_1+t^{2m-2}(x_1+u),\ t^{2m-2}(x_1+u)\}\Big\}.$$

*Then there is a Room square of side* $4m-1$.

Archbold used this theorem to construct Room squares of sides 7,11,19 and 23, and proves that it cannot be used to construct a Room square of side 15. We shall not prove the theorem yet, because we present a more general version as theorem 6.4 below.

We first show that the given construction is in fact quite limited.

THEOREM 6.3. *The conditions of theorem 6.2 cannot be satisfied unless* $4m-1$ *is prime.*

PROOF. We write $4m-1 = p$, and denote the set of positive integers less than p and prime to p by P. If the distinct primes dividing p are $\pi_1,\pi_2,\dots,\pi_\alpha$, then it is a well known result of number theory that P has order $\phi(p)$, where

$$\phi(p) = (1-\pi_1^{-1})(1-\pi_2^{-1})\dots(1-\pi_\alpha^{-1})p.$$

Since p is odd, $\phi(p)$ will be even. It is also known that if $x \notin P$ then no power of x can be congruent to 1 modulo p, and that for $x \in P$,

$$x^{\phi(p)} \equiv 1.$$

The integer t of theorem 6.2 is required to satisfy $t^{2m-1} \equiv 1$, so $t \in P$ and $2m-1 | \phi(p)$. This is possible when $\phi(p) = p-1 = 4m-2$, which occurs if and only if p is prime. But when p is composite, $\phi(p) < 4m-2$, so the only case where $2m-1$ divides $\phi(p)$ is

$$\phi(p) = 2m-1;$$

but this is impossible since $\phi(p)$ is even.

In view of theorem 6.3, there would be no added restriction in theorem 6.2 if we specified that the difference set X was to be in the additive group of integers modulo a prime $4m-1$; that is, in the additive group of the Galois field of prime order $4m-1$. In view of this, the following theorem is a generalization of theorem 6.2:

THEOREM 6.4. *Suppose $p = 4m-1$ is a prime power, and suppose $X$ is a set of $2m$ elements of the Galois field $G$ of order $4m-1$. Suppose there is an element $t$ of $G$, $t \neq 0$ and $t \neq 1$, which fixes $X$ and which satisfies*

$$t^{2m-1} = 1, \quad t^a \neq 1 \text{ for } 0 < a < 2m-1. \tag{6.5}$$

*Let $x_1$ be a non-zero member of $X$, and suppose further that there exists an element $u$ of $G$ such that both $u$ and $u+x_1$ are non-zero and such that if*

$$S = \{0,-x_1\} \cup \{t^i u: \ 0 < i < 2m\} \cup \{t^i(u+x_1)-x_1: \ 0 < i < 2m\} \tag{6.6}$$

*then $S$ is a proper subset of $G$.*

*Then there is a starter in $G$ with the set of all negatives of non-zero elements of $X$ as an adder - that is, the adder is*

$$A = \{-x: \ x \in X, \ x \neq 0\}.$$

PROOF. Since $x_1 \neq 0$ the $2m-1$ elements

$$x_1, tx_1, t^2x_1, \ldots, t^{2m-2}x_1$$

are all different. They are all members of X, since t fixes X. Let us write $x_i = t^{i-1}x_1$. It is clear that the remaining member of X, $x_0$, say, must satisfy $tx_0 = x_0$ (since $x_i \mapsto tx_i$ is one-to-one), so $x_0$ or $t-1$ is zero; as $t \neq 1$, $x_0 = 0$.

$G \setminus S$ is non-empty, as S is a proper subset of G. Select $v \in G \setminus S$. Write Y for the set of all unordered pairs

$$\{-t^{i-1}u, -t^{i-1}v\},$$

where $0 < i < 2m$. We show that Y is the required starter.

To prove that the pairs in Y contain all non-zero members of G - that is

$$\{-t^{i-1}u, -t^{i-1}v: \ 0 < i < 2m\} = G \setminus \{0\},$$

we observe that the $-t^{i-1}u$ are all distinct and non-zero, as are the $-t^{i-1}v$, since u and v are non-zero; and if

$$t^{j-1}v = t^{i-1}u$$

for some i and j then

$$v = t^{i-j}u$$

and $v \in S$, which is not allowed. We have $2(2m-1) = p-1$ distinct non-zero elements in the set, so it equals $G \setminus \{0\}$.

The set of all differences of the pairs in Y is

$$\{\pm t^{i-1}(v-u): \ 0 < i < 2m\};$$

v-u is non-zero because $u \in S$, so the elements $t^{i-1}(v-u)$ are all distinct and non-zero, as are the elements $-t^{i-1}(v-u)$. This means that Y is a starter unless

$$t^{i-1}(v-u) = -t^{j-1}(v-u) \tag{6.7}$$

for some i and j. But if (6.7) were true then

$$t^{j-1}(t^{i-j}+1)(v-u) = 0,$$

and consequently $t^{i-j}+1 = 0$. However,

$$(t^{i-j}+1)(t^{(2m-2)(i-j)}-t^{(2m-1)(i-j)}+\ldots+1) = t^{(2m-1)(i-j)}+1 = 1+1;$$

1 +1 is non-zero in GF(p), since p is odd, so $t^{i-j}+1$ is also. So Y is a starter.

We now prove that the set of negatives of non-zero elements of X is an adder for Y. We attach the elements $-x_i$ to the pair $\{-t^{i-1}u,-t^{i-1}v\}$. We must verify that

$$\{-t^{i-1}u-x_i,-t^{i-1}v-x_i:\ 0 < i < 2m\} = G\setminus\{0\}.$$

Since $x_i = t^{i-1}x_1$, it is sufficient to show that the elements of form $t^{i-1}(u+x_1)$ and $t^{i-1}(v+x_1)$ are all distinct and non-zero. Now

$$t^{i-1}(u+x_1) = 0 \Rightarrow u+x_1 = 0.$$

$$t^{i-1}(u+x_1) = t^{j-1}(u+x_1) \Rightarrow (T^{i-j}-1)(u+x_1) = 0$$

$$\Rightarrow u+x_1 = 0 \quad \text{if } i \neq j,$$

so the $t^{i-1}(u+x_1)$ are distinct and non-zero. Similar calculations with u replaced by v yield the corresponding result, $v+x_1 = 0$ being impossible unless $v \in S$. If

$$t^{i-1}(u+x_1) = t^{j-1}(v+x_1)$$

then

$$v = t^{i-j}(u+x_1)-x_1 \in S,$$

which is not allowed. So the required set is an adder.

As an application, we prove

THEOREM 6.8. *If 4m - 1 is a prime power greater than 3, then there is a skew adder for the patterned starter in the elementary abelian group G of order 4m - 1.*

PROOF. Without loss of generality we may assume that G is the additive group of GF(4m-1). There will be a primitive root r in GF(4m-1) - that is, an element r such that $r^{4m-2} = 1$ but $r^i \neq 1$ for $0 < i < 4m-2$. Select $t = r^2$; then t satisfies (6.5). Write

$$X = \{0,t,t^2,\ldots,t^{2m-1}\}.$$

Obviously t fixes X. As t belongs to X we can set $x_1 = t$ and in general $x_i = t^i$ for $0 < i < 2m$; $x_0$ is of course 0.

We now show that, provided $m > 1$, we can find a non-zero element u in G which satisfies

$$u + x_1 \neq 0, \quad -u \notin S, \tag{6.9}$$

where S is defined as in (6.6). We then choose $v = -u$. The starter Y of theorem 6.4 is

$$\{(-t^{i-1}u, t^{i-1}u): \; 0 < i < 2m\},$$

which is the patterned starter. The adder given by theorem 6.4 is the set Q of negatives of non-zero quadratic elements of GF(4m-1); it is known that -1 is not a quadratic element of GF(4m-1), so if $x \in Q$ then $-x \notin Q$; hence the adder is skew.

The restrictions (6.9) can be restated as

$$u \neq -t \tag{6.10}$$

$$u \notin \{0,t\} \cup \{-t^i u: \; 0 < i < 2m\} \tag{6.11}$$

$$t-u \notin \{t^i(t+u): \; 0 < i < 2m\}. \tag{6.12}$$

If $u = -t^i u$ then $t^i = -1$, so -1 is a quadratic element, which is impossible. Therefore, (6.10) and (6.11) are satisfied provided

$$u \notin \{0,t,-t\}.$$

This will also imply that u is non-zero, as required. Now $t - u \in \{t^i(t+u)\}$ occurs if and only if $(t-u)(t+u)^{-1}$ is one of the $2m-1$ elements $1, t, t^2, \ldots, t^{2m-2}$. If

$$(u-t)(u+t)^{-1} = (w-t)(w+t)^{-1}$$

then

$$(u-w)t = (w-u)t,$$

which is impossible unless $u = w$. So the $4m-1$ elements $(u-t)(u+t)^{-1}$ formed as u varies through G are all different, and the condition that (6.12) is true is satisfied by 2m values of u. These values may contain 0,t or -t, but since $m > 1$, there will be at least one suitable u.

Patterned Room squares of various sizes were constructed by Byleen [5], and his results include those of theorem 6.8. However, Byleen's methods are different from what we have given here. Even they do not give patterned starters in groups of non-prime-power order. There are no adders for the patterned starters in the groups of order 9, but computer results make it seem likely that there are patterned starters in all larger *cyclic* groups at least.

## 6.3 CONSTRUCTION FROM PAIRWISE BALANCED DESIGNS.

A pairwise balanced design is a way of selecting subsets (blocks) from a set of v objects in such a way that every pair of distinct objects occur together in precisely λ of the subsets, for some fixed λ; the constant λ is called the *index* of the design. If k is a set of positive integers such that the number of objects in each block is a member of K, then we shall refer to the design as a PB(v;K;λ).

THEOREM 6.13. *Suppose there is a PB(v;K;1). If there is a Room square of side k for every* $k \in K$, *then there is a Room square of side v with subsquares of sides k for every* $k \in K$; *if the original squares were skew then the square of side v will be skew and the subsquares will be skew subsquares. If there are complementary Room squares of side k for every* $k \in K$, *then there will be complementary Room squares of side v.*

PROOF. Suppose the set of objects on which the PB(v;K;1) is based is $\{1,2,\ldots,v\}$. For each block B of size k in the design, label the members of B $a_1,a_2,\ldots,a_k$ in some order; write $\mathcal{R}_k$ for a standardized Room square of side k. We shall construct a standardized Room square $\mathcal{R}$ of side v.

Suppose p and q are any two distinct integers from 1 to v. There will be exactly one block of the pairwise balanced design which contains both p and q; say it is $B = \{a_1,a_2,\ldots,a_k\}$, and say $p = a_i$, $q = a_j$. If the (i,j) cell of $\mathcal{R}_k$ is empty, then leave the (p,q) cell of $\mathcal{R}$ empty. If the (i,j) cell of $\mathcal{R}_k$ contains $\{x,y\}$, then place $\{a_x,a_y\}$ in position (p,q) of $\mathcal{R}$. Finally, place $\{0,p\}$ in position (p,p) of $\mathcal{R}$ for every p.

Consider the unordered pair $\{p,q\}$, where $1 \le p \le v$ and $1 \le q \le v$. The entry $\{p,q\}$ arises in $\mathcal{R}$ only from some block B where $p \in B$ and $q \in B$. As there is exactly one such block, $\{p,q\}$ will arise exactly once. The pairs $\{0,p\}$ are found once each in $\mathcal{R}$, on the diagonal. So $\mathcal{R}$ contains every unordered pair of entries from $\{0,1,2,\ldots,v\}$, once each.

Now we examine all the entries of row p. An entry will be inserted in row p only if it comes from a block B which contains p. If $p = a_i$ and $B = \{a_1,a_2,\ldots,a_k\}$, then the entries of $\mathcal{R}$ in row p coming from B will be all the $\{a_x,a_y\}$ where $\{x,y\}$ is in row i of $\mathcal{R}_k$, off the diagonal. Since $\mathcal{R}_k$ is standardized, its ith row contains the objects

$$\{1,2,\ldots,i-1,i+1,\ldots,k\}$$

off its diagonal, and the entries in row p contributed by B are

$$\{a_1,a_2,\ldots,a_{i-1},a_{i+1},\ldots,a_k\} = B \setminus \{p\}.$$

So the entries in row p contain 0, p and the members of every $B \setminus \{p\}$ where $p \in B$. For any q satisfying $1 \le q \le v$, or q = b, there is precisely one block containing p and also containing q. So every number from 0 to v is represented precisely once in row p.

A similar proof applies to columns. So $\mathcal{R}$ is a Room square. If $B = \{a_1,a_2,\ldots,a_k\}$ is any block of the pairwise balanced design, then $\mathcal{R}_k$ appears as a subsquare in $\mathcal{R}$ as the intersection of rows $a_1,a_2,\ldots,a_k$ with columns $a_1,a_2,\ldots,a_k$.

Suppose the initial squares $\mathcal{R}_k$ were all skew. The block containing $\{p,q\}$ is

$B = \{a_1,a_2,\ldots,a_k\}$ and $p = a_i$, $q = a_j$. Then exactly one of cells (i,j) and (j,i) is occupied in $\mathcal{R}_k$, so exactly one of cells (p,q) and (q,p) will be occupied in $\mathcal{R}$. So $\mathcal{R}$ is skew.

The square $\mathcal{R}_k$ is skew, and is symmetrically situated as a subsquare of $\mathcal{R}$, so it is a skew subsquare.

Now, instead of assuming skewness, assume that each square $\mathcal{R}_k$ has a complement $\mathcal{S}_k$; and let $\mathcal{S}$ be the Room square constructed from the pairwise balanced design using the $\mathcal{S}_k$ instead of the $\mathcal{R}_k$. If $B = \{a_1,a_2,\ldots,a_k\}$ is the block containing $\{p,q\}$, where $p = a_i$ and $q = a_j$, then the (p,q) cell of $\mathcal{R}$ will be occupied if and only if the (i,j) cell of $\mathcal{R}_k$ is occupied and the (p,q) cell of $\mathcal{S}$ will be occupied if and only if the (i,j) cell of $\mathcal{S}_k$ is occupied. Both of these cannot happen, since $\mathcal{R}_k$ and $\mathcal{S}_k$ are complementary. So $\mathcal{R}$ and $\mathcal{S}$ are complementary.

6.4 NOTES ON THE METHOD OF PAIRWISE BALANCED DESIGNS. The method of the preceding section should be applicable in finding Room squares with specified subsquares. The main shortcoming is our lack of knowledge of families of pairwise balanced designs. In particular, in many of the known constructions for pairwise balanced designs, there is at least one small member of K, and the construction cannot be used when K contains a number smaller than 7.

The construction can sometimes yield skew squares with skew subsquares which cannot be found in other ways. For example, Lawless [16] points out that there is a PB(151;{25,7};1). Since there are skew Room squares of sides 25 and 7, there is a skew square of side 151 with skew subsquares of sides 7 and 25. We can construct a square of side 151 using

$$151 = 25(7-1)+1$$

and theorem 3.3 or 3.4, but the square constructed in that way would not have the skew subsquares.

On the other hand, there are many Room squares which cannot be constructed by the method:

LEMMA 6.14. *Suppose there is a PB(v;K;1) where h and k are the smallest and largest members of K respectively. Then*

$$v \geq k(h-1)+1$$

*(except for the trivial cases where k = v or where the design actually contains no block of size k).*

PROOF. Let $B_0 = \{a_1,a_2,\ldots,a_k\}$ be a block of the PB(v;K;1) of size k. As $v \neq k$, there is at least one of the v objects which is not in $B_0$; call it $a_0$. Denote by $B_i$ the (unique) block containing $\{a_0,a_i\}$, and write $A_i = B_i \setminus \{a_0\}$. Since the intersection of any two blocks has at most one element, $A_i \cap A_j$ is empty when $i \neq j$. Each $A_i$ has at least $h-1$ elements, so the union of the $A_i$ has at least $k(h-1)$ elements, and $a_0$ is not included. So

$$v \geq k(h-1)+1.$$

COROLLARY 6.15. *The construction of theorem 6.13 cannot be used to find a Room square of side $v$ with a subsquare of side $k$ unless $v \geq 6k+1$.*

The squares constructed in theorem 3.5 cannot be gotten from theorem 6.8, for example.

## CHAPTER VII. THE EXISTENCE PROBLEM

7.1 FERMAT NUMBERS. We now prove that there is a skew Room square of side $f_k$, where

$$f_k = 2^{2^k} + 1,$$

for all k except 0,1 and 3. Our main tools are theorems 3.1 and 3.2.

LEMMA 7.1. *If there exists a skew Room square of side $2^d + 1$ and a skew Room square of side $2^{a+d} + 1$ with a skew subsquare of side $2^a + 1$ then there is a skew Room square of side $2^{a+md} + 1$ with a skew subsquare of side $2^{a+(m-1)d} + 1$ for $m = 1,2,3,\ldots$*

PROOF. We proceed by induction on m. The case m = 1 is included in the data. Suppose the case m = k is true, so that there is a skew Room square of side $2^{a+kd} + 1$ with a skew subsquare of side $2^{a+(k-1)d} + 1$.

$$2^{a+(k+1)d} + 1 = (2^d+1)[(2^{a-kd}+1)-(2^{a+(k-1)d}+1)] + 2^{a+(k-1)d} + 1,$$

so by theorem 3.1 the case m = k + 1 is true.

LEMMA 7.2. *If there exist skew Room squares of sides $x+1$, $x-1$ and $x^2+1$, then there exists a skew Room square of side $x^{2n+3} + 1$ with a skew subsquare of side $x^2+1$, for $n = 1,2,3,\ldots$*

PROOF. We define $s_n$ by

$$s_0 = x^2-x+1,$$

$$s_n = x^{2n+2}-x^{2n+1}+s_{n-1},$$

so that

$$x^{2n+3}+1 = (x+1)s_n.$$

If we prove that there is a skew Room square of side $s_n$ with a skew subsquare of side $x^2+1$ for $n = 1,2,3,\ldots$ then we have the result. We shall in fact prove by induction that there is a skew Room square of side $s_n$ with skew subsquares of sides $s_{n-1}$ and $x^2+1$ for $n \geq 1$.

We first observe that

$$s_0 = (x-1)[(x+1)-1] + 1,$$

so that there is a skew Room square of side $s_0$, and that

$$s_1 = (x^2+1)(s_0-1) + 1,$$

so that the result holds for the case n = 1 (using theorem 3.1). In general we can write

$$s_{k+1} = x^{2k+4}-x^{2k+3}+x^{2k+2}-x^{2k+1}+s_{k-1}$$

$$= (x^2+1)(s_k-s_{k-1})+s_{k-1}.$$

So, by theorem 3.1 again, the existence of a skew square of side $s_k$ with a skew subsquare of side $s_{k-1}$ will imply the existence of a square of side $s_{k+1}$ with the desired properties.

LEMMA 7.3. *If $f_k$ is not prime then there is a skew Room square of side $f_n$. In particular there are skew Room squares of sides $f_5, f_6, f_7$ and $f_8$.*

PROOF. We can write

$$f_k = f_{k-1}f_{k-2}\ldots f_1f_0+2.$$

So $f_k$ is not divisible by any $f_i$ other than itself. If $f_k$ is not prime, and has prime factor decomposition

$$f_k = p_1p_2\ldots p_t,$$

then none of the $p_i$ is a Fermat prime, and by theorem 2.7 there is a skew Room square of side $p_i$ for every i. So there is a skew Room square of side $f_k$.

All of $f_5, f_6, f_7$ and $f_8$ are known to be composite.

LEMMA 7.4. *There are skew Room squares of sides $f_2, f_4$ and $f_5-2$.*

PROOF. A skew Room square of side $f_2 = 17$ has been found by computer. Using the equations

$$77 = 7.11$$

$$989 = 13\ (77-1)+1$$

$$f_4 = 65537 = 67\ (989-11)+11$$

and the existence of skew squares of sides 7 and 11 we obtain successively a skew square of side 77 with a skew subsquare of side 11, a skew square of side 989 with a skew subsquare of side 77 which in turn has a skew subsquare of side 11, and a skew square of side $f_4 = 65537$. Since

$$257.5.3 = 3855 = 47\ (83-1)+1$$

$$f_5-2 = 65537.17.(257.5.3)$$

there is a skew square of side $f_5-2$.

THEOREM 7.5. *There is a skew Room square of side $f_k$ unless $k = 0, 1$, or $3$.*

PROOF. From lemmas 7.3 and 7.4 we need only discuss $k > 8$.

Write $x = 2^{32}$. Then $f_k = x^{4.2^{k-7}}+1$. In particular $x^2+1 = f_6$, $x+1 = f_5$ and $x-1 = f_5-2$, so from lemmas 7.3 and 7.4 there are skew Room squares of sides

$x^2+1$, $x+1$ and $x-1$. So, by lemma 7.2, there is a skew Room square of side $x^{2n+3}+1$ with a skew subsquare of side $x^2+1$ for every $n \geq 1$.

If $a = 64$ and $d = 64n + 32$, then lemma 7.1 says that the existence of a skew Room square of side $x^{2n+1}+1$ and a skew Room square of side $x^{2n+3}+1$ with a skew subsquare of side $x^2+1$ imply the existence of a skew Room square of side $x^{2+(2n+1)m}+1$ for every $m \geq 0$. In the case $m = 2$, the skew square constructed will have order $x^{4(n+1)}+1$. From the preceding paragraph there is a skew square of side $x^{2n+1}+1$ when $n \geq 2$, and the other required square exists when $n \geq 1$. So there is a skew square of side $f_k$ whenever $2^{k-7} \geq 3$; that is, whenever $k \geq 9$.

## 7.2 GENERAL EXISTENCE RESULTS.

We now prove that there is a Room square of every side except 3,5 and possibly 257.

Suppose r is an odd integer. We can write r as

$$r = 3^a 5^b 257^c m$$

where m is not divisible by 3,5 or 257. There is a skew Room square of every prime power side except 3,5,9 or 257 (theorems 7.5 and 2.7), so by the multiplication theorem (theorem 3.2) there is a skew Room square of side r unless $a = 1$ or 2, or $b = 1$, or $c = 1$.

There are Room squares of side 9 and 15 (for example, those constructed in [39]). There are Room squares of sides 3r and 5r whenever there is a Room square of side $r > 1$ (theorem 4.3). Side 1 is trivial. It follows that there is a Room square of side r unless $r = 3,5$ or 257n where n is prime to 257. Side 257n is covered by theorem 4.3 provided there is a Room square of side n and $n > 257$.

We can in fact restrict our attention to finitely many squares. For, if $n > 257$ and there is no square of side n, then n is a multiple of 257. So 257n is divisible by $257^2$, and the square of side 257n exists.

We now examine 257n, for odd n less than 257. First, observe that if $n = pq$, where there are Room squares of sides p and 257q, then there will be a Room square of side 257n by the multiplication theorem; this reduces the number of cases to be considered.

If there is a Room square of side r ($r > 1$) then there is a Room square of side $4r+1$ (theorem 3.5). When

$$257n = 4r+1,$$

257 and r will be coprime, and r will certainly be greater than 5. So a square of side 257n can be constructed in this way for every odd r. This covers every case where $257n \equiv 5 \pmod 8$; that is, $n \equiv 5 \pmod 8$. To summarize,

*(A) There is a Room square of side 257n whenever*

$$n \equiv 5 \pmod 8.$$

If there is a skew Room square of side $r > 1$, then there are Room squares of sides $2r-1$ and $2r+1$ (Chapter V). If

$$257n = 2r \pm 1$$

then r will be prime to 257, so there will be a skew Room square of side r provided neither 3 nor 5 divides r, so there is a skew square when

$$r \equiv 1,2,4,7,8,11,13,14 \pmod{15};$$

since r must be odd, this list means

$$r \equiv 1,7,11,13,17,19,23,29 \pmod{30}.$$

So there is a Room square of side 257n whenever (case $257n = 2r-1$)

$$257n \equiv 1,13,21,25,33,37,45,57 \pmod{60}$$

or (case $257n = 2r+1$)

$$257n \equiv 3,15,23,27,35,39,47,59 \pmod{60}.$$

This means

*(B) There is a Room square of side 257n whenever*

*$n \equiv 5,7,9,15,19,21,27,29,31,33,39,41,45,51,53,55 \pmod{60}$.*

We now use various applications of theorem 3.1. In each case we factorize 257n in the form $r(s-1)+1$, where Room squares of sides r and s are known to exist.

| | | | | | |
|---|---|---|---|---|---|
| *(C1)* | *3.257* | = | *771* | = | *77(11-1)+1* |
| *(C2)* | *11.257* | = | *2827* | = | *9(315-1)+1* |
| *(C3)* | *17.257* | = | *4369* | = | *7(625-1)+1* |
| *(C4)* | *23.257* | = | *5971* | = | *591(11-1)+1* |
| *(C5)* | *25.257* | = | *6425* | = | *11(585-1)+1* |
| *(C6)* | *43.257* | = | *11051* | = | *1105(11-1)+1* |
| *(C7)* | *47.257* | = | *12079* | = | *9(1343-1)+1* |
| *(C8)* | *59.257* | = | *15163* | = | *7(2167-1)+1* |
| *(C9)* | *73.257* | = | *18761* | = | *469(41-1)+1* |
| *(C10)* | *83.257* | = | *21331* | = | *2133(11-1)+1* |
| *(C11)* | *97.257* | = | *24929* | = | *779(33-1)+1* |
| *(C12)* | *103.257* | = | *26471* | = | *2647(11-1)+1* |
| *(C13)* | *131.257* | = | *33667* | = | *181(187-1)+1* |
| *(C14)* | *137.257* | = | *35209* | = | *9(3913-1)+1* |
| *(C15)* | *163.257* | = | *41891* | = | *4189(11-1)+1* |
| *(C16)* | *167.257* | = | *42919* | = | *933(47-1)+1* |
| *(C17)* | *179.257* | = | *46003* | = | *11(4183-1)+1* |
| *(C18)* | *191.257* | = | *49087* | = | *9(5455-1)+1* |

| | | | | | |
|---|---|---|---|---|---|
| (C19) | $193.257$ | = | $49601$ | = | $31(1601-1)+1$ |
| (C20) | $223.257$ | = | $57311$ | = | $5731(11-1)+1$ |
| (C21) | $227.257$ | = | $58339$ | = | $7(8335-1)+1$ |
| (C22) | $239.257$ | = | $61423$ | = | $1059(59-1)+1$ |
| (C23) | $241.257$ | = | $61937$ | = | $3871(17-1)+1$ |
| (C24) | $251.257$ | = | $64507$ | = | $13(4963-1)+1$ |

Finally, we observe that

$$71.257 = 18247 = 6.3041+1,$$

$$107.257 = 27499 = 6.4583+1.$$

Since 3041 and 4583 are prime, there are skew Room squares of sides 3041 and 4583. So, by theorem 3.3 or theorem 3.4,

(D) *There are Room squares of sides 71.257 and 107.257.*

In Figure 7.1 we list the odd numbers from 1 to 255. The letter after n refers to the statement which shows there is a Room square of side 257n. (The reference $M_q$ means that $n = pq$, where Room squares of sides p and 257q are known.) From the table we see that every side 257n has been established except when n = 1. So we can summarize the existence position as follows:

THEOREM 7.6. *There is a Room square of every side except 3,5 and 257. Sides 3 and 5 are impossible, and 257 remains in doubt.*

| | | | | | | | | | | | |
|---|---|---|---|---|---|---|---|---|---|---|---|
| 1 | | 3 | C1 | 5 | A | 7 | B | 9 | B | 11 | C2 |
| 13 | A | 15 | B | 17 | C3 | 19 | B | 21 | M3 | 23 | C4 |
| 25 | C5 | 27 | M3 | 29 | A | 31 | B | 33 | M3 | 35 | M5 |
| 37 | A | 39 | M3 | 41 | B | 43 | C6 | 45 | M3 | 47 | C7 |
| 49 | M7 | 51 | M3 | 53 | A | 55 | M5 | 57 | M3 | 59 | C8 |
| 61 | A | 63 | M3 | 65 | M5 | 67 | B | 69 | M3 | 71 | D |
| 73 | C9 | 75 | M3 | 77 | M7 | 79 | B | 81 | M3 | 83 | C10 |
| 85 | M5 | 87 | M3 | 89 | B | 91 | M7 | 93 | M3 | 95 | M5 |
| 97 | C11 | 99 | M3 | 101 | A | 103 | C12 | 105 | M3 | 107 | D |
| 109 | A | 111 | M3 | 113 | B | 115 | M5 | 117 | M3 | 119 | M7 |
| 121 | M11 | 123 | M3 | 125 | M5 | 127 | B | 129 | M3 | 131 | C13 |
| 133 | M7 | 135 | M3 | 137 | C14 | 139 | B | 141 | M3 | 143 | M11 |
| 145 | M5 | 147 | M3 | 149 | A | 151 | B | 153 | M3 | 155 | M5 |
| 157 | A | 159 | M3 | 161 | M7 | 163 | C15 | 165 | M3 | 167 | C16 |
| 169 | M13 | 171 | M3 | 173 | A | 175 | M5 | 177 | M3 | 179 | C17 |
| 181 | A | 183 | M3 | 185 | M5 | 187 | M11 | 189 | M3 | 191 | C18 |
| 193 | C19 | 195 | M3 | 197 | A | 199 | B | 201 | M3 | 203 | M7 |
| 205 | M5 | 207 | M3 | 209 | M11 | 211 | B | 213 | M3 | 215 | M5 |
| 217 | M7 | 219 | M3 | 221 | M13 | 223 | C20 | 225 | M3 | 227 | C21 |
| 229 | A | 231 | M3 | 233 | B | 235 | M5 | 237 | M3 | 239 | C22 |
| 241 | C23 | 243 | M3 | 245 | M5 | 247 | M13 | 249 | M3 | 251 | C24 |
| 253 | M11 | 255 | M3 | | | | | | | | |

FIGURE 7.1

# CHAPTER VIII. ROOM SQUARES OF SIDE 7

8.1 ONE-FACTORIZATIONS OF $K_8$. We shall write $K_8$ for the complete graph with vertices {0,1,2,3,4,5,6,7}.

If $\mathcal{F}$ is a one-factorization of K we shall denote the factors by A,B,C,D,E, F,G and take them in that order (so that 01 ε A, 02 ε B, etc.).

No one-factorization of $K_8$ can contain a 4-division. If $\mathcal{F}$ contains a 3-division we can assume it is {A,B,C}, and write

A = 01 23 45 67

B = 02 13 46 57

C = 03 12 47 56

(where the notation means that the edges in A are 01,23,45 and 67). Every edge in D,E,F and G will be of the form xy where x ε {0,1,2,3} and y ε {4,5,6,7}. D contains 04; if it also contains 17 then we can carry out the permutation (23)(67), which leaves {A,B,C} unchanged and maps {04,17} into {04,16}. So we can assume $D = D_1, D_2, D_3$ or $D_4$, where

$D_1$ = 04 15 26 37 $\qquad$ $D_3$ = 04 16 27 35

$D_2$ = 04 16 25 37 $\qquad$ $D_4$ = 04 15 27 36.

However, the permutation (13)(57) takes $\{A,B,C,D_4\}$ into $\{A,B,C,D_2\}$, so we can ignore $D_4$.

In each of the three remaining cases we can extend to a one-factorization of $K_8$ in precisely four ways. If $D = D_1$,E,F and G can be

| | | | |
|---|---|---|---|
| 05 14 27 36 or | 05 16 27 34 or | 05 14 27 36 or | 05 17 24 36 |
| 06 17 24 35 | 06 17 24 35 | 06 17 25 34 | 06 14 27 35 |
| 07 16 25 34 | 07 14 25 36 | 07 16 24 35 | 07 16 25 34, |

and we denote the resulting factorizations by $\mathcal{F}_1$, $\mathcal{F}_2$, $\mathcal{F}_{21}$ and $\mathcal{F}_{22}$ respectively. If $D = D_2$ we can have

| | | | |
|---|---|---|---|
| 05 14 27 36 or | 05 17 24 36 or | 05 17 26 34 or | 05 17 24 36 |
| 06 17 24 35 | 06 14 27 35 | 06 14 27 35 | 06 15 27 34 |
| 07 15 26 34 | 07 15 26 34 | 07 15 24 36 | 07 14 26 35, |

and we denote the resulting factorizations by $\mathcal{F}_{23}$, $\mathcal{F}_{24}$, $\mathcal{F}_3$ and $\mathcal{F}_{31}$ respectively. With $D = D_3$ the possibilities are

| | | | | | | |
|---|---|---|---|---|---|---|
| 05 14 26 37 | or | 05 17 24 36 | or | 05 17 26 34 | or | 05 17 26 34 |
| 06 17 25 34 | | 06 14 25 37 | | 06 14 25 37 | | 06 15 24 37 |
| 07 15 24 36 | | 07 15 26 34 | | 07 15 24 36 | | 07 14 25 36 |

and we denote the resulting factorizations by $\mathcal{F}_{32}$, $\mathcal{F}_{33}$, $\mathcal{F}_4$ and $\mathcal{F}_{34}$ respectively.

A calculation shows that $\mathcal{F}_1$, $\mathcal{F}_2$, $\mathcal{F}_3$ and $\mathcal{F}_4$ are non-isomorphic, as

$\mathcal{F}_1$ has $d_3 = 7$ and $d_2 = 0$,

$\mathcal{F}_2$ has $d_3 = 3$ and $d_2 = 4$,

$\mathcal{F}_3$ has $d_3 = 1$ and $d_2 = 4$,

$\mathcal{F}_4$ has $d_3 = 1$ and $d_2 = 6$.

It is found that if we apply the permutations (12)(56), (13)(57), (02)(57) and (157236) respectively to $\mathcal{F}_2$ we obtain $\mathcal{F}_{21}$, $\mathcal{F}_{22}$, $\mathcal{F}_{23}$ and $\mathcal{F}_{24}$, so these five factorizations are all isomorphic. Similarly, if we map $\mathcal{F}_3$ by (03)(47), (123)(457), (45)(67) and (0427)(1536) we obtain $\mathcal{F}_{31}$, $\mathcal{F}_{32}$, $\mathcal{F}_{33}$ and $\mathcal{F}_{34}$ respectively. Therefore, $\mathcal{F}_1$, $\mathcal{F}_2$, $\mathcal{F}_3$ and $\mathcal{F}_4$ are a complete set of non-isomorphic one-factorizations of $K_8$ which contain a 3-division.

By similar arithmetic we can show that there are exactly two one-factorizations with no 3-divisions. We shall denote them by $\mathcal{F}_5$ and $\mathcal{F}_6$. $\mathcal{F}_5$ has $d_2 = 4$; $\mathcal{F}_6$ has $d_2 = 0$.

A table of all six one-factorizations is given in Figure 8.1. The table lists a one-factorization, the values of $d_3$ and $d_2$, and all 3-divisions and maximal 2-divisions.

The structure of a factorization is reflected in its set of automorphisms, so we now calculate the automorphism group $G_i$ of each $\mathcal{F}_i$. When speaking of transitivity, etc., we are referring to the natural faithful representation of $G_i$ as a permutation group on {0,1,2,3,4,5,6,7}. We write $A_n$, $Z_n$ and $D_n$ for the alternating group on n symbols and the cyclic and dihedral groups of order n respectively; AGL (3,2) is the collineation group of the affine 3-space over GF(2). A cross denotes direct product. G(x) is the stabilizer of the symbol x in G.

THEOREM 8.1 (i) *$G_1$ has order 1344 and is triply transitive. $G_1$ is the group AGL(3,2) represented as a permutation group on the points of the 3-space, and may be generated by (0123)(4567), (1245736) and (23)(45);*

(ii) *$G_2$ has order 64 and is transitive of rank 4; $G_2(0)$ has orbits {0},{1},{2,3,4,5} and {6,7}. It is the split extension of the $Z_2 \times Z_2 \times Z_2 \times Z_2$ generated by (01)(67), (06)(17), (23)(45) and (24)(35), by the $Z_2 \times Z_2$ generated by (02)(13)(46)(57) and (45)(67);*

$\mathcal{F}_1$ $d_3 = 7$, $d_2 = 0$

| | | | | | | |
|---|---|---|---|---|---|---|
| A | = | 01 | 23 | 45 | 67 | ABC |
| B | = | 02 | 13 | 46 | 57 | ADE |
| C | = | 03 | 12 | 47 | 56 | AFG |
| D | = | 04 | 15 | 26 | 37 | BDF |
| E | = | 05 | 14 | 27 | 36 | BEG |
| F | = | 06 | 17 | 24 | 35 | CDG |
| G | = | 07 | 16 | 25 | 34 | CEF |

$\mathcal{F}_2$ $d_3 = 3$, $d_2 = 4$

| | | | | | | |
|---|---|---|---|---|---|---|
| A | = | 01 | 23 | 45 | 67 | ABC |
| B | = | 02 | 13 | 46 | 57 | ADE |
| C | = | 03 | 12 | 47 | 56 | AFG |
| D | = | 04 | 15 | 26 | 37 | BD |
| E | = | 05 | 14 | 27 | 36 | BE |
| F | = | 06 | 17 | 25 | 34 | CD |
| G | = | 07 | 16 | 24 | 35 | CE |

$\mathcal{F}_3$ $d_3 = 1$, $d_2 = 4$

| | | | | | | |
|---|---|---|---|---|---|---|
| A | = | 01 | 23 | 45 | 67 | ABC |
| B | = | 02 | 13 | 46 | 57 | CD |
| C | = | 03 | 12 | 47 | 56 | CF |
| D | = | 04 | 16 | 25 | 37 | DF |
| E | = | 05 | 17 | 26 | 34 | EG |
| F | = | 06 | 14 | 27 | 35 | |
| G | = | 07 | 15 | 24 | 36 | |

$\mathcal{F}_4$ $d_3 = 1$, $d_2 = 6$

| | | | | | | |
|---|---|---|---|---|---|---|
| A | = | 01 | 23 | 45 | 67 | ABC |
| B | = | 02 | 13 | 46 | 57 | DE |
| C | = | 03 | 12 | 47 | 56 | DF |
| D | = | 04 | 16 | 27 | 35 | DG |
| E | = | 05 | 17 | 26 | 34 | EF |
| F | = | 06 | 14 | 25 | 37 | EG |
| G | = | 07 | 15 | 24 | 36 | FG |

$\mathcal{F}_5$ $d_3 = 0$, $d_2 = 3$

| | | | | | | |
|---|---|---|---|---|---|---|
| A | = | 01 | 23 | 45 | 67 | AB |
| B | = | 02 | 13 | 47 | 56 | CF |
| C | = | 03 | 14 | 27 | 56 | EG |
| D | = | 04 | 16 | 25 | 37 | |
| E | = | 05 | 17 | 26 | 34 | |
| F | = | 06 | 12 | 35 | 47 | |
| G | = | 07 | 15 | 24 | 36 | |

$\mathcal{F}_6$ $d_3 = 0$, $d_2 = 0$

| | | | | | | |
|---|---|---|---|---|---|---|
| A | = | 01 | 23 | 45 | 67 | |
| B | = | 02 | 14 | 36 | 57 | |
| C | = | 03 | 16 | 25 | 47 | |
| D | = | 04 | 17 | 26 | 35 | |
| E | = | 05 | 12 | 37 | 46 | |
| F | = | 06 | 15 | 27 | 34 | |
| G | = | 07 | 13 | 24 | 56 | |

Table of one-factorizations of $K_8$. The divisions are shown next to the factorizations.

FIGURE 8.1

(iii) *$G_3$ has order 16 and is transitive of rank 6; $G_3(0)$ has orbits {0},{1},{2},{3},{46} and {57}. $G_3$ is $D_8 \times Z_2$ where $D_8$ is generated by (0614)(2735) and (01)(23) and $Z_2$ is generated by (03)(12)(47)(56);*

(iv) *$G_4$ has order 96 and is transitive of rank 3; $G_4(0)$ has orbits {0},{1,2,3} and {4,5,6,7}. It is the split extension of the $Z_2 \times Z_2 \times Z_2 \times Z_2$ generated by (01)(23), (02)(13), (45)(67) and (46)(57), by the $S_3$ generated by (04)(15)(27)(36) and (123)(567);*

(v) *$G_5$ has order 24 and is transitive of rank 4; $G_5(0)$ has orbits {0},{4},{1,3,5} and {2,6,7}. $G_5$ is $A_4 \times Z_2$ where $A_4$ is generated by (135)(267) and (067)(134) and $Z_2$ is generated by (04)(16)(25)(37);*

(vi) *$G_6$ has order 42 and fixes 2; it is the sharply doubly transitive group of degree 7, and is generated by (0156374) and (165437).*

PROOF. The first five groups are easily found if the division structure is used. For example, in $\mathcal{F}_2$, A is the only factor in three 3-divisions and F and G are the only factors in no 2-division. So any $\phi$ in $G_2$ must send A to A and {F,G} to {F,G}; for example, if $0\phi = 0$ then $1\phi = 1$ and $\{6\phi, 7\phi\} = \{6,7\}$. This sort of information makes a "complete" search quite short.

In the case of $\mathcal{F}_6$, first observe that if we have three factors of $K_8$ of which no two are a division, then the union of the three factors is one of the graphs in Figure 8.2. The graph contains either one or two triangles. In the one-triangle case, call the set of three vertices forming the triangle a *triad*. Clearly, "being a triad" is invariant under automorphism. In $\mathcal{F}_6$ there are fourteen triads, namely:

016,017,034,037,045,056,135,145,147,346,357,467,567,136.

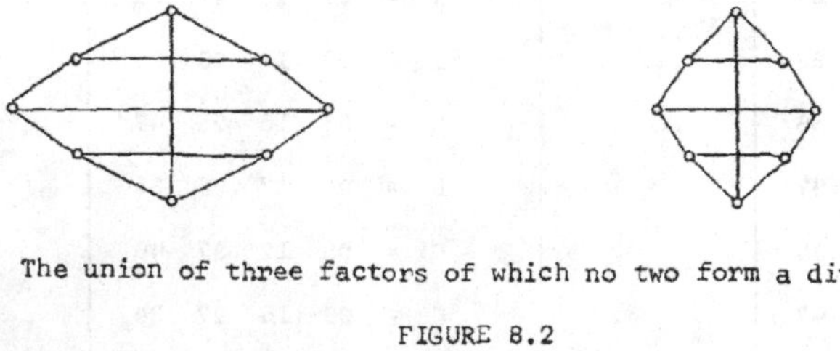

The union of three factors of which no two form a division.

FIGURE 8.2

Using this information $G_6$ is easy to find.

8.2 ROOM SQUARES OF SIDE 7. Suppose $\mathcal{F}$ and $\mathcal{G}$ are factorizations of $K_8$ which together form a Room square. We may choose a permutation {0,1,2,3,4,5,6,7} which will send $\mathcal{F}$ into one of the six canonical forms, but not one which will also send $\mathcal{G}$

into canonical form. To find all Room squares of side 7, we find all factorizations $\mathcal{G}_i = \{T,U,V,W,X,Y,Z\}$ which are orthogonal to $\mathcal{F}_i$, for i = 1,2,3,4,5,6.

Although we cannot restrict $\mathcal{G}_i$ to canonical form, we can cut down on the search. For example, suppose $T_1U_1V_1$ and $T_2U_2V_2$ are possible sets of the first three factors in $\mathcal{G}_i$.

If there is an automorphism $\phi$ of $\mathcal{F}_i$ which maps

$$\{T_1,U_1,V_1\} \mapsto \{T_2,U_2,V_2\}$$

then the Room square resulting from any extension of $T_2U_2V_2$ will be isomorphic to one of the Room squares resulting from extensions of $T_1U_1V_1$ so the case $T_2U_2V_2$ can be ignored.

We denote a Room square as $\mathcal{R}_{ij}$ if its row factorization is $\mathcal{F}_i$ and its column factorization is $\mathcal{F}_j$. This notation is adequate because it is found that there is no example of two non-isomorphic Room squares of side 7 whose row factorizations are isomorphic and whose column factorizations are isomorphic. The non-isomorphic Room squares of side 7 are

$$\mathcal{R}_{11}, \mathcal{R}_{14}, \mathcal{R}_{15}, \mathcal{R}_{16}, \mathcal{R}_{41}, \mathcal{R}_{44}, \mathcal{R}_{45}, \mathcal{R}_{51}, \mathcal{R}_{54}, \mathcal{R}_{61};$$

in every case when $i \neq j$, $\mathcal{R}_{ij}$ and $\mathcal{R}_{ji}$ are equivalent (one is isomorphic to the transpose of the other). Therefore, there are six inequivalent Room squares of side 7; a table is given in Figure 8.3. The calculations are all straightforward; we shall go through the case $\mathcal{F} = \mathcal{F}_4$ as an example.

If $\mathcal{F}_4$ and $\mathcal{G} = \{T,U,V,W,X,Y,Z\}$ form a Room square then T must be one of

$T_1$ = 01 24 37 56 $\quad$ $T_2$ = 01 25 36 47

$T_3$ = 01 26 35 47 $\quad$ $T_4$ = 01 27 34 56.

$T_1 = T_2(45)(67) = T_3(46)(57) = T_4(47)(56)$, and by theorem 8.1 all of these permutations are automorphisms of $\mathcal{F}_4$. So we can assume $T = T_1$. Then U can be $U_1$,$U_2$ or $U_3$, where

$U_1$ = 02 14 35 67, $U_2$ = 02 15 34 67, $U_3$ = 02 17 36 45;

however, $(T,U_3) = (U_2,T)(0526)(1437)$, and this permutation is in $O_4$, so we can ignore $U_3$. Similarly, V is one of

$V_1$ = 03 14 26 57, $V_2$ = 03 15 27 46, $V_3$ = 03 17 25 46.

$U_1$ and $V_1$ cannot occur together, nor can $U_2$ and $V_2$; and since

$$(T,U_2,V_3) = (T,V_2,U_1)(06)(15)(24)(37)$$

we need only try to extend

$$TU_1V_2,\ TU_1V_3,\ TU_2V_1.$$

| | | | | | | | | | | | | | | | |
|---|---|---|---|---|---|---|---|---|---|---|---|---|---|---|---|
| $\mathcal{R}_{11}$ | 01 | - | 45 | 67 | - | - | 23 | $\mathcal{R}_{16}$ | 01 | 67 | - | - | - | 45 | 23 |
| | 57 | 02 | - | - | - | 13 | 46 | | 57 | 02 | 46 | - | 13 | - | - |
| | - | 56 | 03 | 12 | - | 47 | - | | - | - | 03 | - | 47 | 12 | 56 |
| | - | 37 | - | 04 | 26 | - | 15 | | - | 15 | - | 04 | 26 | 37 | - |
| | 36 | 14 | 27 | - | 05 | - | - | | 36 | - | - | 27 | 05 | - | 14 |
| | 24 | - | - | 35 | 17 | 06 | - | | 24 | - | 17 | 35 | - | 06 | - |
| | - | - | 16 | - | 34 | 25 | 07 | | - | 34 | 25 | 16 | - | - | 07 |
| $\mathcal{R}_{14}$ | 01 | - | 45 | 67 | - | 23 | - | $\mathcal{R}_{44}$ | 01 | 67 | - | - | 23 | - | 45 |
| | 57 | 02 | - | 13 | - | - | 46 | | - | 02 | 46 | - | - | 57 | 13 |
| | - | 56 | 03 | - | - | 47 | 12 | | 56 | - | 03 | - | 47 | 12 | - |
| | - | 37 | - | 04 | 26 | 15 | - | | - | 35 | 27 | 04 | 16 | - | - |
| | 36 | 14 | 27 | - | 05 | - | - | | - | - | - | 17 | 05 | 34 | 26 |
| | 24 | - | - | - | 17 | 06 | 35 | | 37 | 14 | - | 25 | - | 06 | - |
| | - | - | 16 | 25 | 34 | - | 07 | | 24 | - | 15 | 36 | - | - | 07 |
| $\mathcal{R}_{15}$ | 01 | - | - | - | 67 | 23 | 45 | $\mathcal{R}_{45}$ | 01 | 67 | - | - | 23 | 45 | - |
| | 57 | 02 | 46 | - | - | - | 13 | | - | 02 | 57 | - | - | 13 | 46 |
| | - | 56 | 03 | - | 12 | 47 | - | | 56 | - | 03 | - | 47 | - | 12 |
| | - | 37 | - | 04 | - | 15 | 26 | | - | - | - | 04 | 16 | 27 | 35 |
| | 36 | 14 | - | 27 | 05 | - | - | | - | 34 | 26 | 17 | 05 | - | - |
| | 24 | - | 17 | 35 | - | 06 | - | | 37 | - | 14 | 25 | - | 06 | - |
| | - | - | 25 | 16 | 34 | - | 07 | | 24 | 15 | - | 36 | - | - | 07 |

Table of inequivalent Room squares of side 7.

FIGURE 8.3

It is found that each of these can be extended to a one-factorization orthogonal to $\mathcal{F}_4$ in precisely one way. The three one-factorizations are isomorphic to $\mathcal{F}_4$, $\mathcal{F}_1$ and $\mathcal{F}_5$ respectively, and give rise to Room squares $\mathcal{R}_{44}$, $\mathcal{R}_{41}$ and $\mathcal{R}_{45}$.

It is easy to calculate the automorphism groups of the Room squares. By theorem 1.7 the automorphisms are precisely those permutations which are automorphisms of the row and column factorizations simultaneously. So we calculate the two automorphism groups and find their intersections.

THEOREM 8.2. *Write $G_{ij}$ for the automorphism group of $\mathcal{R}_{ij}$. Then:*

*(i) $G_{11}$ is the simple group PGL(2,2) of order 168 generated by (1254637) and (02)(15)(37)(46), and is doubly transitive;*

*(ii) $G_{14}$ is the group $S_4$ of order 24 generated by (05)(16)(27)(34), (05)(17)(24)(36) and (01)(27)(36)(45), and is transitive of rank 4;*

*(iii) $G_{15}$ is the group $A_4$ of order 12 generated by (057)(146) and (134)(275); it is intransitive;*

*(iv) $G_{16}$ is the group of order 21 which is a split extension of the $Z_7$ generated by (0147352) by the $Z_3$ generated by (134)(275); it leaves 6 invariant but is transitive of degree 7;*

*(v) $G_{44}$ is the group $D_8$ of order 8 generated by (04)(15)(27)(36) and (0537)(1426), and is sharply transitive;*

*(vi) $G_{45}$ is the group $A_4$ of order 12 generated by (013)(476) and (123)(567), and is intransitive.*

8.3 SKEW SQUARES OF SIDE 7. We now investigate skew Room squares of side 7, with a view to finding out how many non-isomorphic skew squares of that side exist.

$\mathcal{R}_{11}$ is skew. Since $G_{11}$ is transitive, corollary 1.5 tells us that every square isomorphic to $\mathcal{R}_{11}$ is skew. Similarly, since $G_{14}$ and $G_{44}$ are transitive, no isomorph of $\mathcal{R}_{14}$ or $\mathcal{R}_{44}$ could be skew.

{0,1,2,3,4,5,6,7} contains two orbits under each of $G_{15}$, $G_{16}$ and $G_{45}$, namely:

{0,2,5,7}, {1,3,4,6},

{0,1,2,3,4,5,7}, {6},

and

{0,1,2,3}, {4,5,6,7}

respectively. So, by corollary 1.4, it is sufficient to consider $\mathcal{R}_{15}\psi_1$, $\mathcal{R}_{16}\psi_6$ and $\mathcal{R}_{45}\psi_4$ respectively, where $\psi_x$ is the cycle (0x). It is found that the second and third of these are skew, but the first is not.

In summary: there are three equivalent skew Room squares of side 7; they are equivalent to $\mathcal{R}_{11}$, $\mathcal{R}_{16}$ and $\mathcal{R}_{45}$. Consequently there are five non-isomorphic squares, isomorphic to $\mathcal{R}_{11}$, $\mathcal{R}_{16}$, $\mathcal{R}_{45}$, $\mathcal{R}_{54}$ and $\mathcal{R}_{61}$. The inequivalent squares are tabulated in Figure 8.4.

We can now answer the question asked in section 1.5. We say a Room square based on R is *standardized with respect to x* if the map from R to {0,1,...,r} involved maps x → 0. The question was whether Room squares exist which are skew when standardized with respect to one symbol but not when standardized with respect to another. The answer is yes; for example, when $\mathcal{R}_{16}$ (as given in Figure 8.3) is standardized with respect to 6 the result is skew; when it is standardized with respect to any other symbol it is not.

$\mathcal{R}_{11}$

| | | | | | | |
|---|---|---|---|---|---|---|
| 01 | - | 45 | 67 | - | - | 23 |
| 57 | 02 | - | - | - | 13 | 46 |
| - | 56 | 03 | 12 | - | 47 | - |
| - | 37 | - | 04 | 26 | - | 15 |
| 36 | 14 | 27 | - | 05 | - | - |
| 24 | - | - | 35 | 17 | 06 | - |
| - | - | 16 | - | 34 | 25 | 07 |

$\mathcal{R}_{15}\psi_6$

| | | | | | | |
|---|---|---|---|---|---|---|
| 01 | - | - | 25 | 67 | - | 34 |
| 46 | 02 | - | - | - | 37 | 15 |
| 27 | 56 | 03 | - | 14 | - | - |
| - | 13 | 57 | 04 | - | - | 26 |
| - | 47 | - | 36 | 05 | 12 | - |
| 35 | - | 24 | 17 | - | 06 | - |
| - | - | 16 | - | 23 | 45 | 07 |

$\mathcal{R}_{45}\psi_4$

| | | | | | | |
|---|---|---|---|---|---|---|
| 01 | 37 | - | 25 | 46 | - | - |
| - | 02 | 15 | 36 | - | 47 | - |
| 26 | - | 03 | 17 | - | - | 45 |
| - | - | - | 04 | 27 | 35 | 16 |
| - | 14 | 67 | - | 05 | - | 23 |
| 57 | - | 24 | - | 13 | 06 | - |
| 34 | 56 | - | - | - | 12 | 07 |

Table of inequivalent skew Room squares of side 7.

FIGURE 8.4

## CHAPTER IX. ROOM DESIGNS OF HIGHER DIMENSION

9.1 THE NUMBER OF ORTHOGONAL SYMMETRIC LATIN SQUARES. By analogy with the corresponding problem for orthogonal Latin squares, it is interesting to ask the value of $\nu(r)$, the maximum number of pairwise orthogonal symmetric Latin squares of side r.

We restrict ourselves to odd r. For r even, a symmetric Latin square with diagonal $(1,2,\ldots,r)$ is impossible, since it would imply the existence of a one-factorization of $K_{r+1}$, and $r+1$ would be odd. We could remove the restriction on the diagonal - making possible such Latin squares as

$$\begin{matrix} 1 & 2 & 3 & 4 \\ 2 & 1 & 4 & 3 \\ 3 & 4 & 2 & 1 \\ 4 & 3 & 1 & 2 \end{matrix}$$

which is of even side - but this would lose track of the relation to Room squares. Consequently, when we write $\nu(r)$, it is assumed that r is odd.

The case of $\nu(1)$ is special, since a symmetric Latin square of side r can be orthogonal to itself if and only if $r = 1$. Clearly

$$\nu(1) = 1,$$

and the existence of two (distinct) orthogonal symmetric Latin squares of side r is equivalent to the existence of a Room square of side r when $r > 1$. So

$$\nu(3) \leq 1, \quad \nu(5) \leq 1,$$

$$\nu(r) \geq 2 \quad \text{when } r \notin \{3,5,257\},$$

(Theorem 7.6). Moreover, it is known [10,p.87] that there are one-factorizations of all complete graphs of even order, so

$$\nu(r) \geq 1 \quad \text{all } r.$$

We therefore know that $\nu(3)$ and $\nu(5)$ both equal 1; we shall soon show that $\nu(7) = 3$.

Suppose F and G are orthogonal symmetric Latin squares of side r; write $F = (f_{ij})$ and $G = (g_{ij})$. Then

$$f_{12} \neq 1, \quad f_{12} \neq 2,$$

$$g_{12} \neq 1, \quad g_{12} \neq 2,$$

$$f_{12} \neq g_{12}.$$

There are $r-2$ possible choices for $f_{12}$, so

$$\nu(r) \le r-2.$$

However, this bound is probably not very good; we conjecture that $\nu(r) \le \frac{1}{2}(r-1)$.

9.2 CONSTRUCTION FROM STARTERS. Since orthogonal starters give rise to orthogonal one-factorizations, they give rise to orthogonal symmetric Latin squares. Theorem 2.4 can be re-stated as

THEOREM 9.1. *A starter in an abelian group of order r implies the existence of a symmetric Latin square of side r. If there are two orthogonal starters in a group then the corresponding Latin squares are orthogonal symmetric.*

The construction of the Latin square $F = (f_{ij})$ corresponding to a starter X in a group $H = \{h_1, h_2, \ldots, h_r\}$ is quite simple. Given two elements $h_i$ and $h_j$ of H, where $h_i \ne h_j$, there are a unique member $\{x,y\}$ of X and a unique member $h_k$ of H such that

$$\{h_i, h_j\} = \{x+h_k, y+h_k\}.$$

Then $f_{ij} = k$; $f_{ii} = i$.

COROLLARY 9.2. *If there are k pairwise orthogonal starters in a group of order r, then*

$$\nu(r) \ge k.$$

*In particular, if there is a strong starter in a group of order r then*

$$\nu(r) \ge 3.$$

PROOF. Theorems 9.1 and 2.6.

THEOREM 9.3. *If $r = 2n+1$ is a prime power congruent to 3 modulo 4, then*

$$\nu(r) \ge n.$$

PROOF. Let H be the finite field with r elements, considered as an additive group; and partition its non-zero elements into the sets of quadratic elements Q and non-quadratic elements P. Write

$$X_p = \{\{q, pq\} : q \in Q\}$$

for each $p \in P$. As q runs through Q, pq runs through P, so $X_p$ contains each element of $H \setminus \{0\}$ in one of its pairs. The set of differences $\pm(q-pq)$ also precisely cover $H \setminus \{0\}$, since otherwise

$$q-pq = \pm(q_1-pq_1)$$

for some q, $q_1 \in Q$, whence

$$(1-p)(q \mp q_1) = 0;$$

$1 \in Q$, so $p \neq 1$ and we can divide by $1-p$; $q = \pm q_1$, and $q = q_1$ since $-1 \notin Q$. (Notice that we have used the fact $r \equiv 3 \pmod 4$ here; if $r \equiv 1 \pmod 4$ then $-1 \in Q$.) So each $X_p$ is a starter.

Now assume $X_m$ and $X_p$ are not orthogonal, where m and p are distinct members of P. If $X_m$ and $X_p$ have a pair in common, say

$$\{q,pq\} = \{q_1,mq_1\},$$

then either $q = q_1$ and $pq = mq_1$, whence $p = m$, or $q = mq_1$ and $pq = q_1$, whence $m = qq_1^{-1}$ which is in Q. So $X_m$ and $X_p$ have no common element. Consequently, if the pair $\{q,pq\}$ determines the pair $\{q_1,mq_1\}$ and the element $d_q$ under one of the relations

$$d_q = q_1-q = mq_1-pq$$

or

$$d_q = q_1-pq = mq_1-q_1,$$

$d_q$ will be non-zero. So orthogonality can only fail because $d_q = d_s$ for some $q \neq s$. Therefore, for some $q_1$ and $s_1$ in Q, one of the following cases occurs:

(i) $q_1-q = mq_1-pq = s_1-s = ms_1-ps$

(ii) $q_1-q = mq_1-pq = s_1-ps = ms_1-s$

(iii) $q_1-pq = mq_1-q = s_1-s = ms_1-ps$

(iv) $q_1-pq = mq_1-q = s_1-ps = ms_1-s$.

In case (i) we have

$$q_1-mq_1 = q-pq$$
$$s_1-ms_1 = s-ps,$$

whence $(1-m)(1-p)^{-1} = qq_1^{-1} = ss_1^{-1}$; substituting into

$$s(q_1-q) = s(s_1-s)$$

we have

$$q(s_1-s) = s(s_1-s)$$

and $q = s$, which is a contradiction. In case (iv) we again obtain

$$(1-m)(p-1)^{-1} = qq_1^{-1} = ss_1^{-1},$$

and if we divide both sides of

$$s(q_1-pq) = s(s_1-ps)$$

by $s_1-ps$ we get $q = s$ again.

Case (ii) gives

$$(1-m)(1-p)^{-1} = qq_1^{-1} = -ss_1^{-1},$$

which is impossible since -1 is a non-residue, and case (iii) yields the same result.

This means that the n starters are orthogonal, so $\nu(r) \geq n$.

9.3 ROOM DESIGNS OF HIGHER DIMENSION. Just as a Room square could be defined starting from two orthogonal symmetric Latin squares, we can define a higher-dimensional array from a larger set of squares.

A Room t-design of side r shall mean a t-dimensional array of size $r \times r \times \ldots \times r$ which is constructed from an ordered set of t orthogonal symmetric Latin squares in the following way:

(i) cell $(i,i,\ldots,i)$ contains $\{0,i\}$;

(ii) when $i \neq j$, if the $(i,j)$ entries of the Latin squares are $x_1,x_2,\ldots,$ and $x_t$ respectively, then cell $(x_1,x_2,\ldots,x_t)$ contains $\{i,j\}$;

(iii) all other cells are empty.

Room 2-designs are simply Room squares. If the $(i,j)$ projection of a t-dimensional array is defined as the 2-dimensional array formed by putting the entry from cell $(k_1,k_2,\ldots,k_t)$ of the old array into cell $(k_i,k_j)$ of the new, then the $(i,j)$ projection of a Room t-design is a Room square. More generally, any projection of a Room t-design onto s dimensions is a Room s-design.

One can define a sub-t-design by analogy with a subsquare. Given a t-design $\mathcal{R}$ of side r, we say $\mathcal{S}$ is a sub-t-design of $\mathcal{R}$ of side s if $\mathcal{S}$ consists of all entries in the intersection of certain sets of s rows, s columns, and so on. (We take one set for each dimension.) We can prove

THEOREM 9.4. *If there are a t-design of side $r_1$ and a t-design of side $r_2$ with a sub-t-design of side $r_3$, and there are t pairwise orthogonal Latin squares of side $r_2-r_3$, then there is a t-design of side*

$$r_1(r_2-r_3)+r_3$$

*with sub-t-designs of sides $r_1,r_2$ and $r_3$.*

The proof is quite analogous to the proof of the corresponding theorem for Room squares. We can assume that any t-design has sub-t-designs of side 1 and 0.

As an illustration of the proof we consider the case $r_3 = 0$. Our notation is as follows:

$L_1,L_2,\ldots,L_t$: pairwise orthogonal symmetric Latin squares of side $r_1$;

$M_1,M_2,\ldots,M_t$: pairwise orthogonal symmetric Latin squares of side $r_2$;

$N_1,N_2,\ldots,N_t$: pairwise orthogonal Latin squares of side $r_2$;

$M_{ik},N_{ik}$ : arrays obtained from $M_i$ and $N_i$ by replacing an entry x by $x_k$.

We construct an array $S_i$ as an $r_1 \times r_1$ array of $r_2 \times r_2$ blocks. The $(x,x)$ block of $S_i$

is $M_{ix}$. If the (x,y) entry of $L_i$ is k, then the (x,y) block of $S_i$ is

$$N_{ik} \quad \text{if } x < y$$

$$N_{ik}^{T} \quad \text{if } x > y.$$

It may be verified that $S_1, S_2, \ldots, S_t$ are orthogonal symmetric Latin squares.

The general method follows what we have just done. A set of one square of each order $r_1, r_2, r_3, r_2 - r_3$ is used to construct a square of side $r_1(r_2 - r_3) + r_3$. The details of the proof are given in [13], in slightly different terminology; the whole proof is very like the proof of theorem 3.1, and of course it reduces to a proof of that theorem in the case t = 2. It is worthwhile to ask whether similar generalizations apply to other Room square constructions.

We can use these results to calculate lower bounds for $\nu(r)$.

9.4 HIGHER DESIGNS OF SIDE 7. In discussing t-designs of side 7 it is convenient to speak in terms of t orthogonal one-factorizations rather than t orthogonal symmetric Latin squares, since we constructed Room squares of side 7 from factorizations.

We denote a Room 3-design, or cube, of side 7 by $\mathcal{R}_{ijk}$ if its three one-factorizations are isomorphic to $\mathcal{F}_i$, $\mathcal{F}_j$ and $\mathcal{F}_k$ in the notation of section 8.1. If an $\mathcal{R}_{ijk}$ exists then $\mathcal{R}_{ij}$, $\mathcal{R}_{ik}$ and $\mathcal{R}_{jk}$ must exist, so ijk must be

111,114,115,116,144,145,444,445,

or some permutation of one of these sets. In particular, {i,j,k} must contain {1,1}, {1,4} or {4,4}. So we can search for all Room cubes of side 7 by trying to extend $\mathcal{R}_{11}$, $\mathcal{R}_{14}$ and $\mathcal{R}_{44}$.

Say $\mathcal{R}_{14}$ and the factorization {J,K,L,M,N,P,Q} form a Room cube. It is found that N is uniquely determined,

N = 05 16 24 37,

and proceeding from here we find successively that

M = 04 17 23 56

Q = 07 13 26 45

P = 06 12 34 57

J = 01 27 35 46

and there is no possible K which extends this set. So there is no cube extending $\mathcal{R}_{14}$. Similarly $\mathcal{R}_{44}$ cannot be extended. It is found that $\mathcal{R}_{11}$ can be extended, but in precisely one way.

The Room cube which we find is of type $\mathcal{R}_{116}$. We can take its three orthogonal symmetric Latin squares as

| | | |
|---|---|---|
| 1 3 2 5 4 7 6 | 1 4 6 2 7 3 5 | 1 6 7 3 2 5 4 |
| 3 2 1 6 7 4 5 | 4 2 7 1 6 5 3 | 6 2 4 5 3 7 1 |
| 2 1 3 7 6 5 4 | 6 7 3 5 4 1 2 | 7 4 3 6 1 2 5 |
| 5 6 7 4 1 2 3 | 2 1 5 4 3 7 6 | 3 5 6 4 7 1 2 |
| 4 7 6 1 5 3 2 | 7 6 4 3 5 2 1 | 2 3 1 7 5 4 6 |
| 7 4 5 2 3 6 1 | 3 5 1 7 2 6 4 | 5 7 2 1 4 6 3 |
| 6 5 4 3 2 1 7 | 5 3 2 6 1 4 7 | 4 1 5 2 6 3 7 |

(the first two squares correspond to our formulation of $\mathcal{R}_{11}$). The cube gives rise to three non-isomorphic cubes, of form $\mathcal{R}_{116}$, $\mathcal{R}_{161}$ and $\mathcal{R}_{611}$, but they are all equivalent.

Suppose there were a Room 4-design of side 7, of type $\mathcal{R}_{ijk\ell}$. Since any three dimensions must give a cube, we have

$$\{i,j,k\} = \{i,j,\ell\} = \{i,k,\ell\} = \{j,k,\ell\} = \{1,1,6\}.$$

This is impossible. So

$$\nu(7) = 3.$$

The cube we exhibited above has an interesting property. Each of the three Room squares which arises from it is skew.

## CHAPTER X. MISCELLANEA

10.1 SOME ISOMORPHISM RESULTS. All our concrete theorems on Room square isomorphism so far deal with the squares of side 7. We saw that there are inequivalent Room squares of side 7, and that some squares of side 7 are isomorphic to their transposes and some are not. We now prove that these results hold more generally.

Suppose G is a finite abelian group of order $r = 2n+1$; we write the elements of G as $g_1, g_2, \ldots, g_r$, where $g_1 = 0$ is the identity element; $g_0$ is an object which is not an element of G. We assume that X is a strong starter in G, where

$$X = \{\{x_1,y_1\},\{x_2,y_2\},\ldots,\{x_n,y_n\}\}.$$

From theorem 2.6,

$$Y = \{\{-x_i,-y_i\}: \{x,y\} \in X\}$$

is a strong starter in G, and

$$Z = \{\{g_i,-g_i\}: g_i \in G\setminus\{0\}\}$$

is a starter, and X,Y and Z are orthogonal. We shall write $\mathcal{F}$, $\mathcal{G}$ and $\mathcal{H}$ for the one-factorizations of the complete graph on $\{g_0,g_1,\ldots,g_r\}$ corresponding to X,Y and Z respectively. The factors in $\mathcal{F}$, $\mathcal{G}$ and $\mathcal{H}$ will be $\{F_i\}$, $\{G_i\}$ and $\{H_i\}$ respectively, where

$$F_i = \{\{x+g_i,y+g_i\}: \{x,y\} \in X\} \cup \{\{g_0,g_i\}\}$$

and similarly for $G_i$ and $H_i$. We write $\mathcal{R}_{XY}$ for the standardized Room square whose row factorization is $\mathcal{F}$ and whose column factorization is $\mathcal{G}$; $\mathcal{R}_{ZX}$ and so on are defined analogously.

Suppose $\phi$ is the permutation $x \mapsto -x$. Then

$$\begin{aligned} F_i\phi &= \{\{-x-g_i,-y-g_i\}: \{x,y\} \in X\} \\ &= \{\{-x-g_i,-y-g_i\}: \{-x,-y\} \in Y\} \\ &= G_j \quad \text{where } g_j = -g_i, \end{aligned}$$

so

$$\mathcal{F}\phi = \mathcal{G};$$

also

$$\mathcal{G}\phi = \mathcal{F}.$$

Therefore $\mathcal{R}_{XY}\phi = \mathcal{R}_{YX}$ in the notation of section 1.5, so $\mathcal{R}_{XY}$ and $\mathcal{R}_{YX}$ are isomorphic. Consequently

THEOREM 10.1. *If there is a strong starter in a group of order r then there is a*

*Room square of side r which is isomorphic to its transpose.*

This theorem proves the existence of a Room square of side r which is isomorphic to its transpose whenever r is a prime power other than a Fermat number or when r is five times one of these numbers, by the results of Chapter II.

Now suppose $\mathcal{F}$ and $\mathcal{H}$ are not isomorphic. We then know that $\mathcal{R}_{XZ}$ and $\mathcal{R}_{ZX}$ cannot be isomorphic, so $\mathcal{R}_{XZ}$ is a Room square not isomorphic to its transpose. If $\mathcal{R}_{XY}$ and $\mathcal{R}_{XZ}$ are equivalent then either they are both isomorphic to their transposes or neither is, so we have a pair of inequivalent squares. To investigate the possibility of isomorphism between $\mathcal{F}$ and $\mathcal{H}$ we consider the division structure.

First, consider $H_i \cup H_j$, where $i \neq j$. This graph has as its edges $\{g_0,g_i\}$, $\{g_0,g_j\}$, and all the $\{g+g_i,-g+g_i\}$ and $\{g+g_j,-g+g_j\}$ where g ranges through $G\setminus\{0\}$. The union, being a regular graph of degree 2, is a union of disjoint cycles. Each of the cycles must contain an even number of edges: if we go around the cycle starting with an edge $\{g_a,g_b\}$ in $H_i$, then the third, fifth, ... edges are in $H_i$, and if the number of edges were odd then the two edges containing $g_a$ would both be in $H_i$ (which is impossible).

We must consider the cycle containing $g_0$ separately from the cycles not containing $g_0$, because the edges containing $g_0$ are defined differently from the others. The cycle which contains $g_0$ must contain the edges $\{g_0,g_i\}$ and $\{g_0,g_j\}$. If the cycle is written as

$$g_0g_i = k_1\ell_1,\ p_1q_1,\ k_2\ell_2,\ p_2q_2,\ \ldots,\ p_zq_z,$$

the edges $k_a\ell_a$ must all belong to $H_i$ and the edges $p_aq_a$ must all belong to $H_j$, and

$$\ell_a = p_a,\quad q_a = k_{a+1},\quad p_z = g_j.$$

We prove by induction that $p_y = (2y-1)g_i - (2y-2)g_j$, where multiplication of a group element by an integer m denotes the sum of m copies of that group element. It is immediate that $p_1 = g_i$ since $p_1 = \ell_1$. If we assume that

$$p_a = (2a-1)g_i - (2a-2)g_j = (2a-1)(g_i-g_j)+g_j,$$

then

$$p_a = -(2a-1)(g_i-g_j)+g_j = 2ag_j-(2a-1)g_i$$

by the definition of $H_j$, and

$$k_{a+1} = q_a = 2a(g_j-g_i)+g_i.$$

We then have

$$p_{a+1} = \ell_{a+1} = -2a(g_j-g_i)+g_i = (2a+1)g_i - 2ag_j,$$

which has the required form. (All of these calculations make the assumption that the edge involving $g_0$ does not occur; this will be valid in the required range $1 \leq a < z$ since $g_0$ only occurs at the beginning $k_1$ and the end $q_z$ of the cycle.)

Now, since $p_z = g_j$,

$$(2z-1)(g_i-g_j) = 0.$$

The order of the non-zero group element $g_i-g_j$ must divide the order of the group, so

$$2z-1 \mid 2n+1.$$

If p is the smallest prime dividing $2n+1$ then the length of the cycle, 2z, must be at least $p+1$. ($2z-1 = 1$ is not allowed because no cycle can have length 2.)

We now consider a cycle which does not contain $g_0$. We can write it as

$$k_1\ell_1,\ p_1q_1,\ k_2\ell_2,\ \ldots,\ p_zq_z,$$

where

$$\ell_a = p_a,\ q_a = k_{a+1};$$

we assume $k_1 = g+g_i$, so that $\ell_1 = -g+g_i$ and $q_z = g+g_i$. We can prove that

$$q_z = (2z-1)(g_j-g_i) + g+g_j$$

(the proof is an induction similar to the one above, and is omitted). The equation $q_z = k$ yields

$$2z(g_j-g_i) = 0;$$

if p is as above, then 2z must be at least 2p.

LEMMA 10.2. *If r is congruent to 1 modulo 6, then the union of two factors of $\mathcal{H}$ contains no cycle of length 4.*

PROOF. The prime p must be at least 5, so no cycle can have length smaller than $p+1 = 6$.

Now recall the following result from the construction of strong starters using Steiner triple systems in section 2.4.

LEMMA 10.3. *If $\mathcal{F}$ is the one-factorization constructed from a Steiner triple system, then the union of any pair of factors contains a cycle of length 4.*

THEOREM 10.4. *If r is a prime power congruent to 1 modulo 6 then there is a Room square of side r which is not isomorphic to its transpose, and there are Room squares of side r which are not equivalent.*

PROOF. Take G to be the additive group of GF(r), and X to be the strong starter constructed in section 2.4 from a strong set of Steiner difference blocks. By the two lemmas the one-factorizations $\mathcal{F}$ and $\mathcal{H}$ have different types of 2-divisions, so they are not isomorphic. So $\mathcal{R}_{XZ}$ is not isomorphic to its transpose. Since X is strong, theorem 10.1 can be applied and $\mathcal{R}_{XY}$ is isomorphic to its transpose. So $\mathcal{R}_{XY}$ and $\mathcal{R}_{XZ}$ are inequivalent.

10.2 HOWELL ROTATIONS. To motivate the discussion of Howell rotations we must introduce the terminology of bridge, and in particular of duplicate bridge tournaments.

In an ordinary bridge game, the relative scores of the two partnerships depend on the luck of the cards to a certain extent. Duplicate bridge attempts to remove this element. The cards are dealt once, into four hands labelled North, East, South and West (N,E,S,W). Several pairs of partnerships play the deal independently, and then the *relative* performance of all the North-South partnerships is compared, and similarly for the East-West partnerships. We shall refer to each deal of the cards as a *board*. If partnership i is NS on a particular board and partnership j plays the EW cards *against* i on that board, then we say that i *plays with* j on the board. If i and j both play NS (or both play EW) on a particular board, we say that i *competes against* j on the board.

The following properties are desirable in a bridge tournament:

(a) for every i and j, the number of times i plays with j is constant, $\mu$ say;

(b) for every i and j, the number of times i competes against j is constant, k say.

In the actual play of a tournament, various partnerships play various boards at the same time, and the set of pairings and the boards they play concurrently is called a *round*. To avoid dealing the same board twice, it is desirable that:

(c) no board is played twice on the same round.

A *balanced Howell rotation* is a design for a tournament in which the conditions (a), (b) and (c) are met in the case $\mu = 1$, and moreover:

(d) every partnership plays every board once.

A *complete* balanced Howell rotation is one in which every partnership plays in every round.

It is clear that (d) can only be satisfied by an even number of partnerships. We shall assume that this number is at least 4 ("duplicate" bridge for two partnerships is meaningless).

THEOREM 10.5. *If there is a complete balanced Howell rotation for 2n partnerships, then n is even.*

PROOF. Let $B_i$ be the set of partnerships which compete against x on board i. Then $B_i$ is of size $n-1$. There are $2n-1$ boards, and $2n-1$ partnerships other than x. By (b), each of these partnerships belongs to k of the $B_i$, so $k = n-1$.

Suppose partnerships y and z compete against x on the same board $\lambda$ times, that is $\{y,z\} \subseteq B_i$ for $\lambda$ values of i. Then

$$y \in B_i, \quad z \notin B_i \quad \text{for } n-1-\lambda \text{ values,}$$

$$y \notin B_i, \quad z \in B_i \quad \text{for } n-1-\lambda \text{ values,}$$

so $y \notin B_i$ and $z \notin B_i$ for $2n-1-\lambda-2(n-1-\lambda) = \lambda+1$ values of i. The cases where y competes against z are those where y and z both belong to $B_i$ or neither does; there are $2\lambda+1$ of these. So

$$k = n-1 = 2\lambda+1$$

$$\lambda = \tfrac{1}{2}n-1,$$

and n must be even.

Observe that we have in fact proven a stronger result: the sets $B_i$ form a symmetric balanced incomplete block design with parameters $(2n-1, n-1, \frac{1}{2}n-1)$. So the existence of a complete balanced Howell rotation implies the existence of one of these designs, or equivalently of an Hadamard matrix of order 2n.

THEOREM 10.6. *If there is a complete balanced Howell rotation for 2n partnerships, then there is a Room square of side 2n - 1.*

PROOF. We assume the partnerships have been numbered $0,1,\ldots,2n-1$. We construct the square array $\mathcal{R}$ of side $2n-1$ which contains $\{x,y\}$ in its $(i,j)$ cell when partnerships x and y meet in round i and play board j there. Then $\mathcal{R}$ is the required Room square: it contains every unordered pair once by (a); every number occurs once per row and once per column because of (c) and (d).

This theorem means that a complete balanced Howell rotation is a Room square whose entries have been ordered in such a way that (b) is true. The term "complete Howell rotation" has been used in bridge to describe a system which satisfies (a), (c) and (d) with $\mu = 1$, so a complete Howell rotation is simply a Room square.

Complete balanced Howell rotations of order $r+1$ have been constructed for every prime power r congruent to 3 modulo 4 [3]. The term "complete balanced Howell rotation" is also used when the number of players is not divisible by 4: when the number is congruent to 2 modulo 4 then the restriction $\mu = 1$ is replaced by $\mu = 2$, and an odd number of partnerships is converted into an even number by introducing one artificial partnership - that is, a bye.

## 10.3 A STATISTICAL APPLICATION.

In this section we assume some familiarity with the general principles of experimental design. A brief sketch of the necessary background is given below.

The usual interpretation of a balanced incomplete block design as an experimental design is best explained by an example. Suppose it is desired to compare the yield of v varieties of grain. It is quite possible that there would be an interaction between the environment (type of soil, rainfall, drainage, etc.) and the variety of grain, so b blocks (sets of experimental plots) are chosen in which the environment is fairly consistent throughout the block. Ideally, there would be b identifiable different types of environment; it is not always possible to include all different types. Often it will be difficult to distinguish between the environment-types, and the experimental area will be divided into blocks according to some arbitrary criterion - for example, physical proximity. Alternatively, it may be true that the environmental differences are irrelevant (either the environment is constant or else there is empirical evidence that it will not affect the yield), and each "block" could be a set of plots to which a given treatment (e.g., a particular fertilizer) is applied. In this way the classification of the experimental plots into blocks and varieties can be used whenever there are two factors which may influence yield.

The obvious technique is to grow every variety in a plot in every block. This requires bv experimental plots. The obvious disadvantage when bv is large is that the experiment may be costly to run or that plots may become so small that chance variations in the yield of a plot become highly significant. To overcome this a balanced incomplete block design is used: the v objects on which the design is constructed are the v varieties, and each block contains k plots which are sown with the k varieties belonging to that block. It is found that balance (the property that $\lambda$ is constant) minimizes the chance effects due to the incompleteness of the blocks. In this case only bk plots must be used.

Suppose $\mathcal{R}$ is an Hadamard Room square of side $r = 4m-1$. We shall show how to interpret $\mathcal{R}$ as a design for a three-factor experiment, with r blocks, r varieties and $r+1$ treatments. Each block contains $r+1$ plots.

We first interpret $\mathcal{R}$ as a balanced incomplete block design for r varieties in r blocks. This is done by ignoring the contents of $\mathcal{R}$. There are m varieties used in each block, and these are chosen as follows: variety i is used in block j if and only if the (i,j) cell of $\mathcal{R}$ is occupied. We shall call the areas which contain one variety a "main plot" in its block.

Now treatments are imposed. Every "main plot" is divided into two "subplots". There are $r+1$ subplots in each block. If $\{x,y\}$ is the entry in the (i,j) cell of $\mathcal{R}$, then the two subplots of block j which contain variety i are treated with treatments x and y respectively. Each treatment occurs once in every block and once on every variety, so the design could be considered as a complete block design for treatments on varieties or treatments in blocks. The whole design uses $r(r+1)$ subplots, whereas a complete design for varieties, blocks and treatments would use $r^2(r+1)$ subplots.

The method of statistical analysis of an experiment based on the design is given by Shah [36]. The design is found to be quite efficient, principally because of the relationship between Hadamard Room squares and the symmetric balanced incomplete block design with $v = 4m-1$, $k = 2m$ and $\lambda = m$, which we mentioned in section 1.4.

10.4 TEN UNSOLVED PROBLEMS. This is no exhaustive list of the new directions waiting to be explored. Rather, it is what came to me when I sat down for an hour to write out some problems. It is also limited in that I decided to stop on reaching ten problems; possibly the eleventh problem to come to mind would have been more interesting than some of the ten.

1. Is there a Room square of side 257? If so, is there a skew one?
2. Suppose there is a Room square of side r with a subsquare of side n. It is necessarily true that $r \geq 3n+2$; however the best known result is $r = 4n+1$. Is it possible that $3n+2 \leq r < 4n+1$? Is there a stronger bound than $r \geq 3n+2$?
3. Is there an abelian group of order 3n with a strong starter for some n prime to 3?
4. Find infinite families of adders for patterned starters in groups which are not of prime power order.
5. Are there skew Room squares of sides 9, 15 or 21? Are there sides for which skew squares (or embedded squares) do not exist?
6. Is there any theorem of the form "if there are n pairwise orthogonal Latin squares of side r then there is a skew Room square of side r"? Is there such a theorem if "skew" is deleted?
7. One could ask: "find out something about isomorphism". More specifically, the text contains four constructions for Room squares of side $r = 6n+1$ when n and r are prime powers. Are they isomorphic?
8. Is there a Room square which has a complement but is not isomorphic to an embedded square? To a skew square? Are there infinite families of them?
9. Suppose G is a group of order $r = 2n+1$, written in multiplicative notation. Write H for $G \setminus \{1\}$. A *starter* in G is defined as a set of ordered pairs

$$X = \{(x_1,y_1), (x_2,y_2),\ldots, (x_n,y_n)\}$$

such that

$$H = \{x_1,x_2,\ldots,x_n,y_1,y_2,\ldots,y_n\}$$
$$= \{x_1y_1^{-1},x_2y_2^{-1},\ldots,x_ny_n^{-1},y_1x_1^{-1},y_2x_2^{-1},\ldots,y_nx_n^{-1}\}.$$

An *adder* $A_x$ for X in G means an ordered set $(a_1,a_2,\ldots,a_n)$ of distinct elements of H satisfying

$$H = \{x_1a_1, x_2a_2, \ldots, x_na_n, y_1a_1, \ldots, y_na_n\}.$$

Construct a square array whose rows and columns are indexed by the elements of G and whose (g,h) entry is

$$\begin{array}{ll} \{x_ia_ig, y_ia_ig\} & \text{if } h = a_ig \in A_xg, \\ \text{empty} & \text{if } h \notin A_xg. \end{array}$$

It is not hard to see that the array is a Room square. So the abelian property is not needed in defining starters and adders. (However, no definition of "strong starter" is very successful.)

Find examples of starters and adders in non-abelian groups.

10. What is the maximum number of pairwise orthogonal symmetric Latin squares of side 9?

## BIBLIOGRAPHIC NOTES

CHAPTER ONE. Most of the literature on Room squares is in research papers. Bruck [4] has a short discussion of Room squares. Two surveys have appeared, one by Stanton and Mullin [40] and one by the author [42]. An analysis of the non-existence of a Room square of side 5, similar to the one we present here, is given in [40]; Room [33] indicates a geometric argument.

The author introduced skew Room squares in [43], but the idea of a skew *adder* had been presented independently - see [6]. Embedded Room squares are used by Mullin [22]. (Beware: Collens and Mullin [6] used "embedded Room squares" as a synonym for "subsquare".) As far as we know, complementary Room squares appear for the first time in this survey. Self-orthogonal Latin squares are discussed by Mendelsohn [19] and Horton [11], and their relationship to Room squares is explained by Mullin and Nemeth [27].

Isomorphisms will be treated by the author in a forthcoming paper, and are discussed by Lindner [17]; but Lindner's definition of an isomorphism is different from the one given here, and more restrictive.

One-factorizations were first brought into the discussion of Room squares by Nemeth [30] under the guise of edge-colourings. Horton [13] uses them. For general discussion of factorizations one should refer to standard works on Graph Theory, such as [10]. The method of divisions is introduced in [46]. Theorem 1.6 is taken from [6].

Orthogonal symmetric Latin squares were introduced by the author [43] in a slightly different form. Bruck [4] and Lindner [17] discuss quasigroups and Room squares.

CHAPTER TWO. Starters and adders were first defined in [39], for cyclic groups. In [26] they are defined for general abelian groups. Most of the results in section 2.1 are from [26]; sections 2.2, 2.3 and 2.5 are based on [10], [25] and [12] respectively.

O'Shaughnessy [31] introduced orthogonal Steiner systems and their relation to Room squares. (That paper contains the first known example of a Room square of side 13.) Further work on these systems may be found in [11], [18], [22], [24] and [26]. The general construction, theorem 2.7, is in [24]. For the construction of a Steiner triple system from a set of Steiner difference blocks, see [9].

CHAPTER THREE. Theorem 3.2 was originally announced by Bruck [4]. However, his proof was erroneous, and the error is pointed out in [23]. A correct multiplication theorem was proven in [37], and a version which shows that skewness is preserved appears in [43].

Theorem 3.1 may be found in [11] and [14], and a proof in more algebraic terms is given in [17]. None of these proofs involves skewness. Skew subsquares are defined in [29], and it is pointed out there that skewness is preserved in theorem 3.1.

CHAPTER FOUR. The results of this chapter are based on [44] and [47].

CHAPTER FIVE. The doubling constructions first appeared in [45] and [43]. However, they are presented here in a slightly more general form: the original proofs assume the existence of a skew Room square rather than a pair of complementary Room squares. The proof of theorem 5.4 given here is much simpler than that used in [43], but is essentially equivalent.

CHAPTER SIX. The construction of section 6.1 is taken from [2]. An alternative proof, essentially the same but phrased in algebraic rather than geometric terms, is given by Bruck [4].

Theorem 6.2 is due to Archbold [1]. The proof of theorem 6.4 is actually modelled on the proof given in [1], but has been rewritten in terms of starters and considerably generalized.

Theorem 6.13 is based on a result of Lawless [16]. His theorem is simply a Room square construction, and does not discuss skew or complementary squares.

CHAPTER SEVEN. The skew Room square of side 17 is given in [6], and the Room square of side 65537 was found by Mullin [22]. The other results in section 7.1 are from [14].

An analysis like that in section 7.2 was used in [29] to show that "most" Room square sides congruent to 3 modulo 4 can be realized. Those results have been extended by similar techniques in [22], [28] and [47]. Theorem 7.6 was reached in [47].

In order to show the existence of Room squares of side congruent to 5 modulo 8, the results of section 3.5 are cited in section 7.2. The author believes that Schellenberg [35] has constructed Room squares of these orders, but has not seen his results.

CHAPTER EIGHT. Most of this material appeared in duplicated lecture notes prepared by the author (University of Newcastle, 1971).

CHAPTER NINE. Horton [13] discusses higher-dimensional Room designs and obtains the results of sections 9.2 and 9.3. His proofs are written in terms of orthogonal one-factorizations.

CHAPTER TEN. Howell rotations are discussed in [32], [4] and [15]. These papers contain constructions of complete balanced Howell rotations, which we consider to be beyond the scope of this monograph.

The statistical analysis of an Hadamard Room square is presented in [2], and a revised version appears in [36].

## REFERENCES

[1] J.W. ARCHBOLD, A combinatorial problem of T.G. Room. *Mathematika* 7 (1960), 50-55.

[2] J.W. ARCHBOLD and N.L. JOHNSON, A construction for Room squares and an application in experimental design. *Ann. Math. Statist.* 29 (1958), 219-225.

[3] E.R. BERLEKAMP and F.K. HWANG, Constructions for balanced Howell rotations for bridge tournaments. *J. Combinatorial Theory, Series A*, 12 (1972), 159-166.

[4] R.H. BRUCK, What is a loop. *Studies in Modern Algebra* (Mathematical Association of America, 1963), 59-99.

[5] K. BYLEEN, On Stanton and Mullin's construction of Room squares. *Ann. Math. Statist.* 41 (1970), 1122-1125.

[6] R.J. COLLENS and R.C. MULLIN, Some properties of Room squares - a computer search. *Proceedings of the First Louisiana Conference on Combinatorics, Graph Theory and Computing* (Baton Rouge, 1970), 87-111.

[7] HARRIET GRIFFIN, *Elementary Theory of Numbers.* (McGraw-Hill, 1954)

[8] ALEX GRONER, *Duplicate Bridge Direction.* (Barclay, New York, 1967)

[9] MARSHALL HALL JR., *Combinatorial Theory.* (Blaisdell, Waltham, 1967)

[10] FRANK HARARY, *Graph Theory.* (Addison-Wesley, Reading, 1969)

[11] J.D. HORTON, Variations on a theme by Moore. *Proceedings of the First Louisiana Conference on Combinatorics, Graph Theory and Computing* (Baton Rouge, 1970), 146-166.

[12] J.D. HORTON, Quintuplication of Room squares. *Aequationes Math.* 7 (1971), 243-245.

[13] J.D. HORTON, Room squares and one-factorizations. (to appear)

[14] J.D. HORTON, R.C. MULLIN and R.G. STANTON, A recursive construction for Room designs. *Aequationes Math.* 6 (1971), 39-45.

[15] F.K. HWANG, Some more contributions on constructing balanced Howell rotations. *Proceedings of the Second Chapel Hill Conference on Combinatorial Mathematics and its Applications* (Chapel Hill, 1970), 307-323.

[16] J.F. LAWLESS, Pairwise balanced designs and the construction of certain combinatorial systems. *Proceedings of the Second Louisiana Conference on Combinatorics, Graph Theory and Computing* (Baton Rouge, 1971), 353-366.

[17] CHARLES C. LINDNER, An algebraic construction for Room squares. *SIAM J. Appl. Math.* (to appear)

[18] CHARLES C. LINDNER and N.S. MENDELSOHN, Construction of perpendicular Steiner quasigroups. *Aequationes Math.* (to appear)

[19] N.S. MENDELSOHN, Latin squares orthogonal to their transposes. *J. Combinatorial Theory (Series A)* 11 (1971), 187-189.

[20] N.S. MENDELSOHN, Orthogonal Steiner systems. *Aequationes Math.* (to appear)

[21] E.H. MOORE, Concerning triple systems. *Math. Ann.* 43 (1893), 271-285.

[22] R.C. MULLIN, On the existence of a Room design of side $F_4$. *Utilitas Math.* (to appear)

[23] R.C. MULLIN and E. NEMETH, A counter-example to a multiplicative construction of Room squares. *J. Combinatorial Theory* 7 (1969), 264-265.

[24] R.C. MULLIN and E. NEMETH, On furnishing Room squares. *J. Combinatorial Theory* 7 (1969), 266-272.

[25] R.C. MULLIN and E. NEMETH, An existence theorem for Room squares. *Canad. Math. Bull.* 12 (1969), 493-497.

[26] R.C. MULLIN and E. NEMETH, On the non-existence of orthogonal Steiner triple systems of order 9. *Canad. Math. Bull.* 13 (1970), 131-134.

[27] R.C. MULLIN and E. NEMETH, A construction for self-orthogonal Latin squares from certain Room squares. *Proceedings of the First Louisiana Conference on Combinatorics, Graph Theory and Computing* (Baton Rouge, 1970), 213-226.

[28] R.C. MULLIN and P.J. SCHELLENBERG, Room designs of small side. *Proceedings of the Manitoba Conference on Numerical Mathematics* (University of Manitoba, 1971), 521-526.

[29] R.C. MULLIN and W.D. WALLIS, On the existence of Room squares of order 4n. *Aequationes Math.* 6 (1971), 306-309.

[30] E. NEMETH, *A Study of Room Squares.* (Thesis, University of Waterloo, 1969).

[31] C.D. O'SHAUGHNESSY, A Room design of order 14. *Canad. Math. Bull.* 11 (1968), 191-194.

[32] E.T. PARKER and A.N. MOOD, Some balanced Howell rotations for duplicate bridge sessions. *Amer. Math. Monthly* 62 (1955), 714-716.

[33] T.G. ROOM, A new type of magic square. *Math. Gazette* 39 (1955), 307.

[34] H.J. RYSER, *Combinatorial Mathematics.* (Carus Mathematical Monograph #14, M.A.A., 1963)

[35] P.J. SCHELLENBERG, *Constructions for (Balanced) Room Squares.* (Thesis, University of Waterloo, 1971)

[36] K.R. SHAH, Analysis of Room's square design. *Ann. Math. Statist.* 41 (1970), 743-745.

[37] R.G. STANTON and J.D. HORTON, Composition of Room squares. *Colloquia Mathematica Societatis János Bolyai, 4: Combinatorial Theory and its Applications* (North-Holland, 1970), 1013-1021.

[38] R.G. STANTON and J.D. HORTON, A multiplication theorem for Room squares. *J. Combinatorial Theory* (to appear).

[39] R.G. STANTON and R.C. MULLIN, Construction of Room squares. *Ann. Math. Statist.* 39 (1968), 1540-1548.

[40] R.G. STANTON and R.C. MULLIN, Techniques for Room squares. *Proceedings of the First Louisiana Conference on Combinatorics, Graph Theory and Computing* (Baton Rouge, 1970), 445-464.

[41] R.G. STANTON and R.C. MULLIN, Room quasigroups and Fermat primes. *J. Algebra* 20 (1972), 83-89.

[42] W.D. WALLIS, Room squares. *Invited and Contributed Papers, 1971 Australasian Statistical Conference* (Sydney, 1971).

[43] W.D. WALLIS, Duplication of Room squares. *J. Austral. Math. Soc.* 14 (1972), 75-81.

[44] W.D. WALLIS, A construction for Room squares. *A Survey of Combinatorial Theory*, ed. J. Srivastava (North-Holland, to appear).

[45] W.D. WALLIS, A doubling construction for Room squares. *Discrete Math.* (to appear)

[46] W.D. WALLIS, On one-factorizations of complete graphs. (to appear)

[47] W.D. WALLIS, On the existence of Room squares. *Aequationes Math.* (to appear)

[48] W.D. WALLIS, Room squares with subsquares. *J. Combinatorial Theory Series A* (to appear).

[49] W.D. WALLIS, On Archbold's construction of Room squares. *Utilitas Math.* (to appear)

[50] L. WEISNER, A Room design of order 10. *Canad. Math. Bull.* 7 (1964), 377-378.

# PART 3

## SUM-FREE SETS

ANNE PENFOLD STREET

# CONTENTS

| | | | |
|---|---|---|---|
| CHAPTER I. | | PRELIMINARIES | 127 |
| | 1.1 | Introduction | 127 |
| | 1.2 | The proof of Ramsey's theorem | 129 |
| CHAPTER II. | | SUM-FREE SETS OF INTEGERS | 131 |
| | 2.1 | Schur's original problem | 131 |
| | 2.2 | An improved lower bound | 133 |
| | 2.3 | Applications to estimates of Ramsey numbers | 136 |
| | 2.4 | References and related topics | 137 |
| CHAPTER III. | | REGULARITY OF SYSTEMS OF EQUATIONS | 138 |
| | 3.1 | Definitions and notation | 138 |
| | 3.2 | The law of regularity | 140 |
| | 3.3 | An important special case | 141 |
| | 3.4 | Some preliminary results | 147 |
| | 3.5 | The main theorems | 150 |
| | 3.6 | References and related topics | 157 |
| CHAPTER IV. | | SCHUR FUNCTIONS FOR MORE GENERAL SYSTEMS | 158 |
| | 4.1 | A system of simultaneous linear equations | 158 |
| | 4.2 | Estimates of Ramsey numbers | 160 |
| | 4.3 | Another system with several equations | 162 |
| | 4.4 | Another generalization | 164 |
| | 4.5 | References and related topics | 167 |
| CHAPTER V. | | ADMISSIBLE SUBSETS OF A SET OF POSITIVE INTEGERS | 168 |
| | 5.1 | Definitions and notation | 168 |
| | 5.2 | The lower bound for g(N) | 168 |
| | 5.3 | An upper bound for g(N) | 169 |
| | 5.4 | Admissible sets of integers modulo a prime | 170 |
| | 5.5 | Some results on the distribution of prime numbers | 171 |
| | 5.6 | A large sieve inequality | 175 |
| | 5.7 | The improved upper bound for g(N) | 178 |
| | 5.8 | References and related topics | 185 |
| CHAPTER VI. | | ADDITION THEOREMS FOR GROUPS | 186 |
| | 6.1 | Notation and basic concepts | 186 |
| | 6.2 | Further results in abelian groups | 192 |
| | 6.3 | Small sum-sets in abelian groups | 195 |
| | 6.4 | References and related topics | 204 |

| | | |
|---|---|---|
| CHAPTER VII. | SUM-FREE SETS IN GROUPS | 205 |
| 7.1 | Basic results | 205 |
| 7.2 | $\|G\|$ divisible by a prime congruent to 2 (modulo 3) | 208 |
| 7.3 | $\|G\|$ divisible by 3 | 208 |
| 7.4 | All divisors of $\|G\|$ congruent to 1 (modulo 3) | 211 |
| 7.5 | Sum-free sets in non-abelian groups | 242 |
| 7.6 | References and related topics | 246 |
| CHAPTER VIII. | SUM-FREE PARTITIONS AND RAMSEY NUMBERS | 247 |
| 8.1 | $R_3(3,2) = 17$ | 247 |
| 8.2 | The colourings of $K_{16}$ | 250 |
| 8.3 | $R_4(3,2) \geq 50$ | 263 |
| 8.4 | References and related topics | 264 |
| CHAPTER IX. | SOME RELATED TOPICS AND SOME UNSOLVED PROBLEMS | 265 |
| 9.1 | Sum-free sets in groupoids and semigroups | 265 |
| 9.2 | Problems related to Schur's | 265 |
| 9.3 | Unsolved problems | 266 |
| REFERENCES | | 267 |

# CHAPTER I. PRELIMINARIES

1.1 INTRODUCTION. Suppose that we have an additive semigroup, G, and that S, T are subsets of G. We define the set

$$S+T = \{s+t \mid s \in S,\ t \in T\}$$

to be the *sum* of S and T. In particular,

$$S+S = \{s_1+s_2 \mid s_1,s_2 \in S\}$$

where $s_1,s_2$ may or may not be equal.

S is said to be a *sum-free set* if and only if $S \cap (S+S) = \emptyset$ or, equivalently, if and only if the equation $x_1+x_2-x_3 = 0$ has no solution with $x_1,x_2,x_3 \in S$.

Sum-free sets have been studied in several contexts but mainly because of their connection with Ramsey numbers. The relevant definitions are given in section 2.7 of part I and will be discussed more fully in section 2 of this chapter, where we also give the proof of Ramsey's theorem. Here we just consider the *Ramsey number* $R_n(3,2)$ as being the smallest positive integer such that colouring the edges of the complete graph on $R_n(3,2)$ vertices in n colours forces the appearance of a monochromatic triangle.

Let $G = Z_5$, the integers modulo 5. Suppose that we partition the non-zero elements of G into two disjoint sum-free sets, $S_1 = \{1,4\}$ and $S_2 = \{2,3\}$, and assign to the set $S_k$ the colour $C_k$ for $k = 1,2$. Now let $K_5$ be the complete graph on the five vertices $v_0,\ldots,v_4$ and colour the edge from $v_i$ to $v_j$ in colour $C_k$ if $i-j \in S_k$. Since $S_k = -S_k$, this induces a well-defined edge-colouring of the graph. Let $v_\ell, v_m$, $v_n$ be any three vertices of $K_5$ and consider the triangle on these vertices. Suppose that two of its edges, say $\{v_\ell,v_m\}$ and $\{v_m,v_n\}$ are coloured $C_k$. This means that $\ell-m$, $m-n \in S_k$. But since $S_k$ is sum-free, we know that $\ell-n = (\ell-m)+(m-n) \notin S_k$ so that the edge $\{v_\ell,v_n\}$ is coloured in the other colour and no monochromatic triangle can occur. This shows that $R_2(3,2) > 5$. (In fact, an easy argument shows that $R_2(3,2) = 6$.)

All the applications of sum-free sets to estimating Ramsey numbers are similar to this example, in that they all depend on partitioning a group or a set of positive integers into a pairwise disjoint union of sum-free sets.

Work on sum-free sets of positive integers has been aimed directly at this partition problem. The *Schur function* f(n) is defined to be the largest positive integer such that the set of integers $\{1,2,\ldots,f(n)\}$ may be partitioned into n sum-free sets. In chapter II, we consider the original problem of Schur and, by studying the equation $x_1+x_2-x_3 = 0$ directly, we find the bounds

$$89^{(n/4)-c\log n} < f(n) < [n!e]-1$$

for some absolute constant c and sufficiently large n. We then apply this result to estimating Ramsey numbers.

The lower bound on f(n) has been improved by considering various generalizations of the problem. If we have a system (S) of simultaneous linear equations, we proceed by partitioning sets of integers into *(S)-free sets*, that is, into sets which contain no solution to the system (S). Just as we defined the Schur function for the equation $x_1+x_2-x_3 = 0$, we now need to define the analogous function for the system (S). But it is not always possible to define such a function for an arbitrary system (S), so in chapter III we determine the properties of (S) which are necessary and sufficient to enable such a function to be defined. In chapter IV, we consider the systems and the (S)-free partitions which improve the lower bound on f(n) and hence the estimates of the Ramsey numbers.

In chapter V, we consider a closely related problem: namely, how to estimate the largest number g(N) such that, from every set of N distinct natural numbers, we can always select a subset of g(N) integers with the property that no sum of two distinct integers of this subset belongs to the original set.

We then consider sum-free sets in finite groups. Very little is known about the partitions of finite groups into sum-free sets; in fact, so little is known about the sets themselves that most of the results we discuss are concerned with counting and characterising them. In order to do this, we need a number of addition theorems for finite groups; these are contained in chapter VI.

If $S \subseteq G$ is a sum-free set and if $|T| \leq |S|$ for every sum-free set T in G, then we call S a *maximal sum-free set* in G and denote by $\lambda(G)$ the cardinality of S. In chapter VII, we consider the values of $\lambda(G)$ for finite groups and the structure of the maximal sum-free sets. Almost all of these results apply to abelian groups only, but we also deal with the very few results concerning sum-free sets in non-abelian groups.

In chapter VIII, we consider sum-free partitions of finite abelian groups and their application to estimating Ramsey numbers.

Much work related to generalizations of Schur's problem and in particular to sum-free sets in semigroups has not been dealt with here, simply because we had to stop somewhere. In chapter IX we outline briefly some of these results. We also point out some of the unsolved problems in the area.

A list of references and acknowledgements for the material in each chapter is given in the last section of the chapter. No attempt has been made to cover all the methods which have been used in attempts to evaluate Ramsey numbers, though the applications of sum-free sets in finding Ramsey numbers are fully discussed.

1.2 THE PROOF OF RAMSEY'S THEOREM. We give here a proof of Ramsey's theorem, which was stated in section 2.7 of part I (theorem 63).

DEFINITION 1.1. Let S be an s-set, i.e. $|S| = s$. Let $\Pi_r(S)$ denote the collection of all r-subsets of S. Further let $\Pi_r(S) = A_1 \cup A_2 \cup \dots \cup A_n$ be a partition of $\Pi_r(S)$ into n mutually disjoint subsets. Suppose that for some $k \geq r$, there exists a k-subset K of S, such that all the r-subsets of K belong to the same $A_i$, for some $i = 1,\dots,n$. Then we call K a *$(k,A_i)$-subset* of S.

THEOREM 1.2. *Let $n,k_1,k_2,\dots,k_n,r$ be positive integers with $k_i \geq r$, $i = 1,\dots,n$. Then there exists a least positive integer $R(k_1,k_2,\dots,k_n,r)$ such that the following statement is true for any $s \geq R(k\ ,\dots,k_n,r)$:*

*For any s-set S and for any partition of $\Pi_r(S)$ into n classes, $A_1,\dots,A_n$, there exists a subset $K_i \leq S$ which is a $(k_i,A_i)$-subset for some i, $i = 1,\dots,n$.*

PROOF. Obviously, $R(k_1,k_2,\dots,k_n,r) \geq r$.

We consider first the case where $n = 2$ and then use induction on n.

(i) Suppose that $n = 2$ and $r = 1$. In this case, $\Pi_r(S)$ is just S itself and a $(k_i,A_i)$-subset of S is just a $k_i$-subset of $A_i$.

If $s \leq k_1+k_2-2$, then we can partition S as $S = A_1 \cup A_2$, where $|A_i| \leq k_i-1$, $i = 1,2$, so that no $(k_i,A_i)$-subset exists. But if $s \geq k_1+k_2-1$, then $|A_i| \geq k_i$ for at least one i. Hence

$$R(k_1,k_2,1) = k_1+k_2-1. \tag{1.1}$$

Now let $k_2 = r > 1$ and suppose that $s \leq k_1-1$. If $A_1 = S$ and $A_2 = \emptyset$, then S contains neither a $(k_1,A_1)$-subset nor an $(r,A_2)$-subset. Hence we need only consider $s \geq k_1$. If $A_2 \neq \emptyset$, then the set S contains an $(r,A_2)$-subset, namely any element of $A_2$. If $A_2 = \emptyset$, then $A_1 = \Pi_r(S)$ and S contains a $(k_1,\Pi_r(S))$-subset. Hence

$$R(k_1,r,r) = k_1 \tag{1.2}$$

and by a similar argument

$$R(r,k_2,r) = k_2. \tag{1.3}$$

We now assume that $1 < r < k_1,k_2$ and use induction. We take as the induction hypothesis the existence of the Ramsey numbers $R(k_1-1,k_2,r)$, $R(k_1,k_2-1,r)$ and $R(k_1',k_2',r-1)$ for all $k_1',k_2'$ such that $1 \leq r-1 \leq k_1',k_2'$. Equations (1.1), (1.2) and (1.3) then form the starting-point of the induction.

By the induction hypothesis, we know that the Ramsey numbers $\ell_1 = R(k_1-1,k_2,r)$, $\ell_2 = R(k_1,k_2-1,r)$ and $R(\ell_1,\ell_2,r-1)$ exist. We prove the existence of $R(k_1,k_2,r)$ and, in the process, show that $R(k_1,k_2,r) \leq R(\ell_1,\ell_2,r-1)+1$.

For let $s \geq R(\ell_1,\ell_2,r-1)+1$ and let S be an s-set. Let $\alpha \in S$ be any fixed element of S and $T = S\setminus\{\alpha\}$. A partition $\Pi_r(S) = A_1 \cup A_2$ of the r-subsets of S induces a partition $\Pi_{r-1}(T) = B_1 \cup B_2$ of the (r-1)-subsets of T in the following way:

Let U be an (r-1)-subset of T. Then $U \in B_i$ if and only if $U \cup \{\alpha\} \in A_i$, $i = 1,2$.

Now $|T| \geq R(\ell_1,\ell_2,r-1)$ so T contains at least one of an $(\ell_1,B_1)$-subset or an $(\ell_2,B_2)$-subset. We assume without loss of generality that Q is an $(\ell_1,B_1)$-subset of T, i.e. that Q is an $\ell_1$-subset of T, all of whose (r-1)-subsets belong to $B_1$. Since $\ell_1 = R(k_1-1,k_2,r)$ and Q is a subset of S, we know that Q contains either a $(k_1-1)$-subset, P, all of whose r-subsets belong to $A_1$, or a $k_2$-subset, P', all of whose r-subsets belong to $A_2$. In the second case, the set P' satisfies our requirements and the theorem is proved for n = 2.

So suppose that Q contains a $(k_1-1)$-subset P, all of whose r-subsets are in $A_1$. Then $P \cup \{\alpha\} = K$ which is a $k_1$-subset of S. We claim that all the r-subsets of K belong to $A_1$. For if J is an r-subset of K, then either

(a) $\alpha \notin J$, in which case J is an r-subset of P and hence an element of $A_1$, or

(b) $\alpha \in J$, in which $J = H \cup \{\alpha\}$, where H is an (r-1)-subset of P and hence of Q. But this implies that $H \in B_1$ and, by the definition of $B_1$, that $J \in A_1$. So K is a $k_1$-subset of S, with all of its r-subsets in $A_1$, and the theorem is proved for n = 2.

(ii) Now suppose that the theorem has been established for n = m-1 and consider the case where n = m. Let

$$\Pi_r(S) = (A_1 \cup \ldots \cup A_{m-1}) \cup A_m$$

be a partition of $\Pi_r(S)$ and let $k = R(k_1,\ldots,k_{m-1},r)$. Choose $s \geq R(k,k_m,r)$ and let S be an s-set. Then S contains either a $(k_m,A_m)$-subset, in which case the theorem is proved, or a $(k,A_1 \cup \ldots \cup A_{m-1})$-subset K. But then, by the choice of k, K must contain a $(k_i,A_i)$-subset J for at least one i, $i = 1,2,\ldots,m-1$, and since J is a subset of S, the theorem is proved.

Ramsey's theorem appeared originally in Ramsey (1930). Other proofs have been given by Erdos and Szekeres (1935), Skolem (1933); for a discussion of the theorem see Ryser (1936) and Hall (1967).

## CHAPTER II. SUM-FREE SETS OF INTEGERS

2.1 SCHUR'S ORIGINAL PROBLEM. The congruence $x^m+y^m \equiv z^m \pmod p$ has been extensively studied in connection with Fermat's last theorem. Dickson, in 1909, showed that this congruence is satisfied by three numbers x,y,z, not divisible by p, as soon as the prime p exceeds a certain limit M, dependent only on m. Dickson's result implies that there is no hope of trying to use this congruence to show that the equation $x^m+y^m = z^m$ has no integral solutions, $m \geq 3$. His proof involved very tedious computation and Schur's original work on sum-free sets was done in order to provide a neat proof of Dickson's result.

Before considering the problem of Schur, we point out the following useful result:

LEMMA 2.1. *For any integer $n \geq 2$,*

$$[n!e] = n! \sum_{j=0}^{n} 1/j! \tag{2.1}$$

*and hence*

$$[(n+1)!e] = (n+1)[n!e]+1. \tag{2.2}$$

PROOF. By Taylor's theorem we see that

$$n!e = n!(1+1+\frac{1}{2!}+\ldots+\frac{1}{n!}+\frac{1}{(n+1)!}+\ldots)$$

$$\leq n!(1+1+\frac{1}{2!}+\ldots+\frac{1}{n!}+\frac{e}{(n+1)!}),$$

since the remainder after (n+1) terms of the expression in brackets, $r_{n+1}$, satisfies $r_{n+1} \leq \frac{e}{(n+1)!}$. Hence for $n \geq 2$, (2.1) follows. (2.2) is an immediate consequence of (2.1).

Now we consider the problem of Schur: What is the largest integer f(n) for which there exists some way of partitioning the set {1,2,...,f(n)} into n sets, each of which is sum-free?

Only four values of f(n) are known: f(1) = 1, f(2) = 4 and f(3) = 13 are easily determined. The value f(4) = 44 was determined by Baumert in 1961, with the aid of a computer. One sum-free partition he found for the set {1,...,44} is as follows:

$S_1$ = {1,3,5,15,17,19,26,28,40,42,44};
$S_2$ = {2,7,8,18,21,24,27,33,37,38,43};
$S_3$ = {4,6,13,20,22,23,25,30,32,39,41};
$S_4$ = {9,10,11,12,14,16,29,31,34,35,36}.

For $n \geq 5$, we have only estimates of $f(n)$. Schur's argument leads to the following result:

THEOREM 2.2.

$$\frac{3^n-1}{2} \leq f(n) \leq [n!e]-1.$$

PROOF. (i) Suppose that the set of integers $\{1,2,\ldots,N\}$ can be partitioned into $n$ sum-free sets, $S_1,S_2,\ldots,S_n$. Without loss of generality, we may assume that

$$m_1 = |S_1| \geq |S_i| \quad \text{for } i = 2,3,\ldots,N$$

and note that

$$N \leq m_1 n. \tag{2.3}$$

Let $x_1 < x_2 < \ldots < x_{m_1}$ be the integers in $S_1$. Now the $(m_1-1)$ differences

$$x_2-x_1, x_3-x_1, \ldots, x_{m_1}-x_1$$

belong to the set $\{1,2,\ldots,N\}$ and since $S_1$ is sum-free, they must be distributed between the $(n-1)$ sets $S_2,\ldots,S_n$. Let $S_2$ be the set which contains the largest number, say $m_2$, of these $(m_1-1)$ differences, namely $x_{i_j}-x_1$, where $j = i_1,i_2,\ldots,i_{m_2}$, and $i_1 < i_2 < \ldots < i_{m_2}$. Again

$$m_1-1 \leq m_2(n-1). \tag{2.4}$$

Now the differences $x_{i_j}-x_{i_1}$ for $j = 2,\ldots,m_2$, must be distributed between the $(n-2)$ sets $S_3,\ldots,S_n$, and again we let $S_3$ be the set which contains the greatest number, say $m_3$, of these $(m_2-1)$ differences. As before,

$$m_2-1 \leq m_3(n-2). \tag{2.5}$$

We continue in this fashion: for each integer $\nu$, we get a number $m_\nu$ such that

$$m_\nu-1 \leq m_{\nu+1}(n-\nu) \tag{2.6}$$

and

$$\frac{m_\nu}{(n-\nu)!} \leq \frac{m_{\nu+1}}{(n-\nu-1)!} + \frac{1}{(n-\nu)!} .$$

Eventually, $\nu = n$ and $m_\mu = 1$ for some $\mu \leq n$, so that using (2.3), (2.6) and adding, we find

$$N \leq n! \left\{\frac{1}{(n-1)!} + \frac{1}{(n-2)!} + \ldots + \frac{1}{(n-\mu)!}\right\}.$$

Hence, by (2.1), since $N$ is an integer, we have

$$N \leq [n!e]-1.$$

(ii) Suppose that we have a partition of the set $\{1,2,\ldots,f(n)\}$ into $n$ sum-free sets, namely

$$S_1 = \{x_{11},x_{12},\ldots,x_{1,\ell_1}\},\ldots,S_n = \{x_{n1},x_{n2},\ldots,x_{n,\ell_n}\} .$$

Then the (n+1) sets

$$S_1 = \{3x_{11}, 3x_{11}-1, 3x_{12}, 3x_{12}-1, \ldots, 3x_{1,\ell_1}, 3x_{1,\ell_1}-1\},$$

$$\vdots \qquad \vdots$$

$$S_n = \{3x_{n1}, 3x_{n1}-1, 3x_{n2}, 3x_{n2}-1, \ldots, 3x_{n,\ell_n}, 3x_{n,\ell_n}-1\},$$

$$S_{n+1} = \{1, 4, 7, \ldots, 3f(n)+1\}$$

form a sum-free partition of $\{1,2,\ldots,3f(n)+1\}$. Hence

$$f(n+1) \geq 3f(n)+1$$

and, since $f(1) = 1$, we have

$$f(n) \geq 1+3+3^2+\ldots+3^{n-1} = \frac{3^n-1}{2}.$$

We can improve this lower bound slightly by using Baumert's result; since $f(4) = 44$, we see that $f(5) \geq 3.44+1$ and in general that

$$f(n) \geq 3^{n-4}.44+3^{n-5}+\ldots+1 \quad \text{for } n \geq 4.$$

Hence

$$f(n) \geq 3^{n-4}.44+\tfrac{1}{2}(3^{n-4}-1) = \tfrac{1}{2}(89.3^{n-4}-1). \tag{2.7}$$

2.2 AN IMPROVED LOWER BOUND. The bound given by (2.7) can be improved further.

DEFINITION 2.3. Let $g(\ell)$ be the smallest number of sum-free sets into which the set of integers $\{1,2,\ldots,\ell\}$ can be partitioned. Equivalently, we say that if $f(n-1) < \ell \leq f(n)$, then $g(\ell) = n$.

LEMMA 2.4. *For $\ell$ sufficiently large,*

$$g(\ell) < \log \ell. \tag{2.8}$$

PROOF. By (2.7),

$$g\big((89.3^{n-4}-1)/2\big) \leq n. \tag{2.9}$$

Choose the smallest $\ell$ such that $\ell+1 > (89.3^{n-4}-1)/2$, i.e.,

$$\ell > \tfrac{1}{2}(89.3^{n-4}-3).$$

If $n = 7$, then

$$e^n = e^7 < \tfrac{1}{2}(89.3^{7-4}-3).$$

By induction on n,

$$e^n < \tfrac{1}{2}(89.3^{n-4}-3) \quad \text{for } n \geq 7.$$

By our choice of $\ell$, this shows that for n large enough, $e^n < \ell$. Hence $n < \log \ell$ and (2.9) implies (2.8)

THEOREM 2.5. *For all positive integers m and k,*

$$f\big(km+g(kf(m))\big) \geq \big(2f(m)+1\big)^k-1. \tag{2.10}$$

PROOF. Let $X = 2f(m)+1$ and write the numbers $1,2,\dots,X^k-1$ in base $X$, so that each integer is represented as

$$a = a_0+a_1X+a_2X^2+\dots+a_{k-2}X^{k-2}+a_{k-1}X^{k-1}$$

where $0 \leq a_i \leq 2f(m)$, for $i = 0,1,\dots,k-1$.

We call an integer *good* if $a_i \leq f(m)$ for each $i$ and *bad* if $a_i \geq f(m)+1$ for at least one $i$.

We know that the set of integers $\{1,2,\dots,kf(m)\}$ can be partitioned into disjoint sum-free sets $A_1,A_2,\dots,A_{g(kf(m))}$. This partition induces a partition of the good integers into $g(kf(m))$ sum-free sets $B_1,\dots,B_{g(kf(m))}$ in the following way:

For every $a$, $1 \leq a \leq X^k-1$, we define

$$\sigma(a) = a_0+a_1+\dots+a_{k-1},$$

the sum of the digits of $a$. If $a$ is a good integer, then $\sigma(a) \leq kf(m)$ and $\sigma(a) \in A_j$ for some $j$, $1 \leq j \leq g(kf(m))$. Now let $a \in B_j$ if and only if $\sigma(a) \in A_j$. We claim that each $B_j$ is sum-free. For if $a,b \in B_j$, then either $a+b$ is a bad integer and belongs to none of the $B_j$, or $a_i+b_i \leq f(m)$ for every $i$. In the latter case, $a+b \in B_n$ say. If $n = j$, then $\sigma(a)$, $\sigma(b)$ and $\sigma(a+b) = \sigma(a)+\sigma(b)$ are integers in the set $\{1,\dots,kf(m)\}$ all belonging to the set $A_j$. But this contradicts the fact that $A_j$ is sum-free.

Now we consider the bad integers. We first partition them into $k$ classes $C_1,\dots,C_k$ by assigning $a$ to class $C_{j+1}$, $j = 0,1,\dots,k-1$, if $a_i \leq f(m)$ for $i = 0,1,\dots,j$ and $a_{j+1} \geq f(m)+1$. Finally we partition each $C_j$ into $m$ disjoint sets $D_{j1},D_{j2},\dots,D_{jm}$ in the following way:

Let $D_1,D_2,\dots,D_m$ be a sum-free partition of the set $\{1,2,\dots,f(m)\}$. If $a \in C_j$, then $f(m)+1 \leq a_{j+1} \leq 2f(m)$, and we assign $a$ to the set $D_{j\ell}$ if and only if $a_{j+1} \equiv -u \pmod X$ for some $u \in D_\ell$. Exactly one such $u$ exists, so the partition is well-defined. We claim that each $D_{j\ell}$ is sum-free. For suppose that there exist $a,b,c \in D_{j\ell}$ such that $a+b = c$. Now

$$a = \sum_{i=0}^{k-1} a_iX^i, \qquad b = \sum_{i=0}^{k-1} b_iX^i, \qquad c = \sum_{i=0}^{k-1} c_iX^i,$$

where

$$a_i,b_i,c_i \leq f(m) \quad \text{for } i = 0,1,\dots,j,$$

$$a_{j+1},b_{j+1},c_{j+1} \geq f(m)+1$$

and

$$a_{j+1} \equiv -u, \quad b_{j+1} \equiv -v, \quad c_{j+1} \equiv -w \pmod X$$

where $u,v,w \in D_\ell$.

Now

$$a_{j+1}+b_{j+1} = X+c_{j+1},$$

so $u+v \equiv w \pmod X$.

But $u,v,w \leq f(m)$, so $u+v = w$, which contradicts the fact that $D_\ell$ is sum-free.

We have now partitioned the bad integers into km sum-free sets. We had already partitioned the good integers into $g(kf(m))$ sum-free sets and the theorem follows.

COROLLARY 2.6. *$f(n) > 89^{(n/4) - c \log n}$ for all sufficiently large n, where c is some positive absolute constant.*

PROOF. Let $m = 4$ in (2.10). Since $f(4) = 44$, we have

$$f(4k+g(44k)) \geq 89^k-1.$$

By (2.8), $g(44k) < \log(44k)$ for sufficiently large k, so

$$f(4k+\log(44k)) \geq 89^k-1,$$

since f is an increasing function.

Now choose n so that

$$4k+\log(44k) \leq n < 4(k+1)+\log(44(k+1)). \qquad (2.11)$$

Then

$$f(n) \geq f(4k+\log(44k)) \geq 89^k-1. \qquad (2.12)$$

From the right-hand inequality of (2.11) we have

$$k > (n - \log 44 - \log(k+1)-4)/4 \qquad (2.13)$$

and from the left-hand inequality of (2.11) we have $k+1 < n$, so that

$$\log(k+1) < \log n. \qquad (2.14)$$

From (2.14), for large n,

$$\frac{1}{4}(\log 44 + \log(k+1)+4) < \frac{1}{3} \log n \qquad (2.15)$$

and by (2.13) and (2.15),

$$k > \frac{n}{4} - \frac{1}{3} \log n.$$

Hence by (2.12)

$$f(n) > 89^{(n/4)-(\log n)/3}-1,$$

from which the corollary follows, with suitable choice of c.

One other property of f(n) also follows from theorem 2.5.

COROLLARY 2.7. *The limit, $\lim_{n\to\infty} f(n)^{1/n}$, exists.*

PROOF. Let

$$\alpha = \liminf_{n\to\infty} f(n)^{1/n} \leq \limsup_{n\to\infty} f(n)^{1/n} = \beta.$$

Suppose that $\beta$ is finite. Choose $\varepsilon > 0$ and let m be the smallest integer for which

$$f(m)^{1/m} > \beta-\varepsilon. \qquad (2.16)$$

By (2.8), for sufficiently large k,

$$g(kf(m)) < \log(kf(m)) = \log k + \log f(m).$$

Hence, for m fixed,

$$(g(kf(m)))/k \to 0 \quad \text{as } k \to \infty,$$

and there exists an integer $k_0 = k_0(\varepsilon)$, such that for $k \geq k_0$, we have

$$km+g((kf(m)) < [km(1+\varepsilon)]. \tag{2.17}$$

Let

$$[km(1+\varepsilon)] \leq n \leq [(k+1)m(1+\varepsilon)]. \tag{2.18}$$

Hence, by (2.10),(2.17) and (2.18),

$$f(n) \geq f([km(1+\varepsilon)]) > f(km+g(kf(m))) \geq (2f(m)+1)^k-1 > f(m)^k.$$

This implies that

$$f(n)^{1/n} > f(m)^{k/n} > (\beta-\varepsilon)^{km/n}$$

by (2.16) and hence, by (2.18),

$$\liminf_{n\to\infty} f(n)^{1/n} \geq (\beta-\varepsilon)^{1/(1+\varepsilon)}.$$

It follows that $\alpha = \beta$.

A similar argument deals with the case where $\beta$ is infinite.

It is not known whether $\lim_{n\to\infty} f(n)^{1/n}$ is finite or infinite.

2.3 APPLICATIONS TO ESTIMATES OF RAMSEY NUMBERS. Consider the second statement of Ramsey's theorem (theorem 64 of part 1). If

$$k_1 = k_2 = \dots = k_n = k \geq 2,$$

then we abbreviate $R(k_1,k_2,\dots,k_n,2)$ to $R_n(k,2)$. Hence $R_n(3,2)$ is the smallest positive integer such that colouring the edges of the complete graph on $R_n(3,2)$ vertices in n colours forces the appearance of a monochromatic triangle.

THEOREM 2.8. *For all sufficiently large n,*

$$R_n(3,2) > 89^{(n/4) - c\log n} + 1$$

PROOF. We prove that

$$R_n(3,2)-1 \geq f(n)+1 \tag{2.19}$$

from which by corollary 2.6, the theorem follows.

To prove (2.19), let $A_1,\dots,A_n$ be a sum-free partition of the set $\{1,2,\dots,f(n)\}$. Let $K = K_{f(n)+1}$ be the complete graph on $f(n)+1$ vertices $x_0,x_1,\dots,x_{f(n)}$. We colour the edges of K in the n colours $C_1,\dots,C_n$ by colouring the edge $\{x_i,x_j\}$ in the colour $C_m$ if $|i-j| \in A_m$. Suppose this gives us a triangle with vertices $x_i,x_j,x_k$, all of whose edges are coloured $C_m$. We assume that $i > j > k$. Then $i-j$, $i-k$, $j-k \in A_m$. But

$(i-j)+(j-k) = i-k$, which contradicts the fact that $A_m$ is sum-free.

We have also an upper bound on $R_n(3,2)$.

THEOREM 2.9. *$R_{n+1}(3,2) \leq (n+1)\big(R_n(3,2)-1\big)+2$.*

PROOF. Let K be the complete graph on $(n+1)\big(R_n(3,2)-1\big)+2$ vertices and consider a colouring of K with $(n+1)$ colours. Choose one vertex, v, of K. Of the $(n+1)\big(R_n(3,2)-1\big)+1$ edges ending at v, at least $R_n(3,2)$ must have the same colour. Suppose these join v to vertices $x_1, x_2, \ldots, x_s$ respectively, where $R_n(3,2)$. Consider the edges $\{x_i, x_j\}$ where $1 \leq i < j \leq s$. If any of them has the original colour, then the triangle $\{x_i, x_j, v\}$ is monochromatic. If none of them has the original colour then the complete graph $K_s$ on $x_1, \ldots, x_s$ must be coloured in the other n colours. But by the choice of s, this forces the appearance of a monochromatic triangle in $K_s$, and hence in K.

COROLLARY 2.10. *Since $R_1(3,2) = 3$, by the definition of Ramsey numbers, the theorem implies that*

$$R_{n+1}(3,2) \leq 3(n+1)!$$

However, this bound can be improved.

COROLLARY 2.11. *$R_n(3,2) \leq [n!e]+1$.*

PROOF. By corollary 2.10,

$$R_2(3,2) \leq 6 = [2!e]+1,$$

which forms the starting point for our induction. If

$$R_n(3,2) \leq [n!e]+1,$$

then by Theorem 2.9 and (2.2),

$$R_{n+1}(3,2) \leq (n+1)[n!e]+2 = [(n+1)!e]+1.$$

We note that theorem 2.8 and corollary 2.11 give an alternative proof of Schur's upper bound in theorem 2.2.

## 2.4 REFERENCES AND RELATED TOPICS.

Dickson's original work and its relation to Fermat's last theorem are discussed in Dickson (1909a, 1909b, 1919) and Vandiver (1946). Schur (1916) gives the initial work on sum-free sets and a simple proof of Dickson's theorem. The work of Ramsey (1930) leads to the idea of Ramsey numbers. Bounds on these numbers are given in Greenwood and Gleason (1955). Abbott and Moser (1966), using a construction related to that of Frasnay (1963), give the improved lower bound on the Schur function and point out the relation between it and the Ramsey numbers. In particular, they show the relationship between their results on the Schur function and the work of Greenwood and Gleason. The evaluation of f(4) is due to Baumert (1961); see also Golomb and Baumert (1965).

## CHAPTER III. REGULARITY OF SYSTEMS OF EQUATIONS

In order to get more information about sum-free sets and the Schur function, we need to study the corresponding problem for more general systems of equations. The question which immediately arises is: which systems are regular? or in other words, for which systems (S) may we define a function $f_S(n)$, such that:

(i) the set of integers $\{1,2,\ldots,f_S(n)\}$ may be partitioned into n sets, none of which contains a solution of the system (S), but

(ii) if the set of integers $\{1,2,\ldots,f_S(n)+1\}$ is partitioned in any manner into n sets, then at least one of these subsets must contain a solution of the system (S).

3.1 DEFINITIONS AND NOTATION. We shall call a set of n consecutive natural numbers a *segment of length n*. In general, the number of terms of an arithmetic progression will be referred to as its *length*. We shall consider arbitrary partitions of the natural numbers, or of some segment $\{1,\ldots,n\}$ of the natural numbers, into finitely many blocks. Such a partition will be denoted by P and two partitions will be regarded as identical if they differ only in the naming of the blocks. We allow empty blocks, so a partition with k blocks may also be regarded as a partition with (k+1) blocks.

If two natural numbers, m and n, belong to the same block of the partition P, we say that m and n are *congruent modulo P*, and denote this by $m \equiv n \pmod{P}$. More generally, if we are considering two sequences $\{m_1,m_2,\ldots,m_\ell\}$ and $\{n_1,n_2,\ldots,n_\lambda\}$, we shall call the sequences *congruent modulo P* if and only if $\ell = \lambda$ and $m_i \equiv n_i \pmod{P}$, for $i = 1,\ldots,\ell$. We shall often use this terminology when dealing with arithmetic progressions. In particular, we note that a partition P, with k blocks, on the natural numbers induces a partition $P^n$, with $k^n$ blocks, on the set of segments of length n. Again two segments belonging to the same block of this induced partition are said to be congruent.

We shall often change from a given partition P to a new partition P' in the following way:

Let $P = K_1 \cup K_2 \cup \ldots \cup K_k$ be a given partition with k blocks and let $f_1,f_2,\ldots,f_N$ be given functions on the natural numbers. Then we define $m \equiv n \pmod{P'}$ to mean that $f_i(m) \equiv f_i(n) \pmod{P}$ for $i = 1,2,\ldots,N$. Hence P' has in general $k^N$ blocks, some of which may be empty. The induced partition P' will be denoted by

$$P' = P[f_1(\xi),f_2(\xi),\ldots,f_N(\xi)]$$

or alternatively

$$P' = P[f_i(\xi),\ 1 \leq i \leq N].$$

Obviously, for any partition P, we have $P = P[\xi]$.

For example, if

$$P = \{1,5,7,8,9,\ldots\} \cup \{2,3,4,6,10,\ldots\} ,$$

then

$$P[2\xi] = \{4,\ldots\} \cup \{1,2,3,5,\ldots\}$$

and

$$P[\xi,\xi+1] = \{7,8,\ldots\} \cup \{1,5,9,\ldots\} \cup \{2,3,\ldots\} \cup \{4,6,\ldots\}.$$

In particular, we use $P_n$ to denote the partition into residue classes modulo n, for any natural number n.

Let S be any system of conditions on the variables $x_1,x_2,\ldots,x_n$, and let $a_1$, $a_2,\ldots,a_n$ be natural numbers. We use the notation $S(a_1,\ldots,a_n) = 0$ to mean that the numbers $a_1,\ldots,a_n$ satisfy the conditions S.

We call such a system of conditions *homogeneous* if and only if $S(a_1,a_2,\ldots,a_n) = 0$ for some natural numbers $a_1,a_2,\ldots,a_n$, implies that $S(a_1a,a_2a,\ldots,a_na) = 0$, for any natural number a.

We recall the definitions given in section 2.7 of part 1. A system of conditions S is said to be *k-regular* if and only if, for every partition P with k blocks, there exist $a_1,\ldots,a_n$ such that $a_1 \equiv a_2 \equiv \ldots \equiv a_n \pmod{P}$, and $S(a_1,\ldots,a_n) = 0$. S is said to be *regular* if and only if it is k-regular for every k.

For example, the equation

$$(x+1-y)(x+2-y) = 0 \tag{3.1}$$

is 2-regular, for if $P = K_1 \cup K_2$ is any partition with two blocks, then two of the three numbers 1,2,3 are in the same block and $x = 1$, $y = 2$ or $x = 1$, $y = 3$, or $x = 2$, $y = 3$ are all solutions of the equation (3.1). But (3.1) is not 3-regular, for if we choose $P = P_3$, no block of $P_3$ contains a solution of (3.1).

On the other hand, the equation

$$x^2-3xy+4y^2-2 = 0 \tag{3.2}$$

is regular, because $x = y = 1$ is a solution of (3.2) and $1 \equiv 1 \pmod{P}$ for every P. In general, $S(x_1,\ldots,x_n)$ is regular if there exists a natural number a such that $S(a,\ldots,a) = 0$.

From the law of regularity, discussed in the next section, we shall see that the regular systems are exactly those for which a Schur function may be defined.

3.2 THE LAW OF REGULARITY. We show that it is not necessary to consider partitions of the entire set of natural numbers; a suitable finite subset of the natural numbers will always suffice.

THEOREM 3.1. *Suppose that the system of conditions*

$$S(x_1,x_2,\dots,x_n) = 0 \tag{3.3}$$

*is k-regular. Then there exists a positive integer N, which is a function only of k and of S, such that the system of conditions (3.3) and*

$$x_1,x_2,\dots,x_n \leq N \tag{3.4}$$

*is k-regular.*

PROOF. We suppose that the theorem is false and hence that the system of conditions (3.3) and (3.4) is not k-regular for any choice of N. Hence, for every positive integer N, there exists a partition $P_{(N)}$, with k blocks, such that if $a_1 \equiv a_2 \equiv \dots \equiv a_n \pmod{P_{(N)}}$ and $a_1,a_2,\dots,a_n \leq N$, then $S(a_1,a_2,\dots,a_n) \neq 0$.

Denote the blocks of $P_{(N)}$ by $K_{N1},K_{N2},\dots,K_{Nk}$ and, for every natural number x, let $f_N(x)$ denote the number of the block to which x belongs. We construct a function $f'(x)$ as follows:

Label as $K_{N1}$ the block to which the number 1 belongs. Then $f_N(1) = 1$ for every N, and we let $f'(1) = 1$. If $2 \in K_{N1}$, then $f_N(2) = 1$. Otherwise, label as $K_{N2}$ the block to which 2 belongs, so that $f_N(2) = 2$. Since $f_N(2)$ is defined for every natural number N, it must take at least one of the values, 1 or 2, infinitely often. If $f_N(2) = 1$ for infinitely many values of N, we let $f'(2) = 1$. Otherwise let $f'(2) = 2$. Now we have $f'(x) = f_N(x)$ for x = 1,2, for infinitely many values of N and we proceed to consider where the number 3 occurs in the partitions. If $3 \in K_{N1}$, then $f_N(3) = 1$. If $2 \in K_{N1}$ but $3 \notin K_{N1}$, label as $K_{N2}$ the block to which 3 belongs, so that $f_N(3) = 2$. If $2,3 \in K_{N2}$, then again $f_N(3) = 2$. If $2 \in K_{N2}$, $3 \notin K_{N1} \cup K_{N2}$, then label as $K_{N3}$ the block to which 3 belongs, so that $f_N(3) = 3$. Now consider the infinite set $H_1$ of natural numbers N for which $f'(x) = f_N(x)$, for x = 1,2. Since $f_N(3)$ is defined for all natural numbers and in particular for all $N \in H_1$, $f_N(3)$ must take at least one of the values 1, 2 or 3, infinitely often, for $N \in H_1$. If $f_N(3) = 1$ for infinitely many $N \in H_1$, let $f'(3) = 1$. If $f_N(3) = 1$ for only finitely many $N \in H_1$, but $f_N(3) = 2$ for infinitely many $N \in H_1$, let $f'(3) = 2$. Otherwise, let $f'(3) = 3$. In any case, we now have $f'(x) = f_N(x)$ for x = 1,2,3, for infinitely many $N \in H_2 \subseteq H_1$. Proceeding in this way, we may construct a function $f'(x)$ such that, given any $m > 0$, there are infinitely many natural numbers N for which

$$f'(x) = f_N(x) \quad \text{for every } x = 1,2,\dots,m. \tag{3.5}$$

We note that for every x, $1 \leq f_N(x) \leq k$ and $1 \leq f'(x) \leq \min(x,k)$.

Now we define a partition

$$P' = K_1' \cup K_2' \cup \ldots \cup K_k'$$

by letting

$$K_j' = \{x | f'(x) = j\}, \quad \text{for } j = 1,2,\ldots,k.$$

Choose n numbers $a_1', a_2', \ldots, a_n'$ such that

$$a_1' \equiv a_2' \equiv \ldots \equiv a_n' \pmod{P'} \tag{3.6}$$

and

$$S(a_1', a_2', \ldots, a_n') = 0. \tag{3.7}$$

Choose $m = \max(a_1', a_2', \ldots, a_n')$. By (3.5) there exists $N \geq m$ for which $f'(x) = f_N(x)$ for every $x = 1,2,\ldots,m$. In particular, $f'(a_i') = f_N(a_i')$ for $i = 1,2,\ldots,n$. Hence, since $a_i' \leq N$ for $i = 1,2,\ldots,n$, (3.6) and (3.7) show that the conditions (3.3) and (3.4) are k-regular and the theorem is proved.

If we take the minimum value for N in theorem 3.1, where S is a regular system, then (N-1) is the Schur function, $f_S(k)$.

## 3.3 AN IMPORTANT SPECIAL CASE.

To deal with regularity of linear systems in general, there is one special system we must consider first.

THEOREM 3.2. *Given natural numbers k and ℓ, the system of conditions*

$$x_1 - x_2 = x_2 - x_3 = \ldots = x_{\ell-1} - x_\ell \neq 0 \tag{3.8}$$

*is k-regular.*

We know by theorem 3.1 that if (3.8) is k-regular, then there exists a natural number $N = n(k,\ell)$ such that the system of conditions (3.8) and

$$x_1, x_2, \ldots, x_\ell \leq n(k,\ell) \tag{3.9}$$

is k-regular, so we need only consider partitions of the set of integers $\{1,2,\ldots,n(k,\ell)\}$. The theorem may be reformulated in the following way:

*For two arbitrary natural numbers, k and ℓ, there exists a natural number n(k, ℓ) such that if the segment {1,2,...,n(k, ℓ)} is partitioned arbitrarily into k blocks, then at least one of the blocks contains an arithmetic progression of length ℓ.*

PROOF. The theorem is certainly true for any k, for $\ell = 2$. For let $n(k,\ell) = k+1$. If we partition the set $\{1,2,\ldots,k+1\}$ into k blocks, then at least one of these blocks contains more than one integer, and any pair of integers forms an arithmetic progression of length 2.

We now assume that for some $\ell \geq 2$ and for arbitrary k, the theorem has been verified and the number $n(k,\ell)$ found. We then prove the theorem for $\ell+1$ by constructing the number $n(k,\ell+1)$. For this purpose, let

$$q_0 = 1, \; n_0 = n(k,\ell) \tag{3.10}$$

and

$$q_s = 2n_{s-1}q_{s-1},\ n_s = n(k^{q_s},\ell),\ s = 1,2,\ldots \qquad (3.11)$$

We claim that $n(k,\ell+1) \leq q_k$, i.e. that if the set of integers $\{1,2,\ldots,q_k\}$ is partitioned arbitrarily into k blocks, then at least one block contains an arithmetic progression of length at least $\ell+1$.

Before giving the general proof, we carry out the construction for the simplest possible case, showing that $n(2,3) \leq 780$. Here

$$q_0 = 1, \qquad n_0 = n(2,2) = 2+1 = 3,$$

$$q_1 = 2n_0q_0 = 6, \qquad n_1 = n(2^{q_1},2) = 2^6+1 = 65,$$

$$q_2 = 2n_1q_1 = 2.6.65 = 780,$$

and we show that if the set $S = \{1,2,\ldots,780\}$ is partitioned into two blocks, so that $P = K_1 \cup K_2$, then at least one of the blocks contains an arithmetic progression of length at least three.

We regard S, the segment $\{1,2,\ldots,780\}$, as a sequence of $2.65 = 130$ subsegments, each of length six. Any natural number belongs either to $K_1$ or to $K_2$. This partition, P, of the natural numbers into two blocks, induces a partition of the set of all segments of length six into $2^6$ blocks, depending on how the six elements of the segment are partitioned between $K_1$ and $K_2$.

The initial half $\{1,2,\ldots,390\}$ of the segment S contains 65 subsegments of length six each, of the form

$$\{6j+1,\ldots,6j+6\} \quad \text{for } j = 0,1,\ldots,64.$$

Since these 65 subsegments are partitioned into 64 blocks, we see that $\{1,2,\ldots,390\}$ contains an arithmetic progression of two congruent subsegments, say

$$S_1 = \{6a+1,\ldots,6a+6\}$$

and

$$S_2 = \{6a+d_1+1,\ldots,6a+d_1+6\},$$

where the difference $d_1$ between the initial numbers of the segments is called the difference of the arithmetic progression, and the fact that $S_1$ is congruent to $S_2$ ($S_1 \equiv S_2$) means that

$$6a+i \equiv 6a+d_1+i \pmod{P}, \quad \text{for } i = 1,\ldots,6.$$

Now let

$$S_3 = \{6a+2d_1+1,\ldots,6a+2d_1+6\}$$

so that $S_3 \subseteq S$ and $\{S_1,S_2,S_3\}$ form an arithmetic progression of segments with difference $d_1$. We do not know whether $S_3 \subseteq \{1,2,\ldots,390\}$ nor do we know to which of the $2^6$ blocks $S_3$ belongs.

In the second step of the construction, we consider the segment $S_1$. The initial half

$$\{6a+1,6a+2,6a+3\}$$

of $S_1$ is a sequence of $n_0 = 3$ consecutive subsegments, each of length $q_0 = 1$. Hence this initial half of $S_1$ must contain a progression of two congruent numbers. Let $d_2$ be their difference; $d_2 = 1$ or 2. We label these numbers $S_{11}, S_{12}$ and define $S_{13} = S_{12}+d_2$. Now $S_{13} \in S_1$ and $S_{11}, S_{12}, S_{13}$ form an arithmetic progression with difference $d_2$. We do not know whether $S_{13}$ belongs to the initial half of $S_1$, nor do we know whether $S_{13}$ belongs to $K_1$ or to $K_2$.

We now carry out the corresponding construction on $S_2$ and $S_3$, obtaining altogether a set of nine numbers indexed as $S_{ij}$, where $1 \leq i,j \leq 3$. We know that

$$S_{11} \equiv S_{12} \equiv S_{21} \equiv S_{22} \pmod P \tag{3.12}$$

and that $S_{1j}, S_{2j}, S_{3j}$ appear in the corresponding positions in the segments $S_1, S_2, S_3$, for $j = 1,2,3$. Since $S_1 \equiv S_2$, it follows that

$$S_{1j} \equiv S_{2j} \pmod P, \text{ for } j = 1,2,3. \tag{3.13}$$

Also, since $S_1$ and $S_2$ are adjacent segments in the first step of the construction, we have

$$S_{2j}-S_{1j} = d_1, \quad \text{for } j = 1,2,3.$$

Similarly,

$$S_{3j}-S_{2j} = d_1, \quad \text{for } j = 1,2,3.$$

Again, at the second step of the construction, $S_{ij}$ and $S_{i,j+1}$ are adjacent, so

$$S_{13}-S_{12} = S_{12}-S_{11} = S_{23}-S_{22} = S_{22}-S_{21} = S_{33}-S_{32} = S_{32}-S_{31} = d_2.$$

Now we consider the following $k+1 = 3$ numbers of the segment S:

$$a_0 = S_{33}; \quad a_1 = S_{13}; \quad a_2 = S_{11}.$$

Since S has been partitioned into two blocks, two of these must belong to the same block. Let these be the numbers $a_r$ and $a_s$, where $0 \leq r < s \leq 2$, and

$$S_{\underbrace{1\ldots1}_{r}\underbrace{3\ldots3}_{2-r}}, \quad S_{\underbrace{1\ldots1}_{s}\underbrace{3\ldots3}_{2-s}}. \tag{3.14}$$

Now we consider the three numbers

$$c_1 = S_{\underbrace{1\ldots1}_{s}\underbrace{3\ldots3}_{2-s}}, \quad c_2 = S_{\underbrace{1\ldots1}_{r}\underbrace{2\ldots2}_{s-r}\underbrace{3\ldots3}_{2-s}}, \quad c_3 = S_{\underbrace{1\ldots1}_{r}\underbrace{3\ldots3}_{2-r}}.$$

The possibilities here are:

(i) $r = 0$, $s = 1$, so that $c_1 = S_{13}$, $c_2 = S_{23}$, $c_3 = S_{33}$;

(ii) $r = 0$, $s = 2$, so that $c_1 = S_{11}$, $c_2 = S_{22}$, $c_3 = S_{33}$;

(iii) $r = 1$, $s = 2$, so that $c_1 = S_{11}$, $c_2 = S_{12}$, $c_3 = S_{13}$.

In each set, we consider the first $\ell = 2$ numbers: we have $c_1 \equiv c_2$ by (3.12) and (3.13); we now consider the $(\ell+1)$-st number: we have $c_1 \equiv c_3$ by (3.14). Hence in each case,

$$c_1 \equiv c_2 \equiv c_3 \pmod P$$

and we have only to check that they are in arithmetic progression, i.e. that $c_3-c_2 = c_2-c_1$. But by our construction, it follows that:

in case (i), $c_3-c_2 = c_2-c_1 = d_1$;

in case (ii), $c_3-c_2 = c_2-c_1 = d_1+d_2$;

in case (iii), $c_3-c_2 = c_2-c_1 = d_2$.

Hence in each case, we have constructed an arithmetic progression of length three, the terms of which are congruent modulo P, and $n(2,3) \leq 780$.

We now proceed with the general proof. We want to show that if the segment $S = \{1,2,\ldots,q_k\}$ is partitioned into k blocks in an arbitrary manner, then at least one of these blocks contains an arithmetic progression of length at least $\ell+1$.

Let $P = K_1 \cup K_2 \cup \ldots \cup K_k$ be the arbitrary partition of the segment S. This induces, on the set of subsegments of S of length m, a partition into $k^m$ blocks, depending on how the m elements of the subsegment are partitioned between the k blocks of P.

Since

$$q_k = 2n_{k-1}q_{k-1},$$

the segment S may be regarded as a sequence of $2n_{k-1}$ disjoint consecutive subsegments, each of length $q_{k-1}$. The initial half of S contains the $n_{k-1}$ subsegments

$$\{jq_{k-1}+1,\ldots,(j+1)q_{k-1}\} \quad \text{for } j = 0,1,\ldots,n_{k-1}-1.$$

Now the partition P induces a partition P' of this set of subsegments into $k^{q_{k-1}}$ blocks. But P' in turn induces a partition of the $n_{k-1} = n(k^{q_{k-1}},\ell)$ initial numbers of the subsegments into $k^{q_{k-1}}$ blocks, and hence a partition of the segment $\{1,2,\ldots,n_{k-1}\}$ into $k^{q_{k-1}}$ blocks. By the definition of $n_{k-1}$, one block of this last partition must contain an arithmetic progression of length at least $\ell$. Corresponding to this progression, there is in one block of the partition P' an arithmetic progression of $\ell$ congruent segments of length $q_{k-1}$. We call these segments $S_1,S_2,\ldots,S_\ell$.

The difference $d_1$ between the initial numbers of $S_i$ and $S_{i-1}$, for $i = \ell,\ldots,2$, we call the difference of the progression of segments. To this progression we add the segment $S_{\ell+1} = S_\ell+d_1$. Then $S_{\ell+1} \subseteq S$ and $\{S_1,S_2,\ldots,S_\ell,S_{\ell+1}\}$ form a progression of length $\ell+1$ with difference $d_1$. We do not know whether $S_{\ell+1}$ is contained in the initial half of the segment S, nor to which block of the partition P' the segment $S_{\ell+1}$ belongs. This completes the first step of the construction.

We proceed to the second step. Consider $S_1$, a segment of length $q_{k-1}$. The procedure that we have just carried out on the segment S we now apply to the segment $S_1$. Since $q_{k-1} = 2n_{k-2}q_{k-2}$, we may regard the initial half of $S_1$ as a sequence of

$n_{k-2}$ consecutive disjoint subsegments each of length $q_{k-2}$. The partition P induces on this set of subsegments a partition into $k^{q_{k-2}}$ blocks. Since $n_{k-2} = n(k^{q_{k-2}},\ell)$, the initial half of $S_1$ must contain a progression of $\ell$ congruent subsegments $S_{11},S_{12},\ldots,S_{1\ell}$, each of length $q_{k-2}$. Let $d_2$ be the difference of this progression, i.e. the difference between initial numbers of $S_{1i}$ and $S_{1,i-1}$, for $i = 2,\ldots,\ell$. To this progression we add the segment $S_{1,\ell+1} = S_{1\ell}+d_2$. Then $S_{1,\ell+1} \cong S_1$ and $\{S_{11},\ldots,S_{1,\ell},S_{1,\ell+1}\}$ form a progression of length $\ell+1$ with difference $d_2$. We do not know whether $S_{1,\ell+1}$ is contained in the initial half of the segment $S_1$, nor do we know to which block of the partition the segment $S_{1,\ell+1}$ belongs.

This second step of the construction has been carried out for $S_1$ only. We now repeat it congruently on all the other segments $S_2,\ldots,S_\ell,S_{\ell+1}$, so that we obtain altogether a set of $(\ell+1)^2$ segments, indexed as $S_{ij}$, $1 \leq i,j \leq \ell+1$, each of length $q_{k-2}$. By the definition of the construction, we see that

$$S_{gh} \equiv S_{ij}, \quad \text{whenever } 1 \leq g,h,i,j \leq \ell. \tag{3.15}$$

We repeat this process altogether k times. After step k, the results of the construction are $(\ell+1)^k$ segments, each of length $q_{k-k} = q_0 = 1$, i.e. numbers in the segment S, but now indexed as

$$S_{i_1 i_2 \ldots i_k}, \quad \text{for } 1 \leq i_1,i_2,\ldots,i_k \leq \ell+1.$$

We have for shorter segments a generalisation of (3.15):

for $1 \leq s \leq k$, $1 \leq i_1,i_2,\ldots,i_s,j_1,j_2,\ldots,j_s \leq \ell$,

we have

$$S_{i_1 i_2 \ldots i_s} \equiv S_{j_1 j_2 \ldots j_s}. \tag{3.16}$$

We note two important properties of the construction. Firstly, if $1 \leq s < k$ and if $1 \leq i_{s+1},i_{s+2},\ldots,i_k \leq \ell+1$, then the number $S_{i_1\ldots i_s i_{s+1}\ldots i_k}$ appears in the same position in the segment $S_{i_1\ldots i_s}$ as the number $S_{j_1\ldots j_s i_{s+1}\ldots i_k}$ does in the segment $S_{j_1\ldots j_s}$. Since these two segments are congruent by (3.16), we see that

$$S_{i_1\ldots i_s i_{s+1}\ldots i_k} \equiv S_{j_1\ldots j_s i_{s+1}\ldots i_k} \tag{3.17}$$

whenever

$$1 \leq i_1,\ldots,i_s,j_1,\ldots,j_s \leq \ell$$

and

$$1 \leq i_{s+1},\ldots,i_{s+k} \leq \ell+1.$$

Secondly, if $1 \leq s \leq k$ and $j_s = i_s+1$, then $S_{i_1\ldots i_{s-1}i_s}$ and $S_{i_1\ldots i_{s-1}j_s}$ are adjacent segments in step s of the construction. Since for $1 \leq i_{s+1},\ldots,i_k \leq \ell+1$, the numbers

$$S_{i_1\ldots i_{s-1}i_s i_{s+1}\ldots i_k} \quad \text{and} \quad S_{i_1\ldots i_{s-1}j_s i_{s+1}\ldots i_k}$$

appear in the same position in two adjacent segments, we have

$$S_{i_1\ldots i_{s-1}j_s i_{s+1}\ldots i_k} - S_{i_1\ldots i_{s-1}i_s i_{s+1}\ldots i_k} = d_s, \tag{3.18}$$

Now we consider the following k+1 numbers of the segment S:

$$a_0 = S_{\ell+1,\ell+1,\ell+1,\ldots,\ell+1,\ell+1},$$
$$a_1 = S_{1,\ell+1,\ell+1,\ldots,\ell+1,\ell+1},$$
$$a_2 = S_{1,1,\ell+1,\ldots,\ell+1,\ell+1},$$
$$\ldots\ldots\ldots\ldots,$$
$$a_{k-1} = S_{1,1,1,\ldots,1,\ell+1},$$
$$a_k = S_{1,1,1,\ldots,1,1}.$$

Since S has been partitioned into k blocks, and we have defined $a_i$ for k+1 values of i, two of the $a_i$'s must be congruent modulo P. Let these be denoted by $a_r$ and $a_s$, where $r < s$, so that

$$a_r = S_{\underbrace{1,\ldots,1}_{r},\underbrace{\ell+1,\ldots,\ell+1}_{k-r}} \equiv S_{\underbrace{1,\ldots,1}_{s},\underbrace{\ell+1,\ldots,\ell+1}_{k-s}} = a_s. \tag{3.19}$$

Finally, we consider the $\ell+1$ numbers

$$c_i = S_{\underbrace{1,\ldots,1}_{r},\underbrace{i,\ldots,i}_{s-r},\underbrace{\ell+1,\ldots,\ell+1}_{k-s}} \tag{3.20}$$

where $1 \leq i \leq \ell+1$. These are the numbers which we claim satisfy the requirements of the theorem.

Firstly, the first $\ell$ numbers of the set (3.20) (namely, $c_1,\ldots,c_\ell$) are congruent by (3.17). But the first and last numbers of (3.20) (namely, $c_1$ and $c_{\ell+1}$) are congruent by (3.19). Hence

$$c_1 \equiv c_2 \equiv \ldots \equiv c_\ell \equiv c_{\ell+1} \pmod{P}$$

and to complete the proof of the theorem we have only to prove that they are in arithmetic progression.

Consider $1 \leq i \leq \ell$, and let $j = i+1$. For $0 \leq m \leq s-r$, define

$$c_{i,m} = S_{\underbrace{1,\ldots,1}_{r},\underbrace{j,\ldots,j}_{m},\underbrace{i,\ldots,i}_{s-r-m},\underbrace{\ell+1,\ldots,\ell+1}_{k-s}}$$

so that

$$c_{i,0} = c_i \quad\text{and}\quad c_{i,s-r} = c_j.$$

Now

$$c_j - c_i = \sum_{m=1}^{s-r} (c_{i,m} - c_{i,m-1}) \tag{3.21}$$

and we look at the differences $c_{i,m} - c_{i,m-1}$. By (3.18),

$$c_{i,m}-c_{i,m-1} = S_{\underbrace{1,\ldots,1}_{r},\underbrace{j,\ldots,j}_{m},\underbrace{i,\ldots,i}_{s-r-m},\underbrace{\ell+1,\ldots,\ell+1}_{k-s}}$$

$$- S_{\underbrace{1,\ldots,1}_{r},\underbrace{j,\ldots,j}_{m-1},\underbrace{i,\ldots,i}_{s-r-m+1},\underbrace{\ell+1,\ldots,\ell+1}_{k-s}}$$

$$= d_{r+m}.$$

So by (3.21) the difference

$$c_j-c_i = c_{i+1}-c_i = d_{r+1}+d_{r+2}+\ldots+d_s.$$

But this is independent of the value of i, for $1 \leq i \leq \ell$. Hence $c_1,c_2,\ldots,c_\ell,c_{\ell+1}$ are in arithmetic progression and the theorem is proved.

## 3.4 SOME PRELIMINARY RESULTS.

We still need several short lemmas before we can prove our main theorems.

LEMMA 3.3. *Let $S(x_1,\ldots,x_n) = 0$ be a homogeneous k-regular system of conditions, M a natural number and P a partition of the natural numbers into k blocks. Then there exist natural numbers $x_{01},x_{02},\ldots,x_{0n},d$, such that*

$$S(x_{01},\ldots,x_{0n}) = 0$$

*and*

$$x_{0i}+jd \equiv x_{01} \pmod P \quad \text{for } i = 1,\ldots,n,$$

*j an integer, $|j| \leq M$.*

PROOF. By theorem 3.1, there exists a natural number N such that the system of conditions

$$S(x_1,\ldots,x_n) = 0 \tag{3.22}$$

and

$$x_1,\ldots,x_n \leq N \tag{3.23}$$

is k-regular.

Let

$$P' = P[i\xi,\ 1 \leq i \leq N]. \tag{3.24}$$

By theorem 3.2, there exist $a',d' > 0$ such that

$$a'+j'd' \equiv a' \pmod{P'} \quad \text{for } j' \text{ an integer, } |j'| \leq MN^{n-1}. \tag{3.25}$$

In other words, there exists an arithmetic progression of length $(2MN^{n-1}+1)$ of natural numbers congruent modulo P', and we have called the middle term of the progression a'.

Let

$$P'' = P[a'\xi]. \tag{3.26}$$

P" has k blocks. As (3.22) and (3.23) are k-regular, there exist natural numbers $x''_1, \ldots, x''_n$ such that

$$S(x''_1, \ldots, x''_n) = 0, \tag{3.27}$$

$$x''_1, \ldots, x''_n \leq N, \tag{3.28}$$

and

$$x''_1 \equiv x''_2 \equiv \ldots \equiv x''_n \pmod{P''}. \tag{3.29}$$

We claim that the numbers

$$x_{0i} = a'x''_i \quad \text{for } i = 1,2,\ldots,n, \tag{3.30}$$

and

$$d = d'x''_1 x''_2 \ldots x''_n \tag{3.31}$$

satisfy the statement of the theorem.

Firstly, by (3.27), (3.30) and the homogeneity of S,

$$S(x_{01}, \ldots, x_{0n}) = 0. \tag{3.32}$$

Secondly, by (3.26) and (3.30), (3.29) is identical with

$$x_{01} \equiv x_{02} \equiv \ldots \equiv x_{0n} \pmod{P}. \tag{3.33}$$

Finally, by (3.24), (3.25) implies that

$$i(a'+j'd') \equiv ia' \pmod{P} \tag{3.34}$$

for $1 \leq i \leq N$, $j'$ an integer, $|j'| \leq MN^{n-1}$.

By (3.28), (3.30) and (3.31), we have

$$\begin{aligned} x_{0m}+jd &= a'x''_m + jd'x''_1 x''_2 \ldots x''_n \\ &= x''_m(a'+jx''_1 \ldots x''_{m-1} x''_{m+1} \ldots x''_n d') \\ &= i(a'+j'd') \end{aligned} \tag{3.35}$$

for $1 \leq m \leq N$, $j$ an integer, $|j| \leq M$, $1 \leq i = x''_m \leq N$ and

$$|j'| = |jx''_1 \ldots x''_{m-1} x''_{m+1} \ldots x''_n| \leq MN^{n-1}.$$

By (3.34) and (3.35),

$$x_{0m}+jd \equiv x_{0m} \pmod{P}, \text{ for } 1 \leq m \leq N,\ j \text{ an integer, } |j| \leq M. \tag{3.36}$$

The result now follows from (3.32), (3.33) and (3.36).

LEMMA 3.4. *Suppose we are given integers $b_{ij}, b_i$, for $i = 1,\ldots,m$, $j = 1,\ldots,n$, associated with the system of simultaneous linear congruences*

$$\left(\sum_{j=1}^{n} b_{ij}x_j\right) + rsb_i \equiv 0 \pmod{st} \tag{3.37}$$

*for $i = 1,\ldots,m$. Suppose that there exist infinitely many triples $(r,s,t)$ of in-integers, with $t$ arbitrarily large, such that $(r,t) = 1$, and for which the system (3.37) has at least one solution in integers, $x_1, x_2, \ldots, x_n$.*

*Then the system of simultaneous linear equations*

$$\left(\sum_{j=1}^{n} b_{ij}x_j\right) + b_i = 0, \tag{3.38}$$

*for $i = 1,\ldots,m$, also has solutions.*

PROOF. (3.38) has a solution if and only if

$$\sum_{i=1}^{m} c_i b_{ij} = 0, \quad \text{for } j = 1,\ldots,n, \tag{3.39}$$

implies that

$$\sum_{i=1}^{m} c_i b_i = 0 \tag{3.40}$$

where $c_1, c_2, \ldots, c_m$ are integers.

Suppose that (3.39) is true. Then by (3.37),

$$\sum_{i=1}^{m} c_i \left(\sum_{j=1}^{m} b_{ij}x_j\right) + \sum_{i=1}^{m} (c_i rsb_i) \equiv 0 \pmod{st},$$

which by (3.39) implies that

$$0 + rs \sum_{i=1}^{m} c_i b_i \equiv 0 \pmod{st}.$$

Hence $r\sum_{i=1}^{m} c_i b_i \equiv 0 \pmod{t}$, and since $(r,t) = 1$, we have

$$\sum_{i=1}^{m} c_i b_i \equiv 0 \pmod{t}. \tag{3.41}$$

Since (3.41) is true for arbitrarily large t, (3.40) follows, which proves the lemma.

LEMMA 3.5. *Let $S(x_1,\ldots,x_n) = 0$ be a homogeneous k-regular system of conditions, and let a be an integer. Then*

$$S(x_1-a,\ldots,x_n-a) = 0 \tag{3.42}$$

*is also k-regular.*

PROOF. Let b be an integer, $b > |a|$. (3.43)

Let P be a partition with k blocks, and define

$$P' = P[a+b\xi]. \tag{3.44}$$

P' is well defined, because by (3.43), $a+b\xi$ is a natural number. P' has k blocks. So there exist natural numbers $x'_1,\ldots,x'_n$ such that

$$S(x'_1,\ldots,x'_n) = 0 \tag{3.45}$$

and

$$x'_1 \equiv x'_2 \equiv \ldots \equiv x'_n \pmod{P'}. \tag{3.46}$$

By (3.44), (3.46) is equivalent to

$$a+bx'_1 \equiv a+bx'_2 \equiv \ldots \equiv a+bx'_n \pmod{P}. \tag{3.47}$$

Now let $x_j = a+bx'_j$, $j = 1,\ldots,n$. Then by (3.45), (3.47) and the homogeneity of S, we have that

$$S(x_1-a,\ldots,x_n-a) = 0,$$

where $x_1 \equiv x_2 \equiv \ldots \equiv x_n \pmod{P}$. Since P is arbitrary, this proves the k-regularity of (3.42).

LEMMA 3.6. *Let $S(x_1,\ldots,x_n) = 0$ be a k-regular system of conditions, and let a be a negative integer. Then*

$$S(x_1+a,\ldots,x_n+a) = 0 \tag{3.48}$$

*is also k-regular.*

PROOF. Let P be a partition with k blocks and define

$$P'' = P[\xi-a]. \tag{3.49}$$

P" is well defined, because $a < 0$, and P" has k blocks. Hence there exist natural numbers $x''_1,\ldots,x''_n$ such that

$$S(x''_1,\ldots,x''_n) = 0 \tag{3.50}$$

and

$$x''_1 \equiv x''_2 \equiv \ldots \equiv x''_n \pmod{P''}. \tag{3.51}$$

By (3.49), (3.51) is equivalent to

$$x''_1-a \equiv x''_2-a \equiv \ldots \equiv x''_n-a \pmod{P}. \tag{3.52}$$

Now let $x_j = x''_j-a$, $j = 1,\ldots,n$. Then by (3.50) and (3.52) we have that

$$S(x_1+a,\ldots,x_n+a) = 0,$$

where $x_1 \equiv x_2 \equiv \ldots \equiv x_n \pmod{P}$. Since P is arbitrary, this proves the k-regularity of (3.48).

Note that in lemma 3.5 we required $S(x_1,\ldots,x_n) = 0$ to be a homogeneous system, and in lemma 3.6 we required the integer a to be negative. If $S(x_1,\ldots,x_n) = 0$ is k-regular but not homogeneous and $a > 0$, then $S(x_1+a,\ldots,x_n+a) = 0$ need not be k-regular, as the following example shows:

Let S be just the one regular equation $x_1 = 1$, and let $a = 1$.

## 3.5 THE MAIN THEOREMS.

THEOREM 3.7. *The system of equations*

$$\sum_{j=1}^{n} a_{ij}x_j = 0 \tag{3.53}$$

*for $i = 1,\ldots,m$, and $a_{ij}$ rational, is regular if and only if its coefficient matrix $A = [a_{ij}]$ has the following property:*

*The set of integers $\{1,\ldots,n\}$ can be partitioned into $q$ blocks $G_1, G_2, \ldots, G_q$ such that:*

(i) $\sum_{j\in G_1} a_{ij} = 0$ for $i = 1,\ldots,m$;

(ii) $\sum_{j\in G_2} a_{ij} + \sum_{j\in G_1} a_{ij}s_j = 0$ for $i = 1,\ldots,m$, for some rational $s_j$;

(iii) $\sum_{j\in G_3} a_{ij} + \sum_{j\in G_1\cup G_2} a_{ij}t_j = 0$ for $i = 1,\ldots,m$, for some rational $t_j$;

........

(q) $\sum_{j\in G_q} a_{ij} + \sum_{j\in G_1\cup\ldots\cup G_{q-1}} a_{ij}z_j = 0$ for $i = 1,\ldots,m$, for some rational $z_j$.

PROOF. (I) Let $A = [a_{ij}]$ have the required property. We show that (3.53) is regular. For $q = 1$, the statement is true, for in this case $x_1 = x_2 = \ldots = x_n = 1$ is a solution of (3.53).

We proceed by induction on $q$, so we now assume that the statement is proved for all smaller values of $q$ and show that (3.53) is k-regular for every $k \geq 1$.

For $k = 1$, this means that we have to show that (3.53) has a solution $x_1,\ldots, x_n$ in positive integers. We see as follows that this is true: we may assume without loss of generality that $G_q = \{n-r+1,\ldots,n\}$. By the induction hypothesis, the system of equations

$$\sum_{j=1}^{n-r} a_{ij}x_j = 0 \quad \text{for } i = 1,\ldots,m \tag{3.54}$$

is regular and hence satisfied by positive integers such that $x_j = g_j > 0$, $j = 1,\ldots, n-r$. By the properties we have assumed concerning the coefficient matrix A, there exist rational numbers $s_1, s_2,\ldots,s_{n-r}$ such that

$$\sum_{j=1}^{n-r} a_{ij}s_j + \sum_{j=n-r+1}^{n} a_{ij} = 0, \quad \text{for } i = 1,\ldots,m. \tag{3.55}$$

Hence for every X and Y,

$$\sum_{j=1}^{n-r} a_{ij}(Xg_j+Ys_j) + \sum_{j=n-r+1}^{n} a_{ij}Y = 0, \quad \text{for } i = 1,\ldots,m. \tag{3.56}$$

If we choose for Y the common denominator H of all the $s_j$'s, and for X a sufficiently large integer, then (3.56) shows that (3.53) has a solution in positive integers, i.e. (3.53) is 1-regular.

The first half of the theorem will be proved if we can show the following: the regularity of (3.54) and the (k-1)-regularity of (3.53) (for $k \geq 2$) imply the k-regularity of (3.53). We now assume that (3.53) is (k-1)-regular and (3.54) is regular.

Because (3.53) is (k-1)-regular, by theorem 3.1 there exists a constant N such that the system of conditions

$$\sum_{j=1}^{n} a_{ij}x_j = 0 \quad \text{for } i = 1,\ldots,m,$$

and

$$x_1,x_2,\ldots,x_n \leq N$$

is (k-1)-regular. Let P be a partition with k blocks.

Suppose that (3.53) is not k-regular. In particular, suppose that there do not exist natural numbers $x_1,x_2,\ldots,x_n$, which are congruent modulo P and satisfy (3.53). As (3.54) is regular, we can apply lemma 3.3 to the system of equations (3.54) and the partition P. As M in the lemma, we choose

$$M = HN \max(s_1,s_2,\ldots,s_{n-r}). \tag{3.57}$$

(H means, as above, the common denominator of the $s_j$'s.) Then by lemma 3.3, there exist natural numbers $x_{01},x_{02},\ldots,x_{0,n-r},d$ such that

$$\sum_{j=1}^{n-r} a_{ij}x_{0j} = 0 \quad \text{for } i = 1,\ldots,m, \tag{3.58}$$

and

$$x_{0j}+ud \equiv x_{01} \pmod{P} \quad \text{for } j = 1,\ldots,n-r, \tag{3.59}$$

u an integer, $|u| \leq M$.

We let $K_0$ denote the block of P to which the number $x_{01}$ belongs; consequently the other numbers of (3.59) belong to $K_0$ also.

By (3.58) and (3.55), we have

$$\sum_{j=1}^{n-r} a_{ij}(x_{0j}+vHs_jd) + \sum_{j=n-r+1}^{n} a_{ij}vHd = 0, \tag{3.60}$$

for $i = 1,\ldots,m$, $v = 1,\ldots,N$.

By (3.57) and the definition of H, it follows that $vHs_j$ is always an integer, for $v = 1,\ldots,N$ and $j = 1,\ldots,n-r$, and that $|vHs_j| \leq M$. Hence by (3.59),

$$x_{0j}+vHs_jd \in K_0 \quad \text{for } v = 1,\ldots,N, \text{ and } j = 1,\ldots,n-r. \tag{3.61}$$

By (3.60), (3.61) and our earlier supposition that (3.53) is not k-regular, we see that

$$vHd \notin K_0, \quad \text{for } v = 1,\ldots,N. \tag{3.62}$$

Let

$$P' = P[\xi Hd]. \tag{3.63}$$

By (3.62), the set of integers $\{1,2,\ldots,N\}$ are partitioned by P' into at most (k-1) blocks. By the definition of N, it follows that there exist natural numbers $x'_1,x'_2,\ldots,x'_n$ such that

$$\sum_{j=1}^{n} a_{ij}x'_j = 0, \quad \text{for } i = 1,\ldots,m, \tag{3.64}$$

$$x'_1,x'_2,\ldots,x'_n \leq N,$$

and

$$x'_1 \equiv x'_2 \equiv \ldots \equiv x'_n \pmod{P'}. \tag{3.65}$$

By (3.64) we have

$$\sum_{j=1}^{n} a_{ij}x'_j Hd = 0, \quad \text{for } i = 1,\ldots,m. \tag{3.66}$$

By (3.63), (3.65) means that

$$x'_1 Hd \equiv x'_2 Hd \equiv \ldots \equiv x'_n Hd \pmod{P}. \tag{3.67}$$

But now (3.66) and (3.67) contradict our supposition that (3.53) is not k-regular. As P was an arbitrary partition into k blocks, this proves that (3.53) is k-regular. The first half of the theorem follows.

(II) Now suppose that (3.53) is regular. We have to partition $\{1,\ldots,n\}$ so that conditions (i) - (q) are satisfied.

Let p be an arbitrary prime number. Then any natural number x is uniquely expressible in the form

$$x = u_p p^{v_p}, \quad \text{where } u_p,\ v_p \text{ are integers and } (u_p,p) = 1.$$

This gives us a well defined function $u_p(x)$ on the natural numbers.

Now let $P = P_p[u_p(\xi)]$ where, as in section 3.1, $P_p$ is the partition of the natural numbers into residue classes modulo p. As (3.53) is regular, there exist natural numbers $x_1, x_2, \ldots, x_n$ satisfying (3.53) and

$$x_1 \equiv x_2 \equiv \ldots \equiv x_n \pmod{P}. \tag{3.68}$$

(3.68) implies the existence of $u_0$, such that $(u_0,p) = 1$ and

$$x_j = (u_0+k_jp)p^{v_j} \tag{3.69}$$

where $k_j$, $v_j$ are integers, for $j = 1,\ldots,n$. We may assume

$$v_1 \leq v_2 \leq \ldots \leq v_n$$

or, more precisely,

$$0 \leq v_1 = v_2 = \ldots = v_{c_1} < v_{c_1+1} = v_{c_1+2} = \ldots = v_{c_2} < v_{c_2+1} = \ldots$$
$$\ldots = v_{c_3} < \ldots = v_{c_q},$$

where $c_q = n$.

For each of the numbers $q, c_1, c_2, \ldots, c_q$, a value must be chosen from the set $\{1,2,\ldots,n\}$. Hence for infinitely many primes p these numbers must agree. From an infinite set of such primes, we choose one particular value, p, and proceed as follows:

By (3.53) and (3.69),

$$\sum_{j=1}^{n} a_{ij}(u_0+k_jp)p^{v_j} = 0, \quad \text{for } i = 1,\ldots,m. \tag{3.70}$$

Consider (3.70) modulo $p^{v_{c_1}+1}$. Then

$$\sum_{j=1}^{n} a_{ij} u_0 p^{v_{c_1}} \equiv 0 \pmod{p^{v_{c_1}+1}} \quad \text{for } i = 1,\dots,m,$$

and hence

$$u_0 \sum_{j=1}^{c_1} a_{ij} \equiv 0 \pmod{p} \quad \text{for } i = 1,\dots,m. \tag{3.71}$$

Since $(p,u_0) = 1$, (3.71) implies that

$$\sum_{j=1}^{c_1} a_{ij} \equiv 0 \pmod{p}.$$

Since p can take infinitely many prime values, we have

$$\sum_{j=1}^{c_1} a_{ij} = 0, \quad \text{for } i = 1,\dots,m. \tag{3.72}$$

Next, consider (3.70) modulo $p^{v_{c_2}+1}$. We see that

$$\sum_{j=1}^{c} a_{ij} x_j + \sum_{j=c_1+1}^{c_2} a_{ij} u_0 p^{v_{c_2}} \equiv 0 \pmod{p^{v_{c_2}+1}}, \quad \text{for } i = 1,\dots,m. \tag{3.73}$$

Now apply lemma 3.4 to (3.73), with $r = u_0$, $s = p^{v_{c_2}}$, $t = p$. The hypotheses of the lemma are obviously fulfilled. This implies that the system of equations

$$\sum_{j=1}^{c_1} a_{ij} y_j + \sum_{j=c_1+1}^{c_2} a_{ij} = 0, \quad \text{for } i = 1,\dots,m, \tag{3.74}$$

has a solution $y_1,\dots,y_{c_1}$.

This step of the argument can be repeated for $p^{v_{c_3}+1},\dots,p^{v_{c_q}+1}$, so that

$$\sum_{j=1}^{c_{h-1}} a_{ij} x_j + \sum_{j=c_{h-1}+1}^{c_h} a_{ij} u_0 p^{v_{c_h}} \equiv 0 \pmod{p^{v_{c_h}+1}}, \quad \text{for } i = 1,\dots,m,\ h = 3,\dots,q, \tag{3.75}$$

and then by lemma 3.4 applied to (3.75), with $r = u_0$, $s = p^{v_{c_h}}$, $t = p$, the system of equations

$$\sum_{j=1}^{c_{h-1}} a_{ij} y_j + \sum_{j=c_{h-1}+1}^{c_h} a_{ij} = 0, \quad \text{for } i = 1,\dots,m,\ h = 3,\dots,q, \tag{3.76}$$

has a solution $y_1,\dots,y_{c_{h-1}}$.

(3.72), (3.74) and (3.76) show that A can be partitioned as stated in conditions (i) - (q). In the same block $G_h$, put only the indices of those columns corresponding to the $x_j$ divisible by exactly the same power of p.

This completes the proof of theorem 3.7. The only other result we need concerning regularity is the corresponding criterion for non-homogeneous systems.

THEOREM 3.8. *The system of equations*

$$\sum_{j=1}^{n} a_{ij}x_j = a_i \quad \text{for } i = 1,\dots,m, \tag{3.77}$$

*and $a_{ij}$, $a_i$ rational, is regular if and only if:*

*(i) the corresponding homogeneous system of equations*

$$\sum_{j=1}^{n} a_{ij}x_j = 0 \quad \text{for } i = 1,\dots,m, \tag{3.78}$$

*is regular, and at least one of the following two conditions is satisfied:*

*(ii) there exists a natural number $a$ such that*

$$\sum_{j=1}^{n} a_{ij}a = a_i, \quad \text{for } i = 1,\dots,m;$$

*(iii) there exists an integer $a \leq 0$ such that*

$$\sum_{j=1}^{n} a_{ij}a = a_i, \quad \text{for } i = 1,\dots,m.$$

PROOF. (I) First we show that the truth of (i) and either (ii) or (iii) implies the regularity of (3.77).

For the case where (i) and (ii) are true, the regularity of (3.77) is obvious, since a is itself a natural number.

For the case where (i) and (iii) are true, we rewrite (3.77) as

$$\sum_{j=1}^{n} a_{ij}(x_j - a) = 0, \quad \text{for } i = 1,\dots,m. \tag{3.79}$$

The regularity of (3.79) follows immediately from lemma 3.5.

(II) Now we suppose that (3.77) is regular and show that (i) and either (ii) or (iii) are true.

First consider the case m = 1. Suppose that the equation

$$\sum_{j=1}^{n} c_j x_j = c, \quad \text{for } c_j, c \text{ rational integers}, \tag{3.80}$$

is regular. Let h be a natural number such that

$$h \,\Big|\, \sum_{j=1}^{n} c_j. \tag{3.81}$$

Let $P = P_h$. Since (3.80) is regular, there exist natural numbers $x_{01}, x_{02}, \dots, x_{0n}$ such that

$$\sum_{j=1}^{n} c_j x_{0j} = c,$$

and

$$x_{01} \equiv x_{02} \equiv \dots \equiv x_{0n} \pmod{P}.$$

This implies the existence of a natural number $x_0$ such that

$$x_{0j} \equiv x_0 \pmod{h} \quad \text{for } j = 1,\dots,m,$$

and

$$\sum_{j=1}^{n} c_j x_0 \equiv c \pmod{h}. \tag{3.82}$$

By (3.81) and (3.82), we see that $h|c$. But by (3.81), every natural number h which divides $\sum_{j=1}^{n} c_j$ also divides c. Hence either

$$\sum_{j=1}^{n} c_j = c = 0 \quad \text{or} \quad \sum_{j=1}^{n} c_j \neq 0, \quad \Big(\sum_{j=1}^{n} c_j\Big)\Big| c. \tag{3.83}$$

Now let m be arbitrary. (3.77) is a regular system of equations, so any linear equation which is a rational linear combination of equations of (3.77) is also regular. For any such equation, (3.83) is true. Hence, for every i such that $\sum_{j=1}^{n} a_{ij} \neq 0$, the integers

$$\frac{a_i}{\sum_{j=1}^{n} a_{ij}}$$

have the same value. For suppose not. Then without loss of generality, we may assume that

$$\sum_{j=1}^{n} a_{1j} \neq 0, \quad \sum_{j=1}^{n} a_{2j} \neq 0,$$

and

$$\frac{a_1}{\sum_{j=1}^{n} a_{1j}} \neq \frac{a_2}{\sum_{j=1}^{n} a_{2j}}.$$

By the regularity of (3.77), we see that

$$\Big(2a_2 - \sum_{j=1}^{n} a_{2j}\Big) \sum_{j=1}^{n} a_{1j}x_j - \Big(2a_1 - \sum_{j=1}^{n} a_{1j}\Big) \sum_{j=1}^{n} a_{2j}x_j$$

$$= \Big(2a_2 - \sum_{j=1}^{n} a_{2j}\Big)a_1 - \Big(2a_1 - \sum_{j=1}^{n} a_{1j}\Big)a_2$$

is a regular equation. But

$$\frac{\Big(2a_2 - \sum_{j=1}^{n} a_{2j}\Big)a_1 - \Big(2a_1 - \sum_{j=1}^{n} a_{1j}\Big)a_2}{\Big(2a_2 - \sum_{j=1}^{n} a_{2j}\Big) \sum_{j=1}^{n} a_{1j} - \Big(2a_1 - \sum_{j=1}^{n} a_{1j}\Big) \sum_{j=1}^{n} a_{2j}} = \frac{1}{2},$$

which contradicts (3.83), for by (3.83), the ratio should be an integer.

We have now proved the existence of an integer a such that

$$\sum_{j=1}^{n} a_{ij}a = a_i, \quad \text{for } i = 1,\ldots,m. \tag{3.84}$$

If $a > 0$, then condition (ii) is satisfied and the regularity of (3.78) follows by lemma 3.5. If $a = 0$, then by (3.84), $a_i = 0$ for $i = 1,\ldots,m$. Hence (3.78) is regular and (iii) is satisfied. Finally, if $a < 0$, then condition (iii) is satisfied and the regularity of (3.78) follows by lemma 3.6. This completes the proof of the theorem.

3.6 REFERENCES AND RELATED TOPICS. Most of the ideas discussed in this chapter are due to Rado (1933, 1945, 1971). These papers also contain many related and more general results of the same type. The special case of the arithmetic progression (theorem 3.2) is due to van der Waerden (1927); the proof given here is that of M.A. Lukomskaya, quoted in Khinchin (1952). For a discussion of other proofs of this theorem, see Moser (1960). Very much better bounds are known for $n(k,\ell)$ than those which follow directly from this proof. For a discussion of lower bounds of $n(k,\ell)$, see Berlekamp (1968); for upper bounds, see Abbott and Liu (1972).

## CHAPTER IV. SCHUR FUNCTIONS FOR MORE GENERAL SYSTEMS

In this chapter, we consider various generalisations of Schur's problem.

### 4.1 A SYSTEM OF SIMULTANEOUS LINEAR EQUATIONS.

DEFINITION 4.1. Consider the following system (S), which consists of $\binom{k-1}{2}$ equations in $\binom{k}{2}$ unknowns:

$$x_{i,j}+x_{j,j+1} = x_{i,j+1}, \quad \text{for } 1 \leq i < j \leq k-1. \qquad (S)$$

We call a set of positive integers *(S)-free* if and only if it contains no solution to the system (S). By theorem 3.7 the system (S) is regular, so we can define the function $f_k(n)$ as follows: $f_k(n)$ is the largest positive integer such that the set of integers $\{1,2,\ldots,f_k(n)\}$ can be partitioned into n (S)-free classes.

Similarly we define $g_k(m)$, generalising $g(m)$; $g_k(m)$ is the smallest number of (S)-free sets into which the set of integers $\{1,2,\ldots,m\}$ can be partitioned. Equivalently, if $f_k(n-1) < m \leq f_k(n)$, then $g_k(m) = n$.

We note that if k = 3, these functions reduce to our previous f(n) and g(m).

THEOREM 4.2. *For all positive integers n and m,*

$$f_k(n+m) \geq (2f_k(m)+1)f_k(n) + f_k(m)$$

PROOF. We partition the set $\{1,2,\ldots,(2f_k(m)+1)f_k(n)+f_k(m)\}$ into two subsets

$$A = \{a(2f_k(m)+1)+b \mid a = 0,1,\ldots,f_k(n);\ b = 1,2,\ldots,f_k(m)\},$$

and

$$B = \{a(2f_k(m)+1)-b \mid a = 1,2,\ldots,f_k(n);\ b = 0,1,\ldots,f_k(m)\}.$$

Next we partition the set $\{1,2,\ldots,f_k(m)\}$ into m (S)-free subsets $C_1,C_2,\ldots,C_m$, and the set $\{1,2,\ldots,f_k(n)\}$ into n (S)-free subsets $D_1,D_2,\ldots,D_n$.

Finally, we partition A into m subsets $A_1,A_2,\ldots,A_m$ by assigning $a(2f_k(m)+1)+b$ to the set $A_i$ if $b \in C_i$, and B into n subsets $B_1,B_2,\ldots,B_n$ by assigning $a(2f_k(m)+1)-b$ to the set $B_i$ if $a \in D_i$.

Obviously we have defined a partition of the set $\{1,2,\ldots,(2f_k(m)+1)f_k(n)+f_k(m)\}$ into (m+n) subsets and to complete the proof of the theorem we have only to show that each of these subsets is (S)-free.

First consider the sets $A_i$, $i = 1,\ldots,m$. Suppose that one of these sets, say $A_h$, contains a solution to (S). It must be of the form

$$a_{i,j}\big(2f_k(m)+1\big)+b_{i,j}+a_{j,j+1}\big(2f_k(m)+1\big)+b_{j,j+1}$$
$$=a_{i,j+1}\big(2f_k(m)+1\big)+b_{i,j+1} \qquad (4.1)$$

where $1 \leq i < j \leq k-1$, $1 \leq b_{s,t} \leq f_k(m)$ and $b_{i,j}$, $b_{j,j+1}$, $b_{i,j+1} \in C_h$. From (4.1), we have

$$(a_{i,j}+a_{j,j+1}-a_{i,j+1})\big(2f_k(m)+1\big) = b_{i,j+1}-b_{i,j}-b_{j,j+1}$$

and hence

$$b_{i,j+1}-b_{i,j}-b_{j,j+1} \equiv 0 \pmod{2f_k(m)+1}.$$

But since $1 \leq b_{s,t} \leq f_k(m)$, for all possible s,t, we have

$$b_{i,j}+b_{j,j+1} = b_{i,j+1}$$

for all $1 \leq i < j \leq k-1$, which contradicts the fact that $C_h$ is (S)-free.

Next consider the sets $B_i$, $i = 1,\ldots,n$. Suppose that one of these sets, say $B_h$, contains a solution to (S). It must be of the form

$$a_{i,j}\big(2f_k(m)+1\big)-b_{i,j}+a_{j,j+1}\big(2f_k(m)+1\big)-b_{j,j+1}$$
$$=a_{i,j+1}\big(2f_k(m)+1\big)-b_{i,j+1} \qquad (4.2)$$

where $1 \leq i < j \leq k-1$, $0 \leq b_{s,t} \leq f_k(m)$ and $a_{i,j}$, $a_{j,j+1}$, $a_{i,j+1} \in D_h$. As in the first case, we find that

$$b_{i,j}+b_{j,j+1} = b_{i,j+1}$$

for all $1 \leq i < j \leq k-1$, and hence from (4.2) we have

$$a_{i,j}+a_{j,j+1} = a_{i,j+1}$$

for all $1 \leq i < j \leq k-1$, contradicting the fact that $D_h$ is (S)-free.

COROLLARY 4.3. *For all positive integers n and m,*

$$f(n+m) \geq \big(2f(m)+1\big)f(n)+f(m).$$

PROOF. Let k = 3 in theorem 4.2.

COROLLARY 4.4. *For $n \geq 4$, and for some absolute positive constant c,*

$$f(n) \geq c\ 89^{n/4}.$$

PROOF. By corollary 4.3., since f(4) = 44, we have $f(n+4) \geq 89f(n)+44$. The corollary now follows by induction, for c = 44/89.

This is an improvement over the bound given by corollary 2.6.

COROLLARY 4.5. *For $n \geq 1$ and for some constant $c_k$, dependent only on k,*

$$f_k(n) \geq c_k(2k-3)^n.$$

PROOF. We note first that $f_k(1) = k-2$. For, by the definition of (S), the values of $x_{1,2},\ldots,x_{1,k}$ must all be distinct, so the set of integers $\{1,2,\ldots,k-2\}$ is certainly (S)-free. But if we let $x_{r,s} = s-r$, for $1 \leq r < s \leq k$, then we have a solution of (S) in the set $\{1,2,\ldots,k-1\}$.

The corollary now follows by induction on n, with $c_k = \frac{k-2}{2k-3}$.

4.2 ESTIMATES OF RAMSEY NUMBERS. We apply the results of the previous section to obtain lower bounds for $R_n(k,2)$ which are effective for small values of k.

THEOREM 4.6. *Let $f_k(n)$ and the system (S) be as defined in definition 4.1. Then*

$$R_n(k,2) \geq f_k(n)+2 \tag{4.3}$$

*and for $n \geq 1$, $k \geq 2$ and for some absolute constant $c_k$, dependent only on k,*

$$R_n(k,2) \geq c_k(2k-3)^n. \tag{4.4}$$

PROOF. We partition the set of integers $\{1,2,\ldots,f_k(n)\}$ into n (S)-free classes $C_1, C_2,\ldots,C_n$. Let G be the complete graph with vertices $P_0,P_1,\ldots,P_{f_k(n)}$. We colour G in n colours by assigning colour $c_r$ to the edge $(P_i,P_j)$ whenever $|i-j| \in C_r$, for $r = 1,\ldots,n$.

Suppose that $P_{i_1},P_{i_2},\ldots,P_{i_k}$, where $i_1 > i_2 > \ldots > i_k$, are the vertices of a complete monochromatic k-gon, coloured $c_r$. This means that $i_t-i_s \in C_r$ for $1 \leq t < s \leq k$. But then

$$(i_t-i_s)+(i_s-i_{s+1}) = (i_t-i_{s+1}), \quad 1 \leq t < s \leq k-1,$$

is a solution to the system (S) in $C_r$, contradicting the definition of $C_r$. Since G has $f_k(n)+1$ vertices, this proves (4.3).

(4.4) now follows from (4.3) and corollary 4.5.

In some special cases we can improve this bound.

THEOREM 4.7. *For $n \geq 4$ and for some absolute constant c,*

$$R_n(3,3) \geq c\, 89^{n/4}.$$

PROOF. This follows from (4.3) and corollary 4.4.

THEOREM 4.8. *For $n \geq 2$ and for some absolute constant c,*

$$R_n(4,2) \geq c\, 33^{\frac{1}{2}n}.$$

PROOF. We partition the set of integers $\{1,2,\ldots,16\}$ into $C_1$ and $C_2$, where $C_1$ consists of the quadratic residues modulo 17 and $C_2$ the non-residues, i.e.

$$C_1 = \{1,2,4,8,9,13,15,16\};$$
$$C_2 = \{3,5,6,7,10,11,12,14\}.$$

This implies that $f_4(2) \geq 16$, since a routine verification shows that $C_1$ and $C_2$ are (S)-free.

The result now follows from (4.3), by applying theorem 4.2 to estimate $f_4(n+2)$.

A similar argument shows that $f_5(2) \geq 37$ and hence that $R_n(5,2) \geq c\ 75^{\frac{1}{2}n}$ for some absolute constant c. But a better estimate than this can be found. We need an additional result concerning Ramsey numbers.

LEMMA 4.9. *For integers $s,t \geq 2$, we have*

$$R_n(st-s-t+2,2) \geq \left(R_n(s,2)-1\right)\left(R_n(t,2)-1\right)+1.$$

PROOF. Let $p = R_n(s,2)-1$ and $q = R_n(t,2)-1$. Consider the complete graphs, $K_p$ and $K_q$, with sets of vertices $\{u_1,u_2,\ldots,u_p\}$ and $\{v_1,v_2,\ldots,v_q\}$ respectively. Colour the edges of $K_p$ and $K_q$ in n colours $C_1,C_2,\ldots,C_n$ in such a way that $K_p$ does not contain a complete monochromatic s-gon, nor $K_q$ a complete monochromatic t-gon.

Now let $K_{pq}$ be the complete graph with vertices $w_{ij}$, $i = 1,2,\ldots,p$, $j = 1,2,\ldots,q$, and colour the edges of $K_{pq}$ in the following way:

Let e be the edge joining $w_{ij}$ and $w_{gh}$. If $i = g$, colour e in the same colour as the edge joining $v_j$ and $v_h$ in $K_q$. If $i \neq g$, colour e in the same colour as the edge joining $u_i$ and $u_g$ in $K_p$.

Let $r = st-s-t+2$ and consider any complete r-gon contained in $K_{pq}$. We show that it is not monochromatic. For suppose we have a complete monochromatic r-gon, $K_r$, and suppose that all its edges are coloured $C_1$. Two cases arise:

(i) For at least s distinct values of i, the vertex $w_{ij}$ is contained in $K_r$. Then from the colouring scheme it follows that $K_p$ contains a complete monochromatic s-gon, which is a contradiction.

(ii) There are at most s-1 distinct values of i such that $w_{ij}$ is a vertex of $K_r$. Suppose that there are at most t-1 distinct values of j such that $w_{ij}$ is a vertex of $K_r$. Then $K_r$ has at most $(s-1)(t-1) = st-s-t+1 < r$ vertices, which is a contradiction. Hence for at least one value of i, $w_{ij}$ must be a vertex of $K_r$ for at least t values of j. Then from the colouring scheme it follows that $K_q$ contains a complete monochromatic t-gon, which again is a contradiction.

COROLLARY 4.10. $R_n(5,2) > c\ 89^{\frac{1}{2}n}$, *for some absolute constant c.*

PROOF. We apply the lemma in the case $s = t = 3$. Then $R_n(5,2) \geq \left(R_n(3,2)-1\right)^2+1$. The result now follows by theorem 4.7.

4.3 ANOTHER SYSTEM WITH SEVERAL EQUATIONS. The proof of theorem 4.2 is a particular example of a general method of proof. Here we discuss another theorem proved by this method. Consider the equation

$$a_1x_1+a_2x_2+\ldots+a_kx_k = 0 \tag{4.5}$$

where $a_1,\ldots,a_k$ are non-zero integers. By theorem 3.7, equation (4.5) is regular if and only if some subset of the coefficients has zero sum. Hence we can define the function $\phi(n)$, analogous to the Schur function, as usual: $\phi(n)$ is the largest positive integer such that the set of integers $\{1,2,\ldots,\phi(n)\}$ can be partitioned into n subsets, none of which contains a solution to (4.5).

We rewrite equation (4.5) in the form

$$a_1x_1+\ldots+a_tx_t = b_{t+1}x_{t+1}+\ldots+b_kx_k \tag{4.6}$$

where $a_1,\ldots,a_t$, $b_{t+1},\ldots,b_k$ are positive integers, and we now assume that (4.6) is regular. We also assume that

$$A = a_1+\ldots+a_t > b_{t+1}+\ldots+b_k = B,$$

since otherwise $\phi(n) = 0$ for all n.

THEOREM 4.11. *Let m be a positive integer. Let M and N be integers satisfying*

$$(A-1)\phi(m) \leq M < N \tag{4.7}$$

*and*

$$A\phi(m)+1 \leq N \leq \left\{\frac{A}{A-1}(M+1)\right\} \tag{4.8}$$

*where* $\{x\}$ *denotes the largest integer less than* $x$*, i.e.*

$$\{x\} = \begin{cases}[x] & \text{if } x \text{ is not an integer,}\\ x-1 & \text{if } x \text{ is an integer.}\end{cases}$$

*Let* $\gamma(M,N)$ *be the least number of subsets into which the set of integers* $\{1,2,\ldots,M\}$ *can be partitioned, no subset containing a solution of any of the equations*

$$a_1x_1+a_2x_2+\ldots+a_tx_t = b_{t+1}x_{t+1}+\ldots+b_kx_k+jN \tag{4.9}$$

*where*

$$j = -B+1,-B+2,\ldots,A-1 \quad \text{if } N < \frac{A}{A-1}M \quad \text{and}$$

$$j = -B+1,-B+2,\ldots,A-2 \quad \text{if } \frac{A}{A-1}M \leq N \leq \left\{\frac{A}{A-1}(M+1)\right\}.$$

*Let* $\Gamma(m) = \min \gamma(M,N)$ *where the minimum is taken over all pairs M,N satisfying (4.7) and (4.8). Then for all positive integers n,*

$$\phi(n+\Gamma(m)) \geq N'\phi(n)+M'$$

*where* $N',M'$ *satisfy (4.7) and (4.8) and* $\gamma(M',N') = \Gamma(m)$.

PROOF. We partition the set of integers $\{1,\ldots,M'\}$ into $\Gamma(m)$ subsets $C_1,C_2,\ldots,C_{\Gamma(m)}$ which satisfy the conditions given in defining $\Gamma(m)$. We partition the set of integers $\{1,2,\ldots,\phi(n)\}$ into n subsets $D_1,D_2,\ldots,D_n$, none of which contains a solution to equation (4.6).

Next we partition the set of integers $\{1,\dots,N'\phi(n)+M'\}$ into two sets E and F, where

$$E = \{uN'+v \mid u = 0,1,\dots,\phi(n),\ v = 1,2,\dots,M'\} \quad \text{and}$$
$$F = \{uN'-v \mid u = 1,2,\dots,\phi(n),\ v = 0,1,\dots,N'-M'-1\}.$$

Finally, we partition E into $\Gamma(m)$ subsets $E_1,\dots,E_{\Gamma(m)}$ by assigning $uN'+v$ to $E_i$ if $v \in C_i$, and we partition F into n subsets $F_1,\dots,F_n$ by assigning $uN'-v$ to $F_i$ if $u \in D_i$.

In this way, we have partitioned the set of integers $\{1,\dots,N'\phi(n)+M'\}$ into $\Gamma(m)+n$ subsets, and we complete the proof of the theorem by showing that none of these subsets contains a solution to (4.6).

First consider the sets $E_1,\dots,E_{\Gamma(m)}$. If any of these sets contains a solution to (4.6), it must be of the form

$$\sum_{i=1}^{t} a_i(u_iN'+v_i) = \sum_{i=t+1}^{k} b_i(u_iN'+v_i)$$

where $0 \leq u_i \leq \phi(n)$ and $1 \leq v_i \leq M'$. Hence we must have

$$\sum_{i=1}^{t} a_iv_i - \sum_{i=t+1}^{k} b_iv_i = jN'$$

where

$$j = \sum_{i=t+1}^{k} b_iu_i - \sum_{i=1}^{t} a_iu_i.$$

Now we consider the possible values of j. We know that

$$1 \leq \sum_{i=1}^{t} a_iv_i \leq AM' \quad \text{and} \quad 1 \leq \sum_{i=t+1}^{k} b_iv_i \leq BM'.$$

Hence we must have $1-BM' \leq jN' \leq AM'-1$.

Since $M' < N'$, $-BN' < 1-BM'$, so the smallest possible value of j is $-B+1$.

In the first case, where $N' < AM'/(A-1)$, we have $(A-1)N' < AM' < AN'$, so the largest possible value of j is $A-1$.

In the second case, where $AM'/(A-1) \leq N' \leq \{A(M'+1)/(A-1)\}$, the largest possible value of j is $A-2$.

By (4.9) and the definition of $\Gamma(m)$, this gives a contradiction. Hence none of the subsets $E_i$, $i = 1,\dots,\Gamma(m)$, contains a solution of (4.6).

Next consider the sets $F_1,\dots,F_n$. If any of these sets contains a solution to (4.6), it must be of the form

$$\sum_{i=1}^{t} a_i(u_iN'-v_i) = \sum_{i=t+1}^{k} b_i(u_iN'-v_i),$$

where $1 \leq u_i \leq \phi(n)$ and $0 \leq v_i \leq N'-M'-1$. Hence

$$N'\left(\sum_{i=1}^{t} a_iu_i - \sum_{i=t+1}^{k} b_iu_i\right) = \sum_{i=1}^{t} a_iv_i - \sum_{i=t+1}^{k} b_iv_i.$$

By the definition of the sets $F_i$,

$$\sum_{i=1}^{t} a_iu_i \neq \sum_{i=t+1}^{k} b_iu_i, \quad \text{so}$$

either

$$\sum_{i=1}^{t} a_iv_i \geq \sum_{i=t+1}^{k} b_iv_i+N'$$

or

$$\sum_{i=1}^{t} a_iv_i+N' \leq \sum_{i=t+1}^{k} b_iv_i.$$

In the first case, we must have

$$A(N'-M'-1) \geq \sum_{i=1}^{t} a_iv_i \geq \sum_{i=t+1}^{k} b_iv_i+N' \geq N'.$$

But this implies that $N' \geq A(M'+1)/(A-1)$, contradicting our hypothesis that $N' \leq \{A(M'+1)/(A-1)\} < A(M'+1)/(A-1)$.

In the second case, we must have

$$N' \leq N' + \sum_{i=1}^{t} a_iv_i \leq \sum_{i=t+1}^{k} b_iv_i \leq B(N'-M'-1) < A(N'-M'-1),$$

which again is false, for the same reason.

COROLLARY 4.12. *$f(n+m) \geq (2f(m)+1)f(n)+f(m)$.*

(So corollary 4.3 follows from this theorem also.)

PROOF. Consider the equation $x_1+x_2 = x_3$, so that $\phi(n) = f(n)$, $A = 2$, $B = 1$, and choose $M' = f(m)$, $N' = 2f(m)+1$. Then $\Gamma(m) = m$.

4.4 ANOTHER GENERALISATION. We consider another problem, closely related to that of Schur: Let m and n be positive integers and denote by $f(m,n)$ the largest positive integer such that the set of integers $\{m,m+1,...,m+f(m,n)\}$ can be partitioned into n sum-free sets. What is the value of $f(m,n)$?

Obviously $f(1,n) = f(n)-1$. Also, since the set $\{m,2m,...,(f(n)+1)m\}$ cannot be partitioned into n sum-free sets, we have

$$f(m,n) \leq mf(n)-1. \tag{4.10}$$

Theorem 4.6, together with (4.10), imply that

$$f(m,n) \leq m[n!e]-m-1.$$

In fact, we have equality in (4.10) for $n = 2$ and $n = 3$, that is $f(m,2) = 4m-1$ and $f(m,3) = 13m-1$. For the set $\{m,\ldots,5m-1\}$ can be partitioned into the two sum-free sets $\{m,\ldots,2m-1,4m,\ldots,5m-1\}$ and $\{2m,\ldots,4m-1\}$. Similarly, the set $\{m,\ldots,14m-1\}$ can be partitioned into the three sum-free sets

$$\{m,\ldots,2m-1,4m,\ldots,5m-1,10m,\ldots,11m-1,13m,\ldots,14m-1\},$$

$$\{2m,\ldots,4m-1,11m,\ldots,13m-1\} \text{ and } \{5m,\ldots,10m-1\}.$$

To establish a lower bound on $f(m,n)$, we study first a different generalisation of the problem.

DEFINITION 4.13. A set of positive integers is said to be *strongly sum-free* if and only if it contains no solution to either of the equations

$$x_1+x_2 = x_3, \quad \text{and} \tag{4.11}$$

$$x_1+x_2+1 = x_3. \tag{4.12}$$

DEFINITION 4.14. For any positive integer $n$, we define $\phi(n)$ to be the largest positive integer such that the set of integers $\{1,\ldots,\phi(n)\}$ may be partitioned into $n$ strongly sum-free sets. $\phi(n)$ is well-defined by theorems 3.7 and 3.8.

THEOREM 4.15. *For any positive integers m and n,*

$$\phi(n+m) \geq 2f(m)\phi(n)+f(m)+\phi(n),$$

*where f is the Schur function for the equation (4.11) alone.*

PROOF. We partition the set $\{1,2,\ldots,f(m)\}$ into $m$ sum-free subsets $A_1,\ldots,A_m$ and the set $\{1,2,\ldots,2f(m)+1\}$ into $m+1$ subsets $B_1,\ldots,B_{m+1}$ in the following way:

$$B_i = \{2a \mid a \in A_i\} \quad \text{for } i = 1,2,\ldots,m$$

and

$$B_{m+1} = \{1,3,5,\ldots,2f(m)+1\}.$$

Clearly $B_1,\ldots,B_m$ are strongly sum-free and $B_{m+1}$ is sum-free.

Now we partition the set $\{1,2,\ldots,\phi(n)\}$ into $n$ strongly sum-free subsets, $C_1,\ldots,C_n$. Finally we construct $m+n$ sets $D_1,\ldots,D_{m+n}$ in the following way: for $j = 1,2,\ldots,m$, let

$$D_j = \{(2a-1)\phi(n)+a+b \mid 2a \in B_j,\ b = 0,1,\ldots,\phi(n)\}$$

and for $j = m+1,\ldots,m+n$, let

$$D_j = \{(2a\phi(n)+a+b) \mid 2a+1 \in B_{m+1},\ b \in C_{j-m}\}.$$

We have a partition of the set $\{1,2,\ldots,2f(m)\phi(n)+f(m)+\phi(n)\}$ into $m+n$ subsets, and to complete the proof of the theorem we need only prove these subsets are strongly sum-free.

First consider the sets $D_j$, $j = 1,\ldots,m$. Suppose one of these sets is not strongly sum-free. Then either

$$(2a_1-1)\phi(n)+a_1+b_1+(2a_2-1)\phi(n)+a_2+b_2 = (2a_3-1)\phi(n)+a_3+b_3 \qquad (4.13)$$

or

$$(2a_1-1)\phi(n)+a_1+b_1+(2a_2-1)\phi(n)+a_2+b_2+1 = (2a_3-1)\phi(n)+a_3+b_3 \qquad (4.14)$$

where in each case $a_1,a_2,a_3 \in A_j$ and $0 \leq b_1,b_2,b_3 \leq \phi(n)$. Equation (4.13) implies that

$$\big(2\phi(n)+1\big)(a_1+a_2-a_3) = \phi(n)+b_3-b_1-b_2. \qquad (4.15)$$

Since $A_j$ is sum-free, $a_1+a_2-a_3 \neq 0$. Hence the left side of equation (4.15) has absolute value at least $2\phi(n)+1$, but the right side has absolute value at most $2\phi(n)$. Hence (4.13) cannot hold. Similarly, (4.14) cannot hold and $D_j$ is strongly sum-free.

Now consider the sets $D_j$, $j = m+1,\ldots,m+n$. Suppose one of these sets is not strongly sum-free. Then either

$$2a_1\phi(n)+a_1+b_1+2a_2\phi(n)+a_2+b_2 = 2a_3\phi(n)+a_3+b_3 \qquad (4.16)$$

or

$$2a_1\phi(n)+a_1+b_1+2a_2\phi(n)+a_2+b_2+1 = 2a_3\phi(n)+a_3+b_3 \qquad (4.17)$$

where in each case, $2a_1+1,2a_2+1,2a_3+1 \in B_{m+1}$ and $b_1,b_2,b_3 \in C_{j-m}$. Equation (4.16) implies that

$$\big(2\phi(n)+1\big)(a_1+a_2-a_3) = b_3-b_1-b_2. \qquad (4.18)$$

Since $b_1,b_2,b_3 \in C_{j-m}$, the right side of (4.18) has absolute value less than $2\phi(n)$. Hence $a_1+a_2-a_3 = 0 = b_3-b_1-b_2$, contradicting the fact that $C_{j-m}$ is strongly sum-free. Similarly, (4.17) cannot hold and $D_j$ is strongly sum-free.

THEOREM 4.16. *For any positive integers m and n,*

$$f(m,n) \geq m\phi(n)-1.$$

PROOF. We partition the set of integers $\{1,2,\ldots,\phi(n)\}$ into n strongly sum-free subsets $C_1,C_2,\ldots,C_n$. Next we partition the set $\{m,m+1,\ldots,m\phi(n)+m-1\}$ into n subsets $D_1,D_2,\ldots,D_n$ by assigning $am+b$ to $D_i$ if $a \in C_i$, where $a = 1,2,\ldots,\phi(n)$ and $b = 0,1,\ldots,m-1$. Now we need only show that each $D_j$ is sum-free and the theorem will be proved.

So suppose that $D_j$ is not sum-free for some $j$, $1 \leq j \leq n$. Then we must have

$$a_1m+b_1+a_2m+b_2 = a_3m+b_3 \qquad (4.19)$$

where $a_1,a_2,a_3 \in C_j$ and $0 \leq b_1,b_2,b_3 \leq m-1$. Equation (4.19) implies that

$$m(a_1+a_2-a_3) = b_3-b_1-b_2. \qquad (4.20)$$

Since $C_j$ is strongly sum-free, either $1 \leq a_1+a_2-a_3$ or $a_1+a_2-a_3 \leq -2$, and hence the left side of (4.20) is either at least m or at most -2m. Since $-2m-2 \leq b_3-b_1-b_2 \leq m-1$, (4.20) cannot hold and $D_j$ is sum-free.

COROLLARY 4.17. *For any positive integers m and n,*

$$f(m,n) \geq m\big(3f(n-1)+1\big)-1.$$

PROOF. Let n = 1 and m = n-1 in theorem 4.15, so that $\phi(n) \geq 3f(n-1)+1$. Now combine this inequality with theorem 4.16.

COROLLARY 4.18. *For any positive integers m and n,*

$$f(m,n) > cm\ 89^{n/4}.$$

PROOF. This follows from corollary 4.4 and corollary 4.17.

## 4.5 REFERENCES AND RELATED TOPICS.

Most of the results of this chapter are given in Hanson (1970) and Abbott and Hanson (1972). The lower bounds on Ramsey numbers which are derived here are effective for small k; earlier related results are due to Frasnay (1963) and Giraud (1968a, 1968b).

Upper bounds on Ramsey numbers are given in Greenwood and Gleason (1955). Lower bounds effective for n small and k large were found by Erdos (1947) and Abbott (1965). They showed that $R_2(k,2) \geq c\,k\,2^{\frac{1}{2}k}$ for some absolute constant c. An argument similar to that of Erdos was used by Hanson (1970) to show that

$$\binom{R_n(k,2)}{k} \geq n^{\binom{k}{2}-1}$$

and hence to derive a lower bound of approximately $kn^{\frac{1}{2}k}$ on $R_n(k,2)$.

Lemma 4.9 is due to Abbott (1972).

The problem of section 4 was first considered by P. Turan, who showed that f(m,2) = 4m-1. His results were contained in a private communication to L. Moser, quoted in Abbott and Hanson (1972). The remainder of the results in section 4 are due to Abbott (1965), Znam (1967), Hanson (1970) and Abbott and Hanson (1972).

Schur functions for some other regular systems have been determined by Salié (1954).

## CHAPTER V. ADMISSIBLE SUBSETS OF A SET OF POSITIVE INTEGERS

In this chapter, we consider a problem closely related to that of sum-free sets: we estimate the largest number g(N) such that, from every set of N distinct natural numbers, we can always select a subset of g(N) integers with the property that no sum of two distinct integers of this subset belongs to the original set.

### 5.1 DEFINITIONS AND NOTATION.

DEFINITION 5.1. Let X and Y be sets of natural numbers. Then X is said to be *admissible relative to Y* if and only if the sum of every pair of distinct elements of X lies outside Y. We shall consider in particular the case where $X \subseteq Y$.

DEFINITION 5.2. Let g(A) be the number of elements in the largest possible subset of A which is admissible relative to A. If we consider all sets A, such that $|A| = N$, then our function g(N) can be defined as

$$g(N) = \min_{|A|=N} g(A).$$

We show that, for some absolute constants c and k and for N sufficiently large,

$$c \log N < g(N) < k\, N^{(2/5)+\varepsilon} \qquad (5.1)$$

for every $\varepsilon > 0$.

### 5.2 THE LOWER BOUND FOR g(N).

In this section, we prove the left-hand inequality of (5.1).

Let $A = \{a_1, a_2, \ldots, a_N\}$ be a set of N distinct natural numbers. We construct $B \subseteq A$, such that $c \log N < |B|$ and B is admissible relative to A.

For every $a \in A$, let s(a) denote the number of solutions of the equation

$$x + a = y, \quad \text{where } x, y \in A \text{ and } x \neq a. \qquad (5.2)$$

Suppose that a is the jth largest integer in A, i.e. there are (j-1) integers in A larger than a. Then there are at most (j-1) choices for y in (5.2), so that $s(a) \leq j-1$. Hence

$$\sum_{\substack{s(a) \leq r \\ a \in A}} 1 \geq r+1, \quad \text{for } r = 0, 1, \ldots, N-1. \qquad (5.3)$$

Without loss of generality, we assume that the elements of A are numbered so that

$s(a_i) \leq s(a_j)$ whenever $i < j$. By (5.3), we have

$$s(a_i)+1 \leq i. \tag{5.4}$$

We define: $B_1 = A$,

$b_1 = a_1$, the element of $B_1$ with the smallest suffix,

$C_1 = \{b_1\}$

and, for $i \geq 2$, $B_i = B_{i-1} \setminus C_{i-1}$,

$b_i$ = the element of $B_i$ with the smallest suffix,

$C_i = \{a \mid a+b_i \in A,\ a \in A,\ a \neq b_i\} \cup \{b_i\}$.

Hence

$$|B_i| \geq |A| - \sum_{j=1}^{i-1} |C_j| = N - \sum_{j=1}^{i-1} |C_j|. \tag{5.5}$$

Now, for $j \geq 2$,

$$|C_j| = s(b_j)+1 \leq \sum_{h=1}^{j-1} |C_h|+1$$

by (5.4) and the definition of $b_j$. Hence

$$\begin{aligned}\sum_{j=1}^{i-1} |C_j| &\leq 1 + 2\sum_{j=1}^{i-2} |C_j| \\ &\leq 1+2+\ldots+2^{i-3}+2^{i-2}|C_1| \\ &\leq 2^{i-1},\end{aligned}$$

which, by (5.5) shows that $|B_i| \geq N-2^{i-1}$. Hence $|B_i| > 0$ for $i \leq [\log N/\log 2]$, and our set $B = \{b_1, b_2, \ldots, b_k\}$ for $k = [\log N/\log 2]$.

This gives the lower bound for $g(N)$.

## 5.3 AN UPPER BOUND FOR $g(N)$.

We first derive an inequality which, though weaker than the right-hand inequality of (5.1), follows from a far simpler construction.

**THEOREM 5.3.** *For some absolute constant $k$ and for all sufficiently large $N$,*

$$g(N) < k\,N^{\frac{1}{2}}. \tag{5.6}$$

PROOF. Let $s = [N^{\frac{1}{2}}]$ and $t = [N/s]$. (5.7)

For $i = 1,\ldots,s$, we define the set $D_i$ by

$$B_i = \{2^{i-1}n \mid t \leq n \leq 2t-1\}, \tag{5.8}$$

so that $B_i \cap B_j = \emptyset$ for $i \neq j$. We also define $B_0$ to be a set of $(N-st)$ natural numbers such that $B_0 \cap B_i = \emptyset$ for $i = 1,\ldots,s$. Certainly

$$|B_0| = N-st \leq 2\,N^{\frac{1}{2}}. \tag{5.9}$$

Now let

$$B = \bigcup_{i=0}^{s} B_i$$

so that

$$|B| = \sum_{i=0}^{s} |B_i| = N-st+st = N.$$

Let $B' \subseteq B$ be a maximal admissible set relative to B. Suppose that for some i such that $1 \leq i \leq s-1$, we have $|B_i \cap B'| \geq 3$. By (5.8), this means that three integers, $2^{i-1}n_1, 2^{i-1}n_2, 2^{i-1}n_3 \in B_i \cap B'$, where $t \leq n_j \leq 2t-1$, for $j = 1,2,3$. Two of these three integers, say $n_1$ and $n_2$, must have the same parity, so that

$$n_1+n_2 = 2n \quad \text{for some } n \text{ such that } \quad t \leq n \leq 2t-1.$$

But this implies that the sum of two distinct elements of B', namely $2^{i-1}n_1$ and $2^{i-1}n_2$, belongs to $B_{i+1}$ and hence to B. This contradicts the definition of B', so that we must have

$$|B_i \cap B'| \leq 2, \quad \text{for } i = 1,\ldots,s-1. \tag{5.10}$$

Then

$$g(N) \leq g(B) = |B'| \leq |B_0| + |B_s| + 2(s-1).$$

So, by (5.7), (5.8) and (5.9), we have

$$g(N) \leq 2N^{\frac{1}{2}} + [N/[N^{\frac{1}{2}}]] + 2[N^{\frac{1}{2}}] - 2$$

$$\leq k\, N^{\frac{1}{2}}.$$

In fact, since $[N/[N^{\frac{1}{2}}]] \leq N^{\frac{1}{2}}+2$, we may take $k = 5$.

In order to improve this upper bound and to prove some interesting related results, we need some information about admissible sets of integers modulo a prime, and two results concerning the distribution of primes.

## 5.4 ADMISSIBLE SETS OF INTEGERS MODULO A PRIME.

LEMMA 5.4. *Let p be an odd prime, $Z_p$ the cyclic group of order p and h an element of $Z_p$. If $S \subseteq Z_p$ is admissible relative to {h}, then $|S| \leq \frac{1}{2}(p+1)$.*

PROOF. Let $S = \{a_1, a_2, \ldots, a_q\}$ where $a_i+a_j \neq h$, for $i \neq j$. At most one of the elements of S satisfies $2a_i = h$, and without loss of generality, we assume that if there is such an element then $2a_1 = h$. Now $h-a_i \neq h-a_j$ for $i \neq j$, and $a_i \neq h-a_j$ for $i = 1,\ldots,q$, $j = 2,\ldots,q$, because that would imply that $a_i+a_j = h$, for $i \neq j$, contradicting the admissibility of S relative to {h}.

Hence we have found (2q-1) distinct elements of $Z_p$, namely $a_1, a_2, \ldots, a_q$, $h-a_2, \ldots, h-a_q$. But this implies that $2q-1 \leq p$ and $|S| = q \leq \frac{1}{2}(p+1)$.

COROLLARY 5.5. *Let p be an odd prime and let h be an integer such that $0 \leq h < p$. Let S be a set of integers admissible relative to the congruence class of h modulo p. Then there exist integers $x_1, x_2, \ldots, x_r$ such that $0 \leq x_i < p$ for $i = 1,2,\ldots,r$, where*

$r \leq \frac{1}{2}(p+1)$, *such that* $S \subseteq \bigcup_{i=1}^{r} X_i$ *where* $X_i$ *denotes the congruence class of* $x_i$ *modulo* $p$.

5.5 SOME RESULTS ON THE DISTRIBUTION OF PRIME NUMBERS. We shall need an estimate of the number of primes in the interval $(n,2n)$.

DEFINITION 5.6. Let $\pi(x) = \sum_{p \leq x} 1$ and $\theta(x) = \sum_{p \leq x} \log p$, where $x$ is a real number and $p$ is prime.

DEFINITION 5.7. Let $n$ be a natural number and $p$ a prime. Then we say that $p|n$ $s$ times if and only if $p^s|n$, but $p^{s+1} \nmid n$.

LEMMA 5.8. *$p|m!$ exactly $[m/p]+[m/p^2]+[m/p^3]+\ldots$ times.*

PROOF. We note that since $[x] = 0$ for every $x$ such that $0 \leq x < 1$, the number of non-zero terms in the series is finite.

Among the set of integers $\{1,2,\ldots,m\}$, there are exactly $[m/p]$ which are divisible by $p$, namely

$$p, 2p, \ldots, [m/p]p. \tag{5.11}$$

Among the integers listed in (5.11), there are the $[m/p^2]$ integers of the set $\{1,2,\ldots,m\}$ which are divisible by $p^2$, namely

$$p^2, 2p^2, \ldots, [m/p^2]p^2. \tag{5.12}$$

Continuing this argument, we see that there are exactly $[m/p^r]-[m/p^{r+1}]$ numbers in the set $\{1,2,\ldots,m\}$ which are divisible by $p$ $r$ times. Hence $p|m!$ exactly $s$ times, where $s = \sum_{r \geq 1} r\{[m/p^r]-[m/p^{r+1}]\} = \sum_{r \geq 1} [m/p^r]$.

LEMMA 5.9. *Let* $N = \binom{2n}{n} = (2n)!/(n!)^2$.

*(i) Let $p$ be a prime, $p \leq 2n$. Then $N$ is divisible by $p$ exactly $\nu_p$ times, where*

$$\nu_p = \sum_{r \geq 1} ([2n/p^r]-2[n/p^r]), \tag{5.13}$$

*and*

$$\nu_p \leq M_p = [(\log 2n)/(\log p)]. \tag{5.14}$$

*(ii)* $$2^{2n}/(2\sqrt{n}) < N < 2^{2n}/(\sqrt{2n}) \tag{5.15}$$

PROOF. (i) By lemma 5.8, the numerator of $N$ is divisible by $p$ exactly $[2n/p]+[2n/p^2]+\ldots$ times, and the denominator of $N$ is divisible by $p$ exactly $2\{[n/p]+[n/p^2]+\ldots\}$ times. Hence, $N$ is divisible by $p$ exactly $\nu_p$ times, where $\nu_p$ is given by (5.13). In this notation, we write

$$N = \prod_{p \leq 2n} p^{\nu_p}.$$

Now, if $2n < p^r$, then $r > [(\log 2n)/(\log p)] = M_p$.

Hence,

$$\nu_p = \sum_{r=1}^{M_p} ([2n/p^r]-2[n/p^r]). \qquad (5.16)$$

But, for every $y \geq 0$, we have $[y] \leq y < [y]+1$, so that

$$2[y] \leq 2y < 2[y]+2, \quad \text{and} \quad [2y] \leq 2y < [2y]+1.$$

Hence

$$-1 < [2y]-2[y] < 2, \quad \text{and} \quad [2y]-2[y] = 0 \text{ or } 1. \qquad (5.17)$$

But (5.17), with (5.16), implies (5.14).

We note that

$$N = \prod_{p\leq 2n} p^{\nu_p} \leq \prod_{p\leq 2n} p^{M_p}. \qquad (5.18)$$

(ii) Let $P = (1.3.5...(2n-1))/(2.4.6...(2n))$. Then

$$P = (1.3.5...(2n-1).2.4.6...(2n))/(2.4.6...(2n).2.4.6...(2n))$$
$$= (2n)!/(2^{2n}(n!)^2),$$

so that

$$2^{2n}P = N. \qquad (5.19)$$

Since $1 > (1-(1/2^2))(1-(1/4^2))....(1-(1/(2n)^2))$, we have

$$1 > \frac{1.3}{2^2}\frac{3.5}{4^2}\frac{5.7}{6^2}\cdots\frac{(2n-1)(2n+1)}{(2n)^2} = (2n+1)P^2.$$

Hence by (5.19), $1 > 2nP^2 = (2nN^2)/2^{4n}$, which proves the right-hand inequality of (5.15). Similarly, $1 > (1-(1/3^2))(1-(1/5^2))...(1-(1/(2n-1)^2))$ so that

$$1 > \frac{2.4}{3^2}\frac{4.6}{5^2}\frac{6.8}{7^2}\cdots\frac{(2n-2)(2n)}{(2n-1)^2} = 1/(4nP^2).$$

Hence by (5.19), $1 > 2^{4n}/(4nN^2)$, which proves the left-hand inequality of (5.15).

LEMMA 5.10. *For every integer $n \geq 1$, $\theta(n) < 2n \log 2$.* (5.20)

PROOF. We proceed by induction, using the right-hand inequality of (5.15). (5.20) is certainly true, for $n = 1$ or $2$.

We show first that if (5.20) is true for some $n > 2$, then $\theta(2n-1) < 2(2n-1) \log 2$ and hence $\theta(2n) = \theta(2n-1) < 4n \log 2$.

Consider

$$\frac{N}{2} = \frac{1}{2}\binom{2n}{n} = \frac{(2n)!}{(n!)}\cdot\frac{n}{2n} = \frac{(2n-1)!}{n!(n-1)!} = \binom{2n-1}{n-1}.$$

This is divisible by every prime $p$ such that $n < p \leq 2n-1$, and hence also by their product. This implies that

$$N/2 \geq \prod_{n<p\leq 2n-1} p,$$

so that

$$\log (N/2) \geq \theta(2n-1)-\theta(n).$$

By (5.15)

$$\log N < 2n \log 2 \; - \tfrac{1}{2}\log 2n$$

and hence

$$\theta(2n-1)-\theta(n) < (2n-1) \log 2 - \tfrac{1}{2} \log 2n.$$

By assumption, (5.20) is true for n and hence $\theta(n) < 2n \log 2$. So

$$\theta(2n-1) < 2n \log 2 + (2n-1) \log 2 - \tfrac{1}{2} \log 2n,$$

and hence, for $n > 2$,

$$\theta(2n-1) < 2(2n-1) \log 2.$$

This shows that if (5.20) is true for every n such that $2^{r-1} < n \leq 2^r$, for some $r > 1$, then it is true for every n such that $2^r < n \leq 2^{r+1}$. But (5.20) is true for $2 < n \leq 2^2$, since $\theta(4) = \theta(3) = \log 2 + \log 3 = \log 6 < 6 \log 2$. Hence, by induction, we have (5.20) for every $n > 2$. Since we have already checked for n = 1 or 2, we have $\theta(n) < 2n \log 2$ for every integer $n \geq 1$.

THEOREM 5.11. *For sufficiently large n,*

$$\pi(n)-\pi(n/2) > n^{2/3} > n^{\frac{1}{2}}.$$

PROOF. We obtain an estimate for $\theta(n)$ and use this to find the result for $\pi(n)$.

Again let

$$N = (2n)!/(n!)^2 = \prod_{p\leq 2n} p^{\nu_p},$$

so that

$$\log N = \sum_{p\leq 2n} \nu_p \log p. \tag{5.21}$$

We consider this sum in four parts:

$$\textstyle\sum_1 = \sum_{n<p\leq 2n} \nu_p \log p; \quad \sum_2 = \sum_{(2n/3)<p\leq n} \nu_p \log p;$$

$$\textstyle\sum_3 = \sum_{\sqrt{2n}<p\leq(2n/3)} \nu_p \log p; \quad \sum_4 = \sum_{p\leq\sqrt{2n}} \nu_p \log p.$$

(We have assumed here that $n \geq 5$, so that $\sqrt{2n} < 2n/3$.)

Consider $\sum_1$. Since $n/p < 1$, we have $[n/p] = 0$. Since $1 \leq 2n/p < 2$, we have $[2n/p] = 1$ and $[2n/p^2] = 0$. Hence $\nu_p = 1$ for $n < p \leq 2n$, so that

$$\textstyle\sum_1 = \sum_{n<p\leq 2n} \nu_p \log p = \sum_{n<p\leq 2n} \log p = \theta(2n)-\theta(n). \tag{5.22}$$

Consider $\sum_2$. Since $1 \leq n/p < 3/2$, we have $[n/p] = 1$ and $[2n/p] = 2$. If $n \geq 3$, we have also $[2n/p^2] = 0$. Hence

$$\textstyle\sum_2 = 0 \text{ for } n \geq 3. \tag{5.23}$$

Consider $\sum_3$. Since $n/p^2 < 2n/p^2 < 1$, we have by (5.17) that $\nu_p = [2n/p]-2[n/p] = 0$ or 1. Hence

$$\sum_3 \leq \sum_{\sqrt{2n}<p\leq 2n/3} \log p = \theta(2n/3)-\theta(\sqrt{2n}).$$

But since

$$\theta(\sqrt{2n}) = \sum_{p\leq\sqrt{2n}} \log p \geq \log 2 \sum_{p\leq\sqrt{2n}} 1 = \pi(\sqrt{2n}) \log 2,$$

we have

$$\sum_3 \leq \theta(2n/3)-\pi(\sqrt{2n}) \log 2. \tag{5.24}$$

Finally, consider $\sum_4$. By (5.14), we have $\nu_p \leq M_p = [(\log 2n)/(\log p)]$, so that

$$\sum_4 \leq \sum_{p\leq\sqrt{2n}} M_p \log p \leq \sum_{p\leq\sqrt{2n}} \frac{\log 2n}{\log p} \log p = \log 2n \sum_{p\leq\sqrt{2n}} 1.$$

Hence $\qquad \sum_4 \leq \pi(\sqrt{2n}) \log 2n. \qquad\qquad$ (5.25)

Now, by (5.21)-(5.25), we have

$$\log N \leq \theta(2n)-\theta(n)+\theta(2n/3)-\pi(\sqrt{2n}) \log 2 + \pi(\sqrt{2n}) \log 2n,$$

$$\theta(2n)-\theta(n) \geq \log N-\theta(2n/3)-\pi(\sqrt{2n}) \log n. \tag{5.26}$$

We now need three inequalities:

By the left-hand inequality of (5.15), we have

$$\log N > 2n \log 2 - \log 2\sqrt{n}\ . \tag{5.27}$$

By (5.20),

$$\theta(2n/3) = \theta([2n/3]) < 2[2n/3] \log 2 \leq 2(2n/3) \log 2\ . \tag{5.28}$$

For $n \geq 8$, $\pi(n) \leq n/2$, since 2 is the only even prime, and for $n \geq 32$, we have $\sqrt{2n} \geq 8$. Hence

$$\pi(\sqrt{2n}) = \pi([\sqrt{2n}]) \leq \tfrac{1}{2}[\sqrt{2n}] \leq \tfrac{1}{2}\sqrt{2n}. \tag{5.29}$$

Hence, by (5.26)-(5.29), we have, for $n \geq 32$,

$$\theta(2n)-\theta(n) \geq 2n \log 2 - \log(2\sqrt{n})-(4n \log 2)/3 - \tfrac{1}{2}(\sqrt{2n} \log n),$$

which implies that

$$\theta(2n)-\theta(n) \geq (\frac{2n}{3} -1) \log 2 - \tfrac{1}{2}(\sqrt{2n}+1) \log n. \tag{5.30}$$

Now

$$\pi(2n)-\pi(n) = \sum_{n<p\leq 2n} 1, \quad \text{and} \quad \theta(2n)-\theta(n) = \sum_{n<p\leq 2n} \log p,$$

so

$$\log 2n \left(\pi(2n)-\pi(n)\right) \geq \theta(2n)-\theta(n).$$

Hence, by (5.30),

$$\pi(2n)-\pi(n) \geq \frac{(2n-3) \log 2}{3 \log 2n} - \frac{(\sqrt{2n}+1) \log n}{2 \log 2n}\ . \tag{5.31}$$

For n sufficiently large, the right-hand side of (5.31) exceeds $n^{2/3}$ and this implies the theorem.

## 5.6 A LARGE SIEVE INEQUALITY.

The last of the results we need before considering the improved upper bound for g(N) is a large sieve inequality.

Let M,N be natural numbers and let $\mathcal{A}$ be a set of A distinct natural numbers, $\mathcal{A} \subseteq \{M+1, M+2, \ldots, M+N\}$. If A is not too small compared with N, and if the prime p is not too large, we may reasonably expect the set $\mathcal{A}$ to be evenly distributed between the residue classes modulo p, and we estimate the extent to which the elements of $\mathcal{A}$ depart from this even distribution.

For any natural number q, define

$$A(q,h) = \sum_{\substack{n\in\mathcal{A} \\ n\equiv h \pmod q}} 1$$

so that

$$\sum_{h=1}^{q} A(q,h) = A.$$

For a given prime p, the variance

$$D_p = \sum_{h=1}^{p} \left(A(p,h)-(A/p)\right)^2$$

measures the way in which $\mathcal{A}$ is distributed among the residue classes modulo p. Hence, what we are looking for is an inequality of the kind

$$\sum_{p\le X} pD_p \le f(N,A,X), \quad \text{for } X < N,$$

where f depends on the size but not the structure of $\mathcal{A}$. We have at once that

$$pD_p = p\sum_{h=1}^{p} A^2(p,h)-A^2 \le p\sum_{h=1}^{p} A^2(p,h) \tag{5.32}$$

and that

$$A(p,h) \le (N/p)+1 \le 2N/p$$

for all $p < N$. Hence

$$pD_p \le p(2N/p)\sum_{h=1}^{p} A(p,h) = 2NA,$$

and this implies that

$$\sum_{p\le X} pD_p \le 2NAX.$$

We need a non-trivial improvement on this inequality.

We transform the problem to an equivalent one concerning exponential sums. Let

$$S(x) = \sum_{n\in\mathcal{A}} e^{2\pi inx}.$$

Then

$$\sum_{a=1}^{p-1} |S(a/p)|^2 = \sum_{a=1}^{p-1} S(a/p).\overline{S(a/p)}$$

$$= \sum_{a=1}^{p-1} \Big(\sum_{n\varepsilon\mathcal{Q}} e^{2\pi ina/p}\Big)\Big(\sum_{n'\varepsilon\mathcal{Q}} e^{-2\pi in'a/p}\Big)$$

$$= \sum_{n\varepsilon\mathcal{Q}} \sum_{n'\varepsilon\mathcal{Q}} \sum_{a=1}^{p-1} e^{2\pi i(n-n')a/p},$$

where the inner sum

$$\sum_{a=1}^{p-1} e^{2\pi i(n-n')a/p} = \begin{cases} p-1 & \text{if } n \equiv n' \pmod p \\ -1 & \text{if } n \not\equiv n' \pmod p \end{cases}.$$

Hence the sum becomes

$$p\Big(\sum_{n\varepsilon\mathcal{Q}} \sum_{\substack{n'\varepsilon\mathcal{Q} \\ n\equiv n' (\mathrm{mod}\ p)}} 1\Big) - A^2 = p\sum_{h=1}^{p}\Big(\sum_{\substack{n\varepsilon\mathcal{Q} \\ n\equiv h(\mathrm{mod}\ p)}} 1\Big)^2 - A^2,$$

so that, by (5.32),

$$pD_p = \sum_{a=1}^{p-1} |S(a/p)|^2. \tag{5.33}$$

We will therefore be concerned with finding an estimate for the sum

$$\sum_{p\leq X} \sum_{a=1}^{p-1} |S(a/p)|^2. \tag{5.34}$$

Obviously, (5.34) does not exceed

$$\sum_{q\leq X} \sum_{\substack{a=1 \\ (a,q)=1}}^{q} |S(a/q)|^2 \tag{5.35}$$

which is a particular case of a sum of the kind

$$\sum_{r=1}^{R} |S(x_r)|^2, \tag{5.36}$$

where the real numbers $x_r$ are distinct modulo 1 and well-spaced in the following sense:

Let $||x||$ denote the distance of x from the nearest integer. Then $\{x_1,x_2,\dots,x_R\}$ are well-spaced numbers if and only if there exists $\delta > 0$ such that $||x_i-x_j|| > \delta$ for $i \neq j$, $i,j = 1,2,\dots,R$.

If we choose the numbers $x_1,x_2,\dots,x_R$ to be the numbers $a/q$, where $1 \leq a \leq q$, $(a,q) = 1$, and there are X such numbers, then $X^{-2}$ is an admissible value of $\delta$. (We have assumed here that $X \geq \sqrt{q}$, which is true for $q \neq 2$ or 6.) In this case, (5.36) reduces to (5.35).

Finally, for $b_n$ any complex numbers, we define

$$S_0(x) = \sum_{n=-U}^{U} b_n e^{2\pi inx} .$$

For N odd, we make the substitutions $U = \frac{1}{2}(N-1)$ and $b_n = a_{n+M+1+U}$; for N even, we make the substitutions $U = \frac{1}{2}N$, $b_n = a_{n+M+1+U}$ for $n = -\frac{1}{2}N,\dots,\frac{1}{2}N-1$, and $a_{N+M+1} = 0$. This gives

$$|S_0(x)| = \left|\sum_{n=M+1}^{M+N} a_n e^{2\pi inx}\right| .$$

(Note that here we are considering absolute values.) Now we choose $a_n = 0$ for $n \notin \mathcal{a}$, $a_n = 1$ for $n \in \mathcal{a}$, so that $|S_0(x)| = |S(x)|$. Now we have transformed the problem into that of finding an estimate of sums of the type

$$\sum_{r=1}^{R} |S_0(x_r)|^2 .$$

THEOREM 5.12. $\sum_{r=1}^{R} |S_0(x_r)|^2 \leq (\delta^{-1}+2\pi U)Z_0$, *where* $Z_0 = \sum_{-U}^{U} |b_n|^2$.

PROOF. We have

$$S_0^2(x)-S_0^2(x_r) = 2\int_{x_r}^{x} S_0(y)S_0'(y)dy$$

and hence

$$|S_0(x_r)|^2 \leq |S_0(x)|^2 + 2\left|\int_{x_r}^{x} |S_0(y)S_0'(y)|dy\right| .$$

We now integrate with respect to x over the interval $(x_r-\frac{1}{2}\delta, x_r+\frac{1}{2}\delta)$, which gives

$$\delta|S_0(x_r)|^2 \leq \int_{x_r-\frac{1}{2}\delta}^{x_r+\frac{1}{2}\delta} |S_0(x)|^2dx + 2\int_{x_r-\frac{1}{2}\delta}^{x_r+\frac{1}{2}\delta}\left|\int_{x_r}^{x} |S_0(y)S_0'(y)|dy\right|dx. \tag{5.37}$$

Suppose that f(y) is positive and continuously differentiable and let k be any positive number. Then

$$\int_{-k}^{k}\left|\int_0^x f(y)dy\right|dx = \int_0^k\int_0^x f(y)dydx - \int_{-k}^{0}\int_0^x f(y)dydx. \tag{5.38}$$

Considering the first term on the right-hand side, we have

$$\int_0^k\int_0^x f(y)dydx = \int_0^k\int_y^k f(y)dxdy \leq \int_0^k kf(y)dy = k\int_0^k f(y)dy.$$

A similar argument applied to the second term of (5.38) shows that

$$\int_{-k}^{k}\left|\int_0^x f(y)dy\right|dx \leq k\int_{-k}^{k} f(y)dy. \tag{5.39}$$

Applying the inequality (5.39) to the last term of (5.37) we see that

$$\delta|S_0(x_r)|^2 \leq \int_{x_r-\frac{1}{2}\delta}^{x_r+\frac{1}{2}\delta} |S_0(x)|^2dx + \delta\int_{x_r-\frac{1}{2}\delta}^{x_r+\frac{1}{2}\delta} |S_0(y)S_0'(y)|dy. \tag{5.40}$$

We now sum (5.40) over r. Because of the choice of $\delta$, the intervals $(x_r-\frac{1}{2}\delta, x_r+\frac{1}{2}\delta)$ are pairwise disjoint, for $r = 1,\ldots,R$, so that

$$\sum_{r=1}^{R} |S_0(x_r)|^2 \leq \delta^{-1}\int_0^1 |S_0(y)|^2 dy + \int_0^1 |S_0(y)S_0'(y)|\,dy. \qquad (5.41)$$

Now

$$Z_0 = \sum_{-U}^{U} |b_n|^2 = \int_0^1 |S_0(y)|^2 dy.$$

Making this substitution and applying Cauchy's inequality to (5.41), we have

$$\sum_{r=1}^{R} |S_0(x_r)|^2 \leq \delta^{-1}Z_0 + Z_0^{\frac{1}{2}}\left(\int_0^1 |S_0'(y)|^2 dy\right)^{\frac{1}{2}}$$

$$\leq \delta^{-1}Z_0 + Z_0^{\frac{1}{2}}(4\pi^2U^2Z_0)^{\frac{1}{2}} = (\delta^{-1}+2\pi U)Z_0$$

which proves the theorem.

COROLLARY 5.13. $$\sum_{p\leq X} pD_p \leq \sum_{q\leq X} \sum_{\substack{a=1 \\ (a,q)=1}}^{q} |S(a/q)|^2 \leq (\pi N+X^2)A\,.$$

## 5.7 THE IMPROVED UPPER BOUND FOR g(N).

The bound given in equation (5.1) follows from theorem 5.16 of this section, which in turn follows from theorem 5.15. Theorem 5.14 is a rather different construction, given here because it is of independent interest.

THEOREM 5.14. *Let n be a natural number, let S be the set of integers in [2n,4n), let B be any subset of S and let $C\subseteq[n,2n)$ be a maximal admissible set relative to B. Let f(n) be defined by*

$$f(n) = \min_{B\subseteq S} (|C|+|B|).$$

*Then for all sufficiently large n and for some absolute constant c, we have*

$$f(n) < cn^{3/4}. \qquad (5.42)$$

PROOF. To prove (5.42), it will be sufficient to construct a set $B\subseteq S$, such that if $C\subset[n,2n)$ is a maximal admissible set relative to B, then

$$|C|+|B| < cn^{3/4}, \qquad (5.43)$$

for all sufficiently large n.

Let $n > n_0$, where $n_0$ is sufficiently large. From the set of integers in the interval $(\frac{1}{2}n^{\frac{1}{2}}, n^{\frac{1}{2}})$ we can, by theorem 5.11, select a set $\mathcal{Q}'$ of $[n^{\frac{1}{4}}]$ distinct primes. Let B be the set of their multiples in [2n,4n), so that

$$|B| < c_1n^{3/4} \qquad (5.44)$$

for some absolute constant $c_1$.

Now we apply corollary 5.13, with $N = 2n$, $\mathcal{A} = C$ and hence $A = |C|$. Our set $\mathcal{A}'$ consists of $[n^{\frac{1}{4}}]$ distinct primes $p$, all satisfying $p \leq X$, where $X = n^{\frac{1}{2}}$. Hence

$$\sum_{p\in\mathcal{A}'} pD_p \leq \sum_{p\leq X} pD_p = \sum_{p\leq X} p \sum_{h=1}^{p} \left(A(p,h)-(A/p)\right)^2$$

$$\leq (2\pi n+n)|C|. \tag{5.45}$$

Finally, we apply corollary 5.5, with $S = C$ and $h = 0$. This shows that for every $p \in \mathcal{A}'$, we have $A(p,h) = 0$ for at least $\frac{1}{2}(p-1)$ values of $h$. Hence

$$\sum_{p\in\mathcal{A}'} p \sum_{h=1}^{p} \left(A(p,h)-(A/p)\right)^2 \geq \sum_{p\in\mathcal{A}'} \tfrac{1}{2}p(p-1)(A/p)^2$$

$$= |C|^2 \sum_{p\in\mathcal{A}'} (p-1)/2p$$

$$\geq (n^{\frac{1}{4}}|C|^2)/3. \tag{5.46}$$

Hence, by (5.45) and (5.46), we have

$$(n^{\frac{1}{4}}|C|^2)/3 \leq (2\pi n+n)|C|,$$

so that

$$|C| < c_2 n^{3/4} \tag{5.47}$$

for some absolute constant $c_2$. (5.44) and (5.47) imply (5.42), for $n > n_0$.

THEOREM 5.15. *Let $n$ and $k$ be natural numbers, $k \geq 2$. Let $S_1$ be the set of integers in $[n,2n)$ and let $S_i$ be a subset of the integers in $[2^{i-1}n, 2^i n)$, for $i = 2,\ldots,k$. Let $T_i \subseteq S_i$ be a maximal admissible set relative to $S_{i+1}$, for $i = 1,\ldots,k-1$. Let $f^{(k)}(n)$ be defined by*

$$f^{(k)}(n) = \min_{S_2,\ldots,S_k} (|T_1|+|T_2|+\ldots+|T_{k-1}|+|S_k|) \tag{5.48}$$

*where the minimum is taken over all possible choices of $S_2,\ldots,S_k$. Then for $n > n_0(k)$, and for some absolute constant $c$,*

$$f^{(k)}(n) < c2^{2k}kn^{(2/3)+(1/2)^k}. \tag{5.49}$$

*We note that $f^{(2)}(n) = f(n)$.*

PROOF. To prove (5.49), it will be sufficient to construct sets $S_2,\ldots,S_k$ such that

$$|T_1|+|T_2|+\ldots+|T_{k-1}|+|S_k| < c2^{2k}kn^{(2/3)+(1/2)^k}, \tag{5.50}$$

for all sufficiently large $n$.

Let $n > n_0(k)$, where $n_0(k)$ is sufficiently large. From the set of integers in the interval $(\frac{1}{2}n^{(\frac{1}{2})^{i-1}}, n^{(\frac{1}{2})^{i-1}})$ we can, by theorem 5.11, select a set $P_i$ of $[n^{(1/3)(1/2)^{i-2}}]$ distinct primes. We choose such a set $P_i$ for $i = 2,\ldots,k$. We then define sets $R_i$ by $R_1 = \{1\}$, and for $i = 2,\ldots,k$,

$$R_i = \{z | z = p_2p_3\ldots p_i,\text{ where } p_j \in P_j,\ j = 2,3,\ldots,i\}. \tag{5.51}$$

For a set S of integers, we let $\{x,y;S\}$ denote the subset of $[x,y)$ consisting of all the multiples of the elements of S. We recall the notation mS which denotes the set $\{ms \mid s \in S\}$.

Now we define $S_i$, for $i = 2,\dots,k$, by

$$S_i = \bigcup_{z\in R_i} z\{\frac{2^{i-1}n}{z}, \frac{2^i n}{z}; \{1\}\} . \tag{5.52}$$

Since for every $j = 2,\dots,i$, $p_j \in (\tfrac{1}{2}n^{(\frac{1}{2})^{j-1}}, n^{(\frac{1}{2})^{j-1}})$, we have $R_i \subseteq [(\tfrac{1}{2})^{i-1}n^a, n^a]$, where

$$a = \sum_{j=2}^{i} (\tfrac{1}{2})^{j-1} = 1-(\tfrac{1}{2})^{i-1}.$$

Hence each of the sets $\{\frac{2^{i-1}n}{z}, \frac{2^i n}{z}; \{1\}\}$ which appear in (5.52) contains less than

$$2^{i-1}n/(\tfrac{1}{2})^{i-1}n^a < 2^{2i}n^{(\frac{1}{2})^{i-1}}$$

elements. We have also that

$$|R_i| \le |P_2||P_3| \dots |P_i| \le n^b, \quad \text{where} \tag{5.53}$$

$b = \frac{1}{3}\sum_{j=2}^{i} (\tfrac{1}{2})^{j-2} = 2(1-(\tfrac{1}{2})^{i-1})/3.$

Hence, for each $i = 2,3,\dots,k$, we have

$$|S_i| < 2^{2i}n^u \tag{5.54}$$

where $u = (2+(\tfrac{1}{2})^{i-1})/3$. If we let $i = k$ in (5.54), we see that

$$|S_k| < 2^{2k}n^{(2/3)+(1/3)(1/2)^{k-1}} < 2^{2k}n^{(2/3)+(1/2)^k}. \tag{5.55}$$

Now let i satisfy $1 \le i \le k-1$. For every $z \in R_i$, we define

$$X_z = \{\frac{2^{i-1}n}{z}, \frac{2^i n}{z}; \{1\}\}$$

and

$$Y_z = \{\frac{2^i n}{z}, \frac{2^{i+1}n}{z}; P_{i+1}\} ,$$

so that by (5.52) we have

$$S_i = \bigcup_{z\in R_i} zX_z \quad \text{and} \quad S_{i+1} = \bigcup_{z\in R_i} zY_z .$$

Hence, if $X_z' \subseteq X_z$ is a maximal admissible set relative to $Y_z$, then

$$|T_i| \le \sum_{z\in R_i} |X_z'| . \tag{5.56}$$

We now apply corollary 5.13 to obtain a bound on the order of $X_z'$. We let $N = 2^i n/z$, $\mathcal{A} = X_z'$ and hence $A = |X_z'|$. Our set $P_{i+1}$ consists of $[n^{(1/3)(1/2)^{i-1}}]$ distinct primes p, all satisfying $p \leq X$, where $X = n^{(\frac{1}{2})^i}$. Hence

$$\sum_{p\varepsilon P_{i+1}} pD_p \leq \sum_{p\leq X} pD_p = \sum_{p\leq X} p \sum_{h=1}^{p} (A(p,h)-(A/p))^2$$

$$\leq ((2^i \pi n/z)+n^{(\frac{1}{2})^{i-1}})|X_z'|. \tag{5.57}$$

Finally we apply corollary 5.5 with $S = X_z'$ and $h = 0$. This shows that for every $p \varepsilon P_{i+1}$, we have $A(p,h) = 0$ for at least $\frac{1}{2}(p-1)$ values of h. Hence

$$\sum_{p\varepsilon P_{i+1}} p \sum_{h=1}^{p} (A(p,h)-(A/p))^2 \geq \sum_{p\varepsilon P_{i+1}} \tfrac{1}{2}p(p-1)(A/p)^2$$

$$= |X_z'|^2 \sum_{p\varepsilon P_{i+1}} (p-1)/2p$$

$$\geq (n^{(1/3)(1/2)^{i-1}}|X_z'|^2)/3. \tag{5.58}$$

Since

$$z \,\varepsilon\, [(\tfrac{1}{2})^{i-1}n^{1-(\frac{1}{2})^{i-1}}, n^{1-(\frac{1}{2})^{i-1}}],$$

we have

$$(n/z) \leq 2^{i-1}n^{(\frac{1}{2})^{i-1}} < 2^i n^{(\frac{1}{2})^{i-1}}. \tag{5.59}$$

(5.57)-(5.59) imply that

$$(n^{(1/3)(1/2)^{i-1}}|X_z'|^2)/3 \leq ((2^i\pi 2^i n^{(\frac{1}{2})^{i-1}}+n^{(\frac{1}{2})^{i-1}}))|X_z'|,$$

and hence that

$$|X_z'| < d2^{2k}n^{(2/3)(1/2)^{i-1}}, \tag{5.60}$$

for some absolute constant d.

By (5.53), there are at most $n^b$ elements in $R_i$. Hence, by (5.56) and (5.60),

$$|T_i| < d2^{2k}n^{(2/3)+(1/2)^k} \tag{5.61}$$

for $i = 1,2,\ldots,k-1$.

(5.55) and (5.61) now imply (5.50) and hence the theorem.

THEOREM 5.16. *Let $g(N)$ be the function given in definition 5.2. Then, for every $\varepsilon > 0$ and $N \geq N_1(\varepsilon)$, we have*

$$g(N) < cN^{(2/5)+\varepsilon}, \tag{5.62}$$

*for some absolute constant c.*

PROOF. Throughout this proof, c and $c_i$ denote absolute positive constants.

(i) We prove first that for every $N \geq N_0(\varepsilon)$ there exists m satisfying

$$N-2N^{3/5} \leq m < N \tag{5.63}$$

such that

$$g(m) < c_1N^{(2/5)+\varepsilon}. \tag{5.64}$$

To prove this statement, we construct a set A of m distinct natural numbers, where m satisfies (5.63), such that

$$g(A) < c_1N^{(2/5)+\varepsilon}. \tag{5.65}$$

The inequality (5.64) will then follow, since $g(m) \leq g(A)$.

So suppose that $N \geq N_0(\varepsilon)$, where $N_0(\varepsilon)$ is sufficiently large. We define a and b by

$$a = [N^{2/5}], \tag{5.66}$$

and

$$b = [N/a]-[N^{1/5}]. \tag{5.67}$$

For every j such that $1 \leq j \leq a$, we define the set $A_j$ by

$$A_j = \{2^{j-1}t;\ b \leq t \leq 2b-1\}. \tag{5.68}$$

We now extend theorem 5.15 so that, instead of considering each $S_i$ to be a subset of the integers in $[2^{i-1}n, 2^in)$, we let it be a subset of $\{t\ell;\ 2^{i-1}n \leq t < 2^in\}$, where $\ell$ is a natural number, independent of i. We choose k to be the least natural number such that

$$2^k > 1/\varepsilon, \tag{5.69}$$

and apply the extended version of theorem 5.15, with this value of k and with $\ell = 2^{a-1}$ and $n = b$.

Now let $S_i$ be as defined in the proof of theorem 5.15, and define $A_{a+i} = 2^{a-1}S_{i+1}$ for $i = 1,\ldots,k-1$. By theorem 5.15, for $i = 1,\ldots,k-1$, we can choose a subset $A_{a+i}$ of the set $\{2^{a-1}t;\ 2^ib \leq t < 2^{i+1}b\}$ such that, if $B_{a+i} \subseteq A_{a+i}$ for $i = 0, 1,\ldots,k-2$, and if $B_{a+i}$ is a maximal admissible set relative to $A_{a+i+1}$, then we have

$$|B_a|+|B_{a+1}|+\ldots+|B_{a+k-2}|+|A_{a+k-1}| < c_2 2^{2k} k b^{(2/3)+(1/2)^k}. \tag{5.70}$$

It also follows from (5.54) that

$$|A_{a+i}| < 2^{2k} b^{5/6} \tag{5.71}$$

for $i = 1,\ldots,k-1$.

It follows from the definition of the sets $A_1, A_2, \ldots, A_{a+k-1}$ that they are disjoint. We define

$$A = \bigcup_{j=1}^{a+k-1} A_j \tag{5.72}$$

so that

$$\left|A\right| = ab + \sum_{i=1}^{k-1} \left|A_{a+i}\right| . \tag{5.73}$$

Hence, by (5.66), (5.67), (5.71) and (5.73), we have

$$N-2N^{3/5} \leq |A| \leq N-[N^{1/5}][N^{2/5}]+k2^{2k+1}N^{\frac{1}{2}} .$$

If we choose $N_0(\varepsilon)$ sufficiently large, we can ensure that $m = |A|$ satisfies (5.63). Now we need only prove (5.65) in order to establish our statement.

Let $B \subseteq A$ be a maximal admissible set relative to A and, for each $j = 1,2,\ldots,$ $a+k-2$, let $B_j \subseteq A_j$ be a maximal admissible set relative to $A_{j+1}$. By (5.72), we see that

$$|B| \leq \sum_{j=1}^{a-1} |B_j| + \sum_{j=a}^{a+k-2} |B_j|+|A_{a+k-1}|. \tag{5.74}$$

By (5.66) and (5.67), $b < N^{3/5}$, so from (5.69) and (5.70)

$$\sum_{j=a}^{a+k-2} |B_j|+|A_{a+k-1}| < c_2 2^{2k} k N^{(2/5)+(3\varepsilon/5)} < c_2\varepsilon^{-3} N^{(2/5)+(3\varepsilon/5)}$$

and, since $N \geq N_0(\varepsilon)$, where $N_0(\varepsilon)$ has been chosen sufficiently large, this shows that

$$\sum_{j=a}^{a+k-2} |B_j|+|A_{a+k-1}| < c_3 N^{(2/5)+\varepsilon}. \tag{5.75}$$

Now we consider $|B_j|$, for $j = 1,2,\ldots,a-1$. Suppose that for some j, where $1 \leq j \leq a-1$, we have $|B_j| \geq 3$. By the definition of $A_j$ in (5.68) and since $B_j \subseteq A_j$, this implies that three distinct elements $2^{j-1}a_1, 2^{j-1}a_2, 2^{j-1}a_3 \in B_j$, where $b \leq a_i \leq 2b-1$, for $i = 1,2,3$. Of the three integers $a_1,a_2,a_3$, two must have the same parity. Let these be $a_1$ and $a_2$. Then $2|(a_1+a_2)$ so that the sum $(2^{j-1}a_1+2^{j-1}a_2)$ is divisible by $2^j$ and hence by (5.68) belongs to $A_{j+1}$. But this contradicts the fact that $B_j$ is admissible relative to $A_{j+1}$. Hence

$$|B_j| \leq 2, \text{ for } j = 1,2,\ldots,a-1. \tag{5.76}$$

From (5.66) and (5.76), it follows that

$$\sum_{j=1}^{a-1} |B_j| < c_4 N^{(2/5)+\varepsilon}. \tag{5.77}$$

By (5.74), (5.75) and (5.77), we have $|B| < (c_3+c_4)N^{(2/5)+\varepsilon}$, which implies (5.65). This completes the first stage of the proof.

(ii) We now show that the statement of (i) in fact implies the theorem. We have shown that for every $N \geq N_0(\varepsilon)$, there exists m satisfying

$$N-2N^{3/5} \leq m < N \tag{5.63}$$

such that

$$g(m) < c_1 N^{(2/5)+\varepsilon}. \tag{5.64}$$

From the definition of $g(m)$, we see that

$$g(x+y) \leqslant g(x)+g(y), \quad \text{for } x \geqslant 1,\ y \geqslant 1. \tag{5.78}$$

Suppose that

$$N \geqslant N_0^{5/2},$$

and let m be a natural number satisfying (5.63), such that (5.64) holds. By (5.78), we have

$$g(N) \leqslant g(m)+g(N-m) < c_1N^{(2/5)+\varepsilon}+g(N-m). \tag{5.79}$$

We now show that

$$g(N-m) < c_5N^{(2/5)+\varepsilon}. \tag{5.80}$$

If

$$N-m \leqslant N^{(2/5)+\varepsilon},$$

then

$$g(N-m) \leqslant N^{(2/5)+\varepsilon},$$

so we may assume that

$$N-m > N^{(2/5)+\varepsilon}.$$

We have

$$N \geqslant N_0^{5/2},$$

so

$$N-m \geqslant N_0.$$

Hence by (i), there exists m' satisfying

$$(N-m)-2(N-m)^{3/5} \leqslant m' < N-m \tag{5.81}$$

such that

$$g(m') < c_6(N-m)^{(2/5)+\varepsilon}.$$

By (5.78), we have

$$\begin{aligned} g(N-m) &\leqslant g(N-m-m')+g(m') \\ &< g(N-m-m')+c_6(N-m)^{(2/5)+\varepsilon} \\ &< g(N-m-m')+c_6N^{(2/5)+\varepsilon}. \end{aligned} \tag{5.82}$$

We know from (5.63) that

$$N-m < c_7N^{3/5},$$

so by (5.81)

$$N-m-m' < c_7(N-m)^{3/5} < c_7{}^2N^{9/25} < c_7{}^2N^{(2/5)+\varepsilon},$$

and hence

$$g(N-m-m') < c_8N^{(2/5)+\varepsilon}. \tag{5.83}$$

(5.82) and (5.83) imply (5.80), which in turn with (5.79) shows that

$$g(N) < cN^{(2/5)+\varepsilon} \qquad (5.62)$$

for

$$N \geq N_1(\varepsilon) = \{N_0(\varepsilon)\}^{5/2}.$$

This completes the proof of the theorem.

5.8 REFERENCES AND RELATED TOPICS. Erdos (1965) discusses the function $\phi(N)$, defined to be the largest number such that, from every set of N distinct positive real numbers, we can always choose a subset of $\phi(N)$ real numbers with the property that no sum of two distinct numbers of this subset belongs to the original set. He gives bounds on $\phi(N)$. The proofs given here are those of Choi (1971), who showed that the function $\phi(N) = g(N)$ (as defined in section 5.1) and worked in terms of $g(N)$. The lower bound was originally found by D.A. Klarner but never published; the proof in section 5.2 is Choi's. The upper bound and the related theorems in section 5.7 are due to Choi, who also discusses some generalisations of the problem.

Of the preliminary results given in this chapter, lemma 5.4 is again from Choi (1971); the results in section 5.5 largely follow Chandrasekharan (1966), who in particular quotes a proof due to S.S. Pillai on which we have based the proof of theorem 5.11; the treatment of the large sieve inequality is that of Halberstam (1968) and the particular version (theorem 5.12) of the inequality which we use is that of Gallagher (1967).

For related problems, see also Straus (1966).

# CHAPTER VI. ADDITION THEOREMS FOR GROUPS

In order to study sum-free sets in groups, we need some results concerning the addition of subsets of group elements.

## 6.1 NOTATION AND BASIC CONCEPTS.

Throughout this chapter, groups are written additively, whether or not they are abelian. Most of our results apply to finite abelian groups, though some are more general.

Let G denote a group; then $A \leq (\subseteq) G$ means that A is a subgroup (subset) of G. $\bar{A}$ denotes the set-theoretic complement of A in G and $|A|$ the cardinality of A. We define the sum A+B by $A+B = \{a+b \mid a \in A, b \in B\}$ where $A,B \subseteq G$, so that in particular $A+\emptyset = \emptyset$. Also, for $A \subseteq G$, we use the notation $-A = \{-a \mid a \in A\}$ and $kA = \{ka \mid a \in A\}$ for k an integer. Since A-B denotes $\{a-b \mid a \in A, b \in B\}$, we use $A \setminus B$ to denote $\{g \mid g \in A, g \notin B\}$.

We denote by $Z_n$ the cyclic group of order n, for any integer n.

THEOREM 6.1. *Let $A,B \subseteq G$, where G is a group. Then either*

$$G = A+B \quad or \quad |G| \geq |A|+|B|.$$

PROOF. If $G \neq A+B$, choose $g \in \overline{A+B}$. Define the set B' by

$$B' = \{g-b \mid b \in B\} = \{g\}-B,$$

so that $|B'| = |B|$, $B' \subseteq G$.

Suppose that $A \cap B' \neq \emptyset$; then there exists $a \in A \cap B'$. Hence $a = g-b$ or $g = a+b$, which is a contradiction. So $A \cap B' = \emptyset$, implying that

$$|G| \geq |A|+|B'| = |A|+|B|.$$

Most of the addition theorems relate to abelian groups and most of them are proved by transformation of the summand sets. Here we use a transform similar to that introduced by Cauchy.

DEFINITION 6.2. Let G be an abelian group, $A,B \subseteq G$, $g \in G$. Then the *transform* of the pair (A,B) by g is the pair $(A^g,B_g)$ defined by

$$A^g = A \cup (B+g), \quad B_g = B \cap (A-g).$$

LEMMA 6.3. *(i)* $|A^g|+|B_g| = |A|+|B|$;

*(ii)* $A^g+B_g \subseteq A+B$ *where, in particular,* $A^g+B_g = \emptyset$ *if* $B_g = \emptyset$.

PROOF. (i) $|A^g|+|B_g| = |A\cup(B+g)| + |B\cap(A-g)|$

$$= |A\cup(B+g)| + |(B+g)\cap A|$$

$$= |A|+|B+g|$$

$$= |A|+|B|.$$

(ii) Let $a\in A^g$, $b\in B_g$. Then $b\in B$ so that, if $a\in A$, we have $a+b\in A+B$. If $a\notin A$, then $a\in B+g$, so $a = \beta+g$ for some $\beta\in B$. But $B_g\subseteq A-g$, so $b = \alpha-g$ for some $\alpha\in A$. Hence $a+b = \beta+\alpha\in A+B$.

We use this transform to prove the simplest of the addition theorems.

THEOREM 6.4. *Let p be a prime, let* $G = Z_p$ *and let* $A,B\subseteq G$. *Then*

$$|A+B| \geq min(p,|A|+|B|-1).$$

PROOF. If $|A|$ or $|B| = 1$, the theorem is obviously true. If $|A|+|B| > p$, then by theorem 6.1 $|A+B| = p$ and again the theorem is true. So we may assume that $|A|+|B| \leq p$, $|A|,|B| \geq 2$ and (without loss of generality) that $0\in B$. We proceed by induction on $|B|$.

(i) We show first that $A \neq A+B$. For since $|B| \geq 2$, we may choose $b\in B$, $b\neq 0$; adding this element b to some $a\in A$, we have $a,a+b,\dots,a+kb\in A$ for every k. Hence $A = Z_p$, which is a contradiction.

(ii) Next we show that for some $a\in A$, $|B|>|B_a|$. For if $|B| = |B_a|$ for every $a\in A$, then $B\subseteq A-a$ or $B+a\subseteq A$ for every $a\in A$. Hence $A+B\subseteq A$. But since $0\in B$, we have $A\subseteq A+B$ and hence $A = A+B$, contradicting (i).

(iii) If $|B| = 2$, we want to show that $|A+B| \geq min(p,|A|+1)$, i.e. that $|A+B| \geq |A|+1$.

Suppose not. Then $|A+B| \leq |A|$. But we know that $|A+B| \geq |A|$, so again we have $A = A+B$ which is impossible by (i).

(iv) Now if the theorem is true for $|B| < n$, we choose B such that $|B| = n$ and $a\in A$ such that $|B| > |B_a|$. Then $|A+B| \geq |A^a+B_a|$ by lemma 6.3,

$\geq |A^a|+|B_a|-1$ by the induction hypothesis,

$= |A|+|B|-1$ by lemma 6.3.

It is often important to know the conditions under which equality holds in theorem 6.4.

DEFINITION 6.5. Let $A,B\subseteq G = Z_p$. Then (A,B) is said to be a *critical pair* if and only if

$$|A+B| = min(p,|A|+|B|-1).$$

DEFINITION 6.6. Let $A\subseteq G$, where G is an abelian group. A is said to be an *arithmetic progression with difference d* or a *standard set with difference d* if and only if

$A = \{a+id \mid i = 0,1,\ldots,|A|-1,$ for some $a,d \in G,\ d \neq 0\}$. (A,B) is said to be a *standard pair with difference d* if and only if both A and B are standard sets with difference d.

We find that, with certain trivial exceptions, critical pairs in $Z_p$ are standard pairs.

LEMMA 6.7. *Let $A,B,C,D \subseteq G$, where G is an abelian group. Suppose that $A-B = C-D$. Then $A \cap B = \emptyset$ if and only if $C \cap D = \emptyset$.*

PROOF. If $g \in C \cap D$, then $0 = g-g \in C-D$. Hence $0 \in A-B$, which implies that there exists $h \in A \cap B$.

COROLLARY 6.8. *(i) $(A+B) \cap C = \emptyset$ if and only if $A \cap (C-B) = \emptyset$.*

*(ii) $(A-B) \cap (C+D) = \emptyset$ if and only if $(A-C) \cap (B+D) = \emptyset$.*

EXAMPLE. We note that $A-B = C$ implies that $A \subseteq B+C$, but need not imply that $A = B+C$, as in the following example: $G = Z_7$, $A = B = \{3,4\}$ and $C = \{6,0,1\}$.

THEOREM 6.9. *Let $A,B \subseteq G = Z_p$, where p is prime. Then (A,B) is a critical pair if and only if one of the following is satisfied:*

*(i) $|A|+|B| > p$;*

*(ii) $\min(|A|,|B|) = 1$;*

*(iii) $A = \overline{g-B}$, for some $g \in G$;*

*(iv) (A,B) is a standard pair.*

PROOF. (a) We check first that any one of the given conditions is sufficient for criticality. The sufficiency of (i) follows from theorem 6.1, and that of (ii) is obvious. If (iii) holds, then $A \cap (g-B) = \emptyset$. Hence from corollary 6.8, we have $(A+B) \cap \{g\} = \emptyset$ so that

$$|A+B| \leq p-1. \tag{6.1}$$

But $|A| = |\overline{B}| = p-|B|$ implying that $|A|+|B| = p$. So by theorem 6.4,

$$|A+B| \geq \min(p,p-1) = p-1. \tag{6.2}$$

Hence by (6.1) and (6.2), $|A+B| = p-1 = |A|+|B|-1$ and (A,B) is a critical pair.

Finally if (iv) holds, we have

$$A = \{a+id \mid i = 0,1,\ldots,|A|-1;\ a,d \in G,\ d \neq 0\}$$

and

$$B = \{b+id \mid i = 0,1,\ldots,|B|-1;\ b,d \in G,\ d \neq 0\},$$

so that

$$A+B = \{a+b+id \mid i = 0,1,\ldots,|A|+|B|-2\}$$

and

$$|A+B| = \min(p,|A|+|B|-1).$$

(b) Now suppose that (A,B) is a critical pair. If $|A|+|B| > p$, condition (i) is satisfied. If $\min(|A|,|B|) = 1$, condition (ii) is satisfied. If $|A|+|B| = p$, then by criticality, we have $|A+B| = p-1$, so that $A+B = \overline{\{g\}}$ for some $g \in G$. Hence $(A+B)\cap\{g\} = \emptyset$ and by corollary 6.8, $A\cap(g-B) = \emptyset$ or $A \subseteq \overline{g-B}$. But $|A| = p-|B| = |\overline{B}| = |\overline{g-B}|$, so $A = \overline{g-B}$ and (iii) is satisfied.

So now we assume that (A,B) is a critical pair with $\min(|A|,|B|) > 1$ and $|A|+|B| < p$, and we show that (iv) holds. The proof is in five steps, working always with these assumptions.

(I) If A is a standard set, then (A,B) is a standard pair.

PROOF. We assume without loss of generality that $A = \{0,1,\ldots,|A|-1\}$ and that $0 \in B$.

If $b, b+n+1 \in B$ but the n elements $b+1,\ldots,b+n \in \overline{B}$, then we say that there is a *gap of length n* in B. We consider the gaps in B. If $b \in B$, then $\{b,b+1,\ldots,b+|A|-1\} \subseteq A+B$. But since $|A+B| < p-1$, there must be some elements of G left out of A+B and hence there must be at least one gap in B of length at least $|A|$.

Suppose B has at least one other gap. Then in A+B we have all the elements of B, together with $|A|-1$ elements from the big gap in B, and at least one element from the second gap. Hence $|A+B| \geq |B|+(|A|-1) = |A|+|B|$, which contradicts the criticality of (A,B).

Hence B has only one gap, that is B is a set of consecutive elements, and (A,B) is a standard pair.

COROLLARY 6.10. *If $\min(|A|,|B|) = 2$, then (A,B) is a standard pair.*

(II) Let $D = \overline{A+B}$. Then (-A,D) is a critical pair.

PROOF. Since $(A+B)\cap D = \emptyset$, we have $B\cap(D-A) = \emptyset$ by corollary 6.8, and hence

$$B \subseteq \overline{D-A} = E. \tag{6.3}$$

But $E\cap(D-A) = \emptyset$ which implies by corollary 6.8 that $(E+A)\cap D = \emptyset$. Hence

$$A + E \subseteq \overline{D} = A + B. \tag{6.4}$$

By (6.3), $B \subseteq E$, so that $A+B \subseteq A+E$. This, with (6.4), implies that $A+B = A+E$. Now

$$|A+B| = |A|+|B|-1 = |A+E| \geq \min(p,|A|+|E|-1).$$

Hence $|B| \geq |E|$. But $B \subseteq E$, so this implies that $B = E$. Hence $\overline{B} = D-A$. Thus $|D-A| = p-|B|$ and $|D| = p-|A+B| = p-|A|-|B|+1$. So $|D-A| = \min(p,|D|+|-A|-1)$ and (-A,D) is a critical pair.

(III) If A+B is a standard set, then (A,B) is a standard pair.

PROOF. If A + B is a standard set, then $D = \overline{A + B}$ is a standard set. Now by (II), (-A, D) is a critical pair. Also

$$|-A| + |D| = |A| + p - |A| - |B| + 1 < p$$

and

$$\min(|-A|, |D|) > 1.$$

So by (I), (-A, D) is a standard pair. This implies that -A is a standard set. Hence A is also a standard set and, again by (I), (A, B) is a standard pair.

(IV) If $|B| \geq 3$ and $0 \in B$, then there exists $a \in A$ such that

$$|B| > |B_a| \geq 2.$$

PROOF. (i) Let $Y = \{a \in A \mid |B| > |B_a|\}$. If Y = A, then $|Y| \geq 2$. We show that $|Y| \geq 2$ always.

For if $Y \neq A$, let $Z = A \setminus Y \neq \emptyset$. Since $B_z \subseteq B$ and $|B_z| = |B|$ for all $z \in Z$, we have $B_z = B$. Hence $B \cap (A - z) = B$ for all $z \in Z$. This implies that $B \subseteq A - z$ and hence that $B + z \subseteq A$ for all $z \in Z$. So $B + Z \subseteq A$ and we see that

$$|A| \geq |B + Z| \geq |B| + |Z| - 1 \quad \text{by Theorem 6.4}$$

$$\geq |Z| + 2 \quad \text{since } |B| \geq 3.$$

Hence $|Y| = |A| - |Z| \geq 2$ always.

(ii) Now suppose that, for every $a \in Y$, $|B_a| < 2$, ie. that $B \cap \{A - a\} = \{0\}$. Let $E = B\setminus\{0\}$ so that $E \cap (A - a) = \emptyset$. Then by Corollary 6.8, $(E + a) \cap A = \emptyset$. But this is true for every $a \in Y$, so that $(E + Y) \cap A = \emptyset$.

Now $E + Y \subseteq A + B$ and $A \subseteq A + B$. Hence

$$|E + Y| \leq |A + B| - |A| = |B| - 1 \quad \text{since (A, B) is a critical pair}$$
$$= |E|.$$

But

$$|E + Y| \geq |E| + |Y| - 1 \geq E + 1 \text{ by (i).}$$

Hence $|B_a| \geq 2$ for some $a \in Y$.

(V) To complete the proof, we use induction on $|B|$, with Corollary 6.10 as the first step, where $|B| = 2$. Again we may assume without loss of generality that $0 \in B$. We assume that, under our hypotheses, (A,B) is a standard pair if $2 \leq |B| \leq k$; now consider the case where $3 \leq |B| \leq k + 1$.

By (IV), we may choose $a \in A$ such that $|B| > |B_a| \geq 2$. Hence

$$|A| + |B| - 1 = |A + B| \quad \text{by the criticality of (A,B),}$$

$$\geq |A^a + B_a| \quad \text{by Lemma 6.3,}$$

$$\geq |A^a| + |B_a| - 1 \qquad \text{by Theorem 6.4,}$$

$$= |A| + |B| - 1 \qquad \text{by Lemma 6.3.} \qquad (6.5)$$

So

$$|A^a + B_a| = |A^a| + |B_a| - 1$$

and $(A^a, B_a)$ is a critical pair.

But $2 \leq |B_a| < |B|$ so by the induction hypothesis $(A^a, B_a)$ is a standard pair. Hence $A^a + B_a$ is a standard set.

Now

$$A^a + B_a \subseteq A + B \qquad \text{by Lemma 6.3}$$

and

$$|A^a + B_a| = |A + B| \qquad \text{by (6.5).}$$

Hence

$$A + B = A^a + B_a$$

and is thus a standard set. By (III), (A, B) must be a standard pair and the theorem is proved.

We need also one further result on standard sets in $Z_m$.

LEMMA 6.11. *Let $G = Z_m$ and let $A \subseteq G$ be the set*

$$A = \{a + id \mid i = 0, 1, \dots , s\}$$

*where $(d,m) = 1$ and $1 \leq s \leq m - 3$. If we can also write*

$$A = \{b + ie \mid i = 0, 1, \dots , s\}$$

*then*

$$d \equiv \pm e \pmod m.$$

PROOF. Without loss of generality we may assume that a = 0 and d = 1. The lemma is true for s = 1, so we assume also that $2 \leq s \leq m - 3$.

Now for some j, $0 \leq j \leq s$, we have $0 \equiv b + je \pmod m$ and hence

$$A = \{0, 1, \dots , s\} = \{ie \mid -j \leq i \leq s - j\}$$

and at least one of $e, -e \in A$. Hence

$$A = \{ic \mid -k \leq i \leq s - k\},$$

where $1 \leq c \leq s$ and $c \equiv \pm e \pmod m$.

Suppose that $1 < c$. Then $1 \leq s + 1 - c < s + 2 - c \leq s$, and hence $s + 1 - c, s + 2 - c \in A$. But we know that $s + 1, s + 2 \notin A$. Hence each of the elements $s + 1 - c$, $s + 2 - c$ must be the last term of the arithmetic progression so

$$s + 1 - c = s + 2 - c = (s - k)c$$

which is a contradiction.

Hence $c = 1$ and $e \equiv \pm 1 \pmod m$.

**6.2 FURTHER RESULTS IN ABELIAN GROUPS.** Theorems 6.4 and 6.9 have many generalisations. In order to derive those which have been used in studying sum-free sets, we need several preliminary results.

DEFINITION 6.12. Let G be an abelian group and let $C \subseteq G$. If $H \leq G$ is a non-trivial subgroup such that $C + H = C$, then C is a union of cosets of H in G and is said to be *periodic* with period H. H is not uniquely determined by this definition; we let $H(C) = \{g \in G \mid C + g = C\}$ denote the largest period of C.

If $C + H = C$ implies $H = \{0\}$, then C is said to be *aperiodic*.

If $C = C' \cup C''$, where

$$C' \cap C'' = \emptyset,\ C' \neq \emptyset,$$

$$C' + H = C' \text{ for some } H \neq \{0\}$$

and $C' \subseteq c + H$ for some $c \in C''$, then C is said to be *quasiperiodic*, with quasiperiod H.

LEMMA 6.13. *Let G be an abelian group and let $C \subseteq G$, where C is finite. Suppose that*

$$C = C_0 \cup C_1 \cup \ldots \cup C_n, \qquad n \geq 1,$$

*where $\emptyset \neq C_i \neq C$ and*

$$|C| < |C_i| + |H(C_i)| \quad \text{for } i = 0, 1, \ldots, n. \tag{6.6}$$

*Then (i)*

$$|C| + |H(C)| \geq |C_i| + |H(C_i)| \tag{6.7}$$

*for at least one value of $i = 0, 1, \ldots, n$, and*

*(ii) either C is quasiperiodic or there exist $c \in C$ and finite subgroups $H_1, H_2 \leq G$ such that*

$$|H_1| = |H_2|, \quad H_1 \cap H_2 = \{0\}$$

*and*

$$C - c = H_1 \cup H_2.$$

PROOF. We proceed by induction on n.

(i) If $n = 1$, we have

$$C = C_0 \cup C_1 \tag{6.8}$$

where $\emptyset \neq C_0, C_1 \neq C$. We denote by $H_i$ the largest period $H(C_i)$ of $C_i$ for $i = 0, 1$; we denote $H_0 \cap H_1$ by H and we denote by $H^*$ the subgroup $\langle H_0, H_1\rangle$ generated by $H_0$ and $H_1$. We also let $h = |H|$ and $m_i = [H^* : H_i]$. Hence by the isomorphism theorem,

$$|H^*| = m_0 m_1 h, \quad |H_0| = m_1 h \text{ and } |H_1| = m_0 h.$$

By (6.8), neither $\overline{C_0} \cap C_1$ nor $C_0 \cap \overline{C_1}$ is empty. Since $C_0$, $C_1$ are unions of cosets of $H_0$, $H_1$ respectively, each of $\overline{C_0} \cap C_1$ and $C_0 \cap \overline{C_1}$ must be a union of cosets of H. By (6.8),

$$|C| = |C_0 \cup C_1| = |C_0| + |C_1| - |C_0 \cap C_1|$$

and by (6.6)

$$|C| < |C_0| + |H_0|,$$

so that

$$|\overline{C_0} \cap C_1| = |C_1| - |C_0 \cap C_1| < |H_0| = m_1 h.$$

Since we are working with cosets of H, this implies that

$$|\overline{C_0} \cap C_1| \leq (m_1 - 1)h.$$

So we have

$$0 < |\overline{C_0} \cap C_1| \leq (m_1 - 1)h$$

and by a similar argument (6.9)

$$0 < |C_0 \cap \overline{C_1}| \leq (m_0 - 1)h.$$

From (6.8), we know that C is a union of cosets of H and hence

$$|H(C)| \geq h.$$

So to prove (6.7), it will be sufficient to show that

$$|\overline{C_i} \cap C_j| = (m_j - 1)h \qquad (6.10)$$

for at least one of (i,j) = (0, 1) or (1, 0).

Without loss of generality, we assume

$$|H_0| \geq |H_1|$$

ie. $m_0 \leq m_1$. From (6.9), we may choose some $c_0 \in C_0 \cap \overline{C_1}$. Let $D = c_0 + H^*$. Then $c_0 \in C_0 \cap \overline{C_1} \cap D$, so that

$$0 < |C_0 \cap \overline{C_1} \cap D|. \qquad (6.11)$$

Now $C_i \cap D$ is the union of $u_i$ cosets of $H_i$, where $0 \leq u_i \leq m_i$, so $\overline{C_i} \cap D$ is the union of $(m_i - u_i)$ cosets of $H_i$. Since the intersection of a coset of $H_0$ with a coset of $H_1$ is a coset of H, we have

$$|\overline{C_0} \cap C_1 \cap D| = (m_0 - u_0)u_1 h. \qquad (6.12)$$

Also

$$\begin{aligned} 0 &< |C_0 \cap \overline{C_1} \cap D| \quad \text{by (6.11)} \\ &= (m_1 - u_1)u_0 h, \quad \text{by an argument similar to that for (6.12),} \\ &\leq (m_0 - 1)h \quad \text{by (6.9)} \\ &\leq (m_1 - 1)h \quad \text{since } m_0 \leq m_1. \end{aligned} \qquad (6.13)$$

From the first and second inequalities of (6.13) we have

$$1 \leq u_0 \leq m_0 - 1 \qquad (6.14)$$

and from the first and third inequalities we have

$$1 \leq u_1 \leq m_1 - 1. \qquad (6.15)$$

Now by (6.9), (6.12), (6.13), (6.14) and (6.15), we have

$$\begin{aligned}
0 &\geq -\{(m_1-1)h - |\overline{C}_0 \cap C_1|\} - \{(m_0-1)h - |C_0 \cap \overline{C}_1|\} \\
&\qquad - |\overline{C}_0 \cap C_1 \cap \overline{D}| - |C_0 \cap \overline{C}_1 \cap \overline{D}| \\
&= |\overline{C}_0 \cap C_1 \cap D| - (m_1-1)h + |C_0 \cap \overline{C}_1 \cap D| - (m_0-1)h \\
&= (m_0 - u_0)u_1 h + (m_1 - u_1)u_0 h - (m_0 - 1)h - (m_1 - 1)h \\
&= \{(m_0 - u_0 - 1)(u_1 - 1) + (m_1 - u_1 - 1)(u_0 - 1)\}h \\
&\geq 0
\end{aligned}$$

and hence each of these expressions is zero. This implies:

(a) $|\overline{C}_i \cap C_j \cap D| = (m_j - 1)h$, which means that

$$|\overline{C}_i \cap C_j| = (m_j - 1)h \text{ and } \overline{C}_i \cap C_j \cap \overline{D} = \emptyset.$$

This in turn implies (6.10) and hence (6.7);

(b) neither term in the last expression is negative, ie.

$$(m_i - u_i - 1)(u_j - 1) \geq 0 \quad \text{for } i, j = 0, 1,$$

so both must be zero. Hence either

$$u_0 = m_0 - 1 \text{ or } u_1 = 1$$

and either

$$u_0 = 1 \quad \text{or } u_1 = m_1 - 1.$$

Now let $C' = C \cap \overline{D}$. Since

$$\overline{C}_0 \cap C_1 \cap \overline{D} = \emptyset = C_0 \cap \overline{C}_1 \cap \overline{D}$$

and since $C_0 \neq C \neq C_1$, we have

$$C' = C_0 \cap \overline{D} = C_1 \cap \overline{D}.$$

But $C_0 \cap \overline{D}$ is a union of cosets of $H_0$ and $C_1 \cap \overline{D}$ is a union of cosets of $H_1$. Since these are equal, $C'$ must be a union of cosets of $H^*$, ie.

$$C' + H^* = C'.$$

By (6.9), $m_j \geq 2$ and hence $|H_i| \geq 2$.

If $C' \neq \emptyset$ then $H^*$ is a quasiperiod of C. If $h \geq 2$, then H is a quasiperiod of C. If $u_0 = m_0 - 1$, then all but one of the $m_0$ cosets of $H_0$ contained in D is contained in $C_0 \cap D$; this missing coset may or may not be contained in $C_1$; $C'$ is a union of cosets of $H^*$ and hence $H_0$, so $H_0$ is a quasiperiod of C. Similarly if $u_1 = m_1 - 1$, $H_1$ is a quasiperiod of C.

This leaves us to consider finally the case where

$$C' = \emptyset \text{ and } u_0 = u_1 = h = 1.$$

This implies that $H_0 \cap H_1 = \{0\}$ and that C is the union of one coset of $H_0$ and one coset of $H_1$, both of which are contained in D. Since $C \subseteq D$, we have by (6.10) that

$$|\overline{C}_0 \cap C_1| = m_1 - 1,$$

and by (6.13) that

$$|\overline{C}_0 \cap C_1| = m_0 - 1.$$

Hence

$$m_0 = m_1 \quad \text{and} \quad |H_0| = |H_1|.$$

This proves the lemma for the case $n = 1$.

(ii) Suppose the lemma is true for $1 \leq k \leq n - 1$, and consider $n = k$.

If C can be written as the union of some proper subcollection of $C_0, C_1, \ldots, C_n$, then the statement follows by the induction hypothesis. So we assume that none of these sets can be omitted from the union.

Let $C_1' = C_1 \cup \ldots \cup C_n$, so that

$$C_1, \ldots, C_n \not\subseteq C_1' \quad \text{and} \quad C_0, C_1' \not\subseteq C.$$

Since

$$|C| < |C_i| + |H_i| \text{ for } i = 0,1, \ldots, n,$$

and since

$$|C_1'| < |C|,$$

it follows from the induction hypothesis that

$$|C_1'| + |H(C_1')| \geq |C_i| + |H_i| > |C|,$$

for at least one index i, $1 \leq i \leq n$. But

$$C = C_0 \cup C_1', \text{ and } |C_0| + |H_0| > |C|,$$

so from the result for $n = 1$, we now have the lemma for $n = k$.

COROLLARY 6.14. *Let $C_0, C_1, \ldots, C_n$ be finite non-empty sets such that*

$$|C_i| + |H(C_i)| \geq k, \text{ for } i = 0, 1, \ldots, n.$$

*Let $C = C_0 \cup C_1 \cup \ldots \cup C_n$. Then*

$$|C| + |H(C)| \geq k.$$

6.3 SMALL SUM-SETS IN ABELIAN GROUPS. If $A, B \subseteq G$, then the sum $A + B$ is considered to be small when

$$|A + B| \leq |A| + |B| - 1.$$

In this section we derive the generalisations of Theorems 6.4 and 6.9 which we need for studying sum-free sets.

THEOREM 6.15. *Let A and B be finite subsets of an abelian group G. Suppose that*

$$|A + B| \leq |A| + |B| - 1. \tag{6.16}$$

*Then $H = H(A + B)$ satisfies*

$$|A + B| + |H| = |A + H| + |B + H| \tag{6.17}$$

*and hence $A + B$ is periodic if*

$$|A + B| \leq |A| + |B| - 2.$$

PROOF. Fix $b_i \in B$. Let $(A_i, B_i)$ be a pair of finite subsets of G such that

(i) $A \subseteq A_i$, $b_i \in B_i$, $A_i + B_i \subseteq A + B$ and

$$|A_i| + |B_i| = |A + H| + |B + H|;$$

(ii) $|A_i|$ is maximal subject to (i).

Since $A + B + H = A + B$, we know that $A_i = A + H$, $B_i = B + H$ is a pair satisfying (i).

Now let $C_i = A_i + B_i$. Then

$$A + b_i \subseteq C_i \subseteq A + B. \tag{6.18}$$

Let $a \in A_i$. Then

$$A_i' = A_i \cup (a + B_i - b_i) \text{ and}$$

$$B_i' = B_i \cap (-a + A_i + b_i)$$

is a pair which satisfies (i). But $A_i \subseteq A_i'$ so, by (ii),

$$A_i = A_i'$$

and

$$a + B_i - b_i \subseteq A_i \quad \text{for all } a \in A_i.$$

Hence

$$A_i = A_i + B_i - b_i = C_i - b_i$$

and

$$H(C_i) = H(A_i) \supseteq B_i - b_i.$$

So from (i),

$$|C_i| + |H(C_i)| \geq |A + H| + |B + H|$$

and from (5.18),

$$A + B = \bigcup_{b_i \in B} C_i.$$

We now apply Corollary 6.14. Since $H = H(A + B)$, we have

$$|A + B| + |H| \geq |A + H| + |B + H|.$$

If (6.17) were false, then (since all the terms are multiples of $|H|$), we would have

$$|A + B| \geq |A + H| + |B + H|,$$

contradicting (6.16). Hence (6.17) is true and the theorem is proved.

The preceding theorem generalises Theorem 6.4. Now we consider the generalisations of Theorem 6.9 which give us some information about the conditions under which the order of the sum-set is small compared with the orders of the summand sets, ie. when we have

$$|A + B| \leq |A| + |B| - 1.$$

THEOREM 6.16. *Let G be an abelian group, with finite subsets A, B such that*

$$|A|, |B| \geq 2 \quad and \quad |A + B| \leq |A| + |B| - 1.$$

*Then either A + B is a standard set or A + B is quasiperiodic.*

PROOF. Let A + B = C. We have two cases to consider:

(i) First, suppose that for every $b_i \in B$, we can find $C_i \neq C$ such that $A + B_i \subsetneq C_i \subsetneq C$ and

$$|C_i| + |H(C_i)| > |C|. \tag{6.19}$$

Then C = A + B is the union of proper subsets $C_i$.

Suppose C is not quasiperiodic. Then by Lemma 6.13, there exists an element

$$c_0 = a_0 + b_0, \quad \text{where } a_0 \in A, b_0 \in B,$$

so that $c_0 \in C$, and finite subgroups $H_1$, $H_2$ of equal order such that

$$H_1 \cap H_2 = \{0\}$$

and

$$A + B - c_0 = H_1 \cup H_2.$$

Let $A' = -a_0 + A$ and $B' = B - b_0$. Then

$$A' + B' = H_1 \cup H_2, \quad A' \cup B' \subseteq H_1 \cup H_2 \tag{6.20}$$

and

$$|A'|, |B'| \geq 2.$$

Now choose $a \in A'$, $a \neq 0$. Then $a \in H_1 \cup H_2$ and without loss of generality we assume $a \in H_1$, $a \notin H_2$. Choose $b \in B' \cap H_2$. Then $a + b \in H_1 \cup H_2$ by (6.20). Hence $b = 0$ and $B' \subseteq H_1$. Now choose $b \in B'$, $b \neq 0$. Then a similar argument shows that $A' \subseteq H_1$. Hence

$$H_1 \cup H_2 = A' + B' \subseteq H_1$$

which is impossible.

So in this case C must be quasiperiodic.

(ii) Secondly, we suppose that for some $b_i \in B$, if

$$A + b_i \subseteq C_i \subseteq C, \text{ and } |C_i| + |H(C_i)| > |C|,$$

then

$$C_i = C.$$

In this case, consider $B - b_i$ instead of $B$, so that $0 \in B$ and

$$A \subseteq C_0 \subseteq C, \quad |C_0| + |H(C_0)| > |C|$$

implies

$$C_0 = C. \tag{6.21}$$

Now let $(A_0, B_0)$ be a pair of finite subsets of $G$ such that:

(a) $A \subseteq A_0$, $0 \in B_0$, $A_0 + B_0 \subseteq C$ (hence $A_0 \subseteq C$), $|B_0| \geq 2$ and

$$|A_0| + |B_0| = |C| + 1;$$

(b) $|A_0|$ is maximal subject to (a).

Now $|B| \geq 2$, so if

$$|A + B| = |A| + |B| - 1,$$

then the pair $A = A_0$, $B = B_0$ satisfy (a). If

$$|A| + |B| > |C| + 1,$$

then we can choose some subset $A$ to be $A_0$ and $B = B_0$, satisfying (a).

If $A_0 + B_0 = A_0$, then

$$|H(A_0)| \geq |B_0|.$$

Also $A \subseteq A_0 = A_0 + B_0 \subseteq C$. If we apply (6.21) with $A_0 = C_0$, we have

$$|C_0| + |H(C_0)| = |A_0| + |H(A_0)| \geq |A_0| + |B_0| > |C|.$$

Hence by (6.21), $A_0 = C$. But this implies that $|B| = 1$ which is a contradiction.

Since $A_0 + B_0 \neq A_0$, we consider the set

$$D_1 = \{a \in A_0 \mid a + B_0' \not\subseteq A_0\}, \text{ where } B_0' = B_0 \setminus \{0\}, D_1 \neq \emptyset.$$

From (a), and since $A_0 \subseteq C$,

$$|B_0'| = |B_0| - 1 = |C| - |A_0| = |D_0|, \tag{6.22}$$

where $D_0 = C \cap \overline{A_0}$.

We choose $a \in D_1$ and consider the sets

$$A_1 = A_0 \cup (a + B_0'), \quad B_1 = B_0 \cap (-a + A_0).$$

Now $A \subseteq A_0 \subseteq A_1$, $0 \in B_1$, $|A_1| > |A_0|$. By Lemma 6.3, we have

$$A_1 + B_1 \subseteq A_0 + B_0 \subseteq C.$$

By (a) and Lemma 6.3, we have

$$|A_1| + |B_1| = |A_0| + |B_0| = |C| + 1.$$

By (b), since $|A_1| > |A_0|$, we have $|B_1| = 1$ and hence $|A_1| = |C|$. But

$$A_1 \subseteq A_1 + B_1 \subseteq A_0 + B_0 \subseteq C,$$

so we have $A_1 = C$. Hence by the definition of $A_1$,

$$D_0 = C \cap \overline{A_0} = A_1 \cap \overline{A_0} \subseteq a + B_0'.$$

But by (6.22), $|D_0| = |B_0'|$, so we have

$$D_0 = a + B_0', \text{ for any } a \in D_1. \qquad (6.23)$$

We now extend the definition to $D_m$, for $m > 1$, by defining $D_m$ to be the set of elements $a \in A_0$, such that

$$a + \underbrace{B_0' + \ldots + B_0'}_{m-1 \text{ terms}} \subseteq A_0 \quad \text{but} \quad a + \underbrace{B_0' + \ldots + B_0'}_{m \text{ terms}} \not\subseteq A_0.$$

Let k denote the largest value of m for which $D_m \neq \emptyset$. Finally, let $D_\infty$ be the set of all elements $a \in A_0$, such that

$$a + \underbrace{B_0' + \ldots + B_0'}_{n \text{ terms}} \subseteq A_0 \quad \text{for all } n.$$

By the definition, $D_i \cap D_j = \emptyset$ for $i \neq j$ and, since $D_0 = C \cap \overline{A_0}$, we have partitions of $A_0$ and C defined by

$$A_0 = D_\infty \cup D_k \cup \ldots \cup D_1, \qquad C = D_\infty \cup D_k \cup \ldots \cup D_1 \cup D_0. \qquad (6.24)$$

From the definition,

$$D_\infty + B_0' = D_\infty. \qquad (6.25)$$

We know from (6.23) that if $a \in D_1$, then $D_0 = a + B_0'$, and we now show that in general if $a \in D_m$, then

$$D_{m-1} = a + B_0'. \qquad (6.26)$$

Let $2 \leq m \leq k$, $a \in D_m$ and $b' \in B_0'$. From the definition of $D_m$, we know that

$$a + b' \in D_j \quad \text{for some } j \geq m - 1. \qquad (6.27)$$

Also, there exist elements $b_1, b_2, \ldots, b_{m-1} \in B_0'$ such that

$$a_1 = a + b_1 + \ldots + b_{m-1} \in D_1$$

and

$$(a + b') + b_1 + \ldots + b_{m-1} = a_1 + b' \notin A_0$$

by (6.23). Hence

$$a + b' \in D_j \quad \text{for some } j \leq m - 1. \qquad (6.28)$$

(6.27) and (6.28) together show that $a + b' \in D_{m-1}$, and hence that

$$a + B_0' \subseteq D_{m-1}, \quad m = 2, \ldots, k, \quad \text{for } a \in D_m. \qquad (6.29)$$

But this implies that $D_m + B_0' \subseteq D_{m-1}$ and so $|D_m| \leq |D_{m-1}|$. Hence

$$|B_0'| = |a + B_0'| \leq |D_{m-1}| \leq |D_0| = |B_0'|$$

which, with (6.29), implies (6.26).

Now for $m = 1, \ldots, k$, define the subgroup $F_m$ by

$$F_m = \langle a_i - a_j \mid a_i, a_j \in D_m \rangle.$$

Then $D_m$ is contained in one coset of $F_m$ and $|F_m| \geq 2$ if and only if $|D_m| \geq 2$.

By (6.26), if $a_1, a_2 \in D_m$, then

$$a_1 + B_0' = D_{m-1} = a_2 + B_0',$$

so

$$B_0' + F_m = B_0'.$$

Hence $B_0'$ is the union of cosets of $F_m$ and $F_m$ is finite. Also, again by (6.26), since

$$D_\infty + B_0' = D_\infty,$$

we have

$$D_j + F_m = D_j \quad \text{for } j = 0, 1, \ldots, k-1 \text{ or } j = \infty.$$

We note that (6.26) does not imply that $D_j + F_k = D_j$.

By (6.24) and (6.29), $F_k$ is a quasiperiod of C provided $|D_k| \geq 2$; similarly, $F_m$ is a quasiperiod of C provided $|D_m| \geq 2$ for $m = 1, \ldots, k-1$. If $D_\infty \neq \emptyset$, then by (6.24), (6.25) and (6.26), it follows that $\langle B_0 \rangle$ is a quasiperiod of C.

Hence, if C is not quasiperiodic, we must have:

(a) $|D_k| = 1$, so $D_k = \{c_k\}$ say;

(b) $|D_m| = 1$, for $m = 1, \ldots, k-1$;

(c) $D_\infty = \emptyset$.

If $k = 1$, then by (6.24),

$$|A_0| = |D_1| = 1.$$

But

$$|A_0| \geq |A| \geq 2,$$

so

$$k \geq 2.$$

Now we apply (6.26) for $m \geq 2$. Then

$$|B_0'| = |D_1| = 1,$$

so $B_0' = \{d\}$, where $d \neq 0$. Hence from (6.26),

$$D_{k-m} = \{c_k + md\}, \quad \text{for } m = 0, 1, \ldots, k.$$

Since $D_0, D_1, \ldots, D_k$ are mutually disjoint sets, these k+1 elements are distinct; hence by (6.24), C is in arithmetic progression with difference d and the theorem is proved.

We also need to be able to construct pairs (A, B) such that their sum is small. The next theorem gives the information we need.

THEOREM 6.17. *Let G be an abelian group. Then the following construction gives precisely all the pairs (A, B) of finite non-empty subsets of G such that*

$$|A + B| \leq |A| + |B| - 1. \tag{6.30}$$

*Choose a proper finite subgroup H of G and let σ denote the natural mapping σ : G → G/H. Choose finite, non-empty subsets A*, B* of G/H such that A* + B* is aperiodic and*

$$|A^* + B^*| = |A^*| + |B^*| - 1. \tag{6.31}$$

*Now let A be any subset of $\sigma^{-1}A^*$ and B any subset of $\sigma^{-1}B^*$ such that*

$$|\sigma^{-1}A^* \cap \overline{A}| + |\sigma^{-1}B^* \cap \overline{B}| < |H|. \tag{6.32}$$

*Then this construction gives a pair (A, B) satisfying (6.30), and any pair (A, B) satisfying (6.30) may be constructed in this way.*

PROOF. (i) We show first that the construction gives a pair (A, B) satisfying (6.30). By (6.32), we have

$$\begin{aligned} |A| + |B| &> |\sigma^{-1}A^*| + |\sigma^{-1}B^*| - |H| \\ &= (|A^*| + |B^*| - 1)|H| \\ &= |A^* + B^*|.|H| \quad \text{by (6.31)} \\ &= |\sigma^{-1}(A^* + B^*)|. \end{aligned}$$

But

$$\sigma(A + B) = \sigma A + \sigma B \subseteq A^* + B^*,$$

so

$$\sigma^{-1}(A^* + B^*) \supseteq A + B$$

and hence

$$|A| + |B| > |A + B|.$$

(ii) Next we show that any pair (A, B) satisfying (6.30) may be constructed in this way.

For suppose that (6.30) is true. Then by Theorem 6.15, H = H(A + B) is a finite group such that

$$A + B + H = A + B \tag{6.33}$$

and

$$|A + B| = |A + H| + |B + H| - |H|. \tag{6.34}$$

Let $A^* = \sigma A$, $B^* = \sigma B$. Then from (6.33) and (6.34), we have (6.31).

Also

$$\sigma^{-1}A^* = A + H,$$

so

$$|\sigma^{-1}A^* \cap \overline{A}| = |A + H| - |A|$$

and similarly

$$|\sigma^{-1}B^* \cap \overline{B}| = |B + H| - |B|.$$

By (6.30) and (6.34),

$$|A + H| + |B + H| - |H| \leq |A| + |B| - 1 < |A| + |B|.$$

Hence

$$|\sigma^{-1}A^* \cap \overline{A}| + |\sigma^{-1}B^* \cap \overline{B}| = |A + H| - |A| + |B + H| - |B| < |H|$$

and (6.32) is satisfied.

Finally, if

$$A^* + B^* + x = A^* + B^*,$$

then for $g \in \sigma^{-1}x$, we have

$$A + B + g = A + B.$$

Hence $g \in H$ and $x = 0$.

Lastly, we shall need a little more information about the case where

$$|A + B| \leq |A| + |B| - 1$$

and $A + B$ is in arithmetic progression, with difference $d \neq 0$. Let $n$ denote the order of the element $d$, which is finite. (Analogous statements hold for $d$ of infinite order.)

LEMMA 6.18. *Let $G$ be an abelian group and let $A$, $B$ be non-empty subsets of $G$ such that $A + B$ is in arithmetic progression with difference $d$, and $|A + B| < n$. Then*

$$|A + B| \geq |A| + |B| - 1.$$

PROOF. Since $A + B$ is obviously not periodic, the lemma follows from Theorem 6.15.

LEMMA 6.19. *Let $G$ be an abelian group and let $A$, $B$ be non-empty subsets such that $A + B$ is in arithmetic progression with difference $d$, $|A + B| \leq n - 2$ and*

$$|A + B| \leq |A| + |B| - 1. \tag{6.35}$$

*Then each of $A$ and $B$ is in arithmetic progression with difference $d$ and in fact*

$$|A + B| = |A| + |B| - 1.$$

PROOF. Let $H = \langle d \rangle$, the subgroup of $G$ generated by $d$, so that $|H| = n$. We may replace $A$, $B$ by $-a_0 + A$, $B - b_0$ for some $a_0 \in A$, $b_0 \in B$ and hence assume without loss of generality that

$$0 \in A \cap B, \quad 0 \in A + B \subseteq H, \quad A \subseteq H, \quad B \subseteq H.$$

We are now considering only subsets of $H$, so for the remainder of this proof $\overline{D}$ denotes the complement of $D$ in $H$, and we consider a set $D$ to be in arithmetic progression if and only if it has the form

$$\{id \mid i = i_0, i_0 + 1, \ldots, i_0 + |D| - 1\}.$$

Let $A + B = C$, so that $\overline{C} - B \subseteq \overline{A}$, by Corollary 6.8. Hence by (6.35).

$$|\overline{C} - B| \leq |\overline{A}| \leq |\overline{C}| + |B| - 1. \qquad (6.36)$$

Also, since $C$ is in arithmetic progression, so is $\overline{C}$ and $|\overline{C}| \geq 2$. We show that $\overline{C} - B$ is also in arithmetic progression. For suppose not. Then let

$$\overline{C} - B = \overline{A_1} \cup \ldots \cup \overline{A_k},$$

where (i) each $\overline{A_i}$ is in arithmetic progression with difference $d$ and

(ii) $k$ is minimal subject to (i) and $k \geq 2$. Let $b \in B$ and choose $i, j$ such that $1 \leq i < j \leq k$. Now $\overline{C} - b$ is in arithmetic progression with difference $d$ and $k$ is minimal, so $\overline{C} - b$ cannot have elements in common with both $\overline{A_i}$ and $\overline{A_j}$. So let

$$B_i = \{b \in B \mid \overline{A_i} \cap (\overline{C} - b) \neq \emptyset\}.$$

Then the sets $B_1, \ldots, B_k$ are non-empty mutually disjoint sets and

$$B_1 \cup \ldots \cup B_k = B.$$

Since $\overline{C} - B_i = \overline{A_i}$, it follows from Lemma 6.18, that

$$|\overline{A_i}| \geq |\overline{C}| + |B_i| - 1, \quad \text{for } i = 1, \ldots, k. \qquad (6.37)$$

We now sum the $k$ equations of the form (6.37), showing that

$$k(|\overline{C}| - 1) + |B| \leq |\overline{A}|.$$

This, with (6.36), implies that

$$k(|\overline{C}| - 1) + |B| \leq |\overline{C}| - 1 + |B|.$$

But $|\overline{C}| - 1 \geq 1$ so we must have $k \leq 1$, which is a contradiction. Hence $\overline{C} - B$ is in arithmetic progression.

Since by (6.36),

$$|\overline{C} - B| \leq |\overline{A}| < n,$$

we may now apply Lemma 6.18, to show that

$$|\overline{C} - B| \geq |\overline{C}| + |B| - 1.$$

Hence equality holds in both (6.35) and (6.36).

Also since $\overline{C} - B = \overline{A}$ is in arithmetic progression with difference $d$, so is $A$. A similar argument shows that $B$ is also in arithmetic progression with difference $d$ and completes the proof of the lemma.

COROLLARY 6.20. *Suppose that*

$$|A|, \ |B| \geq 2, \quad |A + B| \leq |A| + |B| - 1$$

*and that every element $g \in G$, $g \neq 0$, has order at least $|A + B| + 2$. Then $A$, $B$, $A + B$ are all in arithmetic progression with difference $d$.*

PROOF. Since G has no subgroup F with

$$2 \leq |F| \leq |A + B|,$$

we know that A + B is not quasiperiodic. Hence from Theorems 6.15 and 6.16, A + B is in arithmetic progression and

$$|A + B| = |A| + |B| - 1.$$

The corollary now follows from Lemma 6.19.

6.4. REFERENCES AND RELATED TOPICS. A general reference for addition theorems for groups is the book by Mann(1965). Theorem 6.1 is due to Mann (1952). Theorem 6.4 was first proved by Cauchy (1813) and subsequently rediscovered by Davenport (1935, 1947). The proof given here is due to Vosper (1956b). Theorem 6.9 is also due to Vosper (1956a and b), whose proof is given here; a different proof, using another transform, is given by Mann (1965). Lemma 6.11 was proved by Mann and Olsen (1967).

Lemma 6.13 was proved by Kemperman (1960); Corollary 6.14 was originally proved by Kneser (1955) by a different method.

Theorem 6.15 is again due to Kneser (1955). All the rest of the results discussed in section 6.3 are from Kemperman (1960), which contains many other more detailed results on the small sum-sets in abelian groups.

Other related results are due to McWorter (1964) and Diananda (1968). See also Kneser (1953, 1956).

## CHAPTER VII. SUM-FREE SETS IN GROUPS

7.1 **BASIC RESULTS.** As usual, we call a subset S of a group G a *sum-free set* in G if and only if $(S + S) \cap S = \emptyset$. If, in addition, $|S| \geq |T|$ for every sum-free set T in G, then we call S a *maximal sum-free set* in G. We denote by $\lambda(G)$ the cardinality of a maximal sum-free set in G.

It follows from results of Erdos that, if G is any finite abelian group, then

$$2|G|/7 \leq \lambda(G) \leq |G|/2 \tag{7.1}$$

and obviously both of these bounds can be attained, since $\lambda(Z_2) = 1$ and $\lambda(Z_7) = 2$. In fact, the upper bound holds for all groups and is attained if and only if G has a subgroup of index 2, as we see from the following theorems.

THEOREM 7.1. *Let* G *be a finite group and let* S *be a maximal sum-free set in* G. *Then*

$$|S| \leq \tfrac{1}{2}|G|.$$

PROOF. By Theorem 6.1, either $G = S + S$ or $|G| \geq 2|S|$. But $G \neq S + S$ since S is sum-free.

THEOREM 7.2. *Let* S *be a finite subset of a group* G. *Then* $|S + S| = |S|$ *if and only if there exists a finite subgroup* H *of* G, *such that*

$$S + H = S = H + S$$

*and*

$$S - S = H = -S + S.$$

PROOF. (i) Let $s_1, s_2 \in S$ and let $H_1 = -s_1 + S$, $H_2 = S - s_2$. Then

$$|H_1 + H_2| = |S + S| = |S| = |H_1| = |H_2| < \infty .$$

But $0 \in H_1 \cap H_2$ and hence

$$H_1 \cup H_2 \subseteq H_1 + H_2.$$

But this implies that $H_1 + H_2 = H_1 = H_2 = H$ say. So we have a finite subgroup $H \leq G$, such that S is both a left and a right coset of H. Hence $H = -s + S = S - s$ for every $s \in S$, and the conclusion follows.

(ii) Since $S + H = S = H + S$, we know that S is a union of left cosets and a union of right cosets of H, and $|S|$ is some multiple of $|H|$. But since

$$|H| = |S - S| \geq |S| = |S + H| = |H + S|$$

and since $|H| \leq |S|$, we must have $|H| = |S|$ and S is both a left and a right coset of H.

Then $S = s + H = H + s$, for every $s \in S$ and

$$|S + S| = |s + H + H + s| = |H| = |S|.$$

COROLLARY 7.3. *Let* $|G| = 2m$. *Then* $\lambda(G) = m$ *if and only if* G *has a subgroup* H *of order* m,*and in this case the maximal sum-free set is the coset other than* H.

To find the lower bound on $\lambda(G)$ given in (7.1) is harder. This bound applies only to abelian groups and, to find it, we have to consider different cases depending on the primes dividing the order of the group. But first we need some information concerning sum-free sets in arbitrary finite abelian groups. So we let G be abelian and let S be a maximal sum-free set in G.

We refer to Definition 6.12. For any subset $C \subseteq G$, we define H(C) to be the subgroup

$$H(C) = \{g \in G \mid C + g = C\},$$

so that

(i) $C + H(C) = C$;

(ii) if $C + K = C$ for some subgroup K, then $K \leq H(C)$;

(iii) $\langle H(C), H(D) \rangle$, the subgroup generated by H(C) and H(D), is contained in $H(C + D)$;

(iv) $H(S) = H(-S)$.

We now apply Theorem 6.15: There exists a subgroup $H = H(S + S) \leq G$ such that $S + S + H = S + S$ and either

$$|S + S| \geq 2|S|$$

or

$$|S + S| = 2|S + H| - |H|. \tag{7.2}$$

In fact, we now show that $H = H(S + S) = H(S) = H(S - S)$.

LEMMA 7.4. *Let* $H = H(S + S)$ *as above. Then* $S + H$ *is also a sum-free set in* G *and hence, since* S *is maximal,* $S + H = S$.

PROOF. By the definition of H,

$$(S + H) + (S + H) = S + S.$$

Suppose that $(S + S) \cap (S + H) \neq \emptyset$. Then for some $s, s_1, s_2 \in S$ and for some $h \in H$, we have $s + h = s_1 + s_2$. But this implies that

$$s = s_1 + s_2 - h \in S + S + H = S + S,$$

which contradicts the sum-freeness of S.

Hence $S + H$ is sum-free and, by the maximality of S, we must have $S + H = S$.

COROLLARY 7.5. $H = H(S + S) = H(S) = H(S - S)$.

PROOF. By (iii) above, $H(S) \leq H(S + S)$. But by (ii), $S + H = S$ implies that $H = H(S + S) \leq H(S)$. Hence $H = H(S)$.

Now $S + H = S$ implies that $S - S + H = S - S$, and hence $H \leqslant H(S - S)$. A proof similar to that of the lemma shows that $S + H(S - S) = S$ so that $H(S - S) \leqslant H$. Hence $H = H(S - S)$.

COROLLARY 7.6. *By equation* (7.2) *and the lemma, we have* $S + S + H = S + S$ *and either*

$$|S + S| \geqslant 2|S|$$

*or*

$$|S + S| = 2|S| - |H|;$$

*similarly,* $S - S + H = S - S$ *and either*

$$|S - S| \geqslant 2|S|$$

*or*

$$|S - S| = 2|S| - |H|.$$

LEMMA 7.7. *Consider the three cases:*

(i) $|G|$ *has at least one prime factor congruent to* 2 (modulo 3) *and we let* $q$ *denote the smallest such factor;*

(ii) $|G|$ *has no prime factor congruent to* 2 (modulo 3) *but is divisible by* 3;

(iii) $|G|$ *is divisible only by primes congruent to* 1 (modulo 3) .

*Then in case* (i), $\lambda(G) \leqslant |G|(q + 1)/3q$;
*in case* (ii), $\lambda(G) \leqslant |G|/3$;
*in case* (iii), $\lambda(G) \leqslant (|G| - 1)/3$.

PROOF. Since S is sum-free, we have

$$|G| \geqslant |S| + |S + S| \geqslant 3|S| - |H| \text{ by (7.2).}$$

Hence

$$|S| \leqslant |H| \frac{1}{3}\left[\left(\frac{|G|}{|H|} + 1\right)\right], \tag{7.3}$$

where $[x]$ denotes the integer part of x. So

$$\lambda(G) \leqslant \max_{d \mid |G|} \frac{|G|}{d}\left[\frac{d + 1}{3}\right]. \tag{7.4}$$

Now,

$$\frac{1}{d}\left[\frac{d+1}{3}\right] = \begin{cases} (d + 1)/3d & \text{if } d \equiv 2 \pmod 3; \\ 1/3 & \text{if } d \equiv 0 \pmod 3; \\ (d - 1)/3d & \text{if } d \equiv 1 \pmod 3). \end{cases}$$

This, with (7.4), implies the stated bounds.

These upper bounds are in fact the exact values in the first and second cases, but not always in the third.

7.2 **|G| DIVISIBLE BY A PRIME CONGRUENT TO 2 (MODULO 3).** For case (i), both the size and the structure of maximal sum-free sets are completely determined.

THEOREM 7.8. *In case* (i), $\lambda(G) = |G|(q + 1)/3q$. *Also, if* S *is a maximal sum-free set in* G, *then* S *is a union of cosets of some subgroup* H *of index* q *in* G, S/H *is in arithmetic progression in* G/H *and* $S \cup (S + S) = G$.

PROOF (i) Let $q = 3k - 1$ and let $G = Z_q$. Then by Lemma 7.7, $\lambda(G) \leq k$. But the set $S = \{k, k + 1, \ldots, 2k - 1\}$ is sum-free and contains k elements, so $\lambda(G) = k$.

By Theorem 6.4, $|S + S| \geq 2k - 1$, and since S is sum-free, $|S + S| \leq 2k - 1$. Hence $|S + S| = 2k - 1$ and, by Theorem 6.9, S is in arithmetic progression. Without loss of generality, we may take the difference of this progression to be 1, and we see that $S = \{k, \ldots, 2k - 1\}$ is the only possible set, up to automorphism.

(ii) In general, G has a subgroup K of index q and an element g of order q such that

$$G = K \cup (K + g) \cup \cdots \cup (K + (q - 1)g).$$

Then the set

$$T = \bigcup_{j=k}^{2k-1} (K + jg)$$

is sum-free in G and has order $|G|(q + 1)/3q$. Hence T is a maximal sum-free set in G and $\lambda(G) = |G|(q + 1)/3q$.

Now let S be any maximal sum-free set in G, so that

$$|S| = |G|(q + 1)/3q. \qquad (7.5)$$

Let H be the subgroup satisfying equation (7.2). Then by equation (7.3), we have

$$|G|(q + 1)/3q \leq |H| \left[\frac{1}{3}\left(\frac{|G|}{|H|} + 1\right)\right],$$

so $|H| = |G|/q$. By Lemma 7.4, S is a union of cosets of H. By equation (7.2) and equation (7.5), we have $|S| + |S + S| \geq |G|$. But since S is sum-free, $|S| + |S + S| \leq |G|$. Hence $S \cup (S + S) = G$, and $|S + S| = 2|S| - H$. So

$$|S/H + S/H| = 2|S/H| - 1.$$

Now $|G/H| = q$ and S/H is a subset of G/H. So by Theorem 6.9, S/H is in arithmetic progression and, as in (i), S/H is isomorphic to $\{k, \ldots, 2k - 1\}$.

7.3 **|G| DIVISIBLE BY THREE.** In case (ii), we have |G| not divisible by any prime congruent to 2 (modulo 3), but |G| divisible by 3. In this case also, the size and structure of the maximal sum-free sets has been determined.

THEOREM 7.9. *In case (ii),* $\lambda(G) = |G|/3$. *Also, if* S *is a maximal sum-free set in* G, *then* S *is a union of cosets of some subgroup* H *of* G, *such that* G/H *is the cyclic group* $Z_{3m}$ *for some* m, S/H *is in arithmetic progression in* G/H *and*

$$|S + S| = 2|S| - |H|.$$

PROOF. (i) G has a subgroup K of index 3. Choose $g \in G$ such that

$$G = K \cup K + g \cup K + 2g.$$

Then the set T = K + g is clearly sum-free and has order $|G|/3$. Hence T is a maximal sum-free set in G by Lemma 7.7, and $\lambda(G) = |G|/3$.

(ii) Now let S be any maximal sum-free set in G, so that $|S| = |G|/3$. Let $H = H(S + S)$. By Lemma 7.4, S is a union of cosets of H, so $|H| = |G|/3m$, for some integer m.

Since S is sum-free,

$$|S + S| \leq |G| - |S| = 2|S|.$$

But by Corollary 7.6, either

$$|S + S| \geq 2|S|$$

or

$$|S + S| = 2|S| - |H|.$$

Hence we have two possibilities:

(a) $|S + S| = 2|S| - |H|$

and

(b) $|S + S| = 2|S|$.

We show that (b) cannot in fact occur.

Let A be a subset of G such that A = -A. Since $|G|$ is odd, $|A|$ is odd if and only if $0 \in A$.

$0 \notin S$ and $|S|$ is odd, so $S \neq -S$. $0 \in S - S = -(S - S)$, so $|S - S|$ is odd. Since S is sum-free,

$$S \cap (S + S) = \emptyset = (S \cup (-S)) \cap (S - S).$$

Using Corollary 7.6, the possibilities for S - S, just as for S + S, are the following·

(a) $|S - S| = 2|S| - |H|$

and

(b) $|S - S| = 2|S|$.

But $|S - S|$ is odd, which rules out (b), so

$$|S - S| = 2|S| - |H|.$$

Now we look at the factor group $G^* = G/H$ of order 3m. $S^* = S/H$ is obviously a maximal sum-free set in $G^*$. Since

$$H = H(S) = H(S + S) = H(S - S),$$

both the set $S^*$ itself and its set of differences $S^* - S^*$ are aperiodic. We know that

$$|S^* - S^*| = 2|S^*| - 1 = 2m - 1 \tag{7.6}$$

and that

$$|S^* \cup (S^* - S^*)| = |G^*| - 1 = 3m - 1. \tag{7.7}$$

By (7.6) and Theorem 6.16, $S^* - S^*$ is either quasiperiodic or in arithmetic progression.

First suppose that $S^* - S^*$ is quasiperiodic. Then

$$S^* - S^* = T' \cup T''$$

where

$$T' = T' + K^*$$

and

$$T'' \subseteq t + K^*,$$

for some subgroup $K^*$ of $G^*$ and for some $t \in T''$. But

$$S^* - S^* = -(S^* - S^*),$$

so that

$$T'' \subseteq K^*.$$

If

$$S^* \cap K^* \neq \emptyset,$$

then the sum-freeness of $S^*$ implies that no complete coset of $K^*$ is contained in $S^*$. But this, together with (7.7) contradicts the quasiperiodicity of $S^* - S^*$.

So

$$S^* \cap K^* = \emptyset.$$

But this, together with (7.7) forces $S^*$ to be periodic with period $K^*$. Again we have a contradiction.

Hence $S^* - S^*$ must be in arithmetic progression with difference d. By (7.6), d must be of order 3m. But now, by Lemma 6.19, $S^*$ is also in arithmetic progression with difference d, so that

$$|S^* + S^*| = 2|S^*| - 1$$

and

$$|S + S| = 2|S| - |H|.$$

Also since $G^*$, of order 3m, contains an element d of order 3m, $G^*$ must be the cyclic group $Z_{3m}$ and the automorphism of $G^*$ which maps d to 1 maps $S^*$ to the set $\{m, m + 1, \ldots, 2m - 1\}$.

The structure of the maximal sum-free sets is completely determined by Theorem 7.9. We may also use the Theorem to count the number of non-isomorphic maximal sum-free sets in a given group. We list a few examples:

COROLLARY 7.10. *Let* $G = Z_{3^n}$ *the cyclic group of order* $3^n$. *Then* G *contains exactly* n *non-isomorphic maximal sum-free sets.*

COROLLARY 7.11. *Let* G *be an elementary abelian 3-group. If* S *is a maximal sum-free set in* G, *then* S *is one coset of a maximal subgroup of* G.

COROLLARY 7.12.

(a) *Let* $G = Z_{3p}$ *where* p *is prime,* $p = 3k + 1$. *If* S *is a maximal sum-free set in* G, *then there are two possibilities for* S:

(i) S *is one coset of the subgroup* H *of order* p;

(ii) S *may be mapped under some automorphism of* G *to the set*

$$\{p, p + 1, \ldots, 2p - 1\}.$$

(b) *Let* $G = Z_3 \oplus Z_3 \oplus Z_p$, *where* p *is prime,* $p = 3k + 1$. *If* S *is a maximal sum-free set in* G, *then there are two possibilities for* S:

(i) S *is one coset of the subgroup* H *of order* 3p;

(ii) S *is a union of* p *cosets of a subgroup* K, *of order* 3, *of* G, *where* G/K *is cyclic and* S/K *is a maximal sum-free set in* G/K.

7.4 **ALL DIVISORS OF |G| CONGRUENT TO 1 (MODULO 3).** We are now left with case (iii) of Lemma 7.7; we consider abelian groups of order divisible only by primes congruent to 1 (modulo 3). This is the case about which least is known; even the size of the maximal sum-free sets is known only in a few special cases.

THEOREM 7.13. *In case* (iii), *let* m *be the exponent of* G. *Then*

$$(m-1)|G|/3m \le \lambda(G) \le (|G| - 1)/3.$$

PROOF. The upper bound we already know from Lemma 7.7.

Let $m = 3n + 1$. To find the lower bound, choose a subgroup K, of index m, in G, and an element $g \notin K$, such that g has order m and

$$G = K \cup (K + g) \cup \ldots \cup (K + 3ng).$$

Clearly

$$T = (K + (n+1)g) \cup \ldots \cup (K + 2ng)$$

is a sum-free set of order $(m-1)|G|/3m$, which gives the lower bound.

COROLLARY 7.14. *If* G *is cyclic, then* $|G| = m$ *and the upper and lower bounds coincide, so that*

$$\lambda(Z_m) = (m-1)/3.$$

COROLLARY 7.15. *Since* m *is a product of primes congruent to 1 (modulo 3), the lowest value of* $(m-1)/3m$ *is 2/7, giving the lower bound of equation* (7.1).

Only for the cyclic and elementary abelian groups do we know the size and structure of the maximal sum-free sets. For other cases, not even the value of $\lambda(G)$ is known.

DEFINITION 7.16. Let G be a group, H a subgroup of G and S a maximal sum-free set in G. Then S is said to *avoid* H if and only if $S \cap H = \emptyset$, and to *cover* H if and only if $S \cap H$ is a maximal sum-free set in H.

THEOREM 7.17. *Let* G *be an elementary abelian p-group, where* p *is prime,* $p = 3k + 1$, $|G| = p^n$. *Then* $\lambda(G) = kp^{n-1}$.

PROOF. We first consider the case where $n = 2$ and then generalise.

(i) Let

$$G = \langle x_1, x_2 \mid px_i = 0,\ i = 1, 2;\ x_1 + x_2 = x_2 + x_1 \rangle .$$

Let $X_i$ denote $\langle x_i \rangle$ and let S be a maximal sum-free set in G.

G has $p + 1$ subgroups of order p, none of which contains more than k elements of S by Corollary 7.14. But $\lambda(G) \geq kp$ and the union of these $p + 1$ subgroups is the whole of G; hence at least one of these subgroups is covered by S. Without loss of generality, we assume this subgroup to be $X_2$.

So

$$G = \bigcup_{i=0}^{p-1} (X_2 + ix_1).$$

We let $S_i$ denote the subset of $X_2$ such that

$$S_i + ix_1 = S \cap (X_2 + ix_1), \text{ for } i = 0,\ldots,p-1.$$

In particular, $|S_0| = k$. If $|S_i| \leq k$ for all $i = 1,\ldots,p-1$, then

$$|S| = |S_0| + |S_1| + \ldots + |S_{p-1}| \leq kp$$

and the theorem follows.

So suppose that $|S_i| > k$ for some i. We may choose $x_1$ so that $|S_1| > k$. Since S is sum-free

$$(S_i + S_j) \cap S_{i+j} = \emptyset , \tag{7.8}$$

and in particular

$$(S_0 + S_i) \cap S_i = \emptyset . \tag{7.9}$$

Hence

$$|S_0 + S_1| \leq p - |S_1| \tag{7.10}$$

and by Theorem 6.4,

$$|S_0| + |S_1| - 1 \leq |S_0 + S_1| . \qquad (7.11)$$

By (7.10) and (7.11),

$$2|S_1| \leq p + 1 - |S_0| ,$$

so that

$$|S_1| \leq k + 1.$$

Since we assumed that $|S_1| > k$, we must have

$$|S_1| = k + 1.$$

If $(S_0,S_1)$ is not a standard pair, then by Theorem 6.9, we have

$$|S_0 + S_1| \geq |S_0| + |S_1| = 2k + 1.$$

But by (7.8)

$$|S_0 + S_1| \leq 2k,$$

a contradiction. Hence $(S_0,S_1)$ is a standard pair with difference d; without loss of generality we may assume that d = 1.

Since $S_0$ is sum-free, we have three possibilities:

$$S_0 = \{k,\ldots,2k-1\} \quad \text{or} \quad \{k+1,\ldots,2k\} \quad \text{or} \quad \{k+2,\ldots,2k+1\}.$$

Since $S_1 = \{\ell,\ell+1,\ldots,\ell+k\}$ for some $\ell \in X_2$, neither k nor 2k + 1 can belong to $S_0$. Hence

$$S_0 = \{k+1+r \mid r = 0,1,\ldots,k-1\} \qquad (7.12)$$

and we may choose $x_1$ so that

$$S_1 = \{k+1+r \mid r = 0,1,\ldots,k\} .$$

Since S is sum-free, (7.8) bounds the range of each $S_i$; more precisely, for each i, there exists $\alpha_i \in S_i$ such that

$$S_i \subseteq \{\alpha_i+r \mid r = 0,1,\ldots,k\}.$$

We call $S_i$ a *small-range set* if for some $m_i > 0$, we have

$$S_i \subseteq \{\alpha_i+r \mid r = 0,1,\ldots,k-1-m_i\}.$$

Similarly, we call $S_i$ a *normal-range set* if

$$S_i \subseteq \{\alpha_i+r \mid r = 0,1,\ldots,k-1\} \text{ and } \alpha_i + k - 1 \in S_i,$$

and a *big-range set* if

$$S_i \subseteq \{\alpha_i+r \mid r = 0,1,\ldots,k\} \text{ and } \alpha_i + k \in S_i.$$

By (7.8), we have

$$S_{i+1} \subseteq \{\alpha_i-m_i+r \mid r = 0,1,\ldots,k+m_i\} \qquad (7.13)$$

when $S_i$ is a small-range set;

$$S_{i+1} \subseteq \{\alpha_i + r \mid r = 0,1,\ldots,k\} \tag{7.14}$$

when $S_i$ is a normal-range set;

$$S_{i+1} \subseteq \{\alpha_i + 1 + r \mid r = 0,1,\ldots,k-1\} \tag{7.15}$$

when $S_i$ is a big-range set.

Now consider the movement of $\alpha_i$ for $i = 1,2,\ldots,p-1$. If $S_i$ is a big-range set then, by (7.15), $\alpha_{i+1} > \alpha_i$. If $S_i$ is a normal-range set then, by (7.14), $\alpha_{i+1} \geq \alpha_i$. If $S_i$ is a small-range set then, by (7.13), $\alpha_{i+1} \geq \alpha_i - m_i$. In this last case, $\alpha_{i+1}$ may be at most $m_i$ steps closer to 0 than $\alpha_i$ is. But then the contribution of $S_i$ to $S$ is $m_i$ elements fewer than the average contribution of k elements. Since $|S| \geq kp$, we must make up these $m_i$ elements, one each from $m_i$ of the big-range sets. But by (7.8) and Theorem 6.4, the cosets corresponding to big-range sets themselves form a sum-free set in $G/X_2$, so there are at most k big-range sets. Hence

$$m = \sum_{i=0}^{p-1} m_i \leq k$$

and

$$\alpha_i \geq \alpha_0 - m \text{ for all } i = 1,\ldots,p-1.$$

But $\alpha_0 = k + 1$ by (7.12) so $\alpha_i \geq 1$ for all i.

A similar argument, using the relation

$$(S_i - S_1) \cap S_{i-1} = \emptyset$$

in place of (7.8), shows that the right-hand end-point of $S_i$ never exceeds $p - 1$ for all i. Hence $0 \notin S_i$, $S \cap X_1 = \emptyset$ and $|S| \leq kp$.

(ii) Now let G be an elementary abelian group of order $p^n$. Then G has $(p^n-1)/(p-1)$ subgroups of order p, none of which contains more than k elements of a maximal sum-free set S. But

$$\lambda(G) \geq kp^{n-1} > (k-1)(p^n-1)/(p-1),$$

so that at least one of these subgroups is covered by S and we denote this subgroup by X. Let Y be a subgroup complementing X in G. Then Y is an elementary abelian group of order $p^{n-1}$ and has

$$(p^{n-1}-1)/(p-1) = \rho$$

subgroups $Y_i$ of order p.

Now $|S \cap X| = k$ and, by (i), $|S \cap (X + Y_i)| \leq kp$ for all $i = 1,\ldots,\rho$. So

$$|S| = \sum_{i=1}^{\rho} |S \cap (X+Y_i)| - (\rho-1)k \leq \rho kp - (\rho-1)k = \rho k(p-1)+k = kp^{n-1}.$$

This completes the proof of the theorem.

Next we determine the structure of the maximal sum-free sets in the group of order p, p = 3k + 1. We need a preliminary lemma.

LEMMA 7.18. *Let* $G = Z_n$, *where* $n = 3k + 1$ *is not necessarily prime. Let* S *be a sum-free set in* G *satisfying*

$$|S| = k, \quad \overline{S} = S + S \quad \text{and} \quad S = -S. \tag{7.16}$$

*Then*

(i) $(S+g) \cap S = \emptyset$ *if and only if* $g \in S$;

(ii) *if* $|(S+g) \cap S| = 1$ *for some* $g \in G$, *then*

$$|(S+g') \cap S| \geq k - 3,$$

*where* $g' = 3g/2$ *and* $\pm g/2 \in S$;

(iii) *if* $|(S+g) \cap S| = \lambda > 1$ *for some* $g \in G$, *then there exists* $g' \in G$ *such that*

$$|(S+g') \cap S| \geq k - (\lambda+1).$$

PROOF. (i) is obvious.

(ii) Let $|(S+g) \cap S| = 1$ for some $g \in G$. Then there exist $s_1, s_2 \in S$ such that $s_1 + g = s_2$. But $S = -S$, hence $-s_2 + g = -s_1 \in S$, so that $s_2 = -s_1$ and $g = -2s_1$. Now

$$S \cap (S-s_1) = (S-s_1) \cap (S-2s_1) = (S-2s_1) \cap (S-3s_1) = \emptyset$$

and

$$|S \cap (S-2s_1)| = |(S-3s_1) \cap (S-s_1)| = 1,$$

so that

$$|S \cap (S-3s_1)| \geq k - 3.$$

Take $g' = -3s_1$ to complete the proof.

(iii) By hypothesis, there exist $s_1, s_2 \in S$ such that $s_1 + g, s_2 + g \in S$ and $s_1 \neq s_2$. Hence

$$\emptyset = (S+s_1) \cap S = (S+s_2) \cap S = (S+g+s_1) \cap S$$
$$= (S+g+s_2) \cap S = (S+g+s_1) \cap (S+g) = (S+g+s_2) \cap (S+g).$$

This implies that

$$|(S+g+s_1) \cap (S+g+s_2)| \geq k - (\lambda+1),$$

with equality only in the case where

$$S \cup (S+g) \cup (S+g+s_1) \cup (S+g+s_2) = G.$$

Choose $g' = s_1 - s_2$ to complete the proof.

THEOREM 7.19 *Let* $G = Z_p$, *where* p *is prime*, $p = 3k + 1$, $p > 7$. *Then any maximal sum-free set* S *may be mapped, under some automorphism of* G, *to one of the following sets:*

$$A = \{k, k+2, \ldots, 2k-1, 2k+1\};$$
$$B = \{k, \ldots, 2k-1\};$$
$$C = \{k+1, \ldots, 2k\}.$$

*Note that if* $p = 7$, *then* $k = 2$ *and sets of type* A *cannot occur.*

PROOF. If S is a standard set then, by taking an automorphism of G if necessary, we may assume the common difference to be 1. This gives two possibilities for S, namely B and C.

If S is not a standard set, then by Theorem 6.9, $|S-S| \geq 2k$. Since $0 \in S - S$ and $S - S = -(S-S)$, we know that $|S-S|$ is odd. But

$$(S \cup (-S)) \cap (S-S) = \emptyset,$$

so

$$|S-S| = 2k + 1, \; S = -S \text{ and } S + S = S - S = \overline{S}.$$

Now S fulfils the conditions of Lemma 7.18. If for some $g \in G$, $|(S+g) \cap S| = 1$, then, by (ii) of the lemma,

$$|(S+(3g/2)) \cap S| \geq k - 3.$$

We map $3g/2$ to 1 so that $g = k + 1$. Now

$$|(S+1) \cap S| \neq k - 1$$

since S is not a standard set. If

$$|(S+1) \cap S| = k - 2,$$

then obviously

$$S = \{\pm k/2, \pm(1+k/2), \ldots, \pm(k-1)\}$$

which maps under automorphism to the set A. If

$$|(S+1) \cap S| = k - 3,$$

then

$$S = \{\alpha, \ldots, \alpha+\rho-1, \; k+\rho+1, \ldots, 2k-\rho, \; 3k+2+\alpha-\rho, \ldots, 3k+1-\alpha\}.$$

where $\alpha \leq k$ and $1 \leq \rho < \frac{1}{2}k$. But $-\frac{1}{2}g = k \in S$ and $g = k + 1 \notin S$ by the lemma. Hence $\alpha + \rho - 1 = k$ and

$$S = \{k+1-\rho, \ldots, k, \; k+\rho+1, \ldots, 2k-\rho, \; 2k+1, \ldots, 2k+\rho\}.$$

But

$$(k+1-\rho) + (k+\rho+1) = 2k + 2 \in \overline{S}.$$

Hence $\rho = 1$ and S is the set A.

We are now left with the case where S satisfies the conditions of Lemma 7.18 and $|(S+g) \cap S| \neq 1$ for any $g \in G$. By taking an automorphism of G if necessary, we map S so that $|(S+1) \cap S|$ is maximal. We list the elements of S as follows:

$$S = \{\alpha_1, \ldots, \alpha_1+\ell_1, \; \alpha_2, \ldots, \alpha_2+\ell_2, \ldots, \alpha_h, \ldots, \alpha_h+\ell_h\}, \quad (7.17)$$

where

$$0 < \alpha_1 \leq \alpha_1+\ell_1 < \alpha_2-1 < \alpha_2+\ell_2 < \ldots < \alpha_h-1 < \alpha_h+\ell_h < p,$$

and $\alpha_i,\ldots,\alpha_i + \ell_i$ denotes a string of $\ell_i + 1$ consecutive elements of S. By (7.16),

$$\alpha_{h-i} + \ell_{h-i} = p - \alpha_{i+1} \quad \text{for all } i = 0,\ldots,h-1. \tag{7.18}$$

Also

$$|(S+1) \cap S| = k - h \geq |(S+g) \cap S| \text{ for all } g \in G. \tag{7.19}$$

Hence h is minimal in (7.17). We show that this case cannot occur.

If $h = 1$, then S is in arithmetic progression and must map under automorphism to either B or C, which have been considered previously.

If $h = 2$, the only possible set is

$$S = \{\pm\tfrac{1}{2}k, \pm (1+\tfrac{1}{2}k),\ldots, \pm (k-1)\}.$$

Since

$$|(S+(2k+1)) \cap S| = 1,$$

this set has been also considered previously; it maps under automorphism to the set A.

So now we assume that $h \geq 3$.

Let

$$X = \{\alpha_1,\alpha_2,\ldots,\alpha_h\}$$

and let

$$Y = \{\alpha_1+\ell_1+1,\ldots,\alpha_h+\ell_h+1\} = \{1-\alpha_1,\ldots,1-\alpha_h\} = 1 - X \text{ by (7.18)}.$$

If $h = 3$, then by assumption,

$$|(S+g) \cap S| \geq h - 1 = 2$$

for every $g \in \overline{S}$. Now suppose that $h \geq 4$, and that for some $g \in \overline{S}$, we have

$$2 \leq |(S+g) \cap S| \leq h - 2.$$

Then by Lemma 7.18, there exists $g' \in \overline{S}$ such that

$$|(S+g') \cap S| \geq k - (h-2+1) = k - h + 1.$$

But this contradicts (7.19), so for all $g \in \overline{S}$, we have

$$h - 1 \leq |(S+g) \cap S| \leq k - h, \quad \text{for } h \geq 3.$$

In particular, for $i = 1,\ldots,h$, we know that $\alpha_i - 1 \in \overline{S}$, so we have

$$|(S+\alpha_i-1) \cap S| \geq h - 1.$$

But for any $s_1, s_2 \in S$, $s_1 + \alpha_i - 1 = s_2$ implies that $s_1 \in X$, $s_2 \in -X$ and $s_1 + \alpha_i \in Y$. Hence

$$h \geq |(X+\alpha_i) \cap Y| \geq h - 1 \text{ for all } i = 1,\ldots,h. \tag{7.20}$$

By Theorem 6.4,

$$|X+X| \geq 2h - 1. \tag{7.21}$$

Since $|Y| = h$, $X + X$ contains at least $h - 1$ elements which do not belong to $Y$. By (7.20), $X + \alpha_i$ contains at most one element which does not belong to $Y$. Suppose $h > 2$. Then, for at least $h - 2$ values of $i = 1,\ldots,h$, we have $2\alpha_i \notin Y$. But $2\alpha_i \notin Y$ implies that $1 - \alpha_i \notin X + \alpha_i$, since $Y = 1 - X$. Hence for at least $h - 2$ values of $i$,

$$\{\alpha_1+\alpha_i,\ldots,\alpha_{i-1}+\alpha_i, \alpha_{i+1}+\alpha_i,\ldots,\alpha_h+\alpha_i\}$$

$$= (X+\alpha_i) \cap Y$$

$$= \{1-\alpha_1,\ldots,1-\alpha_{i-1}, 1-\alpha_{i+1},\ldots,1-\alpha_h\},$$

and summing on both sides of this equation, we find that

$$(h-3)\alpha_i \equiv h - 1 - 2 \sum_{j=1}^{h} \alpha_j \pmod p. \tag{7.22}$$

This implies that $h \leq 3$, which in fact leaves only the possibility that $h = 3$.

If $h = 3$, we can list the elements of $S$ in the following way:

$$S = \{\alpha,\ldots,\alpha+\rho-1, k+\rho+1,\ldots,2k-\rho, 3k+2-\alpha-\rho,\ldots,3k+1-\alpha\}$$

where $\alpha \leq k$ and $1 \leq \rho < \frac{1}{2}k$. By (7.22), we have

$$0 \equiv 3 - 1 - 2(\alpha+k+\rho+1-(\alpha+\rho-1)) \pmod p$$

and hence

$$k + 2 \equiv 1 \pmod p,$$

which is not possible.

So this last case cannot occur and any maximal sum-free set in $G$ may be mapped under automorphism to one of $A$, $B$ or $C$.

We now use Theorems 7.17 and 7.19, together with one additional lemma, to characterise the maximal sum-free sets in elementary abelian $p$-groups, $p = 3k + 1$.

LEMMA 7.20. *Let $G = Z_p$ where $p$ is prime, $p = 3k + 1$. Let $S$ be a maximal sum-free set in $G$ such that*

(i) *$S$ is isomorphic to $C = \{k+1,\ldots,2k\}$ and*

(ii) $S \subseteq \{(k/2)+1,\ldots,(5k/2)\}$.

*Then either $S = C$ or*

$$S = C' = \{(k/2)+1,\ldots,k, 2k+1,\ldots,(5k/2)\}.$$

PROOF. We may assume without loss of generality that

$$S = \{x,x+d,\ldots,x+(k-1)d\} \text{ for some } x, d \in Z_p, 1 \leq d \leq 3k/2.$$

Since $S = -S$, we have

$$2x + (k-1)d = 0$$

and hence

$$x = (k+1)d$$

or equivalently

$$3x = 2d. \tag{7.23}$$

We have two cases to consider:

(a) If $(k/2) + 1 \leq x < x + d < \ldots < x + (k-1)d \leq (5k/2)$, then (7.24)

$$(k-1)d \leq 2k - 1$$

and $d = 1$ or $2$. If $d = 2$ then, by (7.23), $x = 2k + 2$ and $S$ is not contained in the given set; if $d = 1$, then $S = C$.

(b) If (7.24) is not satisfied then, for some $\ell$, $1 \leq \ell \leq k - 1$, we have

$$x + \ell d \leq 5k/2 \quad \text{and} \quad \tfrac{1}{2}k + 1 \leq x + (\ell+1)d \ ,$$

so that

$$k + 2 \leq d \leq 3k/2. \tag{7.25}$$

If, for some $s \in S$,

$$k + 1 \leq s \leq 3k/2$$

then, by (7.25),

$$s - d \in \{(5k/2)+2,\ldots,3k,0,1,\ldots,(k/2)-2\};$$

hence

$$s - d \notin S$$

and $s = x$, the first element of the arithmetic progression. Now

$$k + 1 \leq x \leq 3k/2$$

implies that

$$2 \leq 3x \leq (3k/2) - 1$$

but by (7.25),

$$2k + 4 \leq 2d \leq 3k.$$

Hence $3x \neq 2d$, contradicting (7.23). Hence $S \cap C = \emptyset$ and $S = C'$.

THEOREM 7.21. *Let* $G$ *be an elementary abelian p-group,* $|G| = p^n$, $p = 3k + 1$, $p > 7$. *Let* $S$ *be a maximal sum-free set in* $G$. *If* $G$ *is denoted by*

$$G = \{(i_1,\ldots,i_n) \mid i_j \in Z_p,\ j = 1,\ldots,n\}$$

*then, under some automorphism of* $G$, $S$ *can be mapped to one of the following* $2n + 1$ *sets:*

$$A_n^n = \{(i_1,\ldots,i_n) \mid i_n \in A\};$$

$$A_{n-r}^n = \{(i_1,\ldots,i_n) \mid \textit{not all } i_1,\ldots,i_r = 0, i_n \in C\}$$
$$\cup \{(0,\ldots,0, i_{n+1},\ldots,i_n) \mid i_n \in A\} \quad \text{for } r = 1,\ldots, n - 1;$$

$B^n_n \;= \{(i_1,\dots,i_n) \mid i_n \in B\};$

$B^n_{n-r} = \{(i_1,\dots,i_n) \mid$ *not all* $i_1,\dots,i_r = 0,\ i_n \in C\}$

$\cup\{(0,\dots,0,\ i_{r+1},\dots,i_n) \mid i_n \in B\}$ *for* $r = 1,\dots,n-1;$

$C^n_n \;= \{(i_1,\dots,i_n) \mid i_n \in C\} = A^n_0 = B^n_0,$

*where* A, B, C *are the sets defined in Theorem 7.19.*

NOTE. If $p = 7$, then $k = 2$ and sets of type A do not occur. A similar proof shows that, in an elementary abelian 7-group of order $7^n$, there are $n + 1$ nonisomorphic maximal sum-free sets, namely $B^n_{n-r}$, $r = 0,1,\dots,n-1$ and $C^n$.

PROOF. A routine computation shows that $A^n_{n-r}$, $B^n_{n-r}$ and $C^n$ are maximal sum-free sets, $r = 0,1,\dots,n-1$. To prove that no other maximal sum-free sets exist, we consider the case where $n = 2$ and then generalise.

1. Let

$$G = \langle x_1,x_2 \mid px_i = 0,\ i = 1,\ 2;\ \ x_1 + x_2 = x_2 + x_1\rangle$$

and let $X_i = \langle x_i\rangle$. Since $|G| = p^2$, we know by Theorem 7.17 that $|S| = kp$ and hence that S covers at least $2k + 2$ of the $p + 1$ subgroups of order p. We assume, without loss of generality, $|X_2 \cap S| = k$. We denote by $S_i$ the subset of $X_2$ such that

$$S_i + ix_1 = S \cap (X_2+ix_1) \quad \text{for} \quad i = 0,.,\dots,p-1.$$

We make repeated use of the sum-freeness of S in the form

$$(S_i+S_j) \cap S_{i+j} = \emptyset \tag{7.26}$$

and in particular

$$(S_0+S_i) \cap S_i = \emptyset. \tag{7.27}$$

Since $|S_0| = k$, we find from (7.27) and Theorem 6.4 that $|S_i| \le k + 1$. By Theorem 6.9, if $S_0$ and $S_i$ are not in arithmetic progression with the same common difference, then $|S_i| \le k$; since $|S| = kp$, we then have $|S_i| = k$ for all $i = 0,.,\dots,p-1$. If $S_0$ and $S_i$ are in arithmetic progression with the same common difference, and if $|S_i| = k + 1$ for some i then, since S is sum-free, $S_0$ is isomorphic to C.

(a) Suppose that at least one proper subgroup of G intersects S in a set isomorphic to A. Without loss of generality we assume this subgroup to be $X_2$ and choose its generator $x_2$ so that $S_0 = A$. By (7.27),

$$S_i \subseteq \{\alpha_i,\dots,\alpha_i+k-1,\alpha_i+k+1\} \text{ for some } \alpha_i \in X_2,$$

and not both of $\alpha_i + 1$, $\alpha_i + k + 1 \in S_i$. Since $|S_i| = k$ for all i, we know that for each i,

$$S_i = \alpha_i - k + A \quad \text{or} \quad S_i = \alpha_i - k - 1 + C.$$

(i) If, for some i, $S_i = \alpha_i - k + A$, then we choose $x_1$, the other generator of G, so that $S_1 = A$. Then $S_1 + S_1 = \overline{A}$ and, by (7.26), $S_2 = A$. By induction, $S_i = A$ for all i and $S = A_2^2$.

(ii) If, for all i, $S_i = \alpha_i - k - 1 + C$, then we choose $x_1$ so that $S_1 = C$. Note that $S_i$ and consequently $S_i + S_j$ are in arithmetic progression with common difference d = 1 for all i, j (except for i = j = 0). From this fact and (7.26), we have:

$$\alpha_i + \alpha_{-i} = 2k + 2 \text{ and in particular } \alpha_{-1} = k + 1; \tag{7.28}$$

$$\alpha_{\frac{1}{2}(p+1)} = k + 1 \text{ or } -\tfrac{1}{2}k \text{ or } -\tfrac{1}{2}k+1; \tag{7.29}$$

$$\alpha_{i+1} = \alpha_i - 1 \text{ or } \alpha_i \text{ or } \alpha_i + 1. \tag{7.30}$$

Suppose that $\alpha_{\frac{1}{2}(p+1)} = -\frac{1}{2}k$ and consider the movement of $\alpha_i$ as i runs from $\frac{1}{2}(p+1)$ to p-1. By (7.28), in these (3k/2) - 1 steps, $\alpha_i$ must either decrease from (5k/2) + 1 by 3k/2 or increase from (5k/2) + 1 by (3k/2) + 1. But by (7.30), $\alpha_i$ can increase or decrease by at most 1 at each step. Hence $\alpha_{\frac{1}{2}(p+1)} \neq -\frac{1}{2}k$. A similar argument shows that $\alpha_{\frac{1}{2}(p+1)} \neq -\frac{1}{2}k + 1$ and hence, by (7.28) and (7.29), that

$$\alpha_{\frac{1}{2}(p+1)} = \alpha_{\frac{1}{2}(p-1)} = k + 1. \tag{7.31}$$

By (7.28), (7.30) and (7.31), $\alpha_i$ differs from $\alpha_0$ by at most $\frac{1}{2}(3k-2)$; hence if $\alpha_i < k + 1$, then $k \in S_i$.

Now let

$$X = \{i \in \langle x_1 \rangle \mid (i,k) \in S\} = \{i \in \langle x_1 \rangle \mid \alpha_i < k+1\}.$$

By (7.26), if i, j ε X, then i + j ε X. Hence $X + X \subseteq X$. Since $0 \in X$, $1 \notin X$, this shows that X is a proper subgroup of $X_1$. Since $|X_1| = p$, we have X = {0}. A similar argument shows that only $S_0$ contains an element greater than 2k. Hence $S_i = C$ for all i ≠ 0 and $S = A_1^2$.

(b) Suppose that no proper subgroup of G intersects S in a set isomorphic to A but that at least one proper subgroup intersects S in a set isomorphic to B. We assume this subgroup to be $X_2$ and choose $x_2$ so that $S_0 = B$. By (7.27),

$$S_i \subseteq \{\alpha_i, \ldots, \alpha_i + k - 1\} \text{ for all i, for some } \alpha_i \in X_2.$$

Since |S| = kp, we have $S_i = \alpha_i - k + B$ for all i; we choose $x_1$ so that $S_1 = B + 1 = C$.

By (7.26), $\alpha_i + \alpha_{-i} = 2k$ or $2k + 1$ or $2k + 2$

and in particular

$$\alpha_{-1} = k - 1 \text{ or } k \text{ or } k + 1. \tag{7.32}$$

Also

$$\alpha_{i+1} = \alpha_i - 1 \text{ or } \alpha_i \text{ or } \alpha_i + 1. \tag{7.33}$$

(i) If $\alpha_{-1} = k - 1$ then, by (7.26),

$$\alpha_{i-1} = \alpha_i - 3 \text{ or } \alpha_i - 2 \text{ or } \alpha_i - 1. \tag{7.34}$$

By (7.33) and (7.34),

$$\alpha_{i+1} = \alpha_i + 1 \text{ for all } i.$$

The automorphism of G which maps $(i_1,i_2)$ to $(i_1,i_2-i_1)$ maps S to $B_2^2$.

(ii) If $\alpha_{-1} = k$ then, by (7.26),

$$\alpha_{i-1} = \alpha_i - 2 \text{ or } \alpha_i - 1 \text{ or } \alpha_i. \tag{7.35}$$

By (7.33) and (7.35),

$$\alpha_{i+1} = \alpha_i \text{ or } \alpha_i + 1 \text{ for all } i. \tag{7.36}$$

Consider the movement of $\alpha_i$ as i runs from 1 to p - 1. In these p - 2 steps, $\alpha_i$ must increase by p - 1, but by (7.36) $\alpha_i$ may increase by at most 1 at each step. Hence $\alpha_i \neq k$.

(ii) If $\alpha_{-1} = k + 1$, we use an argument similar to that of (a (ii)). This shows that:

$$\alpha_{\frac{1}{2}(p+1)} = k + 1 \text{ or } -\tfrac{1}{2}k \text{ or } -\tfrac{1}{2}k + 1; \tag{7.37}$$

$$\alpha_{i+1} = \alpha_i - 1 \text{ or } \alpha_i \text{ or } \alpha_i + 1; \tag{7.38}$$

$$\alpha_{\frac{1}{2}(p+1)} = \alpha_{\frac{1}{2}(p-1)} = k + 1; \tag{7.39}$$

that if $\alpha_i < k + 1$, then $k \in S_i$ and that $k \in S_i$ only if $i = 0$.

Similarly if we let $Y = \{i \in \langle x_1\rangle \mid \alpha_i > k + 1\}$, then $Y + Y \subseteq Y$, $Y \neq X_1$, $Y \neq \{0\}$ and hence $Y = \emptyset$. Hence $S_i = C$ for all $i \neq 0$ and $S = B_1^2$.

(c) Finally suppose that every subgroup of G covered by S intersects S in a set isomorphic to C.

(i) If there exists a proper subgroup, covered by S and having at least one coset with k + 1 elements of S, then we assume this subgroup to be $X_2$ and choose $x_2$ so that $S_0 = C$. By (7.37),

$$S_i \subseteq \{\alpha_i,\ldots,\alpha_i + k\} \text{ for all } i$$

and we choose $x_1$ so that

$$S_1 = \{k+1,\ldots,2k+1\}.$$

We know from the proof of Theorem 7.17 that S avoids $X_1$. Hence S must cover every other subgroup of order p in G. But S intersects each such subgroup in a set isomorphic to C, implying that S = -S. Again as in the proof of Theorem 7.17, we consider the movement of $\alpha_i$ (and similarly of the right-hand end-point of $S_i$) as i runs from 1 to p - 1. The fact that S = -S, combined with the previous proof, shows that $\alpha_i$ can move by at most ½k in either direction. Since $\alpha_1 = k + 1$, the right-hand end-point of $S_{-1}$ is 2k and we have

$$S_i \subseteq \{(k/2) + 1,\ldots,5k/2\} \text{ for all } i.$$

Hence for every subgroup $\langle(\rho,1)\rangle$, the second co-ordinates of $\langle(\rho,1)\rangle \cap S$ belong to $C \cup C'$. Since all these subgroups are covered by S, Lemma 7.20 shows that, for any given ρ, the second co-ordinates of the intersection are either C or C'. Let

$$T_j = \{i \in \langle x_1\rangle \mid (i,j) \in S\} = \{i \in X_1 \mid j \in S_i\}.$$

Let $|T_j| = a$, $j \in C'$ and $|T_j| = b$, $j \in C$. If a > 0, b > 0, then by (7.26), in particular,

$$\left(T_{\frac{1}{2}k+1} + T_{\frac{1}{2}k+1}\right) \cap T_{k+2} = \emptyset.$$

Hence by Theorem 6.4,

$$2a + b \leq p + 1.$$

Similarly

$$2b + a \leq p + 1$$

so that

$$a + b \leq 2(p+1)/3.$$

But a + b = p. Hence either a = 0 or b = 0; we may assume that a = 0 and for each $\langle(\rho,1)\rangle$, the second co-ordinates of its intersection with S form the set C.

This contradicts our assumption that $(1,2k+1) \in S$. Hence every coset of every proper subgroup covered by S contains exactly k elements of S.

(ii) Now let $X_2$ be any subgroup of order p covered by S and choose its generator $x_2$ so that $S_0 = C$. Then

$$S_i \subseteq \{\alpha_i,\ldots,\alpha_i + k\}$$

for all i, for some $\alpha_i \in X_1$ and, since $|S| = k$, four types of sets may occur:

$$S_i = \{\alpha_i,\ldots,\alpha_i+k-1\} \in P;$$

$$S_i = \{\alpha_i,\alpha_i+2,\ldots,\alpha_i+k\} \in Q_1;$$

$$S_i = \{\alpha_i,\ldots,\alpha_i+k-2,\alpha_i+k\} \in Q_2;$$

$$S_i = \{\alpha_i,\ldots,\alpha_i+\ell,\alpha_i+\ell+2,\ldots,\alpha_i+k\} \in R, \text{ where } 2 \leq \ell \leq k-2.$$

If a set of type R occurs, choose $x_1$ so that

$$S_1 = \{k+1,\ldots,k+1+\ell,k+3+\ell,\ldots,2k+1\} \text{ for some } \ell,\ 2 \leq \ell \leq k-2.$$

By (7.26), we find that $\alpha_2 = k+2$, that $\alpha_{i+1} = \alpha_i$ or $\alpha_i+1$ and that $\alpha_{-1} = k$ or $k+1$. Since $\alpha_i$ can never decrease, in the p-3 steps as i runs from 2 to p-1, $\alpha_i$ must increase from k+2 by p-2 or p-1. But $\alpha_i$ can increase by at most 1 at each step. Hence no set of type R can occur.

If a set of type $Q_2$ occurs, choose $x_1$ so that

$$S_1 = \{k+1,\ldots,2k-1,2k+1\}.$$

By (7.26), we find that

$$\alpha_{-1} = k \text{ or } k+1,$$

that

$$\alpha_{\frac{1}{2}(p+1)} = k \text{ or } k+1 \text{ or } -\tfrac{1}{2}k+1,$$

that

$$\alpha_{i+1} = \alpha_i-1 \text{ or } \alpha_i \text{ or } \alpha_i+1$$

and that if

$$\alpha_{i+1} = \alpha_i-1,$$

then

$$\alpha_{i+2} = \alpha_i.$$

Again we use an argument similar to that of (a(ii)), which shows that

$$\alpha_i = k \text{ or } k+1 \text{ for all } i.$$

Hence S avoids $X_1$. Therefore S covers every other proper subgroup each of them in a set isomorphic to C and contained in $\{k,\ldots,2k+1\}$. Hence by Lemma 7.20 every subgroup except $X_1$ intersects S in the set C, contradicting our statement that $(1,2k+1) \in S$. Hence no set of type $Q_2$ (and similarly $Q_1$) can occur.

We now know that every coset of $X_2$ intersects S in a set of type P and we choose $x_1$ so that $S = C$. By (7.26), we find that

$$\alpha_{-1} = k \text{ or } k+1 \text{ or } k+2,$$

and that

$$\alpha_{i+1} = \alpha_i-1 \text{ or } \alpha_i \text{ or } \alpha_i+1. \tag{7.40}$$

If $\alpha_{-1} = k$ then, by (7.26) again,

$$\alpha_{i+1} = \alpha_i \quad \text{or} \quad \alpha_i+1. \tag{7.41}$$

If $\alpha_{-1} = k+2$, then

$$\alpha_{i+1} = \alpha_i-1 \quad \text{or} \quad \alpha_i. \tag{7.42}$$

If $\alpha_{-1} = k$ then, by (7.41), in the p-2 steps as i runs from 1 to p-1, $\alpha_i$ must increase by p-1. But $\alpha_i$ may increase by at most 1 at each step. Hence $\alpha_{-1} \neq k$. A similar argument using (7.42) shows that $\alpha_{-1} \neq k+2$. Hence $\alpha_{-1} = k+1$ and, by (7.26), we see that

$$\alpha_{\frac{1}{2}(p+1)} = -\tfrac{1}{2}k \quad \text{or} \quad -\tfrac{1}{2}k+1 \quad \text{or} \quad k+1.$$

Again we use an argument similar to that of (a(ii)) to show that $\alpha_i = k+1$ and hence that $S_i = C$ for all i. Therefore $S = C^2$.

2. Now let G be an elementary abelian group of order $p^n$.

(a) We show first that any maximal sum-free set S in G avoids exactly one maximal subgroup of G.

By Theorem 7.17, $|S| = kp^{n-1}$, so G has at least one subgroup of order p which is covered by S. Let X be any such subgroup and let Y be a subgroup complementing X in G. Then $|Y| = p^{n-1}$ and

$$Y = \bigcup_{i=1}^{\rho} Y_i,$$

where $Y_i$ is a subgroup of order p and

$$\rho = (p^{n-1} - 1)/(p - 1).$$

Now

$$|S| = kp^{n-1} = \sum_{i=1}^{\rho} |(X + Y_i) \cap S| - (\rho - 1)k.$$

But

$$|(X + Y_i) \cap S| \leq kp \quad \text{for all } i = 1,\ldots,\rho$$

and

$$\sum_{i=1}^{\rho} |(X + Y_i) \cap S| = kp^{n-1} + (\rho - 1)k = kp\rho.$$

Hence

$$|(X + Y_i) \cap S| = kp \quad \text{for all } i = 1,\ldots,\rho.$$

From the proof of (7.8), for each $i = 1,\ldots,\rho$, there exists a subgroup $V_i$ of order p such that $V_i < X + Y_i$ and S avoids $V_i$. These $\rho$ subgroups are distinct for if $V_i = V_j$, then

$$X + Y_i = X + V_i = X + V_j = X + Y_j \text{ and } i = j.$$

Hence S avoids $\rho$ of the $(p^n - 1)(p - 1)$ subgroups of order p in G, and since $|S| = kp^{n-1}$, S covers the $p^{n-1}$ remaining subgroups of order p, which we denote by $X_i$, $i = 1,\ldots,p^{n-1}$.

Suppose that for some h, i, j with $1 \leq h \leq p^{n-1}$, $1 \leq i,j \leq \rho$, we have

$$X_h < V_i + V_j.$$

Then we repeat the proof, choosing $X_h$ as our subgroup X which is covered by S, and show that

$$|(V_i + V_j) \cap S| = kp.$$

But since S avoids both $V_i$ and $V_j$,

$$|(V_i + V_j) \cap S| \leq k(p - 1).$$

Hence for any i, j = $1,\ldots,\rho$ we have

$$V_i + V_j \subseteq \bigcup_{\ell=1}^{\rho} V_\ell.$$

Now

$$\left|\bigcup_{\ell=1}^{\rho} V_\ell\right| = p^{n-1}$$

and

$$\bigcup_{\ell=1}^{\rho} V_\ell \text{ is a subgroup.}$$

For if $v_1, v_2 \in \bigcup_{\ell=1}^{\rho} V_\ell$, then

either $\quad v_1, v_2 \in V_i$ and $v_1 + v_2 \in V_i \subseteq \bigcup_{\ell=1}^{\rho} V_\ell$

or $\quad v_1 \in V_i$, $v_2 \in V_j$ and $v_1 + v_2 \in V_i + V_j \subseteq \bigcup_{\ell=1}^{\rho} V_\ell$.

Hence S avoids a maximal subgroup of G.

(b) Finally we suppose that in elementary p-groups of orders $p^{n-1}$ or less, the maximal sum-free sets have been characterised. If G has order $p^n$ and H, K are subgroups of G, each of order p, then we see from (7.8) that it is impossible to have

$$S \cap H = A \quad \text{and} \quad S \cap K = B.$$

Hence two cases arise:

(i) Subgroups of order p intersect S in sets A or C. If no subgroup of order p intersects S in A, then $S = C^n$. If exactly one subgroup of order p intersects S in A, then $S = A_1^n$. If two subgroups of order p intersect S in A, then the subgroup of order $p^2$ which they generate intersects S in $A_2^2$ so that altogether p subgroups of order p intersect S in A and $S = A_2^n$. By induction, if the subgroups of order p intersecting S in A generate a subgroup of order $p^r$, then $p^{r-1}$ subgroups of order p intersect S in A and $S = A_r^n$. In each case, since S avoids a maximal subgroup of G, S is determined up to automorphism by the order of the subgroup generated by all those subgroups of order p which intersect S in A. Hence n+1 sets are possible.

(ii) Subgroups of order p intersect S in sets B or C. An argument similar to (i) shows that again n+1 sets are possible, namely $C^n, B_1^n, \ldots, B_n^n$.

Since $C^n$ occurs in both cases, we have altogether 2n+1 nonisomorphic sets.

Finally we characterise the maximal sum-free sets in cyclic groups of prime-power order, for primes congruent to 1 (modulo 3). We need one preliminary lemma.

LEMMA 7.22. *Let* $G = Z_n$, *where* $n = 3k + 1$. *Let* S *be a maximal sum-free set in* G *and let* H *be the subgroup of* G *of order* m. *Let* $S_i$ *denote the subset of* H *such that*

$$S_i + i = S \cap (H + i),$$

*where* H+1 *generates* G/H. *Then the cosets of* H, *more than half of whose elements belong to* S, *form a sum-free set in* G/H.

PROOF. Since S is sum-free,

$$(S_i + S_j) \cap S_{i+j} = \emptyset \quad \text{for all } i, j = 0,1,\ldots,(n/m)-1.$$

By Theorem 6.15, for some subgroup $K < H$, we have

$$S_i + S_j + K = S_i + S_j$$

and

$$|S_i + S_j| \geq |S_i + K| + |S_j + K| - |K|.$$

Let $|K| = q$.

Since $|S_i| \geq \frac{1}{2}(m + 1)$ and since $q \mid |S_i + K|$, we must have

$$|S_i + K| \geq \tfrac{1}{2}(m + q)$$

and similarly for $S_j$. Hence

$$|S_i + S_j| \geq 2((m + q)/2) - q = m.$$

Hence

$$S_{i+j} = \emptyset$$

and the lemma is proved.

THEOREM 7.23. *Let* $G = Z_n$, *where* $n = 3k + 1 = p^e$ *and* $p$ *is prime,* $p = 3k' + 1$. *Then any maximal sum-free set* $S$ *may be mapped, under some isomorphism of* $G$, *to one of the following sets:*

$$A = \{k, k+2, \ldots, 2k-1, 2k+1\};$$

$$B = \{k, \ldots, 2k-1\};$$

$$C = \{k+1, \ldots, 2k\}.$$

PROOF. By Corollary 7.14, $|S| = k$. Since $S$ is sum-free,

$$S + S \subseteq \overline{S},$$

so

$$|S + S| \leq 2k + 1.$$

Similarly,

$$|S - S| \leq 2k + 1.$$

Suppose that

$$|S + S| \leq 2k - 2.$$

By Theorem 6.15, there exists a non-trivial subgroup $H = H(S + S)$ of $G$, such that

$$S + S + H = S + S \quad \text{and} \quad |S + S| = 2|S + H| - |H|.$$

By Lemma 7.4, $S + H = S$. But this implies that $|S + S| = 2k - |H|$ and that $|H|\,|\,k$. Now $|H|\,|\,n$ and $(k,n) = 1$, so that $|H| = 1$ which is a contradiction. Hence

$$|S + S| \geq 2k - 1.$$

A similar argument using Corollary 7.5 instead of Lemma 7.4, shows that

$$|S - S| \geq 2k - 1.$$

Now $0 \in S - S$ and $S - S = -(S - S)$, so $|S - S|$ is odd. Hence we have two cases:

$$|S - S| = 2k - 1 \quad \text{and} \quad |S - S| = 2k + 1.$$

(I) If $|S - S| = 2k - 1$, then, by Theorem 6.16, either $S - S$ is in arithmetic progression or $S - S$ is quasiperiodic.

Suppose that $S - S$ is quasiperiodic. Then for some proper subgroup $H$ of $G$, $S - S$ consists of the union of complete cosets of $H$, together with a subset of one other coset of $H$. Now $|H|\,|\,n$, so we let

$$n = 3k + 1 = (3\ell + 1)(3m + 1), \text{ where } |H| = 3m + 1.$$

Then

$$k = 3\ell m + \ell + m,$$

so that

$$|S - S| = 2\ell|H| + 2m - 1.$$

Hence S - S consists of $2\ell$ complete cosets of H, together with 2m - 1 elements contained in one other coset of H. But S - S = -(S - S), so these 2m - 1 elements must be contained in H itself.

Now

$$|(S - S) \cup S| = n - 2$$

so, of the m+2 elements of H not belonging to S - S, at least m must belong to S. As before, let $S_i$ denote the subset of H such that

$$S_i + i = S \cap (H + i),$$

where H+1 generates G/H. Now since S is sum-free, we must have

$$(S_i + S_0) \cap S_i = \emptyset \quad \text{for every } i = 0, 1, \ldots, 3\ell;$$

but $|S_0| \geq m$, and for $2\ell$ values of i, $S_i = H$. So we have a contradiction and S - S cannot be quasiperiodic.

Hence S - S is in arithmetic progression, with difference d. Since $|S - S| = 2k - 1$, this difference d must have order n. Now we may apply Lemma 6.19, which shows that S (and -S) must also be in arithmetic progression with difference d. Hence some automorphism of G will map d to ±1 and S to one of the sets B or C in the statement of the Theorem.

(II) If

$$|S - S| = 2k + 1,$$

then $S - S = \overline{S}$. Hence S = -S and

$$S + S = S - S.$$

Now we may apply Lemma 7.18.

(a) Suppose that, for some $g \in G$,

$$|(S + g) \cap S| = 1.$$

Then by (ii) of the lemma,

$$|(S + (3g/2)) \cap S| \geq k - 3.$$

(i) If (g,n) = 1, then

$$((3g/2),n) = 1$$

also; we map 3g/2 to 1 under some automorphism of G, so that g = k+1. Now

$$|(S + 1) \cap S| \neq k - 1$$

since S is not a standard set.

If $$|(S + 1) \cap S| = k - 2,$$

then obviously

$$S = \{\pm\tfrac{1}{2}k, \pm(1 + \tfrac{1}{2}k), \ldots, \pm(k-1)\},$$

which maps under automorphism to the set A.

If $$|(S + 1) \cap S| = k-3,$$

then

$$S = \{\alpha,\ldots,\alpha+\rho-1,k+\rho+1,\ldots,2k-\rho,3k+2+\alpha-\rho,\ldots,3k+1-\alpha\}$$

where $\alpha \leq k$ and $1 \leq \rho < \tfrac{1}{2}k$. But $-\tfrac{1}{2}g = k \in S$ and $g = k+1 \notin S$ by the lemma. Hence

$$\alpha + \rho - 1 = k$$

and

$$S = \{k+1-\rho,\ldots,k,k+\rho+1,\ldots,2k-\rho,2k+1,\ldots,2k+\rho\}.$$

But

$$(k + 1 - \rho) + (k + \rho + 1) = 2k + 2 \in \overline{S}.$$

Hence $\rho = 1$ and S is the set A.

(ii) If $(g,n) = 3m+1$, then

$$((3g/2),n) = 3m + 1$$

also; we map $3g/2$ to $3m+1$ under some automorphism of G, so that

$$g = (\ell+1)(3m+1),$$

where again

$$n = (3\ell + 1)(3m + 1).$$

Now

$$|(S + 3m + 1) \cap S| \geq k - 3.$$

Let $H = \langle 3m + 1\rangle$, the cyclic group of order $3\ell+1$, generated by $3m+1$. Now

$$S = \bigcup_{i=0}^{3m} (S_i + i) \quad \text{and}$$

$$|(S + 3m + 1) \cap S| = \sum_{i=0}^{3m} |(S_i + 3m + 1) \cap S_i|.$$

Note that $S_i + 3m+1 = S_i$ if and only if $S_i = \emptyset$ or $S_i = H$.

If

$$|(S + 3m + 1) \cap S| = k,$$

then S consists of a union of cosets of H, so that $(3\ell+1)|k$, which is a contradiction. Hence we may assume that

$$k - 3 \leqslant |(S + 3m + 1) \cap S| \leqslant k - 1.$$

We know that

$$|S| = k = 3\ell m + \ell + m = m|H| + \ell,$$

so that there are three possibilities for S:

(1) S consists of a union of m complete cosets of H, together with $\ell$ elements distributed between one, two or three other cosets of H. Since S = -S, if these $\ell$ elements are distributed between an odd number of cosets of H, some of them must be contained in H itself. Hence $S_0 \neq \emptyset$. But now consider $S_i$ for any of the m values of i for which $S_i = H$. Since S is sum-free, we know that

$$(S_i + S_0) \cap S_i = \emptyset,$$

but we have

$$S_i + S_0 = S_i$$

which is a contradiction.

So we need only consider the case where the $\ell$ elements are distributed between two cosets of H. Since S = -S, we have $\frac{1}{2}\ell$ elements in each of the cosets H+j, H-j, for some j = 1,...,3m. Since $S + S = \overline{S}$, the remaining elements of the coset H+j must occur in S + S. But the only proper subsets of cosets which can occur in S + S must be the cosets H+2j, H and H-2j. Hence

$$H + j = H - 2j,$$

so that

$$3j \equiv 0 \pmod{3m+1}.$$

But this is impossible and again no such maximal sum-free set can exist.

(2) S consists of a union of m-1 complete cosets of H, together with $4\ell+1$ elements distributed through two or three cosets of H. Since n is odd, m-1 is odd. Since S = -S, one of the complete cosets must be H itself. But H is certainly not sum-free, so this case cannot occur.

(3) S consists of m-2 complete cosets of H, together with $7\ell+2$ elements distributed through three cosets of H. Since S = -S, one of

these cosets must be H itself and $S_0 \neq \emptyset$. But

$$\lambda(H) = \ell \quad \text{so} \quad |S_0| \leq \ell.$$

Hence we must have $S_0 = \ell$ and the remaining $6\ell+2$ elements fill up two more cosets of H. But now we are back to case (1), which we have already seen is impossible.

Hence

$$|(S + g) \cap S| \neq 1$$

for any g such that $(g,n) > 1$.

(b) Suppose that for some $g \in G$,

$$|(S + g) \cap S| = 2.$$

Then by (iii) of Lemma 7.18, for some $g^* \in G$, we have

$$|(S + g^*) \cap S| \geq k - 3.$$

By (a(ii)), $(g^*,n) = 1$ and hence, under some automorphism of G, we may map $g^*$ to 1. But now we are back to case (a(i)), which has already been dealt with.

(c) So now assume that for all $g \in \overline{S}$,

$$|(S + g) \cap S| \geq 3.$$

We show that this case cannot in fact occur.

(i) Suppose that by taking an automorphism of G, we may arrange that

$$|(S + 1) \cap S| \geq |(S + g) \cap S| \text{ for all } g \in G.$$

We list the elements of S as follows:

$$S = \{\alpha_1,\ldots,\alpha_1+\ell_1,\alpha_2,\ldots,\alpha_2+\ell_2,\ldots,\alpha_h,\ldots,\alpha_h+\ell_h\} \qquad (7.43)$$

where $0 < \alpha_1 \leq \alpha_1+\ell_1 < \alpha_2-1 < \alpha_2+\ell_2 < \ldots < \alpha_h-1 < \alpha_h+\ell_h < n$ and $\alpha_i,\ldots,\alpha_i+\ell_i$ denotes a string of $\ell_i+1$ consecutive elements of S. Since $S = -S$,

$$\alpha_{h-i} + \ell_{h-i} = n - \alpha_{i+1}, \qquad (7.44)$$

for all $i = 0, 1, \ldots, h-1$. Also

$$|(S + 1) \cap S| = k-h \geq |(S + g) \cap S| \quad \text{for all } g \in G. \qquad (7.45)$$

Hence h is minimal in (7.43).

If $h = 1$, then S is in arithmetic progression and must map under automorphism to either B or C, which have been considered previously.

If $h = 2$, the only possible set is

$$S = \{\pm\tfrac{1}{2}k, \pm(1+\tfrac{1}{2}k), \dots , \pm(k-1)\}.$$

Since

$$\left|\left(S + (2k+1)\right) \cap S\right| = 1,$$

this set has also been considered previously; it maps under automorphism to the set A.

So now we assume that $h \geq 3$.

Let

$$X = \{\alpha_1 ,\alpha_2, \dots ,\alpha_h\}$$

and let

$$Y = \{\alpha_1+\ell_1+1, \dots , \alpha_h+\ell_h+1\}$$

$$= \{1-\alpha_1, \dots ,1-\alpha_h\}$$

$$= 1 - X \quad \text{by (7.44)}.$$

If $h = 3$ or $4$, then by assumption,

$$|(S + g) \cap S| \geq 3 \geq h - 1$$

for every $g \in \overline{S}$. Now suppose that $h \geq 5$, and that for some $g \in \overline{S}$, we have

$$3 \leq |(S + g) \cap S| \leq h - 2.$$

Then by Lemma 7.18, there exists $g' \in \overline{S}$ such that

$$|(S + g') \cap S| \geq k - (h - 2 + 1) = k - h + 1.$$

But this contradicts (7.45), so for all $g \in \overline{S}$, we have

$$h - 1 \leq |(S + g) \cap S| \leq k - h, \quad \text{for } h \geq 3.$$

In particular, for $i = 1, \dots , h$, we know that $\alpha_i-1 \in \overline{S}$, so we have

$$|(S + \alpha_i - 1) \cap S| \geq h - 1.$$

But for any $s_1, s_2 \in S, s_1+\alpha_i-1 = s_2$ implies that $s_1 \in X$, $s_2 \in -X$ and $s_1+\alpha_i \in Y$. Hence

$$h \geq \left|(X + \alpha_i) \cap Y\right| \geq h - 1 \quad \text{for all } i = 1, \dots ,h. \tag{7.46}$$

This implies that

$$h \leq |X + X| \leq 2h. \tag{7.47}$$

Also $Y \subseteq \overline{S} = S + S$. But $s_1+s_2 = \alpha_i+\ell_i+1$ implies that $s_1, s_2 \in X$. So

$$Y \subseteq X + X \quad \text{and} \quad Y \cap \left[(S + S)\setminus(X + X)\right] = \emptyset.$$

We consider the different possible cases in view of (7.46) and (7.47). If $|X + X| = 2h$, then $2\alpha_i \notin Y$ for any $i = 1, \ldots, h$. But $2\alpha_i \notin Y$ implies that $1-\alpha_i \notin X+\alpha_i$, since $Y = 1-X$. For each value of i, we have

$$\{\alpha_1+\alpha_i, \ldots, \alpha_{i-1}+\alpha_i, \alpha_{i+1}+\alpha_i, \ldots, \alpha_h+\alpha_i\}$$

$$= (X + \alpha_i) \cap Y$$

$$= \{1-\alpha_1, \ldots, 1-\alpha_{i-1}, 1-\alpha_{i+1}, \ldots, 1-\alpha_h\},$$

and summing on both sides of this equation, we find that

$$(h - 3)\alpha_i \equiv h - 1 - 2 \sum_{j=1}^{h} \alpha_j \pmod{n}. \tag{7.48}$$

Since (7.48) holds for all i, we see that

$$(h - 3)(\alpha_i - \alpha_j) \equiv 0 \pmod{n} \tag{7.49}$$

for all $i,j = 1,\ldots,h$.

If $(h-3,n) = 1$, then $\alpha_i$ is uniquely determined and only one string of elements of S is possible. This reduces to the case where S = B or C which have been considered previously.

If $(h-3,n) = 3s+1 > 1$, then $(3r+1)|(\alpha_i-\alpha_j)$ for all i and j, where $n = (3s+1)(3r+1)$. Hence X is contained in one coset of the subgroup $H = \langle 3r+1 \rangle$, of order $3s+1$. But this implies that $h \leq 3s+1 \leq h-3$, which is a contradiction.

If $|X + X| = 2h-1$, then by (7.46), $2\alpha_i \notin Y$ for at least h-2 values of $i = 1,\ldots,h$. Hence (7.48) and (7.49) will follow for at least h-2 values of i.

If $(h-3,n) = 1$, then only one value of $\alpha_i$ is possible for h-2 values of i. Hence $h \leq 3$. But we have already assumed that $h \geq 3$, so now we have $h = 3$. We can list the elements of S in the following way:

$$S = \{\alpha,\ldots,\alpha+\rho-1, k+\rho+1,\ldots,2k-\rho, 3k+2-\alpha-\rho,\ldots,3k+1-\alpha\},$$

where $\alpha \leq k$ and $\rho < \frac{1}{2}k$. From (7.48), we have

$$0 \equiv 3 - 1 - 2\big(\alpha + k + \rho + 1 - (\alpha + \rho - 1)\big) \pmod{n}$$

or

$$1 \equiv k + 2 \pmod{n},$$

which is not possible.

If $(h-3,n) = 3s+1 > 1$, then $(3r+1)|(\alpha_i-\alpha_j)$ for at least h-2 values of i and j, where $n = (3s+1)(3r+1)$. Hence these h-2 elements of X are contained in one coset of the subgroup $H = \langle 3r+1 \rangle$, of order 3s+1. But this implies that $h-2 \leq 3s+1 < h-3$, which is a contradiction.

Finally, suppose that $|X + X| \leq 2h-2$. By Theorem 6.15, X + X is periodic and for some subgroup $H < G$, we have

$$X + X + H = X + X \quad \text{and} \quad |X + X| = 2|X + H| - |H|.$$

Let $|H| = 3s+1$. Now by Theorem 6.17, we can construct all the possible sets X. We choose a subset X* of G/H such that X* + X* is aperiodic in G/H and

$$|X^* + X^*| = 2|X^*| - 1.$$

If $\sigma$ denotes the natural mapping of G to G/H, then X can be any subset of $\sigma^{-1}X^*$, such that

$$|\sigma^{-1}X^* \cap \overline{X}| \leq 3s/2.$$

Hence any coset of H which contains the first element of a string of elements of S must contain the first elements of at least $\frac{1}{2}(3s+2)$ strings of S.

Since $Y \subseteq X + X$, we can describe the distribution of the strings of S. Suppose

$$X^* = \{H+i_1,\dots,H+i_t\} \text{for some } i_1,\dots,i_t \in \{0,1,\dots,3r\},$$

where

$$n = (3s+1)(3r+1).$$

In each coset $H + i_j$, more than half of the elements of the coset are starting-points of strings of S. Since S = -S, the strings finish in the cosets of -X*. If a string finishes in $H - i_j$, then the next coset $H + (1-i_j)$ contains an element of Y and hence is contained in X + X. So no string can continue into this coset and similarly no string could pass through $H + (i_j-1)$.

Hence any coset which contains an element of S contains at least $\frac{1}{2}(3s+2)$ elements of S. But by Lemma 7.22 these cosets must form a sum-free set in G/H and therefore S is contained in at most r cosets of H. But this means that

$$k \leq r(3s+1)$$

which is a contradiction.

(ii) Finally we consider the case where it is impossible to take an automorphism of G such that

$$|(S+1) \cap S| \geq |(S+g) \cap S|$$

for all $g \in G$. So we take an automorphism of G such that

$$|(S+g) \cap S| \geq |(S+f) \cap S| \text{ for all } f \in G,$$

where $g > 1$ and $g|n$, and

$$|(S+g) \cap S| > |(S+f) \cap S| \text{ for all } f|g,\ f \in \overline{S}.$$

Let

$$n = (3r+1)(3s+1)$$

as before, where $g = 3r + 1$ and the subgroup $H = \langle 3r+1 \rangle$ has order $3s + 1$. Then S consists of a union of complete cosets of H, together with subsets of cosets of H. More precisely,

$$S = \bigcup_{j=1}^{t} (H+i_j) \cup T,$$

where

$$T = \{\alpha_1, \alpha_1+g, \ldots, \alpha_1+g\ell_1, \ldots, \alpha_h, \ldots, \alpha_h+g\ell_h\}, \tag{7.50}$$

$$\ell_i < 3s \text{ for all } i,$$

and

$$\alpha_i, \ldots, \alpha_i+g\ell_i$$

denotes a set of $\ell_i + 1$ consecutive elements of a coset of H, which we call an *H-string* of S. Now

$$|(S+g) \cap S| = k - h \geq |(S+f) \cap S| \text{ for all } f \in G. \tag{7.51}$$

Hence h is minimal in (7.50).

If $h = 1$, we see that since $S = -S$, we must have $T \subseteq H$. Hence since S is sum-free, no complete coset of H can be contained in S, so $S = T$. But $|T| \leq s < k$ which is a contradiction.

If $h = 2$, then T consists of two H-strings. Since $S = -S$, either $T \subseteq H$, which we have just seen is impossible, or T is contained in the two cosets $H + i$, $H - i$ for some i. Since $S + S = \overline{S}$, the remaining elements of these cosets must be contained in $T + T$, that is

$$\Big(\big((H+i) \cup (H-i)\big) \backslash T\Big) \subseteq T + T.$$

But

$$T + T \subseteq (H+2i) \cup H \cup (H-2i),$$

so that

$$H + i = H - 2i$$

and

$$3i \equiv 0 \pmod{3s+1}, \text{ where } i = 1,\dots,3s.$$

But this is not possible.

So now we assume that $h \geq 3$.

Again we let

$$X = \{\alpha_1,\alpha_2,\dots,\alpha_h\}$$

and let

$$Y = \{\alpha_1 + (g+1)\ell_1,\dots,\alpha_h+(g+1)\ell_h\}$$

$$= g - X \quad \text{since } S = -S$$

$$= \{g-\alpha_1,\dots,g-\alpha_h\}.$$

We apply Lemma 7.18 as we did in (i), to show that for all $f \in \overline{S}$, we have

$$h - 1 \leq |(S+f) \cap S| \leq k - h, \quad \text{for } h \geq 3.$$

In particular, for $i = 1,\dots,h$, we have'

$$h \geq |(X+\alpha_i) \cap Y| \geq h - 1 \quad \text{as before.} \tag{7.52}$$

This implies that

$$h \leq |X+X| \leq 2h. \tag{7.53}$$

Also

$$Y \subseteq \overline{S} = S + S.$$

But

$$s_1 + s_2 = \alpha_i + (g+1)\ell_i$$

implies that $s_1, s_2 \in X$. So $Y \subseteq X + X$ and

$$Y \cap \left((S+S)\setminus(X+X)\right) = \emptyset.$$

We consider again the different possible cases. If

$$X + X = 2h,$$

then $2\alpha_i \notin Y$ for any $i = 1,\dots,h$. But now $2\alpha_i \notin Y$ implies that

$$g - \alpha_i \notin X + \alpha_i,$$

since $Y = g - X$. So for each value of $i$, we have

$$\{\alpha_1+\alpha_i,\dots,\alpha_{i-1}+\alpha_i, \alpha_{i+1}+\alpha_i,\dots,\alpha_h+\alpha_i\}$$

$$= (X+\alpha_i) \cap Y$$

$$= \{g-\alpha_1,\dots,g-\alpha_{i-1}, g-\alpha_{i+1},\dots,g-\alpha_h\}.$$

and summing on both sides of this equation, we find that

$$(h-3)\alpha_i \equiv (h-1)g - 2 \sum_{j=1}^{h} \alpha_j \pmod{n}. \tag{7.54}$$

Since (7.54) holds for all i, we see that

$$(h-3)(\alpha_i - \alpha_j) \equiv 0 \pmod{n}, \tag{7.55}$$

for all $i, j = 1,\ldots,h$.

If $(h-3,n) = 1$, then $\alpha_i$ is uniquely determined and only one H-string of elements of S is possible. Hence $T \subseteq H$, no complete coset of H can be contained in S and

$$|S| = |T| \leq s < k,$$

which is impossible.

If

$$(h-3,n) = 3u + 1 > 1,$$

then

$$(3v+1) \mid (\alpha_i - \alpha_j)$$

for all i and j, where

$$n = (3u+1)(3v+1).$$

Hence X is contained in one coset of the subgroup $K = \langle 3v+1 \rangle$, of order $3u + 1$. But this implies that

$$h \leq 3u + 1 \leq h - 3,$$

which is a contradiction.

If

$$|X+X| = 2h - 1,$$

then by (7.46), $2\alpha_i \notin Y$ for at least $h - 2$ values of $i = 1,\ldots,h$. Hence (7.54) and (7.55) will follow for at least $h - 2$ values of i.

If

$$(h-3,n) = 1,$$

then only one value of $\alpha_i$ is possible for $h - 2$ values of i. Hence $h \leq 3$. But we have already assumed that $h \geq 3$, so we have $h = 3$. Since $S = -S$, at least one of these three H-strings must be contained in H itself. Hence no complete coset of H can be contained in S. Now since S is sum-free, we know that

$$(S_0 + S_i) \cap S_i = \emptyset \quad \text{for all } i,$$

where $H + 1$ is a generator of $G/H$, and $S_i$ is the subset of H such that

$$S \cap (H+i) = S_i + i.$$

Since $S_0 \neq \emptyset$, we see that $S_i \leq 3s/2$ for each i. Hence in our three H-strings, we have at most s elements of H and at most 3s/2 in each of two other cosets. Hence

$$|S| = k \leq 4s.$$

But

$$k = 3rs + r + s,$$

so this is impossible.

If

$$(h-3,n) = 3u + 1 > 1,$$

then

$$(3v+1) \mid (\alpha_i - \alpha_j)$$

for at least h - 2 values of i and j, where

$$n = (3u+1)(3v+1).$$

Hence these h - 2 elements of X are contained in one coset of the subgroup K = <3v+1>, of order 3u + 1. But this implies that

$$h - 2 \leq 3u + 1 \leq h - 3,$$

which is a contradiction.

Finally, suppose that $|X+X| \leq 2h - 2$. By Theorem 6.15, X + X is periodic and for some subgroup K < G, we have

$$X + X + K = X + X \text{ and } X + X = 2|X+K| - |K|.$$

Let

$$|K| = 3u + 1.$$

Since $n = p^e$, either

$$K \leq H \quad \text{or} \quad H \leq K.$$

By Theorem 6.17, we can again construct all of the possible sets X. We choose a subset X* of G/K such that X* + X* is a periodic in G/K and

$$|X^* + X^*| = 2|X^*| - 1.$$

If σ denotes the natural mapping of G to G/K, then X can be any subset of $\sigma^{-1}X^*$, such that

$$|\sigma^{-1}X^* \cap \overline{X}| \leq 3u/2.$$

Hence any coset of K which contains the first element of an H-string of elements of S must contain the first elements of at least $\frac{1}{2}(3u+2)$ H-strings of S.

If $K \geq H$, then any coset of K is a union of cosets of H. Hence there exists at least one coset of H, more than half of whose elements are starting-points of H-strings of S. This is impossible, so S must consist of a union of complete cosets of H. But this implies that

$$k = 3rs + r + s \leq (3s+1)r,$$

which is a contradiction.

If $K \leq H$, then any coset of H is a union of cosets of K. Consider a coset of H which contains H-strings of S. Suppose these H-strings start in the cosets

$$K + i_1, K + i_2, \ldots, K + i_t,$$

all of which are contained in the coset $H + i_1$. Suppose one of the H-strings finishes in the coset $K + j_1$. Then the next coset, $K + j_1 + g$, contains an element of Y and hence is contained in $X + X$. This means that no H-string can pass through this coset and similarly no H-string could pass through $K + i_1 - g$.

Hence any coset of K which contains an element of S contains at least $\frac{1}{2}(3u+2)$ elements of S. But by Lemma 7.22, these cosets must form a sum-free set in G/K and therefore S is contained in at most v cosets of K. But this means that

$$k \leq v(3u+1)$$

which is a contradiction.

The assumption that n is a prime power is used only at the very last stage of the proof, where we have to have either

$$K \leq H \quad \text{or} \quad H \leq K.$$

We have attempted (unsuccessfully) to avoid making this assumption by using the following argument, after establishing equation (7.53).

Suppose that for $i \neq j$, we have

$$X + \alpha_i = X + \alpha_j.$$

Then

$$X + \alpha_i - \alpha_j = X,$$

so that X is a union of cosets of a subgroup

$$H \geq \langle \alpha_i - \alpha_j \rangle.$$

Hence $X + \alpha_f$, for each f, and $Y = 1 - X$, must also be unions of cosets of H. This implies that

$$|(X+\alpha_f) \cap Y| \neq h - 1 \text{ for any } f,$$

so that

$$X + \alpha_i = Y = 1 - X \text{ for every } i = 1,\ldots,h.$$

Let

$$\langle X-X \rangle = K.$$

Then

$$X + K = X$$

and since

$$|K| \geq |X|,$$

we have

$$X = a + K \text{ for some } a \in X,$$

say $a = \alpha_1$. Now let

$$n = 3k + 1 = (3r+1)(3s+1),$$

where

$$K = \langle 3s+1 \rangle$$

so that

$$|K| = 3r + 1.$$

Then

$$\alpha_1 = a, \quad \alpha_2 = a + 3s + 1,\ldots,\alpha_h = a + 3r(3s+1).$$

Now

$$Y = 1 - X = X + X = 2a + K,$$

so that

$$1 - a \equiv 2a \pmod{K}.$$

Hence

$$3a - 1 \in K \text{ and } (3a-1)(3r+1) \equiv 0 \pmod{n}.$$

But this implies that $a = 2s + 1$, so that

$$S = \{2s+1,\ldots,2s+1+\ell_1,\ldots,2s+1+3r(3s+1),\ldots,2s+1+3r(3s+1)+\ell_h\}.$$

Hence by (7.44), with $i = 0$, we must have

$$2s + 1 + 2s + 1 + 3r(3s+1) + \ell_h = n = (3r+1)(3s+1).$$

Hence

$$s + 1 + \ell_h = 0$$

which is impossible.

So

$$X + \alpha_i \neq X + \alpha_j \text{ for any } i \neq j.$$

This gives us two possibilities:

either

$$X + \alpha_i = Y \text{ for one value of } i, \text{ and } |(X+\alpha_j) \cap Y| = h - 1 \text{ for all } j \neq i;$$

or

$$|(X+\alpha_j) \cap Y| = h - 1 \text{ for all } j = 1,\dots,h.$$

This shows that

$$h + 1 \leq |X+X| \leq 2h.$$

We have been unable to use this information to complete the proof.

7.5 SUM-FREE SETS IN NON-ABELIAN GROUPS. Very little is known about the non-abelian case. Of the results discussed so far, only Theorems 7.1 and 7.2 and Corollary 7.3 apply to non-abelian groups. The following theorem gives one additional result.

THEOREM 7.24. *Let* G *be a non-abelian group of order* 3p, *where* p *is prime,* p = 3k + 1. *If* S *is a maximal sum-free set in* G, *then* S *is a coset of the subgroup* H *of order* p.

PROOF. We know that

$$G = \langle a,b \mid 3a = 0 = pb,\ b + a = a + rb\rangle,$$

for some r such that

$$r^2 + r + 1 \equiv 0 \pmod p,$$

and that H = ⟨b⟩ is the only subgroup of order p in G.

Let

$$H_0 = H,\ H_1 = a + H,\ H_2 = 2a + H.$$

Let S be a maximal sum-free set in G, and let $S_i$ denote the subset of H such that

$$ia + S_i = S \cap H_i \quad \text{for } i = 0,1,2.$$

(i) Since H is sum-free,

$$\lambda(G) \geq p.$$

(ii) If S is contained in one coset of H, then $S_0$ and one of $S_1$ and $S_2$ are empty. This is the case considered in (i).

(iii) If S is contained in two cosets of H, then

$$S_0 = \emptyset \text{ and } |S_1| + |S_2| \geq 3k + 1.$$

Now

$$(a+S_1) + (a+S_1) = 2a + rS_1 + S_1.$$

Since S is sum-free

$$(rS_1+S_1) \cap S_2 = \emptyset.$$

But by Theorem 6.4,

$$|rS_i+S_i| \geq 2|S_i| - 1.$$

This implies that

$$2|S_1| + |S_2| \leq 3k + 2.$$

A similar argument shows that

$$2|S_2| + |S_1| \leq 3k + 2.$$

Adding these inequalities, we find that

$$3(|S_1| + |S_2|) \leq 6k + 4,$$

so that

$$|S_1| + |S_2| \leq 2k + 1.$$

But by (i),

$$|S_1| + |S_2| \geq 3k + 1$$

which is a contradiction. So this case cannot occur.

(iv) Now assume that no $S_i$ is empty. By Corollary 7.14,

$$|S_0| \leq k.$$

Hence

$$|S_1| + |S_2| \geq 2k + 1$$

and, without loss of generality, we may assume that

$$|S_1| \geq k + 1.$$

By a similar argument to that of (iii),

$$p \geq |S_1| + |rS_1| - 1 + |S_2| \geq k + |S_1| + |S_2| \geq |S_0| + |S_1| + |S_2| - |\varepsilon|.$$

Hence $\lambda(G) = p$.

(v) Finally, we show that the case considered in (iv) cannot in fact occur. For suppose it does. Since S is sum-free,

$$(rS_i+S_j) \cap S_{i+j} = \emptyset,$$

and in particular

$$(rS_0+S_i) \cap S_i = \emptyset.$$

But by Theorem 6.4

$$|S_i| + |S_j| - 1 \leqslant |S_i+S_j|.$$

Hence we have

$$2|S_1| + |S_2| \leqslant p + 1 = 3k + 2, \tag{7.56}$$

$$2|S_2| + |S_1| \leqslant p + 1 = 3k + 2, \tag{7.57}$$

$$2|S_1| + |S_0| \leqslant p + 1 = 3k + 2, \tag{7.58}$$

and $$2|S_2| + |S_0| \leqslant p + 1 = 3k + 2. \tag{7.59}$$

Now adding (7.56) and (7.57), we have

$$|S_1| + |S_2| \leqslant 2k + 1.$$

But we know that

$$|S_0| \leqslant k, \text{and } |S_0| + |S_1| + |S_2| = 3k + 1.$$

Hence we must have

$$|S_0| = k \text{ and } |S_1| + |S_2| = 2k + 1.$$

By 7.58 and (7.59),

$$|S_1| \leqslant k + 1, \ |S_2| \leqslant k + 1.$$

So we now assume that

$$|S_1| = k + 1 \text{ and } |S_2| = k.$$

Now by Theorem 6.4,

$$|rS_0+S_1| \geqslant 2k.$$

Since S is sum-free, we know that

$$(rS_0+S_1) \cap S_1 = \emptyset,$$

so that

$$|rS_0+S_1| \leqslant 2k.$$

Hence

$$|rS_0+S_1| = 2k$$

and, by Theorem 6.9, $(rS_0,S_1)$ must be a standard pair, with difference d. Without loss of generality, we assume that $d = 1$. Then

$$S_1 = \{mb, (m+1)b,\ldots,(m+k)b\}$$

for some m such that $0 \leqslant m \leqslant p-1$.

Now

$$((a+S_1)+(a+S_1)) \cap (2a+S_2) = \emptyset,$$

since S is sum-free.

But

$$(a+S_1) + (a+S_1) = 2a + rS_1 + S_1,$$

which implies that

$$|rS_1+S_1| = 2|S_1| - 1.$$

Hence by Theorem 6.9, $(rS_1,S_1)$ must be a standard pair. This implies that

$$rd \equiv \pm d \pmod{p},$$

and hence by Lemma 6.11,

$$r \equiv \pm 1 \pmod{p}.$$

But this contradicts the definition of G.

Hence S must consist of one coset of H.

From Corollary 7.15, we see that for an abelian group $G \neq \{0\}$,

$$2|G|/7 \leq \lambda(G).$$

No such lower bound is known when G is not abelian. Obviously if the commutator subgroup G' is smaller than G then

$$\lambda(G) \geq \lambda(G/G') \cdot |G'| \geq 2|G|/7.$$

But if G' = G, we have no lower bound on $\lambda(G)$ and there is a conjecture that no such lower bound exists. More specifically, there is a conjecture that for $\mathfrak{A}_n$, the alternating group of degree n, we have

$$\lambda(\mathfrak{A}_n) = \tfrac{1}{2}(n-1)!$$

Since any coset of a proper subgroup is sum-free, we certainly have

$$\lambda(\mathfrak{A}_n) \geq \tfrac{1}{2}(n-1)!$$

But we cannot hope to restrict ourselves to cosets, as the following examples show.

Let n = 5, so that we are looking for sum-free sets of at least 12 elements in $\mathfrak{A}_5$. Besides the obvious cosets, we can find

$$S_1 = \left\{\begin{matrix} (12345), & (15432), & (12543), & (13452) \\ (10423), & (15243), & (13245), & (15423) \\ (12453), & (13542), & (12435), & (15342) \end{matrix}\right\} \quad \text{and}$$

$$S_2 = \left\{\begin{matrix} (14)(23), & (15)(24), & (15)(23), & (14)(35) \\ (12)(35), & (25)(34), & (14)(25), & (123) \\ (13)(45), & (12)(34), & (24)(35), & (245) \end{matrix}\right\} .$$

7.6 **REFERENCES AND RELATED TOPICS.** The bounds of equation (7.1) were given by Erdos (1965). Diananda and Yap (1969) gave these bounds by a different argument and also proved most of the results in Sections 7.1 and 7.2, parts of Theorems 7.9 and 7.24 as well as Theorem 7.13 and its corollaries. This extended earlier work by Yap (1968); other papers by Yap (1969, 1970a) are also relevant. The proof of Theorem 7.9 was completed by Street (1972a,b); from this final version of the theorem the results of Yap (1970b, 1971a) immediately follow. Theorems 7.17, 7.19, 7.20 and the related lemmas are due to Rhemtulla and Street (1970, 1971). Theorem 7.22, characterising maximal sum-free sets in cyclic groups of prime-power order is due to Street (1971); for a discussion of the structure of maximal sum-free sets in cyclic groups of arbitrary order, see Yap (1971b). The proof of Theorem 7.24 was completed by Yap (1970b).

The conjecture discussed at the end of Section 7.5 was told to the author by L. Moser, to whom it is due, in 1969, and dates back to about 1950. The examples are due to the author.

## CHAPTER VIII. SUM-FREE PARTITIONS AND RAMSEY NUMBERS

In this chapter, we consider in detail a few particular cases where by partitioning an abelian group into a disjoint union of sum-free sets, we may find a lower bound for a Ramsey number.

8.1 $R_3(3, 2) = 17$. In section 1.1, using a sum-free partition of $Z_5$, we showed that $R_2(3, 2) > 5$ or, equivalently, that the edges of $K_5$, the complete graph on five vertices, can be coloured in two colours without the appearance of a monochromatic triangle. In fact, $R_2(3, 2) = 6$. For suppose we colour the edges of $K_6$ in two colours, red and blue. Consider the vertex $v_1$. At least three of the edges incident with $v_1$ must be coloured with the same colour; suppose that the edges $\{v_1, v_2\}$, $\{v_1, v_3\}$ and $\{v_1, v_4\}$ are all coloured red. If any of the edges joining two of $v_2$, $v_3$ and $v_4$ is coloured red, then we have a red triangle. If none of these edges is coloured red, then all of them must be coloured blue and we have a blue triangle. So if we edge-colour the complete graph on six vertices in two colours, we force the appearance of a monochromatic triangle.

By Theorem 2.9 we see that

$$R_3(3, 2) \leq 3(R_2(3, 2) - 1) + 2 = 17$$

So if we can edge-colour $K_{16}$ in three colours without the appearance of a monochromatic triangle, we can say that

$$R_3(3, 2) > 16$$

and hence that

$$R_3(3, 2) = 17.$$

In fact, two non-isomorphic edge-colourings with this property exist, and each of them is induced by a sum-free partition of an abelian group of order 16, one of $Z_2 \oplus Z_2 \oplus Z_2 \oplus Z_2$, the elementary group, and the other of $Z_4 \oplus Z_4$.

Suppose that we have a colouring of $K_{16}$ in three colours, red, blue and green. Choose any vertex, v, of $K_{16}$. At least five of the 15 edges incident with v must be coloured in the same colour, say red. Suppose these edges join v with vertices $v_1, \ldots, v_h$ respectively where $h \geq 5$, and consider the complete subgraph on these h vertices. If any edge of this subgraph is coloured red, then we have a red triangle in the original graph $K_{16}$. Hence this subgraph must be coloured in blue and green, and must be free of monochromatic triangles. Since $R_2(3, 2) = 6$, we see that $h \leq 5$. Hence $h = 5$, and for each vertex, exactly five of the 15 edges incident with that vertex must be coloured in each colour.

Suppose that we have an abelian group G of order 16 and that we can partition its non-zero elements into three sets, so that

$$G \setminus \{0\} = S_1 \cup S_2 \cup S_3,$$

where $S_i = -S_i$ is a sum-free set for each $i = 1, 2, 3$ and $S_i \cap S_j = \emptyset$ for $i \neq j$. Label each of the vertices of $K_{16}$ with an element of G and assign to each set $S_i$ a corresponding colour $C_i$. Now consider the edge joining the vertices a and b. Since $S_i = -S_i$, the elements a-b and b-a must belong to the same set of the partition, say $S_j$. Colour the edge {a, b} with the colour $C_j$. This colouring will be free of monochromatic triangles for if the triangle with vertices a, b, c is monochromatic, then the elements a-b, b-c, a-c must all belong to the same set $S_i$. But

$$(a - b) + (b - c) = a - c,$$

contradicting the sum-freeness of $S_i$.

If the group G is an elementary abelian 2-group, then the requirement that $S_i = -S_i$ is no restriction at all. In other cases, we have to check that this requirement is satisfied so that the colouring is well-defined.

Consider the complete subgraph on the vertices belonging to $S_1$. Since the difference of any two elements in $S_1$ must belong to $S_2 \cup S_3$, this subgraph must be coloured in $C_2$ and $C_3$ only, without the appearance of a monochromatic triangle. Hence $|S_1| \leq 5$, and similarly $|S_2| \leq 5$, $|S_3| \leq 5$. But

$$|S_1| + |S_2| + |S_3| = 15$$

so

$$|S_i| = 5 \quad \text{for } i = 1, 2, 3.$$

EXAMPLE 1. Let $G_1$ be the elementary abelian group of order 16. We regard it as the additive group of the field $GF[2^4]$ and let the generator x of the field be a root of $x^4 = x + 1 \pmod 2$. Then

$$S_1 = \{x^3, x^3 + x^2, x^3 + x, x^3 + x^2 + x + 1, 1\}$$

is the set of cubic residues in the field, and we take $S_2$ and $S_3$ to be the cosets of $S_1$ in the multiplicative group of the field, so that

$$S_2 = \{x, x + 1, x^3 + x + 1, x^2 + x + 1, x^3 + x^2 + 1\}$$

and

$$S_3 = \{x^2, x^2 + x, x^2 + 1, x^3 + x^2 + x, x^3 + 1\}.$$

Each of these sets is sum-free.

EXAMPLE 2. Let $G_2 = Z_4 \oplus Z_4$. We write this group as the set of ordered pairs of integers modulo 4. Thus

$$G_2 = \{(i, j) \mid i, j \in Z_4\}.$$

$G_2$ contains three elements of order two and 12 elements of order four. We must partition $G_2 \setminus \{0\}$ into three sets $S_1$, $S_2$, $S_3$, in such a way that each set contains five elements and each element belongs to the same set as its negative.

Hence each $S_i$ must contain one element of order two and two pairs $\{x, -x\}$ of elements of order four.

The elements of order two are (0, 2), (2, 0) and (2, 2) and for convenience we label each set by the element of order two which it contains, so that

$$S_1 = S_{(0,\ 2)},\ S_2 = S_{(2,\ 0)} \text{ and } S_3 = S_{(2,\ 2)}.$$

The six pairs of elements of order four we list as follows:

$$A_1 = \{(0, 1), (0, 3)\}; \quad B_1 = \{(1, 0), (3, 0)\}; \quad C_1 = \{(1, 1), (3, 3)\};$$

$$A_2 = \{(2, 1), (2, 3)\}; \quad B_2 = \{(1, 2), (3, 2)\}; \quad C_2 = \{(1, 3), (3, 1)\}.$$

Since

$$(0, 1) + (0, 2) = (0, 3),$$

we see that

$$A_1 \not\subseteq S_{(0,\ 2)}.$$

Similarly,

$$A_2 \not\subseteq S_{(0,\ 2)},\ B_1,\ B_2 \not\subseteq S_{(2,\ 0)}$$

and

$$C_1,\ C_2 \not\subseteq S_{(2,\ 2)}.$$

Again, since

$$(1, 0) + (0, 2) = (1, 2),$$

$$(B_1 \cup B_2) \not\subseteq S_{(0,\ 2)}.$$

Similarly,

$$(C_1 \cup C_2) \not\subseteq S_{(0,\ 2)},$$
$$(A_1 \cup A_2) \not\subseteq S_{(2,\ 0)},$$
$$(C_1 \cup C_2) \not\subseteq S_{(2,\ 0)},$$
$$(A_1 \cup A_2) \not\subseteq S_{(2,\ 2)},$$

and

$$(B_1 \cup B_2) \not\subseteq S_{(2,\ 2)}.$$

Finally, by direct computation we see that the eight isomorphic partitions given in Table 1 are sum-free.

| $S_{(0,\ 2)}$ | $S_{(2,\ 0)}$ | $S_{(2,\ 2)}$ |
|---|---|---|
| $\{(0, 2)\} \cup B_1 \cup C_1$ | $\{(2, 0)\} \cup A_1 \cup C_2$ | $\{(2, 2)\} \cup A_2 \cup B_2$ |
| $\{(0, 2)\} \cup B_1 \cup C_2$ | $\{(2, 0)\} \cup A_1 \cup C_1$ | $\{(2, 2)\} \cup A_2 \cup B_2$ |
| $\{(0, 2)\} \cup B_2 \cup C_1$ | $\{(2, 0)\} \cup A_1 \cup C_2$ | $\{(2, 2)\} \cup A_2 \cup B_1$ |
| $\{(0, 2)\} \cup B_2 \cup C_2$ | $\{(2, 0)\} \cup A_1 \cup C_1$ | $\{(2, 2)\} \cup A_2 \cup B_1$ |
| $\{(0, 2)\} \cup B_1 \cup C_1$ | $\{(2, 0)\} \cup A_2 \cup C_2$ | $\{(2, 2)\} \cup A_1 \cup B_2$ |
| $\{(0, 2)\} \cup B_1 \cup C_2$ | $\{(2, 0)\} \cup A_2 \cup C_1$ | $\{(2, 2)\} \cup A_1 \cup B_2$ |
| $\{(0, 2)\} \cup B_2 \cup C_1$ | $\{(2, 0)\} \cup A_2 \cup C_2$ | $\{(2, 2)\} \cup A_1 \cup B_1$ |
| $\{(0, 2)\} \cup B_2 \cup C_2$ | $\{(2, 0)\} \cup A_2 \cup C_1$ | $\{(2, 2)\} \cup A_1 \cup B_1$ |

TABLE 1.

8.2 THE COLOURINGS OF $K_{16}$. Consider the complete graph, $K_n$, on n vertices. Suppose its edges are coloured in two different ways and that each colouring involves m colours, $C_1, \dots, C_m$. These two colourings are said to be *isomorphic* if and only if there exists a 1-1 mapping of the set of vertices $\{v_1, \dots, v_n\}$ onto itself, such that each edge of colour $C_i$ in the first colouring is mapped to an edge of colour $C_{\sigma(i)}$ in the second colouring, where $\sigma$ is a permutation of the set $\{1, \dots, m\}$.

Less formally, two edge-colourings are isomorphic if and only if one can be obtained from the other by renaming the vertices and colours.

The edge-colourings discussed here involve at most three colours, which we consider to be red, blue and green. In the diagrams, red edges are represented by solid lines, blue edges by broken lines and green edges by dotted lines.

An edge-colouring of $K_n$ in m colours is said to be a *proper* colouring if and only if it is free of monochromatic triangles.

If all the proper edge-colourings of $K_n$ in m colours are isomorphic then we say that the colouring is *unique*. For example, the proper colouring of $K_5$ in two colours is unique (*v.* figure 1) and has two important properties: each vertex is incident with exactly two edges in each colour and the set of edges in either of the colours forms a pentagon. These properties and the uniqueness of the colouring are used repeatedly in the following proof that there are exactly two non-isomorphic proper edge-colourings of $K_{16}$ in three colours.

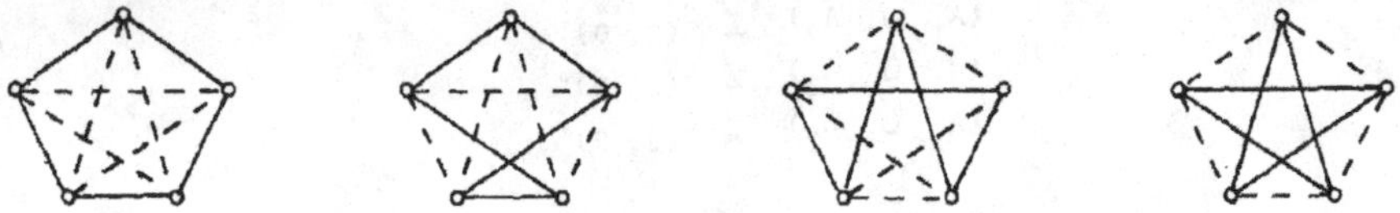

FIGURE 1.

LEMMA 8.1. *Let 1, 2 be two vertices joined by an edge of colour $C_i$ in a proper colouring of $K_{16}$ in 3 colours. Then at most two vertices can be joined to both 1 and 2 by edges of colour $C_j$, $j \neq i$.*

PROOF. Without loss of generality, we assume that the colours $C_i$ and $C_j$ are blue and red respectively.

(a) See figure 2. We suppose first that four or more vertices are joined to both 1 and 2 by red edges; number these vertices 3, 4, 5, 6. We know from the discussion in section 8.1 that each vertex is incident with five edges in each of the three colours. Since the edge {1, 2} is blue, there must be four other vertices joined to 1 by blue edges; number these vertices 7, 8, 9, 10.

Consider the edges {2, 7}, {2, 8}, {2, 9}, {2, 10}. None of these edges can be blue, for if say {2, 7} is blue, then the triangle {1, 2, 7} is blue. Since the four edges {2, 3}, {2, 4}, {2, 5}, {2, 6} are already coloured red, only one other red edge is incident with 2, so at most one of the edges {2, 7}, {2, 8}, {2, 9}, {2, 10} can be red and at least three of them must be green. Suppose that {2, 8}, {2, 9} and {2, 10} are green. To avoid blue and green triangles, we must colour the edges {8, 9},{9, 10},{8, 10} in red. But now we have a green triangle. So this case cannot occur.

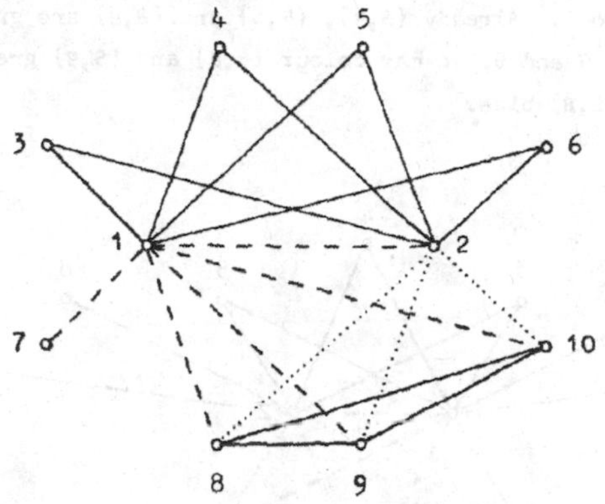


FIGURE 2.

(b) We now suppose that three vertices are joined to both 1 and 2 by red edges; number these vertices 3, 4 and 5. Since each vertex is incident with five edges in each of the three colours, there must be two other vertices, say 6 and 7, joined to 1 by red edges and similarly two other vertices, say 8 and 9, joined to 2 by red edges.

Now consider the remaining vertices 10, ... , 16. See Figure 3. Five vertices must be joined to 1 by blue edges; so far, only one of these has been considered, namely 2. Suppose that three of the vertices 10, ... , 16 are joined to 1 by blue edges, say 10, 11, 12. To avoid a blue triangle, we must colour {2, 10}, {2, 11} and {2, 12} green. To avoid forming blue or green triangles, we must colour {10, 11},{11, 12} and {10, 12} red. But now we have a red triangle.

To avoid this, at most two of the vertices 10,...,16 may be joined to 1 by blue edges. Hence {1,8} and {1,9} must be blue and we number the remaining vertices so that {1,10} and {1,11} are also blue. Similarly, {2,6}, {2,7}, {2,15} and {2,16} are blue. The remaining edges from 1 and 2 are green. See figure 4.

The edge {6,7} must be green, since otherwise the triangle {1,6,7} would be red or the triangle {2,6,7} would be blue. Similarly, {8,9} is green. The vertices 3,4,5,6,7 are joined by red edges to 1; hence the complete subgraph on these five vertices must be properly coloured in blue and green. This proper colouring is unique; the edges in each colour form a pentagon. By symmetry we choose the green edges to be {3,4}, {4,5}, {5,6} and {7,3}. So the blue edges are {3,5}, {4,6}, {5,7}, {6,3} and {7,4}. We apply a similar argument to the vertices 3,4,5,8,9 which are joined by red edges to 2. Already {3,4}, {4,5} and {8,9} are green and {3,5} is blue. By the symmetry of 8 and 9, we may colour {3,8} and {5,9} green. This leaves {3,9}, {4,9}, {4,8} and {5,8} blue.

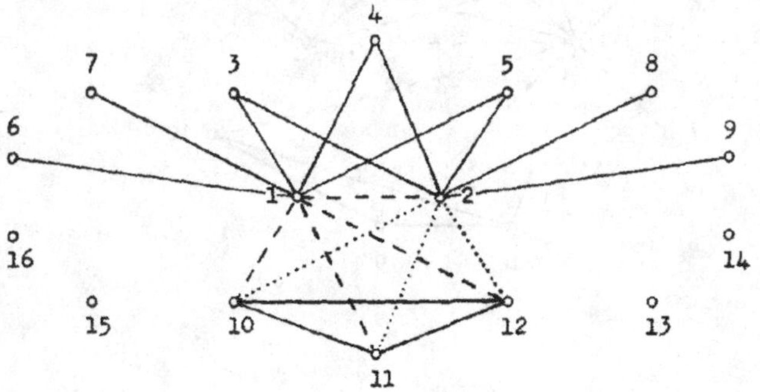


FIGURE 3

The edge {6,9} must be red, since otherwise the triangle {4,6,9} would be blue or the triangle {6,8,9} would be green. Similarly {7,8} is red. Now the four vertices 6,7,8 and 9, together with some other vertex x, are joined to 4 by blue edges, so the complete subgraph on these vertices 6,7,8,9,x must be properly coloured in red and green. The red edges form a pentagon, so one of {6,8} and {7,9} is red and the other is green. By the symmetry of the figure, we may interchange 6 with 7, 8 with 9, 3 with 5, so we choose to colour {7,9} red and {6,8} green.

Similar arguments determine the colouring of the edges in the lower half of figure 4. Red and green are interchanged in the colouring.

We now consider the edges joining a vertex in the set {3,4,...,9} to one in the set {10,11,...,16}.

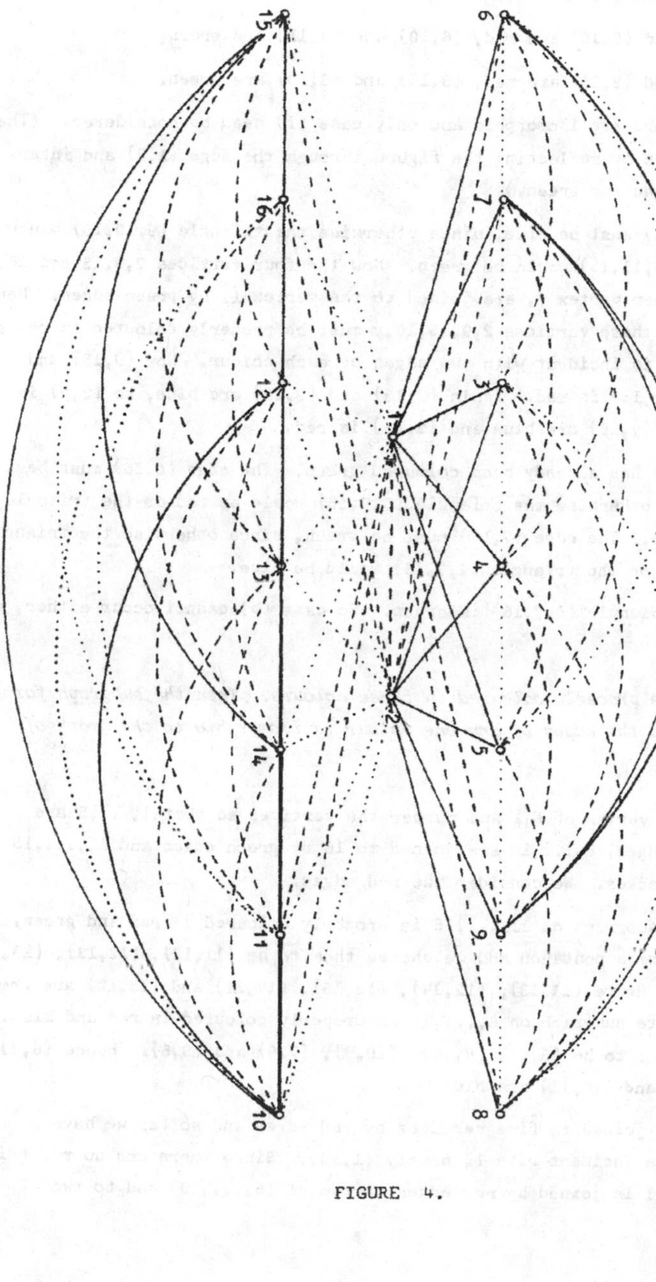


FIGURE 4.

The vertices 2,8,9,10,11 are joined to 1 by blue edges. Hence the complete subgraph on these five vertices must be properly coloured in red and green. The edges {2,8}, {2,9} and {10,11} are already coloured red, and the red edges must form a pentagon, so we have two possibilities:

(i) {8,11} and {9,10} are red, {8,10} and {9,11} are green;

(ii) {8,10} and {9,11} are red, {8,11} and {9,10} are green.

But these two colourings are isomorphic and only case (i) need be considered. (The isomorphism may be seen by reflecting the figure through the edge {1,2} and interchanging the colours red and green.)

The edge {9,15} must be blue, since otherwise the triangle {9,10,15} would be red or the triangle {9,11,15} would be green. Now the four vertices 2,9,15 and 16, together with some other vertex y, are joined to the vertex 11 by green edges. Hence the complete graph on these vertices 2,9,15,16,y must be properly coloured in red and blue, and each vertex is incident with two edges of each colour. Now {9,15} and {2,15} are blue, so {y,15} is red. Again {2,15} and {2,16} are blue, so {2,y} is red. Hence {y,9} and {y,16} are blue and {9,16} is red.

The edge {6,7} has already been coloured green. The edge {6,16} must be coloured green, since otherwise the triangle {6,9,16} would be red or the triangle {2,6,16} would be blue. The edge {7,16} must be green, since otherwise the triangle {7,9,16} would be red or the triangle {2,7,16} would be blue.

But now the triangle {6,7,16} is green. So case (b) cannot occur either, and the lemma is proved.

LEMMA 8.2. *Let* $K_{16}$ *be properly coloured in three colours. Then the subgraph formed by the 16 vertices and the edges of any one colour is isomorphic to the graph of Figure 5.*

PROOF. Let 16 be any vertex of $K_{16}$ and number the vertices so that 1,...,5 are joined to 16 by red edges, 6,...,10 are joined to 16 by green edges and 11,...,15 are joined to 16 by blue edges. We consider the red edges.

The complete subgraph on 11,...,15 is properly coloured in red and green. The red edges must form a pentagon and we choose them to be {11,12}, {12,13}, {13,14}, {14,15} and {15,11}. Hence {11,13}, {12,14}, {13,15}, {14,11} and {15,12} are green. Similarly, the complete subgraph on 6,...,10 is properly coloured in red and blue. We choose the red edges to be {6,8}, {8,10}, {10,7}, {7,9} and {9,6}. Hence {6,7}, {7,8}, {8,9}, {9,10} and {10,11} are blue.

Now 1 must be joined to five vertices by red edges and so far we have coloured only one edge incident with 1, namely {1,16}. Since there are no red triangles in the graph, 1 is joined by red edges to two of {6,...,10} and to two of

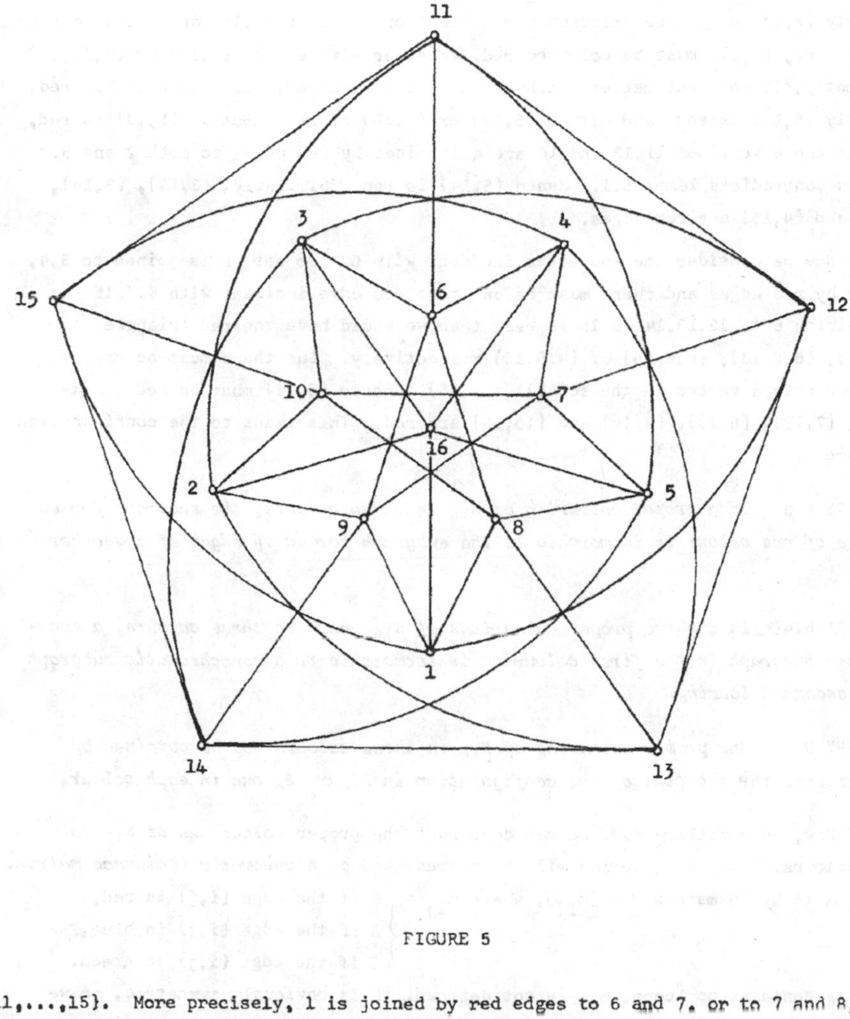


FIGURE 5

{11,...,15}. More precisely, 1 is joined by red edges to 6 and 7, or to 7 and 8, or to 8 and 9, or to 9 and 10, or to 10 and 6. Similarly, 1 is joined by red edges to 11 and 13, or to 13 and 15, or to 15 and 12, or to 12 and 14, or to 14 and 11.

But if we consider the red edges incident with 2,3,4 and 5 we have exactly the same possibilities. Suppose that both 1 and 2 are joined by red edges to both 6 and 7. Then the three vertices 6,7 and 16 are all joined by red edges to both 1 and 2. But this contradicts lemma 8.1. By the symmetry of the set {1,2,3,4,5}, choose {1,8}, {1,9}, {2,9}, {2,10}, {3,10}, {3,6}, {4,6}, {4,7}, {5,7} and {5,8} to colour red. Again, by the symmetry of the set {11,12,13,14,15}, choose {1,15} and {1,12} to colour red. Suppose that {2,12} is coloured red. Then the three vertices 9,12 and 16 are all joined by red edges to both 1 and 2. But this contradicts lemma 8.1. Similarly {2,15} cannot be coloured red. Since only one of {2,13} and {2,14} can be coloured red, {2,11} must be coloured red, together with either {2,13} or {2,14}. By symmetry, it does not matter which of these edges is red; we colour {2,13} red. Similarly {5,11} is red, and either {5,13} or {5,14} is red. But if {5,13} is red, then the three vertices 11,13 and 16 are all joined by red edges to both 2 and 5. But this contradicts lemma 8.1. Hence {5,14} is red. Similarly, {3,12}, {3,14}, {4,13} and {4,15} are red edges.

Now we consider the red edges incident with 6. So far, 6 is joined to 3,4, 8 and 9 by red edges and there must be one more red edge incident with 6. If the edge joining 6 to 12,13,14 or 15 is red, then we would have the red triangle {3,6,12}, {6,8,13}, {6,9,14} or {4,6,15} respectively. But there must be one red edge from 6 to a vertex in the set {11,...,15}. Hence {6,11} must be red. Similarly, {7,12}, {8,13}, {9,14} and {10,15} are red. This leads to the configuration of Figure 5.

COROLLARY 8.3. *In a proper colouring of* $K_{16}$ *in three colours, the subgraph formed by edges of one colour is isomorphic to the subgraph formed by edges of any other colour.*

COROLLARY 8.4. *In any two proper colourings of* $K_{16}$*, each in three colours, a monochromatic subgraph in the first colouring is isomorphic to a monochromatic subgraph in the second colouring.*

COROLLARY 8.5. *Any proper colouring of* $K_{16}$ *in three colours may be obtained by superimposing three copies of the configuration in Figure 5, one in each colour.*

Now, by corollary 8.5, we can construct the proper colourings of $K_{16}$ in three colours. Such a colouring will be represented by a *chromatic incidence matrix.* This is a 16 by 16 matrix $M = [m_{ij}]$, where

$$m_{ij} = \begin{cases} R \text{ if the edge } \{i,j\} \text{ is red,} \\ B \text{ if the edge } \{i,j\} \text{ is blue,} \\ G \text{ if the edge } \{i,j\} \text{ is green.} \end{cases}$$

Since $K_{16}$ contains no loops, $m_{ii}$ is not defined. M is obviously symmetric, so we record only the elements $m_{ij}$ for $i > j$.

The incidence matrix for the configuration of lemma 8.2 is given in Table 2.

All the red edges were coloured, in the process of finding the configuration of Figure 5, so the blank spaces in the matrix represent edges whose colours have not yet been assigned and all of these will be either blue or green. The four 5 by 5 submatrices labelled I, II, III, IV in Table 2 contain between them all the blank spaces.

| | 1 | 2 | 3 | 4 | 5 | 6 | 7 | 8 | 9 | 10 | 11 | 12 | 13 | 14 | 15 | 16 |
|---|---|---|---|---|---|---|---|---|---|---|---|---|---|---|---|---|
| 1 | | | | | | | | | | | | | | | | |
| 2 | | | | | | | | | | | | | | | | |
| 3 | | I | | | | | | | | | | | | | | |
| 4 | | | | | | | | | | | | | | | | |
| 5 | | | | | | | | | | | | | | | | |
| 6 | | | R | R | | | | | | | | | | | | |
| 7 | | II | | R | R | B | | | | | | | | | | |
| 8 | R | | | | R | R | B | | | | | | | | | |
| 9 | R | R | | | | R | R | B | | | | | | | | |
| 10 | | R | R | | | B | R | R | B | | | | | | | |
| 11 | | R | | | R | R | | | | | | | | | | |
| 12 | R | | R | | | | R | | IV | | R | | | | | |
| 13 | III | R | | R | | | | R | | | G | R | | | | |
| 14 | | | R | | R | | | | R | | G | G | R | | | |
| 15 | R | | | R | | | | | | R | R | G | G | R | | |
| 16 | R | R | R | R | R | G | G | G | G | G | B | B | B | B | B | |

TABLE 2

**We consider the proper colourings of $K_5$ in two colours, say blue and green. The colouring is unique (up to isomorphism) and the four ways in which the vertices and edges may be labelled are shown in Table 3 (or equivalently in Figure 1).**

(a)

| | 1 | 2 | 3 | 4 | 5 |
|---|---|---|---|---|---|
| 1 | | | | | |
| 2 | G | | | | |
| 3 | B | G | | | |
| 4 | B | B | G | | |
| 5 | G | B | B | G | |

(b)

| | 1 | 2 | 3 | 4 | 5 |
|---|---|---|---|---|---|
| 1 | | | | | |
| 2 | G | | | | |
| 3 | B | B | | | |
| 4 | B | G | G | | |
| 5 | G | B | G | B | |

(c)

| | 1 | 2 | 3 | 4 | 5 |
|---|---|---|---|---|---|
| 1 | | | | | |
| 2 | B | | | | |
| 3 | G | G | | | |
| 4 | G | B | B | | |
| 5 | B | G | B | G | |

(d)

| | 1 | 2 | 3 | 4 | 5 |
|---|---|---|---|---|---|
| 1 | | | | | |
| 2 | B | | | | |
| 3 | G | B | | | |
| 4 | G | G | B | | |
| 5 | B | G | G | B | |

TABLE 3

Now submatrix I of the incidence matrix of $K_{16}$ is the matrix of a proper colouring of the complete graph on vertices 1,2,3,4,5 in two colours, blue and green. Hence the submatrix I must be one of the four in Table 3. We show that if I is the matrix (a) of Table 3, we are led to a proper colouring of $K_{16}$, isomorphic to that induced by the sum-free partition of the elementary abelian group of order 16 into three sets and that if I is the matrix (b) of Table 3, we are led to a proper colouring of $K_{16}$, isomorphic to that induced by the sum-free partition of $Z_4 \oplus Z_4$ into three sum-free sets, and that these two colourings are not isomorphic to each other. We also show that if I is either of the matrices (c) or (d) of Table 3, then no proper colouring of $K_{16}$ can be derived. Hence there are exactly two non-isomorphic proper colourings of $K_{16}$ in three colours, and both of these are induced by sum-free partitions of appropriate abelian groups.

Before looking at the four cases which arise from the four choices of submatrix I, we consider the other submatrices II, III and IV.

Submatrix II gives the edges joining vertices 1,2,3,4 and 5 to vertices 6,7, 8,9 and 10. We have applied lemma 8.2 to the red edges in the colouring; now we apply it to the green edges. Figure 5, with the vertices relabelled, represents the green edges. Since 6,7,8,9,10 are joined to 16 by green edges, they now correspond to 1,2,3,4,5 in Figure 5. This means that 1,2,3,4,5 must correspond to either 6,7, 8,9,10 or 11,12,13,14,15. Hence each of the vertices 6,7,8,9,10 must be joined by green edges to two of the vertices 1,2,3,4,5 and conversely. Hence in submatrix II, every row and every column must contain two G's.

Submatrix IV gives the edges joining vertices 6,7,8,9 and 10 to vertices 11, 12,13,14 and 15. In our relabelling of Figure 5, vertices 11,12,13,14,15 will correspond to either vertices 11,12,13,14,15 or vertices 6,7,8,9,10. A similar argument shows that in submatrix IV, every row and every column must contain two G's.

Now we apply lemma 8.2 to the blue edges. Again by a similar argument we find that in submatrices III and IV, every row and every column must contain two B's.

We are now ready to consider the four possible choices of submatrix I.

<u>Case (a)</u>: Let the submatrix I be the matrix (a) of Table 3. We fill in the incidence matrix of $K_{16}$ as shown in Table 4, by the following argument.

6 is joined by green edges to two of 1,2 and 5. Already edges {1,2} and {1,5} are green, so if we colour {1,6} green we force either {1,2,6} or {1,5,6} to be a green triangle. So {1,6} must be blue and hence {2,6} and {5,6} must be green. Similar arguments give us the rest of submatrix II.

11 is joined by blue edges to two of 1,3 and 4. Already edges {1,3} and {1,4} are blue, so if we colour {1,11} blue we force either {1,3,11} or {1,4,11} to be a blue triangle. So {1,11} must be green and hence {3,11} and {4,11} must be blue. Similar arguments give us the rest of submatrix III.

| | | | | | | | | | | | | | | | | |
|---|---|---|---|---|---|---|---|---|---|---|---|---|---|---|---|---|
| 1 | | | | | | | | | | | | | | | | |
| 2 | G | | | | | | | | | | | | | | | |
| 3 | B | G | | | | | | | | | | | | | | |
| 4 | B | B | G | | | | | | | | | | | | | |
| 5 | G | B | B | G | | | | | | | | | | | | |
| 6 | B | G | R | R | G | | | | | | | | | | | |
| 7 | G | B | G | R | R | B | | | | | | | | | | |
| 8 | R | G | B | G | R | R | B | | | | | | | | | |
| 9 | R | R | G | B | G | R | R | B | | | | | | | | |
| 10 | G | R | R | G | B | B | R | R | B | | | | | | | |
| 11 | G | R | B | B | R | R | B | G | G | B | | | | | | |
| 12 | R | G | R | B | B | B | R | B | G | G | R | | | | | |
| 13 | B | R | G | R | B | G | B | R | B | G | G | R | | | | |
| 14 | B | B | R | G | R | G | G | B | R | B | G | G | R | | | |
| 15 | R | B | B | R | G | B | G | G | B | R | R | G | G | R | | |
| 16 | R | R | R | R | R | G | G | G | G | G | B | B | B | B | B | |
| | 1 | 2 | 3 | 4 | 5 | 6 | 7 | 8 | 9 | 10 | 11 | 12 | 13 | 14 | 15 | 16 |

TABLE 4

Now {1,7}, {1,10} and {1,11} are green and to avoid a green triangle, we must colour {7,11} and {10,11} blue. Hence {8,11} and {9,11} must be green. Similar arguments give us the rest of submatrix IV and hence complete Table 4 uniquely.

Direct calculation shows that the subgraphs of green and blue edges respectively are isomorphic to the graph of Figure 5. Figure 5 contains no triangles, so there are no monochromatic triangles in the colouring given by Table 4.

To show that this colouring is isomorphic to that induced by the sum-free partition given in Example 1.1, we assign the colour red to the set $S_1$, green to the set $S_2$ and blue to the set $S_3$. The labelling of the vertices is shown in Table 5.

| | | | |
|---|---|---|---|
| 1 | $x^3+x^2+x+1$ | 9 | $x+1$ |
| 2 | $x^3+x^2$ | 10 | $x^3+x^2+1$ |
| 3 | $1$ | 11 | $x^2$ |
| 4 | $x^3+x$ | 12 | $x^3+x^2+x$ |
| 5 | $x^3$ | 13 | $x^2+x$ |
| 6 | $x^3+x+1$ | 14 | $x^3+1$ |
| 7 | $x$ | 15 | $x^2+1$ |
| 8 | $x^2+x+1$ | 16 | $0$ |

TABLE 5

Case (b): Let the submatrix I be the matrix (b) of Table 3. We fill in the incidence matrix of $K_{16}$ as shown in Table 6, by the following argument.

6 is joined by green edges to two of 1,2 and 5. But edges {1,2} and {1,5} are already green, so that if we colour {1,6} green we force either {1,2,6} or {1,5,6} to be a green triangle. So {1,6} must be blue and {2,6} and {5,6} must be green. Again 8 is joined by green edges to two of 2,3 and 4. Already {3,4} and {2,4} are green, so {4,8} must be blue and {2,8} and {3,8} are green. Now since every row and every column of submatrix II must contain two G's, we have all the other entries in this submatrix uniquely determined.

Similar arguments show that {1,11} is green and {3,11}, {4,11} are blue. Again {5,12} is green and {2,12}, {4,12} are blue. Now since every row and every column of submatrix III contains two B's, we have all the other entries in this submatrix uniquely determined.

| | | | | | | | | | | | | | | | |
|---|---|---|---|---|---|---|---|---|---|---|---|---|---|---|---|
| 1 | | | | | | | | | | | | | | | |
| 2 | G | | | | | | | | | | | | | | |
| 3 | B | B | | | | | | | | | | | | | |
| 4 | B | G | G | | | | | | | | | | | | |
| 5 | G | B | G | B | | | | | | | | | | | |
| 6 | B | G | R | R | G | | | | | | | | | | |
| 7 | G | B | G | R | R | B | | | | | | | | | |
| 8 | R | G | G | B | R | R | B | | | | | | | | |
| 9 | R | R | B | G | G | R | R | B | | | | | | | |
| 10 | G | R | R | G | B | B | R | R | B | | | | | | |
| 11 | G | R | B | B | R | R | B | G | G | B | | | | | |
| 12 | R | B | R | B | G | B | R | G | B | G | R | | | | |
| 13 | B | R | G | R | B | G | B | R | B | G | G | R | | | |
| 14 | B | B | R | G | R | G | G | B | R | B | G | G | R | | |
| 15 | R | G | B | R | B | B | G | B | G | R | R | G | G | R | |
| 16 | R | R | R | R | R | G | G | G | G | G | B | B | B | B | B |
| | 1 | 2 | 3 | 4 | 5 | 6 | 7 | 8 | 9 | 10 | 11 | 12 | 13 | 14 | 15 |

16

TABLE 6

Similar arguments, together with the fact that every row and every column of submatrix IV contains two B's and two G's, determine the entries in this submatrix and hence complete Table 6 uniquely.

Again direct calculation shows that the subgraphs of green and blue edges respectively are isomorphic to the graph of Figure 5 and hence that the colouring given by Table 6 contains no monochromatic triangles.

To show that this colouring is isomorphic to that induced by the sum-free partition given in Example 2 of section 8.1, consider the partition given in the first line of Table 1, and assign the colour red to the set $S_{(2,2)}$, the colour green to the set $S_{(0,2)}$ and the colour blue to the set $S_{(2,0)}$. The labelling of the vertices is shown in Table 7.

| | | | | | | | |
|---|---|---|---|---|---|---|---|
| 1 | (2,2) | 5 | (1,2) | 9 | (1,0) | 13 | (1,3) |
| 2 | (3,2) | 6 | (0,2) | 10 | (1,1) | 14 | (3,1) |
| 3 | (2,3) | 7 | (3,3) | 11 | (2,0) | 15 | (0,3) |
| 4 | (2,1) | 8 | (3,0) | 12 | (0,1) | 16 | (0,0) |

TABLE 7

Case (c): Let the submatrix I be the matrix (c) of Table 3. We attempt to fill in the incidence matrix of $K_{16}$ as shown in Table 8, by the following argument.

| | 1 | 2 | 3 | 4 | 5 | 6 | 7 | 8 | 9 | 10 | 11 | 12 | 13 | 14 | 15 | 16 |
|---|---|---|---|---|---|---|---|---|---|---|---|---|---|---|---|---|
| 1 | | | | | | | | | | | | | | | | |
| 2 | B | | | | | | | | | | | | | | | |
| 3 | G | G | | | | | | | | | | | | | | |
| 4 | G | B | B | | | | | | | | | | | | | |
| 5 | B | G | B | G | | | | | | | | | | | | |
| 6 | B | G | R | R | G | | | | | | | | | | | |
| 7 | G | G | B | R | R | B | | | | | | | | | | |
| 8 | R | B | G | G | R | R | B | | | | | | | | | |
| 9 | R | R | G | G | B | R | R | B | | | | | | | | |
| 10 | G | R | R | B | G | B | R | R | B | | | | | | | |
| 11 | | R | | | R | R | | | | | | | | | | |
| 12 | R | | R | | | | R | | | | R | | | | | |
| 13 | | R | | R | | | | R | | | G | R | | | | |
| 14 | | | R | | R | | | | R | | G | G | R | | | |
| 15 | R | | | R | | | | | | R | R | G | G | R | | |
| 16 | R | R | R | R | R | G | G | G | G | G | B | B | B | B | B | |

TABLE 8

7 is joined by green edges to two of 1,2 and 3. Already {1,3} and {2,3} are green, so that {3,7} must be blue and hence {1,7} and {2,7} must be green. Again {1,4} and {4,5} are green, so that {4,10} must be blue and {1,10} and {5,10} must be green. Since every row and every column of submatrix II contains two G's, the remaining entries of II are uniquely determined. But now 8 and 9 are both joined by green edges to all three of 3,4 and 16. This contradicts lemma 8.1. Hence this

case cannot lead to a proper colouring in three colours.

Case (d): Let the submatrix I be the matrix (d) of Table 3. Again we attempt to fill in the incidence matrix of $K_{16}$, as shown in Table 9, by the following argument.

TABLE 9

| | | | | | | | | | | | | | | | | |
|---|---|---|---|---|---|---|---|---|---|---|---|---|---|---|---|---|
| 1 | | | | | | | | | | | | | | | | |
| 2 | B | | | | | | | | | | | | | | | |
| 3 | G | B | | | | | | | | | | | | | | |
| 4 | G | G | B | | | | | | | | | | | | | |
| 5 | B | G | G | B | | | | | | | | | | | | |
| 6 | G | | R | R | | | | | | | | | | | | |
| 7 | | G | | R | R | B | | | | | | | | | | |
| 8 | R | | G | | R | R | B | | | | | | | | | |
| 9 | R | R | | G | | R | R | B | | | | | | | | |
| 10 | | R | R | | G | B | R | R | B | | | | | | | |
| 11 | B | R | | | R | R | | | | | | | | | | |
| 12 | R | B | R | | | | R | | | | R | | | | | |
| 13 | | R | B | R | | | | R | | | G | R | | | | |
| 14 | | | R | B | R | | | | R | | G | G | R | | | |
| 15 | R | | | R | B | | | | | R | R | G | G | R | | |
| 16 | R | R | R | R | R | G | G | G | G | G | B | B | B | B | B | |
| | 1 | 2 | 3 | 4 | 5 | 6 | 7 | 8 | 9 | 10 | 11 | 12 | 13 | 14 | 15 | 16 |

6 is joined by green edges to two of 1,2 and 5. Since {2,5} is green, either {2,6} or {5,6} is blue, and {1,6} is green. Similarly, {2,7}, {3,8}, {4,9} and {5,10} are green; {1,11}, {2,12}, {3,13}, {4,14} and {5,15} are blue. Now each column of submatrix II contains two G's, so either {3,7} or {3,9} is green and we consider these two cases.

(d) (i) {3,7} is green and {3,9} is blue.

Suppose that {3,11} is green. Then {7,11} and {8,11} must be blue, for otherwise triangles {3,7,11} and {3,8,11} would be green. But then {7,8,11} is a blue triangle.

So {3,11} must be blue. Now each column of submatrix III contains exactly two B's, so {3,15} must be green. Hence {7,15} and {8,15} must be blue, for otherwise triangles {3,7,15} and {3,8,15} would be green. But then {7,8,15} is a blue triangle.

(d) (ii) {3,7} is blue and {3,9} is green.

Suppose that {3,11} is green. Then {8,11} and {9,11} must be blue, for otherwise triangles {3,8,11} and {3,9,11} would be green. But then {8,9,11} is a blue triangle.

So {3,11} must be blue. As before {3,15} must be green. Hence {8,15} and {9,15} must be blue, for otherwise triangles {3,8,15} and {3,9,15} would be green. But then {8,9,15} is a blue triangle.

In either case we have a contradiction and case (d) cannot lead to a proper colouring in three colours.

All that we have to prove now is that the two colourings of cases (a) and (b) are not isomorphic.

Suppose that (a) and (b) are isomorphic. By symmetry, we may assume that the isomorphism of (a) onto (b) maps each edge onto one of the same colour, and that vertex 16 in (a) maps to vertex 16 in (b), which we denote by 16a ≡ 16b.

The vertices joined to 16 by blue edges in (a) must be mapped onto the vertices joined to 16 by blue edges in (b). Since (a) is symmetrical in 11,12,13,14,15, we take 11a ≡ 11b. The mapping preserves red edges, so {15a,12a} ≡ {15b,12b}. Again by the symmetry of (a), we take 15a ≡ 15b, 12a ≡ 12b. Since {13a,12a} is red and {13a,16a} is blue, we must have 13a ≡ 13b. By similar arguments, we find that each vertex of (a) maps to the vertex of (b) with the same number. The colours of the edges are preserved, so that the incidence matrices must be identical. But comparing Tables 4 and 6, we see this is not the case. Hence we have a contradiction, and the two colourings are not isomorphic.

Hence we have proved:

THEOREM 8.6. *There are exactly two non-isomorphic proper colourings of* $K_{16}$ *in three colours.*

8.3 $R_4(3,2) \geq 50$. Table 10 gives a partition of $Z_7 \oplus Z_7$ into four disjoint sum-free sets $S_i$, $i = 1,2,3,4$, such that for each $i$, $S_i = -S_i$. Hence by labelling the vertices of $K_{49}$ with the 49 group elements, and by assigning colour $C_i$ to the edge {a,b} whenever $a-b \in S_i$, we have a proper colouring of $K_{49}$ in four colours.

| $S_1$ | $S_2$ | $S_3$ | $S_4$ |
|---|---|---|---|
| (0,1),(0,6) | (1,1),(6,6) | (2,0),(5,0) | (0,2),(0,5) |
| (1,0),(6,0) | (1,6),(6,1) | (2,5),(5,2) | (1,3),(6,4) |
| (1,2),(6,5) | (0,3),(0,4) | (3,1),(4,6) | (1,4),(6,3) |
| (2,1),(5,6) | (3,0),(4,0) | (3,3),(4,4) | (2,2),(5,5) |
| (2,6),(5,1) | (3,2),(4,5) | (3,5),(4,2) | (2,3),(5,4) |
| (6,2),(1,5) | (2,4),(5,3) | (3,6),(4,1) | (3,4),(4,3) |

TABLE 10

We note that the quartic residues of $GF[7^2]$ are not sum-free.

8.4 REFERENCES AND RELATED TOPICS. The sum-free partition of $GF[2^4]$ is due to Greenwood and Gleason (1955), and that of $Z_4 \oplus Z_4$ is given by Whitehead (1971a). The colourings of $K_{16}$ were determined by Kalbfleisch and Stanton (1968). Greenwood and Gleason (1955) also gave a sum-free partition of GF[41] into four sets $S_i$, such that $S_i = -S_i$, for i = 1,2,3,4, and thus showed that $R_4(3,2) \geq 42$. The improved bound is again due to Whitehead (1972a); this paper gives the sum-free partition shown above as Table 10. For the proof that $R_4(3,2) \leq 65$, see Folkman (1967) and Whitehead (1971b). Whitehead (1972) discusses the algorithms used in finding such partitions.

# CHAPTER IX. SOME RELATED TOPICS AND SOME UNSOLVED PROBLEMS

In this chapter, we first outline some related results which we do not have space to discuss fully and then, to finish, list some of the unsolved problems.

## 9.1 SUM-FREE SETS IN GROUPOIDS AND SEMIGROUPS.

Both Mullin (1961) and Doyle and Warne (1963) studied sum-free sets in groupoids. (Mullin called these sets *mutants*; Doyle and Warne called them *antigroupoids*.)

More generally, Iseki (1962) made the following definition: a subset M of a semigroup A is an *(m,n) mutant* of A if and only if $M^m \subseteq A \setminus M^n$. He proved that if M,N are (m,n) mutants of semigroups A,B respectively then M×N is an (m,n) mutant of A×B.

Kim (1969) proved that in a topological semigroup A, for any non-idempotent $a \in A$, there exists a maximal open mutant containing a. He also showed that no semigroup can be partitioned into a disjoint union of two or three mutants and conjectured that no semigroup could be partitioned into any finite number of mutually disjoint mutants. Kim (1971) studied mutants in symmetric semigroups.

## 9.2 PROBLEMS RELATED TO SCHUR'S.

Many related problems have been suggested by Erdos and Moser.

Theorem 4.11 has been applied by Abbott and Hanson (1972) to improve the estimates of the Schur functions for several other regular equations.

In the same paper various analogous problems are considered. For instance, the question of product-free sets: what is the largest positive integer $\ell(n)$ such that the set of integers $\{2,\ldots,\ell(n)\}$ can be partitioned into n classes, no class containing a solution of the equation $x_1x_2 = x_3$?

Similarly, we have a problem, in set theory, of union-free sets: for any positive integer n, what is the minimum number, k(n), such that the $2^n$ subsets of a set S of n elements can be partitioned into k(n) union-free classes? That is, such that no class contains a solution of $A \cup B = C$, where A,B,C are distinct. This problem was studied earlier by Kleitman (1966) and most recently by Erdos and Komlos (1969).

Another similar question has been asked for abelian groups: Let G be an abelian group of order n and let f(G) denote the minimum number of disjoint sum-free sets into which the non-zero elements of G can be partitioned. Now define f(n) to be the maximum of f(G), as G runs through all abelian groups of order n. Abbott and

Hanson (1972) give the bounds

$$(c_1 \log n)/(\log \log n) < f(n) < c_2 \log n,$$

where $c_1, c_2$ are positive absolute constants.

9.3 UNSOLVED PROBLEMS. The unsolved problems which immediately spring to mind are:

(1) how to improve both the upper and lower bounds on the Schur function, $f(n)$, either by working directly with the equation $x+y = z$ or by working indirectly, using some related equation or system of equations;

(2) how to shorten some of the painfully long proofs of the addition theorems for groups and the (even more painful) proofs which have been used to characterise the maximal sum-free sets.

Many other unsolved problems remain. We list a few of them.

(3) Most of the addition theorems have been proved by using transformations of the summand sets. These transformations are of no use for non-abelian groups. Do there exist transformations which could be used with non-abelian groups? Is it essential to use transformations to prove these theorems or are there other proofs which could be generalised to give analogous theorems for non-abelian groups?

(4) Diananda and Yap (1969) conjectured that the lower bound of theorem 7.13 is in fact the exact value, i.e. that if G is an abelian group of order divisible only by primes congruent to 1 (modulo 3) and of exponent m, then $\lambda(G) = (m-1)|G|/3m$. This question is still undecided.

(5) For an abelian group of order divisible only by primes congruent to 1 (modulo 3) the structure of the maximal sum-free sets is again not known (except for the cyclic and elementary abelian groups).

(6) The lower bound $2|G|/7 \leq \lambda(G)$ was derived in chapter VII for finite abelian groups, but no lower bound is known for finite groups in general. It has been conjectured that for non-abelian finite groups, $\lambda(G)/|G|$ may be arbitrarily small. This conjecture which is still undecided is discussed in section 7.5.

(7) Even after the maximal sum-free sets in a group have been determined, the problem of how to partition the group into as few disjoint sum-free sets as possible is still not settled. For instance, the elementary abelian group of order 16 can be partitioned into three disjoint sum-free sets, each containing five elements. If, however, we partition this group by taking the largest possible sum-free set at each step, we end up with a partition of the group into four sets, containing eight, four, two and one elements respectively. In order to estimate the Ramsey numbers, we need to know something about the smaller sum-free sets, not just the maximal ones.

## REFERENCES


[1] H.L. ABBOTT, Ph.D. thesis, University of Alberta, 1965.

[2] H.L. ABBOTT, Lower bounds for some Ramsey numbers. *Discrete Mathematics*, (1972), to appear.

[3] H.L. ABBOTT and D. HANSON, A problem of Schur and its generalizations. *Acta Arith.*, (1972), to appear.

[4] H.L. ABBOTT and A.C. LIU, On partitioning integers into progression free sets. *J. Comb. Theory*, (1972), to appear.

[5] H.L. ABBOTT and L. MOSER, Sum-free sets of integers. *Acta Arith.*, XI, (1966), 393-396.

[6] L.D. BAUMERT, Sum-free sets. J.P.L. Research Summary, No. 36-10, Vol. 1, Sept. 1, (1961), 16-18.

[7] E.R. BERLEKAMP, A construction for partitions which avoid long arithmetic progressions. *Canad. Math. Bull.*, 11, (1968), 409-414.

[8] A.L. CAUCHY, Recherches sur les nombres. *J. Ecole polytechn.*, 9, (1813), 99-116.

[9] K. CHANDRASEKHARAN, *Einführung in die Analytische Zahlentheorie*. Springer-Verlag, Lecture Notes in Mathematics, 29, Berlin-Heidelberg-New York, (1966).

[10] S.L.G. CHOI, On a combinatorial problem in number theory. *Proc. London Math. Soc.*, (3) 23, (1971), 629-642.

[11] H. DAVENPORT, On the addition of residue classes. *J. London Math. Soc.*, 10, (1935), 30-32.

[12] H. DAVENPORT, A historical note. *J. London Math. Soc.*, 22, (1947), 100-101.

[13] P.H. DIANANDA, Critical subsets of finite abelian groups. *J. London Math. Soc.*, 43, (1968), 479-481.

[14] PALAHENEDI HEWAGE DIANANDA and HIAN POH YAP, Maximal sum-free sets of elements of finite groups. *Proc. Japan Acad.*, 45, (1969), 1-5.

[15] L.E. DICKSON, On the congruence $x^n+y^n+z^n \equiv 0 \pmod p$. *J. fur reine und angew. Mathematik*, 135, (1909a), 134-141.

[16] L.E. DICKSON, Lower limit for the number of sets of solutions of $x^e+y^e+z^e \equiv 0 \pmod p$. *J. fur reine und angew. Math.*, 135, (1909b), 181-188.

[17] L.E. DICKSON, History of the theory of numbers. *Carnegie Institute of Washington*, 2, (1919), 763.

[18] P.H. DOYLE and R.J. WARNE, Some properties of groupoids. *Amer. Math. Monthly*, (1963), 1051-1057.

[19] P. ERDÖS, Some remarks on the theory of graphs. *Bull. Amer. Math. Soc.*, 53, (1947), 292-294.

[20] P. ERDÖS, Extremal problems in number theory. *Proc. Sympos. Pure Math.*, 8, Amer. Math. Soc., Providence, Rhode Island, (1965), 181-189.

[21] P. ERDÖS and J. KOMLÓS, On a problem of Moser, *Combinatorial Theory and its Applications*, Colloq. Math. Soc. János Bolyai, 4, I, (1969), 365-367.

[22] P. ERDÖS and G. SZEKERES, A combinatorial problem in geometry. *Compositio Mathematica*, 2, (1935), 463-470.

[23] J. FOLKMAN, Notes on the Ramsey number N(3,3,3,3). Manuscript, Rand Corporation, Santa Monica, California, 1967.

[24] CLAUDE FRASNAY, Sur des fonctions d'entiers se rapportant au théorème de Ramsey. *C.R. Acad. Sci., Paris, Ser. A.*, 256, (1963), 2507-2510.

[25] P.X. GALLAGHER, The large sieve. *Mathematika*, 14, (1967), 14-20.

[26] GUY R. GIRAUD, Une généralisation des nombres et de l'inégalité de Schur, *C.R. Acad. Sci., Paris, Ser. A.*, 266, (1968a), 437-440.

[27] GUY R. GIRAUD, Minoration de certains nombres de Ramsey binaires par les nombres de Schur généralisés, *C.R. Acad. Sci., Paris, Ser. A.*, 266, (1968b), 481-483.

[28] SOLOMON W. GOLOMB and LEONARD D. BAUMERT, Backtrack Programming. *Journal of the Assoc. for Computing Machinery*, 12, (1965), 516-524.

[29] R.E. GREENWOOD and A.M. GLEASON, Combinatorial Relations and Chromatic graphs. *Canad. J. Math.*, 7, (1955), 1-7.

[30] HEINI HALBERSTAM, Sieve methods and applications. *Seminaire Delange-Pisot-Poitou (Théorie des Nombres)*, 9e année, 1967/68, no. 7.

[31] M. HALL JR., Combinatorial Theory. Blaisdell, Waltham, Massachusetts, (1967).

[32] DENIS HANSON, Studies in Combinatorial analysis. Ph.D. thesis, University of Alberta, 1970.

[33] K. ISEKI, On mutant sets in semigroups. *Proc. Japan Acad.*, 38, (1962), 478-479.

[34] J.G. KALBFLEISCH and R.G. STANTON, On the maximal triangle-free edge-chromatic graphs in three colors. *J. Comb. Theory*, 5, (1968), 9-20.

[35] J.H.B. KEMPERMAN, On small sumsets in an abelian group. *Acta Math.*, 103, (1960), 63-88.

[36] A.Y. KHINCHIN, Three pearls of number theory. *OGIZ*, Moscow, 1948; English translation by F. Bagemihl, H. Komm, W. Seidel, Graylock Press, Rochester, N.Y. (1952).

[37] JIN BAI KIM, Mutants in semigroups. *Czechoslovak Math. Journal*, 19 (94), (1969), 86-90.

[38] JIN BAI KIM, Mutants in the symmetric semigroups. *Czechoslovak Math. Journal*, 21 (96), (1971), 355-363.

[39] D. KLEITMAN, On a combinatorial problem of Erdös. *Proc. Amer. Math. Soc.*, 17, (1966), 139-141.

[40] M. KNESER, Abschätzung der asymptotischen Dichte von Summenmengen. *Math. Z.*, 58, (1953), 459-484.

[41] M. KNESER, Ein Satz über abelschen Gruppen mit Anwendungen auf die Geometrie der Zahlen. *Math. Z.*, 61, (1955), 429-434.

[42] M. KNESER, Summenmengen in lokalcompakten abelschen Gruppen. *Math. Z.*, 66, (1956), 88-110.

[43] WILLIAM A. McWORTER, On a theorem of Mann. *American Math. Monthly*, 71, (1964), 285-286.

[44] HENRY B. MANN, *Addition theorems: the addition theorems of group theory and number theory.* Interscience Tracts in pure and applied mathematics, Number 18, John Wiley, New York-London-Sydney, (1965).

[45] H.B. MANN, On products of sets of group elements. *Canad. J. Math.*, 4, (1952), 64-66.

[46] H.B. MANN and JOHN OLSEN, Sums of sets in the elementary abelian group of type (p,p). *J. Comb. Theory*, 2, (1967), 275-284.

[47] L. MOSER, Notes on number theory II. *Canad. Math. Bull.*, 3, (1960), 23-25.

[48] A.A. MULLIN, Properties of mutants. *Bull. Amer. Math. Soc.*, 67, (1961), 82.

[49] R. RADO, Studien zur Kombinatorik. *Math. Z.*, 36, (1933), 424-480.

[50] R. RADO, Note on combinatorial analysis. *Proc. Lond. Math. Soc.*, (2) 48, (1945), 122-160.

[51] R. RADO, Some partition theorems. *Combinatorial Theory and its Applications*, (ed. Erdös, Rényi, Sós), Colloq. Math. Soc. János Bolyai 4, III (1971), 929-936.

[52] F.P. RAMSEY, On a problem in formal logic. *Proc. Lond. Math. Soc.*, (2) 30, (1930), 264-286.

[53] A.H. RHEMTULLA and ANNE PENFOLD STREET, Maximal sum-free sets in finite abelian groups. *Bull. Austral. Math. Soc.*, 2, (1970), 289-297.

[54] A.H. RHEMTULLA and ANNE PENFOLD STREET, Maximal sum-free sets in elementary abelian p-groups. *Canad. Math. Bull.*, 14, (1971), 73-80.

[55] H.J. RYSER, *Combinatorial Mathematics*. Carus Math. Mono., 14, Wiley, New York, (1963).

[56] H. SALIÉ, Zur Verteilung natürlichler Zahlen auf elermentfremde Klassen. *Berichte über die Verhandlungen der Sachsischen, Akademie der Wissenschaften zu Leipzig*, 4, (1954), 1-26.

[57] I. SCHUR, Uber die Kongruenz $x^m+y^m \equiv z^m$ (mod p). *Jahresbericht der Deutschen Mathematiker Vereinigung*, 25, (1916), 114-117.

[58] TH. SKOLEM, Ein kombinatorischer Satz mit Anwendung auf ein logisches Entscheidungsproblem. *Fundamenta Mathematicae*, 20, (1933), 254-261.

[59] E.G. STRAUS, On a problem in combinatorial number theory. *J. Math. Sci.*, (Delhi), 1, (1966), 77-80.

[60] ANNE PENFOLD STREET, Maximal sum-free sets in cyclic groups of prime-power order. *Bull. Austral. Math. Soc.*, 4, (1971), 407-418.

[61] ANNE PENFOLD STREET, Maximal sum-free sets in abelian groups of order divisible by three. *Bull. Austral. Math. Soc.*, 6, (1972), 439-441.

[62] ANNE PENFOLD STREET, Maximal sum-free sets in abelian groups of order divisible by three: Corrigendum. *Bull. Austral. Math. Soc.*, 7, (1972), 320.

[63] H.S. VANDIVER, Fermat's last theorem. *Amer. Math. Monthly*, 53, (1946), 555-578.

[64] A.G. VOSPER, The critical pairs of subsets of a group of prime order. *J. Lond. Math. Soc.*, 31, (1956a), 200-205.

[65] A.G. VOSPER, Addendum to "The critical pairs of subsets of a group of prime order." *J. Lond. Math. Soc.*, 31, (1956b), 280-282.

[66] B.L. VAN DER WAERDEN, Beweis einer Baudetschen Vermetung. *Niew archuf voor wiskunde*, 15, (1927), 212-216.

[67] E.G. WHITEHEAD JR., Algebraic structure of chromatic graphs associated with the Ramsey number N(3,3,3;2). *Discrete Mathematics*, 1, (1971a), 113-114.

[68] E.G. WHITEHEAD JR., Ramsey numbers of the form N(3,...,3;2). Ph.D. dissertation, University of Southern California, Los Angeles, (1971b).

[69] E.G. WHITEHEAD JR., The Ramsey number N(3,3,3,3;2). *Discrete Mathematics*, (1972a), to appear.

[70] E.G. WHITEHEAD JR., Algorithms associated with Ramsey's theorem. (1972b), to appear.

[71] H.P. YAP, The number of maximal sum-free sets in $C_p$. *Nanta Math.*, 2, (1968), 68-71.

[72] H.P. YAP, Maximal sum-free sets of group elements. *J. Lond. Math. Soc.*, 44, (1969), 131-136.

[73] H.P. YAP, Structure of maximal sum-free sets in $C_p$. *Acta Arith.*, 17, (1970a), 29-35.

[74] HIAN POH YAP, Structure of maximal sum-free sets in groups of order 3p. *Proc. Japan Acad.*, 46, (1970b), 758-762.

[75] H.P. YAP, Maximal sum-free sets in finite abelian groups. *Bull. Austral. Math. Soc.*, 4, (1971a), 217-223.

[76] H.P. YAP, Maximal sum-free sets in finite abelian groups II. *Bull. Austral. Math. Soc.*, 5, (1971), 43-54.

[77] S. ZNÁM, On k-thin sets and n-extensive graphs. *Matematický Časopis*, 17, (1967), 297-307.

PART 4

# HADAMARD MATRICES

JENNIFER SEBERRY WALLIS

# CONTENTS


| | | |
|---|---|---|
| CHAPTER I. | PRELIMINARIES | 279 |
| 1.1 | Hadamard matrices and SBIBD's | 279 |
| 1.2 | Difference sets and incidence matrices | 280 |
| 1.3 | Skew-Hadamard matrices | 292 |
| 1.4 | Symmetric conference matrices | 293 |
| 1.5 | Complex Hadamard matrices | 295 |
| 1.6 | Amicable and special Hadamard matrices | 296 |
| 1.7 | Circulant complex matrices | 296 |
| CHAPTER II. | AMICABLE HADAMARD MATRICES | 300 |
| 2.1 | Constructions | 300 |
| CHAPTER III. | ORTHOGONAL MATRICES WITH ZERO DIAGONAL - SKEW-HADAMARD AND SYMMETRIC CONFERENCE MATRICES | 306 |
| 3.1 | Orthogonal matrices with zero diagonal | 306 |
| 3.2 | Constructions for symmetric conference matrices | 313 |
| 3.3 | Paley matrices | 313 |
| 3.4 | Equivalence of symmetric conference matrices | 316 |
| CHAPTER IV. | CONSTRUCTIONS FOR SKEW-HADAMARD MATRICES | 320 |
| 4.1 | Constructions using Szekeres difference sets | 320 |
| 4.2 | The Williamson type | 325 |
| 4.3 | The Goethals-Seidel type | 329 |
| 4.4 | An adaption of Wallis-Whiteman | 333 |
| 4.5 | Constructions using amicable Hadamard matrices | 337 |
| CHAPTER V. | CONSTRUCTIONS FOR SYMMETRIC HADAMARD MATRICES | 339 |
| 5.1 | Constructions using symmetric conference matrices | 339 |
| 5.2 | Strongly regular graphs | 340 |
| 5.3 | Symmetric Hadamard matrices with constant diagonal | 341 |
| CHAPTER VI. | COMPLEX HADAMARD MATRICES | 347 |
| 6.1 | Constructions using amicable Hadamard matrices | 348 |
| 6.2 | Using special Hadamard matrices | 349 |
| 6.3 | Other constructions | 350 |
| CHAPTER VII. | HADAMARD ARRAYS | 354 |
| 7.1 | The Hadamard arrays of Williamson and Baumert-Hall | 354 |
| 7.2 | The arrays of Goethals-Seidel | 355 |
| 7.3 | Baumert-Hall-Welch arrays | 356 |
| 7.4 | Arrays with each element repeated once | 361 |
| 7.5 | Arrays with order divisible by 8 | 364 |
| 7.6 | An announcement of Turyn | 366 |

| | | |
|---|---|---|
| CHAPTER VIII. | CONSTRUCTIONS FOR HADAMARD MATRICES | 367 |
| 8.1 | Using complex Hadamard matrices | 368 |
| 8.2 | Constructions for Hadamard matrices using skew-Hadamard and amicable Hadamard matrices | 368 |
| 8.3 | Using complex Hadamard matrices and quaternion matrices | 372 |
| 8.4 | Using symmetric conference matrices | 375 |
| 8.5 | Using skew-Hadamard matrices | 378 |
| 8.6 | The Williamson type Hadamard matrices | 381 |
| 8.7 | The Goethals-Seidel type Hadamard matrices and an adaption | 391 |
| 8.8 | Using Hadamard arrays | 394 |
| 8.9 | The results of Yang | 395 |
| 8.10 | A computer search | 398 |
| CHAPTER IX. | GENERALIZED HADAMARD MATRICES | 401 |
| 9.1 | The results of Butson | 401 |
| 9.2 | Delsarte and Goethals results | 404 |
| CHAPTER X. | EQUIVALENCE OF HADAMARD MATRICES | 408 |
| 10.1 | Z-equivalence: the Smith normal form | 410 |
| 10.2 | Z-equivalence: the number of small invariants | 411 |
| 10.3 | Z-equivalence: matrices derived from skew-Hadamard matrices and symmetric conference matrices | 415 |
| 10.4 | H-equivalence: orders 1,2,4,8,12,20 | 418 |
| 10.5 | H-equivalence: order 16 | 419 |
| 10.6 | H-equivalence: order 24 | 421 |
| 10.7 | Z-equivalence: order 32 | 421 |
| 10.8 | Order 36 and S-equivalence | 422 |
| 10.9 | Higher powers of 2 | 423 |
| CHAPTER XI. | USES OF HADAMARD MATRICES | 426 |
| 11.1 | Balanced incomplete block designs | 426 |
| 11.2 | Integral solutions to the incidence equation for finite projective planes | 427 |
| 11.3 | Strongly regular graphs | 429 |
| 11.4 | Design graphs, $(v,k,\lambda)$-graphs, $G_2(\lambda)$ graphs | 429 |
| 11.5 | Tournaments | 429 |
| 11.6 | Orthogonal arrays and projective planes | 430 |
| 11.7 | Automorphism and simple groups | 430 |
| 11.8 | Weighing designs | 431 |
| 11.9 | Modular Hadamard matrices | 433 |
| 11.10 | Sequences | 434 |
| 11.11 | Belevitch on 2n-terminal networks | 437 |
| 11.12 | Walsh functions | 437 |
| 11.13 | Coding theory | 440 |
| 11.14 | Pairwise statistical independence | 442 |
| CHAPTER XII. | UNANSWERED QUESTIONS | 444 |

APPENDICES

| | | |
|---|---|---|
| Appendix A: | Known classes of Hadamard matrices | 447 |
| Appendix B: | Known classes of skew-Hadamard matrices | 451 |
| Appendix C: | Known classes of symmetric conference matrices | 452 |
| Appendix D: | Known classes of symmetric Hadamard matrices | 453 |
| Appendix E: | Known classes of symmetric Hadamard matrices with constant diagonal | 454 |
| Appendix F: | Known classes of complex Hadamard matrices | 455 |
| Appendix G: | Known classes of amicable Hadamard matrices | 457 |
| Appendix H: | A table of orders ≡ 0 (mod 4) < 4,000 for which Hadamard, skew-Hadamard and complex Hadamard matrices are known | 458 |
| Appendix I: | A table of orders ≡ 2 (mod 4) < 4,000 for which symmetric conference and complex Hadamard matrices are known | 467 |
| Appendix J: | A table of orders for which amicable Hadamard matrices are known | 470 |
| Appendix K: | Skew-Hadamard matrices of order ≤ 100 | 471 |

REFERENCES 478

In this monograph the author has endeavoured to survey the constructions and equivalence of Hadamard matrices.

We have not tried to study automorphism groups of Hadamard matrices, although we mention this aspect in discussing uses of Hadamard matrices.

The basic concepts of group theory and linear algebra have been assumed as have the definitions of Part 1.

We use *** to denote the end of a proof. Equation (3.71) denotes the equation is after theorem (lemma-corollary-definition) 3.7 and before theorem 3.8. Theorem (lemma-corollary-definition) numbers without decimal points refer to Part 1.

## CHAPTER I. PRELIMINARIES

1.1 **HADAMARD MATRICES AND SBIBD's.** Every Hadamard matrix H of order 4t is associated in a natural way with an SBIBD with parameters (4t-1,2t-1,t-1), and with its complement, a (4t-1,2t,t)-configuration. To obtain the SBIBD we first normalize H and write the resultant matrix in the form

$$\begin{bmatrix} 1 & 1 \ldots\ldots 1 \\ 1 & \\ \cdot & \\ \cdot & A \\ \cdot & \\ 1 & \end{bmatrix}.$$

Then

$$AJ = JA = -J \quad \text{and} \quad AA^T = 4tI-J.$$

So $B = \frac{1}{2}(A+J)$ satisfies

$$BJ = JB = (2t-1)J \quad \text{and} \quad BB^T = tI + (t-1)J.$$

Thus B is a (0,1)-matrix satisfying the equations for the incidence matrix of an SBIBD with parameters (4t-1,2t-1,t-1). Similarly $C = \frac{1}{2}(J-A)$ is the incidence matrix of an SBIBD with parameters (4t-1,2t,t). Clearly if we start with the incidence matrix of an SBIBD with parameters (4t-1,2t-1,t-1) or (4t-1,2t,t) and replace all the 0 elements by -1 we form either A or -A. Thus

$$\begin{bmatrix} 1 & 1 \ldots\ldots 1 \\ 1 & \\ \cdot & \\ \cdot & A \\ \cdot & \\ 1 & \end{bmatrix} \quad \text{and} \quad \begin{bmatrix} -1 & -1 \ldots\ldots -1 \\ -1 & \\ \cdot & \\ \cdot & -A \\ \cdot & \\ -1 & \end{bmatrix}$$

are Hadamard matrices of order 4t obtained from these SBIBD.

Thus we have shown

THEOREM 1.1. *There exists an Hadamard matrix of order 4t if and only if there exists a (4t-1,2t-1,t-1)-configuration.*

Since a (4t-1,2t-1,t-1)-difference set yields an SBIBD we have:

COROLLARY 1.2. *If there exists a (4t-1,2t-1,t-1)-difference set then there exists an Hadamard matrix of order 4t.*

Also we can show:

THEOREM 1.3. *A regular symmetric Hadamard matrix of order 4s exists only if s is an integer square.*

PROOF. Let H be a regular Hadamard matrix of order 4s. Then

$$HH^T = 4sI \text{ and } HJ = JH = tJ.$$

Write $A = \frac{1}{2}(H+J)$ for the positive elements of H. Then

$$AA^T = sI + (s+\tfrac{1}{2}t)J, \quad AJ = JA = \tfrac{1}{2}(t+4s)J.$$

But by a theorem of Ryser (see Hall [56,p.104]), the existence of a non-singular A satisfying these conditions implies that

$$(s+\tfrac{1}{2}t)(4s-1) = (\tfrac{1}{2}t+2s)(\tfrac{1}{2}t+2s-1),$$

that is

$$t^2 = 4s;$$

t is an integer, so s must be an integer square. ***

Further another theorem of Ryser (theorem 16) tells us that A is the incidence matrix of an SBIBD with parameters $(t^2,\frac{1}{2}(t+t^2),\frac{1}{2}(t+\frac{1}{2}t^2))$. Then writing $t = \pm 2u$ we have

THEOREM 1.4. *A regular Hadamard matrix, H, of order $4u^2$, satisfying $HJ = \pm 2uJ$, exists if and only if there exists an SBIBD with parameters $(4u^2, 2u^2\pm u, u^2\pm u)$.*

1.2 DIFFERENCE SETS AND INCIDENCE MATRICES. We develop some more specialized concepts that we use later.

For a complete study of difference sets we refer the reader to the excellent book of L.D. Baumert, Cyclic Difference Sets, [7].

We now list the classes of difference sets (both cyclic and group) that we will use later from Marshall Hall [56,p.141]. We leave out those sets which duplicate parameters but note that in considering inequivalent Hadamard matrices the sets that have been omitted become significant.

Type S: (Singer difference sets). These are hyperplanes in PG(n,q), q a prime power. The parameters are

$$v = \frac{q^{n+1}-1}{q-1}, \quad k = \frac{q^n-1}{q-1}, \quad \lambda = \frac{q^{n-1}-1}{q-1}.$$

Type Q: (quadratic residues of v (prime power) ≡ 3 (mod 4)). Here

$$v = 4t-1, \quad k = 2t-1, \quad \lambda = t-1.$$

Type T: (Twin primes). Suppose that p and q = p+2 are twin primes. Of the (p-1)(q-1) residues modulo pq prime to pq, let $a_1,\dots,a_m$, $m = (p-1)(q-1)/2$ be those for which $\left(\frac{a_i}{p}\right) = \left(\frac{a_i}{q}\right)$, and also let $a_{m+1},\ldots,a_{m+p}$ be $0,q,2q,\ldots,(p-1)q$. Here, $m+p = (pq-1)/2 = k$. Then $a_1,\ldots,a_k$ form a difference set modulo $v$, $v = pq$. Here the parameters are

$$v = pq, \quad k = \frac{pq-1}{2}, \quad \lambda = \frac{pq-3}{4}.$$

Necessarily, $pq \equiv 3 \pmod 4$, and we have $v = 4t-1$, $k = 2t-1$, $\lambda = t-1$. This also holds for $GF(p^r)$ and $GF(q^s)$ if $q^s = p^r+2$.

Type B: (biquadratic residues of primes $v = 4x^2+1$, $x$ odd). Here,

$$v = 4x^2+1, \quad k = x^2, \quad \lambda = \frac{x^2-1}{4}.$$

Type BO: (biquadratic residues and zero of primes $v = 4x^2+9$, $x$ odd). Here

$$v = 4x^2+9, \quad k = x^2+3, \quad \lambda = \frac{x^2+3}{4}.$$

Type O: (octic residues of primes $v = 8a^2+1 = 64b^2+9$ with $a,b$ odd). Here we have

$$v, \ k = a^2, \quad \lambda = b^2.$$

Type OO: (octic residues and zero for primes $v = 8a^2+49 = 64b^2+441$, $a$ odd, $b$ even). Here we have

$$v, \ k = a^2+7, \quad \lambda = b^2+7.$$

Type W: (a generalization of type T developed by Whiteman). Let $p$ and $q$ be two primes such that $(p-1,q-1) = 4$, and we write $d = (p-1)(q-1)/4$. Let $g$ be a primitive root of both $p$ and $q$. The difference set consists of

$$1,g,g^2,\ldots,g^{d-1},0,q,2q,\ldots,(p-1)q \text{ modulo } pq.$$

We must have $q = 3p+2$ and $(v-1)/4$ an odd square. Here

$$v = pq = p(3p+2), \quad k = (3p^2+2p-1)/4, \quad \lambda = (3p^2+2p-5)/16,$$

or

$$v = 4(2s+1)^2+1, \quad k = (2s+1)^2, \quad \lambda = s(s+1)$$

where $p(3p+2) = 4(2s+1)^2+1$.

In Table 1 we have the (1,-1)-matrices, $A$, associated with the incidence matrices of the above $(v,k,\lambda)$-group difference sets. We use theorem 16 of part 1 to obtain the equations satisfied by these (1,-1)-matrices.

DEFINITION 1.5. Let $S_1,S_2,\ldots,S_n$ be subsets of $V$, an additive abelian group, containing $k_1,k_2,\ldots,k_n$ elements respectively. Write $T_i$ for the totality of all differences between elements of $S_i$ (with repetitions), and $T$ for the totality of elements of all the $T_i$. If $T$ contains each non-zero element a fixed number of times, $\lambda$ say, then the sets $S_1,S_2,\ldots,S_n$ will be called $n$-$\{v; k_1,k_2,\ldots,k_n; \lambda\}$ *supplementary difference sets*, where $v$ is the order of $V$. The incidence matrix of each set may be defined as in definition 33 of part 1.

| Type of Difference Set | $(v,k,\lambda)$-configuration generating A | Incidence Equation for $AA^T$ | Comment |
|---|---|---|---|
| | $(v,v,v) = J$ | $vJ$ | |
| | $(v,v-1,v-2) = J-2I$ | $4I+(v-4)J$ | |
| S type | $\frac{q^{n+1}-1}{q-1}, \frac{q^{n}-1}{q-1}, \frac{q^{n-1}-1}{q-1}$ | $4q^{n-1}I+\left(\frac{q^{n+1}-1}{q-1} - 4q^{n-1}\right)J$ | $q = p^r$ , p prime |
| Q type | $(4t-1,2t-1,t-1)$ | $4tI-J$ | $4t = p^r+1$; $p^r$ a prime power |
| T type | $(pq, \frac{pq-1}{2}, \frac{pq-3}{4})$ | $(pq+1)I-J$ | $p = q+2$; p,q prime powers |
| B type | $(4x^2+1,x^2,\frac{x^2-1}{4})$ | $(3x^2+1)I+x^2J$ | $p = 4x^2+1$; p prime, x odd |
| BO type | $(4x^2+9,x^2+3,\frac{x^2+3}{4})$ | $(3x^2+9)I+x^2J$ | $p = 4x^2+9$; p prime, x odd |
| O type | $(64b^2+9,8b^2+1,b^2)$ | $4(7b^2+1)I+(36b^2+5)J$ | $p = 8a^2+1 = 64b^2+9$; p prime, a,b odd |
| OO type | $(64b^2+441,8b^2+56,b^2+7)$ | $4(7b^2+49)I+(36b^2+245)J$ | $p = 8a^2+49 = 64b^2+441$; p prime, a odd, b even |
| W type | $\left(4(2s+1)^2+1,(2s+1)^2,s(s+1)\right)$ | $4(3s^2+3s+1)I+(2s+1)^2J$ | $p(3p+2) = 4(2s+1)^2+1$; p and 3p+2 prime, s integer |

TABLE 1

NOTATION. If $k_1 = k_2 = \ldots = k_n = k$ we will write $n$-$\{v;k;\lambda\}$ to denote the $n$ supplementary difference sets.

NOTATION. We shall be concerned with collections in which repeated elements are counted multiply, rather than with sets. If $T_1$ and $T_2$ are two such collections then $T_1 \& T_2$ will denote the results of adjoining the elements of $T_1$ to $T_2$, with total multiplicities retained. We use square brackets [ ] to denote collections and braces { } to denote sets.

EXAMPLE. Let $S_1 = \{1,2,x+1,2x+2\}$, $S_2 = \{0,1,2,x+1,2x+2\}$ be subsets of the additive abelian group G of $GF(3^2)$. Let $T_1$ and $T_2$ be the totality of differences between elements of $S_1$ and $S_2$ then

$$\begin{aligned} T_1 &= [1-2,1-(x+1),1-(2x+2),2-1,2-(x+1),2-(2x+2),\\ &\qquad (x+1)-1,(x+1)-2,(x+1)-(2x+2),(2x+2)-1,(2x+2)-2,(2x+2)-(x+1)]\\ &= [2,2x,x+2,1,2x+1,x,x,x+2,2x+2,2x+1,2x,x+1]\\ T_2 &= [0-S_1,S_1-0,T_1]\\ &= [2,1,2x+2,x+1,1,2,x+1,2x+2] \& T_1 \end{aligned}$$

$\therefore$ $T_1 \& T_2 = 4(G\setminus\{0\})$. Thus $S_1$ and $S_2$ are 2-$\{9;4,5;4\}$ supplementary difference sets.

LEMMA 1.6. *The parameters of $n$-$\{v;k_1,k_2,\ldots,k_n;\lambda\}$ supplementary difference sets satisfy*

$$\lambda(v-1) = \sum_{j=1}^{n} k_j(k_j-1). \qquad (1.21)$$

PROOF. This follows immediately from the definition by counting the differences.

***

LEMMA 1.7. *The elements of $Q = \{x^{2b} : b = 1,\ldots,2m+1\}$, where $x$ is a primitive root of a prime power $p^n = 4m+3$, are a $(4m+3,2m+1,m)$-group difference set.*

PROOF. Since $-1$ is not a square in $GF(p^n)$ when $p^n \equiv 3 \pmod 4$, exactly one of c and $-c$ is a square for $c \in GF(p^n)$. Thus one of $r_1-r_2$, $r_2-r_1$ is a square when $r_1$ and $r_2 \in Q$, say $r_1-r_2 \equiv r$. If s is any element of Q then $sr_1$, $sr_2$, $sr$ are squares also. Thus every equation $r_i-r_j \equiv r$ corresponds with the equation $sr_i-sr_j \equiv sr$ and vice versa. Hence every square of $p^n$ is represented equally often as a difference of squares. Reversing these congruences (e.g., $r_2-r_1 \equiv -r$) yields every non-square with the same number of representations. Hence we have a difference set.

***

LEMMA 1.8. *The elements of $Q = \{x^{2b} : b = 1,\ldots,2m\}$ and $R = xQ$, where $x$ is a primitive root of a prime power $p^n = 4m+1$, are 2-$\{4m+1;2m;2m-1\}$ supplementary difference sets. (That is, the number of solutions of*

$$d = a-b, \quad a,b \in Q$$

*together with*

$$d = c-e, \quad c,e \in R$$

*is a constant 2m-1.)*

PROOF. Consider the equations

$$q_1-q_2 \equiv q \qquad r_1-r_2 \equiv r$$
$$q_3-q_4 \equiv r' \qquad r_3-r_4 \equiv q'$$

where $q_1,q_2,q_3,q_4,q,q' \in Q$ and $r_1,r_2,r_3,r_4,r,r' \in R$. As before: if s is any element of Q then $sq_1,sq_2,sq_3,sq_4,sq \in Q$ and $sr' \in R$. Thus every equation $q_i-q_j \equiv n$ corresponds with the equation $sq_i-sq_j \equiv sn$ and vice versa. Hence every element of Q is represented equally often, say $d_1$ times, as a difference of squares and also every element of R is represented equally often, say $d_2$ times. Both Q and R have 2m elements and so counting two ways

$$2md_1 + 2md_2 = 2m(2m-1), \quad d_1+d_2 = 2m-1.$$

Now every equation $q_i-q_j \equiv n$ has the same number of solutions and $xq_i,xq_j \in R$, so putting $xq_i = r_i$, $xq_j = r_j$ and $xq = r$ we have that

$$r_i-r_j \equiv k$$

has $d_1$ solutions when $k \in R$ and $d_2$ solutions when $k \in Q$.

Thus in the totality of differences from Q and R every element occurs $d_1+d_2 = 2m-1$ times. ***

We now generalize the concepts of circulant and backcirculant matrices by considering two special incidence matrices of subsets of an additive abelian group.

DEFINITION 1.9. Let G be an additive abelian group with elements $z_i$. Let X be a subset of G. We define two types of incidence matrices $M = (m_{ij})$ and $N = (n_{ij})$. First we fix an ordering for the elements of G, then M, defined by

$$m_{ij} = \psi(z_j-z_i), \qquad \psi(z_j-z_i) = \begin{cases} 1 & z_j-z_i \in X, \\ 0 & \text{otherwise,} \end{cases}$$

will be called the *type 1 incidence matrix of X in G,* and N, defined by

$$n_{ij} = \phi(z_j+z_i), \qquad \phi(z_j+z_i) = \begin{cases} 1 & z_j+z_i \in X, \\ 0 & \text{otherwise,} \end{cases}$$

will be called the *type 2 incidence matrix of X in G.*

LEMMA 1.10. *Suppose M and N are type 1 and type 2 incidence matrices of a subset $C = \{c_i\}$ of an additive abelian group $G = \{z_i\}$. Then*

$$MM^T = NN^T.$$

PROOF. The inner products of distinct rows i and k in M and N respectively are given by

$$\sum_{z_j \in G} \psi(z_j - z_i)\psi(z_j - z_k) \qquad \text{and} \qquad \sum_{z_j \in G} \phi(z_j + z_i)\phi(z_j + z_k)$$

$$= \sum_{g \in G} \psi(g)\psi(g + z_i - z_k) \qquad\qquad = \sum_{h \in G} \phi(h + z_i - z_k)\phi(h)$$

since as $z_j$ runs through G so does $z_j - z_i = g$ | since as $z_j$ runs through G so does $z_j + z_k = h$

$$= \sum_{c \in C} \psi(c + z_i - z_k) \qquad\qquad = \sum_{c \in C} \phi(c + z_i - z_k)$$

= number of times $c + z_i - z_k \in C$ | = number of times $c + z_i - z_k \in C$.

For the same row

$$\sum_{z_j \in G} [\psi(z_j - z_i)]^2 \qquad \text{and} \qquad \sum_{z_j \in G} [\phi(z_j + z_i)]^2$$

$$= \sum_{g \in G} [\psi(g)]^2 \qquad\qquad = \sum_{h \in G} [\phi(h)]^2$$

$$= \sum_{c \in C} [\psi(c)]^2 \qquad\qquad = \sum_{c \in C} [\phi(c)]^2$$

= number of elements in C | = number of elements in C.

So $MM^T = NN^T$. ***

Now type 1 and type 2 incidence matrices of X in G are (0,1)-matrices, but we shall on occasion use the corresponding matrices which have elements from a commutative ring. So we extend the definition to

DEFINITION 1.11. Let G be an additive abelian group with elements $z_i$, which are ordered in some convenient way and the ordering fixed. Let $X = \{x_i\}$ be a subset of G, $X \cap \{0\} = \emptyset$. Then two matrices $M = (m_{ij})$ and $N = (n_{ij})$ defined by

$$m_{ij} = \psi(z_j - z_i) \quad \text{and} \quad n_{ij} = \phi(z_j + z_i),$$

where $\psi$ and $\phi$ map G into a commutative ring, will be called *type 1* and *type 2* respectively.

Further if $\psi$ and $\phi$ are defined by

$$\psi(x) = \begin{cases} a & x \in X \\ b & x = 0 \\ c & x \notin X \cup \{0\} \end{cases}, \qquad \phi(x) = \begin{cases} d & x \in X \\ e & x = 0 \\ f & x \notin X \cup \{0\} \end{cases}$$

then M and N will be called *type 1 matrix of* $\psi$ *on X* and *type 2 matrix of* $\phi$ *on X* respectively. But if $\psi$ and $\phi$ are defined by

$$\psi(x) = \begin{cases} 1 & x \in X \\ -1 & x \notin X \end{cases}, \qquad \phi(x) = \begin{cases} 1 & x \in X \\ -1 & x \notin X \end{cases},$$

then M and N will be called *type 1 (1,-1) incidence matrix* and *type 2 (1,-1) incidence matrix* respectively.

EXAMPLE. Consider the additive group $GF(3^2)$, which has elements

$$0,1,2,x,x+1,x+2,2x,2x+1,2x+2.$$

Define the set $X = \{y: y = z^2 \text{ for some } z \in GF(3^2)\}$
$= \{x+1,2,2x+2,1\}$
using the irreducible equation $x^2 = x+1$. Now a type 1 matrix of $\psi$ on X, $A = (a_{ij})$, is determined by a function of the type

$$a_{ij} = \psi(g_j - g_i) \quad \text{where} \quad \psi(x) = \begin{cases} 0 & x = 0 \\ 1 & x \in X \\ -1 & \text{otherwise} \end{cases}.$$

So let us order the elements as we have above and put

$$g_1 = 0,\ g_2 = 1,\ g_3 = 2,\ g_4 = x,\ g_5 = x+1,\ g_6 = x+2,$$

$$g_7 = 2x,\ g_8 = 2x+1,\ g_9 = 2x+2.$$

Then the type 1 matrix of $\psi$ on X is

$$A = \begin{bmatrix} 0 & 1 & 1 & -1 & 1 & -1 & -1 & -1 & 1 \\ 1 & 0 & 1 & -1 & -1 & 1 & 1 & -1 & -1 \\ 1 & 1 & 0 & 1 & -1 & -1 & -1 & 1 & -1 \\ -1 & -1 & 1 & 0 & 1 & 1 & -1 & 1 & -1 \\ 1 & -1 & -1 & 1 & 0 & 1 & -1 & -1 & 1 \\ -1 & 1 & -1 & 1 & 1 & 0 & 1 & -1 & -1 \\ -1 & 1 & -1 & -1 & -1 & 1 & 0 & 1 & 1 \\ -1 & -1 & 1 & 1 & -1 & -1 & 1 & 0 & 1 \\ 1 & -1 & -1 & -1 & 1 & -1 & 1 & 1 & 0 \end{bmatrix}$$

Let the function $\phi(x) = \begin{cases} 0 & x = 0 \\ 1 & x \in X \\ -1 & \text{otherwise} \end{cases}$ and $a_{ij} = \phi(g_i + g_j)$ define a type 2 matrix B. Then keeping the same ordering as above the type 2 matrix of $\phi$ on X is

$$B = \begin{bmatrix} 0 & 1 & 1 & -1 & 1 & -1 & -1 & -1 & 1 \\ 1 & 1 & 0 & 1 & -1 & -1 & -1 & 1 & -1 \\ 1 & 0 & 1 & -1 & -1 & 1 & 1 & -1 & -1 \\ -1 & 1 & -1 & -1 & -1 & 1 & 0 & 1 & 1 \\ 1 & -1 & -1 & -1 & 1 & -1 & 1 & 1 & 0 \\ -1 & -1 & 1 & 1 & -1 & -1 & 1 & 0 & 1 \\ -1 & -1 & 1 & 0 & 1 & 1 & -1 & 1 & -1 \\ -1 & 1 & -1 & 1 & 1 & 0 & 1 & -1 & -1 \\ 1 & -1 & -1 & 1 & 0 & 1 & -1 & -1 & 1 \end{bmatrix}$$

LEMMA 1.12. *Suppose $G$ is an additive abelian group of order $v$ with elements $z_1, z_2, \ldots, z_v$. Say $\phi$ and $\psi$ are maps from $G$ to a commutative ring $R$. Define*

$$A = (a_{ij}), \qquad a_{ij} = \phi(z_j - z_i),$$
$$B = (b_{ij}), \qquad b_{ij} = \psi(z_j - z_i),$$
$$C = (c_{ij}), \qquad c_{ij} = \mu(z_j + z_i).$$

*Then (independently of the ordering of $z_1, z_2, \ldots, z_v$ save only that it is fixed)*

*(i)* $C^T = C$,

*(ii)* $AB = BA$,

*(iii)* $AC^T = CA^T$.

PROOF. (i) $c_{ij} = \mu(z_j + z_i) = \mu(z_i + z_j) = c_{ji}$.

(ii) $(AB)_{ij} = \sum_{g\in G} \phi(g - z_i)\psi(z_j - g)$

putting $h = z_i + z_j - g$, it is clear that as $g$ ranges through $G$ so does $h$, and the above expression becomes

$$\sum_{h\in G} \phi(z_j - h)\psi(h - z_i)$$
$$= \sum_{h\in G} \psi(h - z_i)\phi(z_j - h)$$

(since $R$ is commutative); this is $(BA)_{ij}$.

(iii) $(AC^T)_{ij} = \sum_{g\in G} \phi(g - z_i)\mu(z_j + g)$

$$= \sum_{h\in G} \phi(h - z_j)\mu(z_i + h) \qquad (h = z_j + g - z_i)$$

$$= (CA^T)_{ij}.$$

***

COROLLARY 1.13. *If $X$ and $Y$ are type 1 matrices and $Z$ is a type 2 matrix then*

$$XY = YX$$
$$XZ^T = ZX^T.$$

LEMMA 1.14. *If $X$ is a type $i$, $i = 1,2$, matrix then so is $X^T$.*

PROOF. (i) If $X = (x_{ij}) = \phi(z_j+z_i)$ is type 2 then so is $X^T = (y_{ij}) = \phi(z_i+z_j)$.

(ii) If $X = (x_{ij}) = \psi(z_j-z_i)$ is type 1 then so is $X^T = (y_{ij}) = \mu(z_j-z_i)$ where $\mu$ is the map $\mu(z) = \psi(-z)$.

COROLLARY 1.15. *(i) If $X$ and $Y$ are type 1 matrices then*

$$XY = YX,$$
$$X^TY = YX^T,$$
$$XY^T = Y^TX,$$
$$X^TY^T = Y^TX^T.$$

*(ii) If $P$ is a type 1 matrix and $Q$ is a type 2 matrix then*

$$PQ^T = QP^T,$$
$$PQ = Q^TP^T,$$
$$P^TQ^T = QP,$$
$$P^TQ = Q^TP.$$

LEMMA 1.16. *Let $X$ and $Y$ be type 2 matrices obtained from two subsets $A$ and $B$ of an additive abelian group $G$ and for which*

$$a \in A \Rightarrow -a \in A \qquad b \in B \Rightarrow -b \in B$$

*then*

$$XY = YX \quad \text{and} \quad XY^T = YX^T.$$

PROOF. Since $X$ and $Y$ are symmetric we only have to prove that $XY^T = YX^T$.

Suppose $X = (x_{ij})$ and $Y = (y_{ij})$ are defined by

$$x_{ij} = \phi(z_i+z_j),$$
$$y_{ij} = \psi(z_i+z_j),$$

where $z_1, z_2, \ldots$ are the elements of $G$.

Then

$$(XY^T)_{ij} = \sum_k \phi(z_i+z_k)\psi(z_k+z_j)$$
$$= \sum_k \phi(-z_i-z_k)\psi(z_k+z_j)$$

since $a \in A \Rightarrow -a \in A$,

$$= \sum_{\ell} \phi(z_j+z_\ell)\psi(-z_\ell-z_i-z_j+z_j)$$

$$z_\ell = -z_k-z_i-z_j$$

$$= \sum_{\ell} \phi(z_j+z_\ell)\psi(z_\ell+z_i)$$

since $b \in B \Rightarrow -b \in B$

$$= (YX^T)_{ij}.$$

***

We note that if the additive abelian group in definition 1.9 is the integers modulo p with the usual ordering then

(i) the type 1 incidence matrix is circulant since
$m_{ij} = \psi(j-i) = \psi(j-i+1-1) = m_{1,i-j+1}$

(ii) the type 2 incidence matrix is backcirculant since
$n_{ij} = \psi(i+j) = \psi(i+j-1+1) = n_{1,i+j-1}$.

In any case:

*a type 1 matrix is analogous to a circulant matrix;*
*a type 2 matrix is analogous to a backcirculant matrix.*

In all the theorems stated above the words
"type 1" and "circulant"
may be interchanged as may
"type 2" and "backcirculant".

LEMMA 1.17. *Let $R = (r_{ij})$ be the permutation matrix of order n, defined on an additive abelian group $G = \{g_i\}$ of order n by*

$$r_{\ell,j} = \begin{cases} 1 & \text{if } g_\ell+g_j = 0, \\ 0 & \text{otherwise.} \end{cases}$$

*Let M by a type 1 matrix of a subset $X = \{x_i\}$ of G. Then MR is a type 2 matrix. In particular, if G is the set of integers modulo n then MR is a backcirculant matrix.*

PROOF. Let $M = (m_{ij})$ be defined by $m_{ij} = \psi(g_j-g_i)$ where $\psi$ maps G into a commutative ring. Let $\mu(-x) = \psi(x)$. Then MR is

$$(MR)_{ij} = \sum_k m_{ik}r_{kj} = m_{i\ell} \quad \text{where} \quad g_\ell+g_j = 0$$

$$\psi(g_\ell-g_i) = \psi(-g_j-g_i) = \mu(g_j+g_i)$$

which is a type 2 matrix. ***

LEMMA 1.18. *Suppose that $A_j = \{a_{ji}\}$ $i = 1,2,\ldots,k_j$, $j = 1,2,\ldots,n$ are $n-\{v;k_1,k_2,\ldots,k_n;\lambda\}$ supplementary difference sets in an additive abelian group $G = \{z_i\}$.*

*If the $A_j$ have type 1 incidence matrices $X_j$ respectively (formed from the same ordering of the elements of G) then*

$$\sum_{j=1}^{n} X_j X_j^T = (\sum_{j=1}^{n} k_j - \lambda) I + \lambda J.$$

PROOF. If $A_j$ are $n$-$\{v; k_1, k_2, \ldots, k_n; \lambda\}$ supplementary difference sets and $h \in G$ then the number of solutions of

$$a_{1i} - a_{1j} = h$$
$$a_{2i} - a_{2j} = h$$
$$\vdots$$
$$a_{ni} - a_{nj} = h$$

are respectively

$$m_1, m_2, \ldots, \text{ and } m_n \text{ and } m_1 + m_2 + \ldots + m_n = \lambda.$$

Define a map $\beta$ by $\beta(z_j - z_i) = \begin{cases} 1 & z_j - z_i \in M \\ 0 & \text{otherwise} \end{cases}$,

on a subset M of G. Then, in the type 1 incidence matrix N obtained using $\beta$, the inner product of distinct rows i and k of N is given by

$$\sum_{z_j \in G} \beta(z_j - z_i)\beta(z_j - z_k) \qquad \text{as } z_j \text{ runs through G so does } z_j - z_i = g$$

$$= \sum_{g \in G} \beta(g)\beta(g + z_i - z_k)$$

$$= \sum_{x \in M} \beta(x)\beta(x + z_i - z_k)$$

$$= \sum_{x \in M} \beta(x + z_i - z_k)$$

$$= \text{number of times } x + z_i - z_k \in M \qquad \text{Write } z_k - z_i = h \text{ and let } y \in H.$$

$$= \text{number of times } x - y = h.$$

For the same row

$$\sum_{z_j \in G} [\beta(z_j - z_i)]^2$$

$$= \sum_{g \in G} [\beta(g)]^2$$

$$= \sum_{x \in M} [\beta(x)]^2$$

$$= \text{number of elements in M.}$$

Hence the inner product of distinct rows i and j of $X_1, X_2, \ldots, X_n$ are $m_1, m_2, \ldots, m_n$ respectively. The inner product of row i with itself for each of $X_j$ is $k_j$. Thus

$$\sum_{j=1}^{n} A_jA_j^T = \sum_{j=1}^{n} k_j \text{ on diagonal and } \sum_{j=1}^{n} m_j \text{ elsewhere}$$

$$= \Big(\sum_{j=1}^{n} k_j-\lambda\Big)I+\lambda J. \qquad ***$$

Finally we construct some matrices that we use later:

With $q = p^n$, let $a_0,a_1,\ldots,a_{q-1}$ be the elements of GF(q) numbered so that

$$a_0 = 0$$

$$a_{q-i} = -a_i \qquad i = 1,\ldots,q-1.$$

Now define $Q = (x_{ij})$ by

$$x_{ij} = \chi(a_j-a_i) \tag{1.181}$$

where $\chi$ is the character defined in definition 4. Then Q is a type 1 matrix and

$$QQ^T = \Big(\sum_k \chi(a_k-a_i)\chi(a_k-a_j)\Big)$$

$$= \Big(\sum_y \chi(y)\chi(y+a_i-a_j)\Big) \qquad y = a_k-a_i$$

$$= \begin{cases} q-1 & i=j \\ -1 & i\neq j \end{cases}$$

$$= qI-J.$$

Also $\sum_y \chi(y) = 0$ so $QJ = JQ = 0$.

Now $x_{ij} = \chi(a_j-a_i) = \chi(-1)\chi(a_i-a_j) = \chi(-1)a_{ji}$ and since

-1 is a square for $q \equiv 1 \pmod 4$

-1 is a non-square for $q \equiv 3 \pmod 4$

it follows that

Q is symmetric for $q \equiv 1 \pmod 4$

Q is skew-symmetric for $q \equiv 3 \pmod 4$.

Thus we have shown

LEMMA 1.19. *If $q = p^n$ is a prime power there exists a matrix, Q of order q, with zero diagonal and all other elements ±1 satisfying*

$$QQ^T = qI-J, \quad QJ = JQ = 0.$$

*Further if $q \equiv 1 \pmod 4$ $Q^T = Q$ and if $q \equiv 3 \pmod 4$ then $Q^T = -Q$.*

LEMMA 1.20. *If $X_1,\ldots,X_n$ are the type 1 incidence matrices of $n$-$\{v;k_1,\ldots,k_n;\lambda\}$ supplementary difference sets, then if $Y_i$ is formed from $X_i$ by replacing all its zeros by -1 then*

$$\sum_{j=1}^{n} Y_jY_j^T = 4\Big(\sum_{j=1}^{n} k_j-\lambda\Big)I + \Big[nv-4\Big(\sum_{j=1}^{n} k_j-\lambda\Big)\Big]J.$$

PROOF. From lemma 1.18 $\sum_{j=1}^{n} X_jX_j^T = \left(\sum_{j=1}^{n} k_j-\lambda\right)I+\lambda J$. Now $Y_i = 2X_i-J$, so

$$\begin{aligned}\sum_{j=1}^{n} Y_jY_j^T &= \sum_{j=1}^{n} (4X_jX_j^T-4X_jJ+vJ)\\ &= 4\left(\sum_{j=1}^{n} k_j-\lambda\right)I+4\lambda J - 4\sum_{j=1}^{n} k_jJ+nvJ\\ &= 4\left(\sum_{j=1}^{n} k_j-\lambda\right)I + [nv-4\left(\sum_{j=1}^{n} k_j-\lambda\right)]J.\end{aligned}$$ ***

LEMMA 1.21. *For $q = p^n \equiv 1 \pmod 4$ there exist two type 1 (1,-1) matrices M and N satisfying*

$$MM^T+NN^T = 4(2m+1)I-2J.$$

PROOF. Construct the 2-{4m+1;2m;2m-1} supplementary difference sets in lemma 1.8. Then use lemma 1.20 to obtain the result. ***

## 1.3 SKEW-HADAMARD MATRICES

DEFINITION 1.22. A *skew-Hadamard matrix* H of order $h \equiv 0 \pmod 4$, has every element +1 or -1, and is of the form H = S+I where S is skew-symmetric.

LEMMA 1.23. *If H = S+I is a skew-Hadamard matrix of order h then*

$$SS^T = (h-1)I.$$

PROOF. Follows from property (c) of lemma 25. ***

In Appendices B and H we list the classes of known skew-Hadamard matrices and also the orders < 4000 of all the known skew-Hadamard matrices.

DEFINITION 1.24. A *skew-type matrix* A = T+I has T skew-symmetric.

DEFINITION 1.25. The *core of a skew-Hadamard matrix* H of order h is that matrix W of order h-1 obtained from H by first multiplying the columns so that the first row has only +1 elements and then multiplying the rows so every element in the first column (bar the first) is -1: then H becomes

$$\begin{bmatrix} 0 & e \\ -e^T & W \end{bmatrix} + I$$

where e = [1,1,...,1] is a 1×(h-1) matrix.

LEMMA 1.26. *If W of order h-1 is the core of a skew-Hadamard matrix then W satisfies*

$$WW^T = (h-1)I-J, \quad WJ = 0, \quad W^T = -W.$$

PROOF. Let H be the skew-Hadamard matrix of order h of which W is the core. Then since $HH^T = hI_h$ we have (with J of order h-1):

$$HH^T = \left( \begin{bmatrix} 0 & e \\ -e^T & W \end{bmatrix} + I_h \right) \left( \begin{bmatrix} 0 & -e \\ e^T & W^T \end{bmatrix} + I_h \right)$$

$$= \begin{bmatrix} h-1 & eW^T \\ We^T & J+WW^T \end{bmatrix} + \begin{bmatrix} 0 & 0 \\ 0 & W+W^T \end{bmatrix} + I_h$$

$$= hI_h .$$

Since H is skew-Hadamard $W^T = -W$ and so we have

$$We^T = 0 \text{ which gives } WJ = 0,$$

and
$$J+WW^T+W+W^T+I_{h-1} = hI_{h-1},$$

or
$$WW^T = (h-1)I_{h-1}-J.$$ ***

**DEFINITION 1.27. Any (1,-1) matrix W of order h-1 with zero diagonal satisfying**

$$WW^T = (h-1)I-J, \quad WJ = 0$$

will be called a *core*. ***

1.4 SYMMETRIC CONFERENCE MATRICES. These matrices were first used by Belevitch [14,15] in connection with electrical networks. Later they are discussed by Goethals and Seidel [44] and Turyn [142] who called them symmetric C-matrices and J. Wallis [154,156] who called them n-type matrices. Sloan and Seidel call them conference matrices.

DEFINITION 1.28. A *symmetric conference matrix* N of order $n \equiv 2 \pmod 4$ has every element +1 or -1, and is of the form $N = R+I$ where R is symmetric, $RR^T = (n-1)I$ and $RJ = 0$.

We emphasize that a symmetric conference matrix (as defined here) is a (1,-1) matrix. Other authors may assume a zero diagonal.

Comparison of definitions 1.28 and 1.22 show that symmetric conference matrices are analogous to skew-Hadamard matrices. Analogously to definition 1.25 we now define the core of a symmetric conference matrix.

DEFINITION 1.29. The *core of a symmetric conference matrix* N of order n is that matrix W of order n-1 obtained from N by first multiplying the rows and columns so that the first row and column has only +1 elements; then N becomes

$$\begin{bmatrix} 0 & e \\ e^T & W \end{bmatrix} + I$$

where $e = [1,1,\ldots,1]$ is a $1\times(n-1)$ matrix.

We digress to obtain a result used in the next theorem.

LEMMA 1.30. *(i) If $m = x^2+y^2$, $x,y$ rational, then $m = a^2+b^2$, $a,b$ integer.*

*(ii) $m = a^2+b^2$, $m \equiv 1 \pmod 4$ is equivalent to saying $m$ has all factors of its square free part $\equiv 1 \pmod 4$.*

PROOF. Suppose $m = (x_1/d_1)^2+(x_2/d_2)^2$. Let p be a prime power dividing m. Then $(x_1d_2)^2+(x_2d_1)^2 \equiv 0 \pmod p$. So $(x_1d_2)^2$ and $-(x_1d_2)^2$ are both quadratic residues of p. Therefore $p \equiv 1 \pmod 4$. Hence every prime power factor of m is $\equiv 1 \pmod p$.

Now every prime $p_i \equiv 1 \pmod 4$ can be written as the sum of two squares $p_i = a_i^2+b_i^2$ (see Griffin [50, p.158]) and we note that

$$\begin{aligned} p_ip_j &= (a_i^2+b_i^2)(a_j^2+b_j^2) \\ &= (a_ia_j \pm b_ib_j)^2+(a_ib_j \mp b_ia_j)^2. \end{aligned}$$

So a product of primes $\equiv 1 \pmod 4$ can be written as a sum of two squares. Every prime $\equiv 3 \pmod 4$ must appear as an even power in m. So we have the result. ***

THEOREM 1.31. (Raghavarao, proof by van Lint and Seidel) *A necessary condition for the existence of a square rational matrix Q of order $q \equiv 2 \pmod 4$ satisfying $Q^TQ = mI_q$, m integer, is that $m = a^2+b^2$ for some integers a and b.*

PROOF. By Lagrange's four-square theorem we may write $m = m_1^2+m_2^2+m_3^2+m_4^2$, $m_1,m_2,m_3,m_4$ integers. Write

$$M = \begin{bmatrix} m_1 & -m_2 & -m_3 & -m_4 \\ m_2 & m_1 & -m_4 & m_3 \\ m_3 & m_4 & m_1 & -m_2 \\ m_4 & -m_3 & m_2 & m_1 \end{bmatrix} \quad \text{and} \quad Q = \begin{bmatrix} A & B \\ C & D \end{bmatrix},$$

with A a 4x4 matrix. Now $M^TM = mI$. Any row of M may be negated without changing this property. Since $\det(M) \neq 0$, it is not possible that $\det(A-M) = 0$ for all possible choices of M. Hence we may assume $\det(A-M) \neq 0$. We now prove for $P = D-C(A-M)^{-1}B$ that $P^TP = mI_{q-4}$ by calculating in two ways the matrix product

$$[-B^T(A^T-M^T)^{-1} \mid I_{q-4}] \begin{bmatrix} A^T & C^T \\ B^T & D^T \end{bmatrix} \begin{bmatrix} A & B \\ C & D \end{bmatrix} \begin{bmatrix} -(A-M)^{-1}B \\ I_{q-4} \end{bmatrix}.$$

Denote the factors by X,Y,Z,U, then

$$\begin{aligned} X(YZ)U &= B^T(A^T-M^T)^{-1}m(A-M)^{-1}B+mI_{q-4}, \\ (XY)(ZU) &= [B^T(A^T-M^T)^{-1}A^T-B^T][A(A-M)^{-1}B-B]+P^TP \\ &= B^T(A^T-M^T)^{-1}[A^T-(A^T-M^T)][A-(A-M)](A-M)^{-1}B+P^TP \\ &= B^T(A^T-M^T)^{-1}m(A-M)^{-1}B+P^TP. \end{aligned}$$

So now we have a matrix of order P of order q-4 satisfying $P^TP = mI_{q-4}$. By iteration a matrix of order 2 is obtained, hence m is a sum of two squares of rationals. By the previous lemma m is the sum of two integer squares. ***

Theorem 1.31 tells us immediately that there are no symmetric conference matrices of orders 22,34,58,70,78 and 94.

Appendix I shows the orders < 1000 for which symmetric conference matrices exist and those orders which are excluded by theorem 1.31. Also in appendix I is a list of those orders between 1000 and 4000 for which symmetric conference matrices do exist.

Appendix C gives a list of known classes of symmetric conference matrices.

## 1.5 COMPLEX HADAMARD MATRICES.

R.J. Turyn introduced these matrices and showed how they may be used to construct Hadamard matrices.

NOTATION. We use $A^*$ for the Hermitian conjugate (or transpose, complex conjugate) of A and $i = \sqrt{-1}$.

DEFINITION 1.32. A *complex Hadamard matrix* of order c is a matrix all of whose elements are +1,-1,i or -i and which satisfies $CC^* = cI_c$.

It is conjectured that a complex Hadamard matrix exists for every even order.

DEFINITION 1.33. A *complex skew-Hadamard matrix* C = I+U has $U^* = -U$. The matrices I+iN, where I+N is a symmetric conference matrix are of this type.

It can be shown that complex Hadamard matrices exist for orders for which symmetric conference matrices are excluded by theorem 1.31.

LEMMA 1.34. *Every complex Hadamard matrix has order 1 or divisible by 2.*

PROOF. The matrix of order 1 is [1] or [i]. There are two H-inequivalent complex Hadamard matrices of order 2, $\begin{bmatrix}1 & 1\\ 1 & -1\end{bmatrix}$ and $\begin{bmatrix}i & 1\\ 1 & i\end{bmatrix}$.

Suppose the matrix is of order m > 2. Then the first two rows may be chosen as

$$\begin{matrix} 1 & 1 & \cdots & 1 & 1 & \cdots & 1 & 1 & 1 & \cdots & 1 & 1 & \cdots & 1 \\ \underbrace{i & i & \cdots & i}_{x} & \underbrace{-i & \cdots & i}_{y} & \underbrace{1 & 1 & \cdots & 1}_{z} & \underbrace{-1 & \cdots & -1}_{w} \end{matrix}$$

by suitably multiplying through the columns by i,-i and -1 and then re-arranging the columns - x,y,z,w are the numbers of columns of each type. So we have

$$x = y, \quad z = w \quad \text{and} \quad x+y+z+w = n,$$

and hence 2|n. ***

We now note a small but useful fact.

LEMMA 1.35. *Suppose $AB^* = BA^*$. Then if $C = iA$, $CB^* = -BC^*$. Specifically, if A and B are real and $AB^T = BA^T$, then if $C = iA$, $CB^* = -BC^*$.*

PROOF. $CB^* = iAB^* = B(iA^*) = -BC^*$. ***

**DEFINITION 1.36. If W is a matrix of order n with zero diagonal and non-diagonal elements 1,-1, i or -i, satisfying $WW^* = nI-J$, $WJ^* = 0$, $W^* = \pm W$ then W is the *core of a complex Hadamard matrix* (not all complex Hadamard matrices have cores).**

In appendix F are listed classes of known complex Hadamard matrices. Appendices H and I give those orders < 1000 for which complex Hadamard matrices are known. Naturally there is a complex Hadamard matrix for every order for which there is a real Hadamard matrix.

## 1.6 AMICABLE AND SPECIAL HADAMARD MATRICES.

We now turn our attention to certain pairs of Hadamard matrices which satisfy useful conditions.

DEFINITION 1.37. Two matrices $M = I+U$ and N will be called (complex) *amicable Hadamard matrices* if M is a (complex) skew-Hadamard matrix and N a (complex) Hadamard matrix satisfying

$$N^T = N, \qquad MN^T = NM^T \quad \text{if real,}$$
$$N^* = N, \qquad MN^* = NM^* \quad \text{if complex.}$$

DEFINITION 1.38. Two Hadamard matrices M and N will be called *special Hadamard matrices* if $MN^T = -NM^T$. M and N will be called *complex special Hadamard matrices* if $MN^* = -NM^*$.

THEOREM 1.39. *Special (complex) Hadamard matrices exist for every order for which there exists a (complex) Hadamard matrix.*

PROOF. Let C be any (complex) Hadamard matrix of order c (c even). Let Q be $R_{\frac{1}{2}c} \times \begin{bmatrix} 0 & 1 \\ -1 & 0 \end{bmatrix}$ where $R = (r_{ij})$ is defined by $r_{i,n-i+1} = 1$, $r_{ij} = 0$ otherwise. Then $Q^T = -Q$, $QQ^T = I$. Then if $M = C$ and $N = CQ$,

$$MN^* = CQ^*C^* = CQ^TC^* = -CQC^* = -NM^*.$$

So M and N are (complex) special Hadamard matrices. ***

## 1.7 CIRCULANT COMPLEX MATRICES.

For real Hadamard matrices $(v,k,\lambda)$- group difference sets are used extensively to construct Hadamard matrices. We shall now give some complex analogues of some of these matrices. This area could merit further study.

Let N of order n be the core of a symmetric conference matrix. Then

$$NN^T = nI-J, \quad NJ = 0, \quad N^T = N.$$

Choose

$$X = \tfrac{1}{2}(J-I+N)$$
$$Y = \tfrac{1}{2}(J-I-N). \tag{1.391}$$

Now let W of order h be the core of a skew-Hadamard matrix. Then

$$WW^T = hI-J, \quad WJ = 0, \quad W^T = -W.$$

Choose

$$Z = \tfrac{1}{2}(J-I+W); \tag{1.392}$$

then

$$Z^TZ = ZZ^T = [(h+1)I+(h-3)J]/4, \quad Z+Z^T+I = J. \tag{1.393}$$

In the table we give some matrices with elements i,-i,1,-1 with interesting properties and how they may be formed.

| Ref. No. | Equation | Construction | Order |
|---|---|---|---|
| 1 | $SS^* = [(h+1)I+(h-1)J]/2$ | $S = I+Z+iZ^T$ | $\equiv 3 \pmod 4$ |
| 2 | $SS^* = (n+1)I-J$ | $S = iI+N$ | $\equiv 1 \pmod 4$ |
| 3 | $SS^* = 2I+(n-2)J$ | $S = J-I+iI$ | any integer |
| 4 | $SS^*+RR^* = (n+3)I+(n-3)J$ | $R = I+X+iY,\ S = -I+X+iY$ | $\equiv 1 \pmod 4$ |
| 5 | $SS^*+RR^* = (n+5)I+(n-5)J$ | $R = -I+X+iY,\ S = I+iX-Y$ | $\equiv 1 \pmod 4$ |
| 6 | $SS^*+RR^* = 2(h+1)I-2J,\ RS^* = SR^*$ | $R = iI+W,\ S = iI-W$ | $\equiv 3 \pmod 4$ |
| 7 | $SS^*+RR^* = (h+1)I+(h-1)J$ | $R = -I+Z+iZ^T,\ S = -I+iZ+Z^T$ | $\equiv 3 \pmod 4$ |
| 8 | $SS^*+RR^* = (h+5)I+(h-5)J$ | $R = I+Z+iZ^T,\ S = I+iZ+Z^T$ | $\equiv 3 \pmod 4$ |
| 9 | $SS^*+RR^* = (h+5)I+(h-5)J$ | $R = -I+Z-iZ^T,\ S = -I+Z+iZ^T$ | $\equiv 3 \pmod 4$ |

The existence of other matrices S and R with complex elements satisfying $SS^*+RR^* = aI+bJ$, $SR^* = RS^*$ would be valuable. The following lemma gives some idea of possible methods of construction.

LEMMA 1.40. *If $p = 4m+1$ is a prime power then there is matrix C of order $4m+2$ with zero diagonal and other elements $1,-1,i,-i$ which satisfies $CC^* = pI_{p+1}$.*

PROOF. Let x be a primitive root of GF(p) and generate the cyclic group G of order $p-1 = 4m$. Define

$$C_i = \{x^{4j+i} : 0 \leq j \leq m-1\} \qquad i = 0,1,2,3,$$

and write $H_i$ for the type 1 incidence matrix of $C_i$.

Now $C_0, C_1, C_2$ and $C_3$ are 4-{4m+1;m;m-1} supplementary difference sets since the differences from $C_i$ may be written as

$$\sum_{s=0}^{3} a_s C_{s+i} \qquad i = 0,1,2,3,$$

where the $a_s$ are non-negative integers such that

$$\sum_{s=0}^{3} a_s = m-1.$$

Then by lemma 1.18

$$\sum_{s=0}^{3} H_s H_s^T = (3m+1)I+(m-1)J.$$

If we choose $a_i = x^{i+1}$ then the matrix of (1.181) is $Q = H_0-H_1+H_2-H_3$ and

$$\begin{aligned} QQ^T &= (H_0-H_1+H_2-H_3)(H_0-H_1+H_2-H_3)^T = (4m+1)I-J \\ &= \sum_{s=0}^{3} H_s H_s^T \\ &\quad + H_0(-J+I+2H_2+H_0)^T - H_1(J-I-2H_3-H_1)^T + H_2(-J+I+2H_0+H_2)^T \\ &\quad - H_3(J-I-2H_1-H_3)^T \\ &= 2\sum_{s=0}^{3} H_s H_s^T - (-J+I)^2 + 2\sum_{s=0}^{3} H_s H_{s+2}^T \\ &= (6m+1)I+(-2m-1)J + 2\sum_{s=0}^{3} H_s H_{s+2}^T. \end{aligned}$$

So

$$\sum_{s=0}^{3} H_s H_{s+2}^T = -mI+mJ.$$

Now

$$\begin{aligned} &-(H_0-H_2)(H_1-H_3)^T+(H_1-H_3)(H_0-H_2)^T \\ &= \sum_{s=0}^{3} H_s H_{s+3}^T - \sum_{s=0}^{3} H_s H_{s+1}^T \\ &= \begin{cases} \sum_{s=0}^{3} H_s H_{s+3} - \sum_{s=0}^{3} H_s H_{s+1} & \text{if } p \equiv 1 \pmod 8\text{, since } H_s^T = H_s, \\ \sum_{s=0}^{3} H_s H_{s+1} - \sum_{s=0}^{3} H_s H_{s+3} & \text{if } p \equiv 5 \pmod 8\text{, since } H_s^T = H_{s+2}. \end{cases} \\ &= \pm\left(\sum_{s=0}^{3} H_s H_{s+3} - \sum_{s=0}^{3} H_{s+1} H_s\right) \quad \text{by corollary 1.15,} \end{aligned}$$

$$= \pm\left\{\sum_{s=0}^{3} H_s H_{s+3} - \sum_{t=0}^{3} H_t H_{t+3}\right\}$$

$$= 0. \qquad (1.401)$$

Let

$$B = (H_0-H_2)+i(H_1-H_3) \qquad \text{where } i = \sqrt{-1}$$

then

$$BB^* = (H_0-H_2)(H_0-H_2)^T + (H_1-H_3)(H_1-H_3)^T$$

$$-i(H_0-H_2)(H_1-H_3)^T+i(H_1-H_3)(H_0-H_2)^T$$

$$= (H_0-H_2)(H_0-H_2)^T+(H_1-H_3)(H_1-H_3)^T \quad \text{by } (1.401)$$

$$= \sum_{s=0}^{3} H_s H_s^T - \sum_{s=0}^{3} H_s H_{s+2}^T$$

$$= (3m+1)I+(m-1)J+mI-mJ$$

$$= (4m+1)I-J,$$

and

$$C = \begin{bmatrix} 0 & 1 \dots 1 \\ \pm 1 & \\ \cdot & \\ \cdot & B \\ \cdot & \\ \pm 1 & \end{bmatrix},$$

which satisfies

$$CC^* = pI_{p+1}$$

is the required matrix with zero diagonal. If $p \equiv 1 \pmod 8$ B is symmetric and if $p \equiv 5 \pmod 8$ B is skew-symmetric (*not* skew-Hadamard). ***

## CHAPTER II. AMICABLE HADAMARD MATRICES

Amicable Hadamard matrices are useful in constructing skew-Hadamard matrices: so we now consider their incidence.

We recall definition 1.33 which defines these matrices:

Two matrices $M = I+U$ and $N$ will be called (complex) *amicable Hadamard matrices* if $M$ is a (complex) skew-Hadamard matrix and $N$ a (complex) Hadamard matrix satisfying

$$N^T = N, \quad MN^T = NM^T \quad \text{if real,}$$
$$N^* = N, \quad MN^* = NM^* \quad \text{if complex.}$$

We will only use constructions with real matrices to construct (real) amicable Hadamard matrices but it is obvious that if complex matrices are used then complex amicable Hadamard matrices can be obtained.

### 2.1 CONSTRUCTIONS.

The results of this section are due to Jennifer Wallis [151, 155].

LEMMA 2.1. *If $M = W+I$ and $N$ are amicable Hadamard matrices then*

$$WN^T = NW^T.$$

PROOF. Since $MN^T = NM^T$ we have

$$MN^T = (W+I)N^T = WN^T+N^T = WN^T+N = NM^T = N(W^T+I) = NW^T+N$$

and so

$$WN^T = NW^T. \qquad ***$$

LEMMA 2.2. *If $m = 2^t\Pi(p_i^{r_i}+1)$ where $t$ is a non-negative integer and $p_i^{r_i}$ (prime power) $\equiv 3 \pmod 4$ then there are amicable Hadamard matrices of order $m$.*

PROOF. (i) $M_2 = \begin{bmatrix} 1 & 1 \\ -1 & 1 \end{bmatrix}$ and $N_2 = \begin{bmatrix} 1 & 1 \\ 1 & -1 \end{bmatrix}$ are two suitable matrices of order 2.

(ii) Let $e = (1,\ldots,1)$ be the vector of $q$ 1's. Further let $A$ be the matrix of order $q$ defined in (1.181) for $h-1 = q = p^r \equiv 3 \pmod 4$. Then the matrix

$$B = \begin{bmatrix} 0 & e \\ -e^T & A \end{bmatrix}$$

has the properties

$$B^T = -B, \quad BB^T = (h-1)I_h.$$

Here

$$M = I_h+B$$

is a (1,-1)-matrix satisfying $MM^T = hI_h$ and so is a skew-Hadamard matrix of order h.

Let U be the matrix of order $h-1 = p^r = q$, defined by

$$U = (u_{ij}), \qquad i,j = 0,\dots,q-1, \quad u_{00} = 1,$$

$$u_{i,q-i} = 1, \qquad i = 1,\dots,q-1, \quad u_{ij} = 0, \quad \text{otherwise.}$$

Now define N by the equation

$$N = \begin{bmatrix} 1 & 0 \\ 0 & -U \end{bmatrix} M = \begin{bmatrix} 1 & 0 \\ 0 & -U \end{bmatrix} + \begin{bmatrix} 1 & 0 \\ 0 & -U \end{bmatrix}\begin{bmatrix} 0 & e \\ -e^T & A \end{bmatrix}$$

$$= \begin{bmatrix} 1 & 0 \\ 0 & -U \end{bmatrix} + \begin{bmatrix} 0 & e \\ e^T & -UA \end{bmatrix}.$$

Since $A = (q_{ij})$, $i,j = 0,\dots,q-1$ and $q_{ij} = \chi(a_j - a_i)$, with $UA = (c_{ij})$, then $c_{0j} = \chi(a_j - 0)$ and

$$c_{ij} = \chi(a_j - a_{q-i}) = \chi(-a_i - a_j), \qquad i = 1,\dots,q-1,$$

whence in all cases $c_{ij} = \chi(a_j + a_i)$ and so UA is symmetric, whence also N is symmetric: $N^T = N$.

In addition, noting $U^T = U$ and $U^2 = 1$, we find

$$NN^T = \begin{bmatrix} 1 & 0 \\ 0 & -U \end{bmatrix} MM^T \begin{bmatrix} 1 & 0 \\ 0 & -U \end{bmatrix} = hI_h.$$

So N is a symmetric Hadamard matrix. Now

$$MN^T = MM^T \begin{bmatrix} 1 & 0 \\ 0 & -U \end{bmatrix} = hI_h \begin{bmatrix} 1 & 0 \\ 0 & -U \end{bmatrix} = NM^T.$$

So M and N are two suitable matrices of order $p^r+1 \equiv 0 \pmod 4$, $p^r$ a prime power.

(iii) Let $M_m = W_m + I_m$ and $N_m$ be amicable Hadamard matrices of order m and $M_n = W_n + I_n$ and $N_n$ be amicable Hadamard matrices of order n. Then

$$M_{mn} = I_m \times M_n + W_m \times N_n$$

is a skew-Hadamard matrix of order mn and

$$N_{mn} = N_m \times N_n$$

is a symmetric Hadamard matrix of order mn.

Now

$$M_{mn}N_{mn}^T = (I_m \times M_n + W_m \times N_n)(N_m^T \times N_n^T)$$

$$= N_m^T \times M_n N_n^T + W_m N_m^T \times N_n N_n^T$$

$$= N_m \times N_n M_n^T + N_m W_m^T \times N_n N_n^T \qquad \text{using lemma 2.1}$$

$$= (N_m \times N_n)(I_m \times M_n^T + W_m^T \times N_n^T)$$

$$= N_{mn} M_{mn}^T.$$

So $M_{mn}$ and $N_{mn}$ are amicable Hadamard matrices of order mn.

(iv) Combining the results of (i), (ii) and (iii) we have the lemma for $m > 1$. But the case $m = 1$ is trivial. ***

DEFINITION 2.3. Let G be an additive abelian group of order 2m+1. Then two subsets, M and N of G, which satisfy

(i) M and N are m-sets,

(ii) $a \in M \Rightarrow -a \notin M$,

(iii) for each $d \in G$, $d \neq 0$, the equations

$$d = a_1 - a_2, \quad d = b_1 - b_2$$

have together m-1 distinct solution vectors for

$$a_1, a_2 \in M, \quad b_1, b_2 \in M,$$

will be called *Szekeres difference sets*. Alternatively, 2-{2m+1;m;m-1} supplementary difference sets, M and $N \subset G$, are called *Szekeres difference sets* if $a \in M \Rightarrow -a \notin M$.

THEOREM 2.4. *Suppose there exist (1,-1)-matrices A,B,C,D of order y satisfying*

$$C = I+U, \quad U^T = -U, \quad A^T = A, \quad B^T = B, \quad D^T = D,$$

$$AA^T+BB^T = CC^T+DD^T = 2(y+1)I-2J,$$

*and with* $e = [1,\ldots,1]$ *a* $1\times y$ *matrix*

$$eA^T = eB^T = eC^T = eD^T = e, \quad AB^T = BA^T \quad \text{and} \quad CD^T = DC^T.$$

*Then if*

$$X = \begin{bmatrix} 1 & 1 & e & e \\ 1 & -1 & -e & e \\ e^T & -e^T & A & -B \\ e^T & e^T & -B & -A \end{bmatrix}, \qquad Y = \begin{bmatrix} 1 & 1 & e & e \\ -1 & 1 & e & -e \\ -e^T & -e^T & C & D \\ -e^T & e^T & -D & C \end{bmatrix},$$

*X is a symmetric Hadamard matrix and Y is a skew-Hadamard matrix both of order 2(y+1). Further if*

$$AC^T-BD^T \quad \text{and} \quad BC^T+AD^T$$

*are symmetric, X and Y are amicable Hadamard matrices of order 2(y+1).*

COROLLARY 2.5. *Let G be an additive abelian group. Suppose there exist Szekeres difference sets, M and N in G, where* $a \in M \Rightarrow -a \notin M$. *Further suppose there exist 2-{2m+1;m;m-1} supplementary difference sets P and S in G such that* $x \in X \Rightarrow -x \in X$ *for* $X \in \{N,P,S\}$. *Then there exist*

*(i) a symmetric Hadamard matrix,*

*(ii) a skew-Hadamard matrix,*

*(iii) amicable Hadamard matrices,*

*of order 4(m+1).*

PROOF. Form the type 1 (1,-1)-incidence matrix C of G. Form the type 2 (1,-1)-incidence matrices, D,A,B of N,P,S respectively. Now use lemmas 1.14 and 1.16, corollary 1.15 and the theorem. ***

In these theorems circulant and backcirculant can be used to replace type 1 and type 2 incidence matrices respectively when the orders are prime.

It now remains to show that sets satisfying the conditions of corollary 2.5 exist for orders $y \equiv 1 \pmod 4$. Let $y = 4t+1$ be a prime power and choose $Q = \{x^{2b}: b = 1,2,\ldots,2t\}$ and $R = xQ$, where $x$ is a primitive element of $GF(y)$. Then by lemma 1.4, Q and R are 2-{4t+1;2t;2t-1} supplementary difference sets. Further $y \in Q \Rightarrow -y \in Q$, and $y \in R \Rightarrow -y \in R$ since $-1 = x^{2t}$. So Q and R satisfying the conditions of the corollary exist. To find M and N we use two results of Szekeres [129].

THEOREM 2.6. (Szekeres) *If $q = 4m+3$ is a prime power and G is the cyclic group of order $2m+1$, then there exist Szekeres difference sets M and N in G.*

PROOF. Let $x$ be a primitive root of $GF(q)$, $Q = \{x^{2b}: b = 1,\ldots,2m+1\}$ the set of quadratic residues in $GF(q)$. Define M and N by the rules

$$a \in M \quad \text{iff} \quad x^{2a}-1 \in Q, \tag{2.60}$$

$$b \in N \quad \text{iff} \quad x^{2b}+1 \in Q. \tag{2.61}$$

Since

$$-1 = x^{2m+1} \notin Q,$$

$$x^{2a}-1 \in Q \Rightarrow x^{-2a}-1 = -x^{-2a}(x^{2a}-1) \notin Q$$

so that $a \in M \Rightarrow -a \notin M$, and conditions (i) and (ii) of definition 2.5 are satisfied. Also, writing N' for the complement of N,

$$b' \in N' \quad \text{if} \quad -(x^{2b'}+1) \in Q. \tag{2.62}$$

Suppose that

$$d = \alpha - a \neq 0, \quad a,\alpha \in A \tag{2.63}$$

where

$$x^{2a} = 1+x^{2(i-d)}, \tag{2.64}$$

$$x^{2\alpha} = 1+x^{2j} \tag{2.65}$$

by (2.60) for suitable $i,j \in G$. Then

$$x^{2\alpha} = x^{2(a+d)} = x^{2d} + x^{2i}$$

by (2.63) and (2.64), hence by (2.65)

$$x^{2d}-1 = x^{2j}-x^{2i} \tag{2.66}$$

where $x^{2j}+1 \in Q$ by (2.65).

Similarly, if

$$d = b'-\beta' \neq 0, \quad b',\beta' \in N' \tag{2.67}$$

where by (2.67)

$$-x^{2\beta'} = 1+x^{2(i-d)} \tag{2.68}$$

$$-x^{2b'} = 1+x^{2j} \tag{2.69}$$

for some $i,j \in G$, we get

$$-x^{2b'} = -x^{2(d+\beta')} = x^{2d}+x^{2i}$$

hence again

$$x^{2d}-1 = x^{2j}-x^{2i}$$

with $-(x^{2j}+1) \in Q$ by (2.69).

Conversely to every solution $i,j \in G$ of equation (2.66) we can determine uniquely $\alpha \in M$ or $b \in N'$ from (2.65) or (2.69) depending on whether $1+x^{2j} = x^{2d}+x^{2i}$ is in Q or not, hence a or $\beta$ from (2.63), (2.67) so that also (2.64) or (2.68) be satisfied, implying $a \in M$, $\beta \in N'$. Thus the total number of solutions of (2.63) and (2.67) is equal to the number of solutions of (2.66) which is m by lemma 1.7. ***

LEMMA 2.7. *There exist amicable Hadamard matrices of order $2(t+1)$ whenever $t \equiv 1 \pmod 4$ and $2t+1$ are prime powers.*

PROOF. With $q = 4m+3 = 2t+1$ we form Szekeres difference sets M and N of order $2m+1 = t+1$. Using the notation of theorem 2.6

$$b \in N \Rightarrow x^{2b}+1 \in Q \Rightarrow x^{-2b}+1 = x^{-2b}(1+x^{2b}) \in Q \Rightarrow -b \in N$$

and so M and N are as required by corollary 2.5.

Choose P = Q and S = xQ then as observed above they too satisfy the conditions of the theorem and we have the result. ***

LEMMA 2.8. *Let $q = 4t+1 = 2m+1$ be a prime $\equiv 5 \pmod 8$. Let $x$ be a primitive root of $GF(q)$. Denote by $H_i = \{x^{4j+i} : j = 0,1,\ldots,t-1\}$, $i = 0,1,2,3$ the subgroup of index 4 in the group of powers of $x$ and its cosets $H_1, H_2$ and $H_3$. Choose*

$$M = H_0 \cup H_1 \quad \text{and} \quad N = H_0 \cup H_2.$$

*Then for $q = 5, 13, 29$ and $53$, M and N satisfy (i), (ii), (iii) of definition 2.3 and are Szekeres difference sets, further $y \in N \Rightarrow -y \in N$.*

PROOF. (i) and (ii) follow since $-1 \in H$. Now from the following table

| prime | differences from M | differences from N | total |
|---|---|---|---|
| 5 | $H_0$ & $H_2$ | $H_1$ & $H_3$ | $-I+J$ |
| 13 | $3H_0$ & $2H_1$ & $3H_2$ & $2H_3$ | $2H_0$ & $3H_1$ & $2H_2$ & $3H_3$ | $-5I+5J$ |
| 29 | $7H_0$ & $6H_1$ & $7H_2$ & $6H_3$ | $6H_0$ & $7H_1$ & $6H_2$ & $7H_3$ | $-13I+13J$ |
| 53 | $13H_0$ & $12H_1$ & $13H_2$ & $12H_3$ | $12H_0$ & $13H_1$ & $12H_2$ & $13H_3$ | $-25I+25J$ |

we see that in each case the differences give m-1 distinct solution vectors of the equations

$$d = a_1 - a_2, \quad d = b_1 - b_2$$

for $a_1, a_2 \in M$, $b_1, b_2 \in N$ as required.

Since $-1 \in H_2$ we have $y \in N \Rightarrow -y \in N$. ***

LEMMA 2.9. *There exist amicable Hadamard matrices of order 2(q+1) for q = 5,13,29 and 53.*

It seems most probable that this is a special case of a more general result.

Thus we have the following list of orders for which amicable Hadamard matrices exist:

| | | |
|---|---|---|
| AI | 2 | |
| AII | $p^r+1$ | $p^r$ (prime power) $\equiv 3 \pmod 4$; |
| AIII | $2(q+1)$ | $2q+1$ is a prime power, $q$ (prime power) $\equiv 1 \pmod 4$; |
| AIV | S | where S is a product of the above orders. |

It would be interesting to find more of these matrices, both real and complex. As yet no complex amicable Hadamard matrices are known for orders other than those for which real ones exist.

## CHAPTER III. ORTHOGONAL MATRICES WITH ZERO DIAGONAL - SKEW-HADAMARD AND SYMMETRIC CONFERENCE MATRICES

We recall that if I+S is a skew-Hadamard matrix or symmetric conference matrix of order s then

$$SS^T = (s-1)I_s \tag{3.01}$$

and

$$S^T = -S \quad \text{or} \quad S^T = S \tag{3.02}$$

respectively. Thus S is the orthogonal matrix with zero diagonal of the title.

Further if W of order s-1 is the core of S

$$WW^T = (s-1)I-J, \quad WJ = 0. \tag{3.03}$$

These matrices are most valuable in constructing both Hadamard and complex Hadamard matrices. Recently Goethals, Seidel, Szekeres, Turyn, Wallis and Whiteman have studied them, though elegant results were noticed previously by Belevitch, Ehlich, Paley and Williamson.

### 3.1 ORTHOGONAL MATRICES WITH ZERO DIAGONAL.

We first consider some results about these matrices:

THEOREM 3.1. (Delsarte-Goethals-Seidel, [36]) *Every orthogonal matrix S of order s with zero diagonal and ±1 elsewhere, satisfying* $SS^T = (s-1)I$, *is H-equivalent (see chapter 10) to a matrix B satisfying*

$$BB^T = B^TB = (s-1)I,$$

$$BJ = B^TJ = 0,$$

$$B^T = (-1)^{\frac{1}{2}(s-2)}B.$$

*This normal matrix B is unique up to multiplication by -I.*

PROOF. The core of S is W which satisfies

$$WW^T = W^TW = (s-1)I-J,$$

$$WJ = W^TJ = 0.$$

Write

$$W^TW-(s-1)I = -xx^T,$$

with $x^T = (x_1,x_2,\ldots,x_{s-1})$. Now since W is zero on the diagonal and +1 or -1 elsewhere, all diagonal elements of $W^TW$ are s-2, and hence $x_i^2 = 1$ for i = 1,2,...,s-1.

Therefore, we may write $x = De$, where $D$ is a diagonal matrix with entries +1 or -1, and $e$ is the $s-1\times 1$ matrix of 1's. Thus $B = WD$ has zero diagonal and other entries +1 or -1 and satisfies

$$BB^T = B^TB = (s-1)I-J.$$

So

$$BB^TBe = B(B^TB)e = (s-1)Be-BJe = (s-1)Be-(s-1)Be = 0.$$

$$BB^TBe = B(B^TB)e = (s-1)Be-BJe = (s-1)Be-B(s-1)e = 0,$$

$$= (BB^T)Be = (s-1)Be-JBe.$$

Thus $(s-1)Be = JBe = JWDe = 0$. Similarly considering $B^TBB^T$ we see $B^Te = 0$.

Now it remains to show $B^T = (-1)^{\frac{1}{2}(s-2)}B$. Let $(b_{i,k})$ and $(b_{j,k})$, $k = 1,2,\ldots,s-1$ be the ith and jth rows of $B$. Write $n_0,n_1,n_2,n_3$, respectively for the number of times the ordered pair $(b_{i,k},b_{j,k})$ equals (1,1), (1,-1), (-1,1), (-1,-1) respectively. Then, from the equations for $B$, we have

$$\begin{aligned} n_0+n_1+n_2+n_3 &= s-3, && \text{from the order of } B,\\ n_0+n_1-n_2-n_3 &= -b_{i,j}, && \text{from } Be = 0,\\ n_0-n_1+n_2-n_3 &= -b_{j,i}, && \text{from } Be = 0,\\ n_0-n_1-n_2+n_3 &= -1, && \text{from } BB^T = (s-1)I-J. \end{aligned}$$

Thus

$$\begin{aligned} 4n_0 &= s-4-b_{i,j}-b_{j,i}\\ 4n_1 &= s-2-b_{i,j}+b_{j,i}\\ 4n_2 &= s-2+b_{i,j}-b_{j,i}\\ 4n_3 &= s-4+b_{i,j}+b_{j,i}. \end{aligned}$$

Now $b_{i,j} = \pm b_{j,i}$. So the above equations become

$$\left.\begin{aligned} 4n_0 &= s-4\mp 2\\ 4n_1 &= s-2\\ 4n_2 &= s-2\\ 4n_3 &= s-4\pm 2 \end{aligned}\right\} \text{for } b_{i,j} = b_{j,i} = \pm 1 \qquad \left.\begin{aligned} 4n_0 &= s-4\\ 4n_1 &= s-2\mp 2\\ 4n_2 &= s-2\pm 2\\ 4n_3 &= s-4 \end{aligned}\right\} \text{for } b_{i,j} = -b_{j,i} = \pm 1.$$

Thus $B^T = B$ for $s-2 \equiv 0 \pmod 4$ and $B^T = -B$ for $s-4 \equiv 0 \pmod 4$.
So $B^T = (-1)^{\frac{1}{2}(s-2)}B$. ***

DEFINITION 3.2. Let $A = [a_{ij}]$ and $C = [c_{ij}]$ be two matrices of order $n$. The *Hadamard product* $A*C$ of $A$ and $C$ is given by

$$A*C = [a_{ij}c_{ij}].$$

LEMMA 3.3. *When $A,B,C,D$ are matrices of the same order*

$$(A\times B)*(C\times D) = (A*C)\times(B*C).$$

PROOF.
$$\begin{aligned}(A\times B)*(C\times D) &= [a_{ij}B]*[c_{ij}D]\\ &= [a_{ij}c_{ij}B*D]\\ &= (A*C)\times(B*D).\end{aligned}$$
***

THEOREM 3.4. *Suppose $I, J$ and $W$ are of order $h$, where $W$ is a matrix with zero diagonal and $+1$ or $-1$ elsewhere. Suppose $A$ is a matrix of order $q = h^p$ which is of the form*

$$A = B_1 + B_2 + \ldots + B_k$$

*where each $B_i$ is a Kronecker product of $p$ terms, $I, J$ or $W$ in some order, such that*

- (a) *each $B_i$ contains at least one term $W$,*
- (b) *for any two summands $B_i$ and $B_j$ there is a position $r$ such that one of the summands has $I$ in the $r$th position and the other has $W$ in that position,*
- (c) *for any two summands $B_i$ and $B_j$ there is a position $s$ such that one of the summands has $J$ in the $s$th position and the other has $W$ in that position,*

*and suppose $A$ satisfies*

$$AA^T = qI_q - J_q.$$

*Then $A$ has zero diagonal and $+1$ or $-1$ in every other place and*

$$AA^T = \sum_{i=1}^{k} B_iB_i^T.$$

PROOF. Clearly

$$W*I = I*W = 0;$$

so, by part (b) of the hypothesis and lemma 3.3

$$B_i*B_j = B_j*B_i = 0$$

whenever $i \neq j$.

Hence no two $B_i$ have non-zero elements in the same position, so each non-zero element of $A$ comes from exactly one of the $B_i$; since each $B_i$ is a $(0,1,-1)$ matrix it follows that $A$ is a $(0,1,-1)$ matrix. But

$$AA^T = qI_q - J_q,$$

so if $A = (a_{ij})$

$$\sum_j a_{ij}^2 = q-1$$

for any $i$: therefore at most one element in any row of $A$ is zero. $W$ appears in each $B_i$, and $W$ has zero diagonal, so each $B_i$ - and consequently $A$ - has zero diagonal.

$$AA^T = \sum_{i=1}^{k} B_iB_i^T + \sum_{j=1}^{k}\sum_{\substack{\ell=1\\ \ell\neq j}}^{k} B_\ell B_j^T$$

but $WJ = JW = 0$ so by part (c) $B_\ell B_j^T = 0$ for $\ell \neq j$. Hence we have the result.

***

An example of terms which satisfy the conditions of theorem 3.4 are

$$W\times W\times W + J\times W\times I + I\times J\times W + W\times I\times J \tag{3.41}$$

for 3,

$$W\times W\times W\times W\times W + \Sigma' I\times J\times[W\times W+I\times J]\times W \tag{3.42}$$

for 5 and

$$W\times W\times W\times W\times W\times W\times W + \Sigma' I\times J\times[W\times W+I\times J]\times[W\times W+I\times J]\times W \tag{3.43}$$

for 7.

It was exactly this pattern as we shall soon see that Turyn generalized. But first a few results, originally due to Belevitch:

LEMMA 3.5. *Let $W$ of order $s$ be a core; then*

$$B = W\times W + I\times J - J\times I$$

*is a core of order $s^2$.*

PROOF. We first check the matrix $I+B$ is a $(1,-1)$ matrix.

$$I+B = \begin{bmatrix} J-I & \pm W-I & \pm W-I & \cdots \\ \pm W+I & J-I & \pm W-I & \cdots \\ \pm W-I & \pm W-I & J-I & \\ \vdots & \vdots & & \end{bmatrix} + I,$$

which is clearly a $(1,-1)$ matrix. Now $B^T = B$ and

$$\begin{aligned} BB^T &= (sI-J)\times(sI-J) + I\times sJ + sJ\times I - 2J\times J \\ &= s^2 I\times I - J\times J. \end{aligned}$$

***

COROLLARY 3.6. *If there is*

*(i) a symmetric conference matrix, or*

*(ii) a skew-Hadamard matrix,*

*of order $s+1$, then there is a symmetric conference matrix of order $s^2+1$.*

PROOF. First construct the core of order $s$. Use lemma 3.5 to obtain the core of order $s^2$. Now border the core of order $s^2$ to get the result. ***

The following theorem is a strengthening of a theorem of Ehlich by Goethals and Seidel [44].

THEOREM 3.7. *If there is a core of order $s$ and one of order $s+2$ then there is an Hadamard matrix of order $(s+1)^2$.*

PROOF. Let Q be the core of order s and G be that of order s+2, then one is symmetric and one is skew-symmetric. Consider

$$K = Q\times G - I\times J + J\times I + I\times I$$

$K(J\times J) = +J\times J$ and

$$\begin{aligned} KK^T &= QQ^T\times GG^T - Q\times GJ^T + QJ^T\times G + Q\times G \\ &\quad - Q^T\times JG^T + JQ^T\times G^T + Q^T\times G^T \\ &\quad - 2J\times J - 2I\times J + 2J\times I + I\times I \\ &\quad + I\times(s+2)J + sJ\times I \\ &= (sI-J)\times\big((s+2)I-J\big) - 2(J\times J) + sI\times J + (s+2)J\times I + I\times I \\ &= [s(s+2)+1]I\times I - J\times J, \end{aligned}$$

where the first matrix of every pair $A\times B$ is of order s and the second is of order s+2.

The matrix

$$H = \begin{bmatrix} -1 & e \\ e^T & K \end{bmatrix}$$

consists of ±1's and satisfies $HH^T = (s+1)^2I$, so H is the required Hadamard matrix.

***

We will now proceed to show that the orders of a core may be raised to any odd power, u, and a core of order $s^u$ exists. Where if the core of order s is $W_1$

$$W_1W_1^T = sI-J, \quad W_1^T = eW_1, \quad e = \pm 1, \quad W_1J = 0$$

and the core of order $s^u$ is $W_u$

$$W_uW_u^T = s^uI-J, \quad W_u^T = eW_u, \quad e = \pm 1, \quad W_uJ = 0.$$

Goldberg [48] showed that this result was true for e = -1, n = 3. Goethals and Seidel [44] pointed out an early result of Belevitch [14], the theorem for e = ±1, n = 2. J. Wallis [154, 156] showed that both these results hold for e = ±1 and gave a proof for n = 5, n = 7, e = ±1. Turyn [142] then generalized the construction of J. Wallis. First a few preliminary results.

DEFINITION 3.8. A *G-string* is a sequence of p matrices (p an odd integer), separated by Kronecker products, each of which is I,J or W such that each I is followed by a J and each J preceded by an I, where the last matrix is considered to be followed by the first one for the purpose of deciding which sequences are G-strings.

(3.41), (3.42) and (3.43) are G-strings.

LEMMA 3.9. *If $G_1$ and $G_2$ are different G-strings of p matrices then there is a position in which*

*(i) $G_1$ has W and $G_2$ has J or vice-versa;*

*(ii) $G_1$ has W and $G_2$ has I or vice-versa.*

PROOF. (i) If all the positions which have a W in either string have a W in the other, the two strings are identical, since each must be completed uniquely from the set of W's in it by adding consecutive pairs I,J in the vacant places. Thus assume that $G_1$ has a W in a position (which we may write as the first) in which $G_2$ does not have a W. If $G_2$ has a J there, we are finished, so assume

$$G_1 = W\times\ldots\text{ and}$$
$$G_2 = I\times J\times\ldots,$$

since a J must follow an I. The second and third positions of $G_1$ may be $W\times W$ (then we are through) or $I\times J$. We now have

$$G_1 = W\times I\times J\times\ldots\text{ and}$$
$$G_2 = I\times J\times\ldots,$$

so that the third position of $G_2$ must have an I, etc. Since each G-string has at least one W, p being odd, $G_2$ has a W somewhere. The smallest index for which $G_1$ or $G_2$ has a W corresponds to a J in the other.

(ii) Proceed as before interchanging I and J in the argument and working from right to left. ***

LEMMA 3.10. *If G is the sum of all the possible G-strings, $G_i$*

*(i)* $G^T = eG$

*(ii)* $GG^T = \sum_i G_iG_i^T$

*(iii) $G+I$ is a (1,-1) matrix.*

PROOF. (i) This follows because $W^T = eW$ and each $G_i$ has an odd number of W's. (ii) and (iii) follow by using lemmas 3.10 and 3.4. ***

THEOREM 3.11. (Turyn) *Let $W_1$ be a core of order s satisfying $W_1{}^T = eW_1$ then $W_p$ the sum of all the G-strings, $G_i$, of order p, an odd integer, is a core of order $s^p$ satisfying $W_p{}^T = eW_p$.*

PROOF. Write $I_p$ and $J_p$ for the matrices of order $s^p$. We have $W_1W_1{}^T = sI-J$, $JJ^T = sJ$, $II^T = I$. It is clear that $W_pW_p{}^T$ can be expressed as a linear combination of the various p-fold Kronecker products of I and J. We know that $W_pW_p{}^T = \sum_i G_iG_i{}^T$ and that $G_1 = W\times W\times\ldots\ldots\times W$ contributes $s^pI_p-J_p$ (plus other terms) to $W_pW_p{}^T$, and that $I_p$ and $J_p$ cannot arise in any other product $G_iG_i{}^T$.

We now ask how any other p-fold product P of I and J, one containing at least one I and at least one J, can arise from $G_iG_i{}^T$. If P contains J(I) in position j it cannot appear in a product $G_iG_i^T$ if $G_i$ has an I(J) in position j. This is the only type of restriction there is. Thus, assume P has exactly b blocks of consecutive J's, $b > 0$, and that it has a total of c J's which are preceded by J. Then P contains b I's followed by J and p-c-2b I's followed by I; if P occurs in $GG^T$ we have

| | P | G | $GG^T$ |
|---|---|---|---|
| b: | I×J | W×W | (sI-J)×(sI-J) |
| | | or I×J | I×sJ |
| c: | (J)J | W | sI-J |
| p-c-2b: | I(I) | W | sI-J. |

There are $\binom{b}{j}$ of all the possible G-strings which have I×J pairs coinciding with j of the pairs I×J in P, so that the coefficient of P in $\Sigma G_iG_i^T$ is

$$\Big\{ \sum_j \underbrace{\tbinom{b}{j}(-s)^{b-j}}_{\text{from the } W\times W} \underbrace{s^j}_{\text{from the } I\times J} \Big\} \underbrace{s^{p-c-2b}}_{\text{from the } I(I)} \underbrace{(-1)^c}_{\text{from the } (J)J}$$

$$= \{-s+s\}^b s^{p-c-2b}(-1)^c$$

$$= 0.$$

Hence $W_pW_p^T = \sum_i G_iG_i^T = s^pI_p - J_p$. By lemma 3.4 $W_pJ_p = 0$, and since each $G_i$ has an odd number of $W_1$, $W_p^T = eW_p$.

Thus $W_p$ is a core of order $s_p$. ***

Then by bordering $W_p$ and using lemma 3.5 we have:

COROLLARY 3.12. (Turyn) *If h is the order of a skew-Hadamard matrix and n is the order of a symmetric conference matrix, then*

*(i)* $(h-1)^u+1$, *u any odd integer, is the order of a skew-Hadamard matrix;*

*(ii)* $(n-1)^v+1$, *v any integer, is the order of a symmetric conference matrix;*

*(iii)* $(h-1)^t+1$, *t any even integer, is the order of a symmetric conference matrix.*

Turyn points out that the construction is still valid if we use non-isomorphic W. Suppose W,W',W" are non-isomorphic, then the core of order $p^3$ can be obtained from

$$W\times W'\times W'' + J\times W'\times I + I\times J\times W'' + W\times I\times J,$$

which will be different from

$$W\times W\times W + J\times W\times I + J\times J\times W + W\times I\times J.$$

**3.2 CONSTRUCTIONS FOR SYMMETRIC CONFERENCE MATRICES.** There are very few constructions for symmetric conference matrices and we already know they can only exist for orders $1+a^2+b^2$ where a and b are integers.

THEOREM 3.13. (Paley) *If $p^h \equiv 1 \pmod 4$ is a prime power there is a symmetric conference matrix of order $p^h+1$.*

PROOF. Let Q be the matrix of §1.4 and e be the $1 \times p^h$ matrix of 1's, then

$$\begin{bmatrix} 0 & e \\ e^T & Q \end{bmatrix} + I$$

is the required matrix. ***

The only other known method for constructing these matrices is the results of corollary 3.12. Appendices C and I list the known classes of symmetric conference matrices.

The remainder of this chapter, from [44], gives the results of Goethals and Seidel.

**3.3 PALEY MATRICES.** Let V be a vector space of dimension 2 over $GF(p^k)$, $q = p^k$, p prime, $p \neq 2$. Let $\chi$ denote the Legendre symbol. Consider the function $\chi$ *det* where the determinant *det* denotes any alternating bilinear form on V. The q+1 one-dimensional subspaces of V, which are the q+1 projective points of the projective line PG(1,q), are represented by the vectors $x_0, x_1, \ldots, x_q$, no two of which are dependent. Then

DEFINITION 3.14. The *Paley matrix* C of the q+1 vectors $x_0, x_1, \ldots, x_q$ is defined by

$$C = [\chi \det(x_i, x_j)], \quad i,j = 0,1,\ldots,q.$$

The *line matrix* (analogous to core) S of the q vectors $y_1, y_2, \ldots, y_q$, which are on a line not through the origin is defined as

$$S = [\chi \det(x_i, x_j)], \quad i,j = 0,1,\ldots,q.$$

DEFINITION 3.15. If a square matrix B can be obtained from a square matrix A by a sequence of the operations

(1) multiply row i *and* column i by -1,

(2) interchange two rows and simultaneously interchange the corresponding columns,

the A will be said to be *Seidel-equivalent* or *S-equivalent* to B. If only (2) is used A and B will be said to be *permutation equivalent.*

THEOREM 3.16. (Goethals and Seidel) *To the projective line PG(1,q) there is attached a class of S-equivalent Paley matrices C of order q+1, symmetric if* $q+1 \equiv 2 \pmod 4$ *and skew if* $q+1 \equiv 0 \pmod 4$*, with elements* $c_{ii} = 0$, $c_{ij} = \pm 1$ *for* $i,j = 0,1,\dots,q$*, satisfying* $CC^T = qI$.

PROOF. The operations (1) and (2) on Paley matrices are effected by multiplication of any vector by a non-square element of GF(q) and by interchange of any two vectors respectively. Therefore, all Paley matrices of order q+1 are equivalent. We only need to prove the property $CC^T = qI$ for one matrix C. To that end we consider the Paley matrix of the vectors x and $y+a_ix$, where x and y are independent and $a_i$ runs through GF(q):

$$C = \chi \det(x,y) \begin{bmatrix} 0 & e \\ e^T\chi(-1) & \chi(a_i-a_j) \end{bmatrix} \qquad i,j = 1,\dots,q.$$

The desired property then follows from Jacobthal's formula (see lemma 5 of part 1)

$$\Sigma_{a \in GF(q)} \chi(a)\chi(a+x) = -1, \qquad x \in GF(q), \qquad x \neq 0. \qquad ***$$

THEOREM 3.17. (Goethals and Seidel) *All line matrices S of order q are permutation equivalent. They satisfy*

$$SS^T = qI-J, \qquad SJ = JS = 0.$$

*They are permutation equivalent to a multicirculant matrix of the form*

$$\begin{bmatrix} 0 & e & -e \\ \chi(-1)e^T & A & B \\ -\chi(-1)e^T & \chi(-1)B^T & -A \end{bmatrix}$$

*with circulant matrices A and B of order* $\frac{1}{2}(q-1)$.

PROOF. The line matrix of the vectors $y+a_1x,\dots,y+a_qx$, where $a_1,\dots,a_q$ denote the elements of GF(q), is

$$S = \chi \det(x,y)\, [\chi(a_i-a_j)], \qquad i,j = 1,\dots,q.$$

If $\chi \det(x,y) = -1$, then for some non-square z all $a_iz$ are distinct and S is permutation equivalent to $[\chi(a_i-a_j)]$. Hence all line matrices are permutation equivalent. The relations for S follow from

$$qI = CC^T = \begin{bmatrix} 0 & e \\ e^T\chi(-1) & S \end{bmatrix} \begin{bmatrix} 0 & e\chi(-1) \\ e^T & S \end{bmatrix}.$$

The multicirculant form is obvious for q = p and readily follows for $q = p^k$; cf [14]. The last standard form is obtained by taking $\chi \det(y,x) = 1$ and by arranging the vectors as follows:

$$y, y+xt^2, y+xt^4, \dots, y+xt^{q-1}, y+xt, y+xt^2, \dots, y+xt^{q-2}$$

where t denotes any primitive element of GF(q). ***

Any linear mapping u: V→V satisfies

$$\det\big(u(x),u(y)\big) = \det u.\det(x,y)$$

for all x,y ε V. We define linear mappings v and w, which will be used in the proof of the next theorem. Let z be any primitive element of $GF(q^2)$, the quadratic extension of GF(q). We choose any basis in V. With respect to this basis, v is defined by the matrix

$$(v) = \tfrac{1}{2}\begin{bmatrix} z^{q-1}+z^{1-q} & (z^{q-1}-z^{1-q})z^{\frac{1}{2}(q+1)} \\ (z^{q-1}-z^{1-q})z^{-\frac{1}{2}(q+1)} & z^{q-1}+z^{1-q} \end{bmatrix},$$

which indeed has its element in GF(q). Then det (v) = 1 and the eigenvalues of v are $z^{q-1}$ and $z^{1-q}$, both elements of $GF(q^2)$ whose $\frac{1}{2}(q+1)$th power, and no smaller, belongs to GF(q). Hence v acts on PG(1,q) as a permutation with period $\frac{1}{2}(q+1)$, without fixed points, which divides the points of PG(1,q) into two sets of transitivity each containing $\frac{1}{2}(q+1)$ points. In addition, w is defined by the matrix

$$(w) = \begin{bmatrix} 0 & z^{q+1} \\ 1 & 0 \end{bmatrix}.$$

Then $\chi$ det (w) = $-\chi(-1)$. The eigenvalues of w are $\pm z^{\frac{1}{2}(q+1)}$, elements of $GF(q^2)$ whose square is in GF(q). Hence w acts on PG(1,q) as a permutation with period 2, which maps any point of one set of transitivity, defined above by v, into the other set. Indeed, for i = 1,...,$\frac{1}{2}(q+1)$, the mapping $v^i w$ has no eigenvalue in GF(q). Finally note vw = wv.

THEOREM 3.18. (Goethals and Seidel) *The S-equivalence class of Paley matrices of order q+1 ≡ 2 (mod 4) contains a member of the form*

$$\begin{bmatrix} A & B \\ B & -A \end{bmatrix}$$

*with square symmetric circulant submatrices A and B.*

PROOF. Represent the q+1 points of PG(1,q) by the following q+1 vectors in V:

$$x,v(x),v^2(x),\dots,v^{\frac{1}{2}(q-1)}(x),w(x),vw(x),v^2w(x),\dots,v^{\frac{1}{2}(q-1)}w(x).$$

Observing that, for i,j = 0,1,...,$\frac{1}{2}(q-1)$,

$$\det\big(v^iw(x),v^jw(x)\big) = \det(w)\cdot\det\big(v^i(x),v^j(x)\big) = \det(w)\cdot\det\big(x,v^{j-i}(x)\big),$$

$$\det\big(v^i(x),v^jw(x)\big) = -\det\big(v^iw(x),v^j(x)\big) = \det\big(v^j(x),v^iw(x)\big),$$

$$\det\big(v^i(x),v^j(x)\big) = -\det\big(v^{\frac{1}{2}(q+1)+i}(x),v^j(x)\big),$$

we conclude that the Paley matrix belonging to these vectors has the desired form.

***

3.4 EQUIVALENCE OF SYMMETRIC CONFERENCE MATRICES. We recall that a necessary condition for the existence of a symmetric conference matrix of order $n \equiv 2 \pmod 4$ is that $n-1 = a^2+b^2$, a and b integers.

LEMMA 3.19. (Goethals and Seidel) *Any symmetrically partitioned symmetric conference matrix of order n satisfies*

$$C = \begin{bmatrix} A & B \\ B^T & D \end{bmatrix} = \begin{bmatrix} P(A+aI) & PB+bQ \\ QB^T+bP & Q(D-aI) \end{bmatrix}^{-1} \begin{bmatrix} aI & bI \\ bI & -aI \end{bmatrix} \begin{bmatrix} P(A+aI) & PB+bQ \\ QB^T+bP & Q(D-aI) \end{bmatrix}$$

*for any real square matrices P,Q and real numbers a,b with $n-1 = a^2+b^2$, for which the transformation matrix is regular.*

PROOF. For any square matrices C,D,R of equal order, which satisfy $C^2 = D^2 = (n-1)I$, we have $(RC+DR)C = D(RC+DR)$. This gives the result. ***

Goethals and Seidel remark that if P,Q,a,b are rational we obtain rational representations of symmetric conference matrices.

THEOREM 3.20. (Belevitch) *Any symmetric conference matrix is permutation equivalent to a matrix of the form*

$$\sqrt{n-1}\begin{bmatrix} 2(I+NN^T)^{-1}-I & -2(I+NN^T)^{-1}N \\ -2(I+N^TN)^{-1}N^T & I-2(I+N^TN)^{-1} \end{bmatrix}$$

*for some square matrix N whose elements have the form $r+s\sqrt{n-1}$, r and s rationals.*

PROOF. By suitable symmetric permutation we may write

$$C = \begin{bmatrix} A & B \\ B^T & D \end{bmatrix} = \begin{bmatrix} U & V \\ W & X \end{bmatrix} \begin{bmatrix} \sqrt{n-1}I & 0 \\ 0 & -\sqrt{n-1}I \end{bmatrix} \begin{bmatrix} U^T & W^T \\ V^T & X \end{bmatrix} \begin{bmatrix} U & X \\ W & X \end{bmatrix} \begin{bmatrix} U^T & W^T \\ V^T & X^T \end{bmatrix} = I,$$

with non-singular $2\sqrt{n-1}\,UU^T = A+I\sqrt{v-1}$.

Then also

$$2\sqrt{n-1}\,XX^T = \sqrt{n-1}\,I-D$$

is non-singular. Applying lemma 3.19 for $a = \sqrt{n-1}$, $b = 0$,

$$P = \{A+I\sqrt{n-1}\}^{-1}, \quad Q = \{D-\sqrt{n-1}\,I\}^{-1},$$

and calling $PB = -N$, we obtain $QB^T = N^T$ and

$$C = \begin{bmatrix} A & B \\ B^T & D \end{bmatrix} = \sqrt{n-1}\begin{bmatrix} I & -N \\ N^T & I \end{bmatrix}^{-1} \begin{bmatrix} I & 0 \\ 0 & -I \end{bmatrix} = \sqrt{n-1}\begin{bmatrix} -I & 0 \\ 0 & I \end{bmatrix} + 2\sqrt{n-1}\begin{bmatrix} I & N \\ N^T & -I \end{bmatrix}^{-1}$$

which is the matrix of the statement of the theorem. ***

Goethals and Seidel remark: for symmetric conference matrices of the form of the theorem, which satisfy the special property

$$NJ = N^T J = kJ \qquad (3.201)$$

for some real number k, it follows that

$$JA = AJ = -JD = -DJ = \frac{1-k^2}{1+k^2}\sqrt{n-1}\,J,$$

$$JB = BJ = \frac{2k}{1+k^2}\sqrt{n-1}\,J.$$

In both formulae the coefficient of J,

$$a = \frac{1-k^2}{1+k^2}\sqrt{n-1}, \quad b = \frac{2k}{1+k^2}\sqrt{n-1},$$

respectively, is an integer. In fact, a is even and b is odd. Since $a^2+b^2 = n-1$, an interpretation of the integers a and b occurring in theorem 3.16 is obtained for symmetric conference matrices with property (3.201). In addition, if any such matrix satisfies a = 0, then all row sums are equal. This applies to symmetric conference matrices with property (3.201) of order

$$v = p^{2k}+1, \quad p \equiv 3 \pmod 4), \quad p \text{ prime},$$

since in this case the decomposition of v-1 as a sum of two squares of integers is unique. We do not know any example of a symmetric conference matrix which is not S-equivalent to a conference matrix with property (3.201).

THEOREM 3.21. (Goethals and Seidel) *Any symmetric Paley matrix of order q+1 is S-equivalent to a matrix of the form*

$$\begin{bmatrix} A & B \\ B & -A \end{bmatrix} = \begin{bmatrix} I & -N \\ N & I \end{bmatrix}^{-1} \begin{bmatrix} aI & bI \\ bI & -aI \end{bmatrix} \begin{bmatrix} I & -N \\ N & I \end{bmatrix}$$

*with symmetric, circulant and rational N, where a and b are any rationals satisfying* $a^2+b^2 = q$ *and let* $(A+aI) \neq 0$.

PROOF. Using the matrix obtained in theorem 3.18 we apply lemma 3.19 for $P = -Q = (A+aI)^{-1}$. Putting $(A+aI)^{-1}(B-bI) = N$, we observe that N is symmetric and circulant because A and B have that property. This gives the result. ***

Seidel and Bussemaker have studied the S-inequivalence of symmetric conference matrices and found that for orders ≤ 18 there is only one equivalence class. These results will be the subject of an Eindhoven University Technical Report.

Currently four S-inequivalent symmetric conference matrices of order 26 are known. We are grateful to Dr. D.E. Taylor for indicating their construction to us, he attributes the constructions to Seidel.

To obtain the four S-inequivalent conference matrices of order 26 we proceed as follows.

(a) Take a Steiner triple system on 13 points, join two blocks whenever they have a point in common. This gives a strong graph on 26 points whose adjacency matrix is a conference matrix of order 26. This gives two S-inequivalent conference matrices as there are two isomorphism classes of Steiner triple systems on 13 points.

(b) Form the Latin square graph on 25 points using 3 mutually orthogonal Latin squares of order 25. The adjacency matrix of this graph is the core of a conference matrix of order 26. This method appears to give just two S-inequivalent conference matrices of order 26.

EXAMPLE: Goethals and Seidel [15] give two essentially different symmetric conference matrices of order 26 corresponding to

$$25 = 4^2+3^2 \quad \text{and} \quad 25 = 0^2+5^2.$$

They are both given in the form

$$\begin{bmatrix} A & B \\ B^T & -A \end{bmatrix}$$

with A and B circulant of order 13. The matrix with first row

0-+++-++-+++- , -+--++++++--+

is a Paley matrix with a = 4, b = 3. The corresponding line matrix of order 25 is permutation equivalent to the matrix S which consists of the cyclically permuted blocks

$$I-J,\ J-2I-2P,\ J-2I-2P^2,\ J-2I-2P^3,\ J-2I-2P^4.$$

Here P, of order 5, is defined by $p_{ij} = 1$ if $j-i \equiv 1 \pmod 5$, $p_{ij} = 0$ otherwise.

We call the *exceptional* conference matrix of order 26 the matrix with first row

0-+--++++--+- , --+-+++++-+++ .

It has $a = 0$, $b = 5$, $A^2 = 13I-J$, $BB^T = 12I+J$.

THEOREM 3.22. (Goethals and Seidel) *The exceptional conference matrix and the Paley matrix of order 26 are not equivalent.*

PROOF. By equivalence operations the first and second rows and columns of the exceptional conference matrix of order 26 are transformed into

0++...++...+ , +0+...+-...- .

Then four submatrices of order 12 arise. Now S-equivalence with the Paley matrix would imply, in view of theorem 3.17, that each of these submatrices can be permuted into a circulant matrix. Hence all rows of the square of any submatrix would have to consist of the same set of numbers. However, by inspection this is not the case.

***

In [36] Delsarte, Goethals and Seidel consider *negacyclic C-matrices* which are polynomials in the matrix

$$P = \begin{bmatrix} 0 & 1 & 0 & \cdot & \cdot & \cdot & 0 & 0 \\ 0 & 0 & 1 & \cdot & \cdot & \cdot & 0 & 0 \\ \cdot & \cdot & \cdot & \cdot & \cdot & \cdot & \cdot & \cdot \\ \cdot & \cdot & \cdot & \cdot & \cdot & \cdot & \cdot & \cdot \\ 0 & 0 & 0 & \cdot & \cdot & \cdot & 0 & 1 \\ -1 & 0 & 0 & \cdot & \cdot & \cdot & 0 & 0 \end{bmatrix}$$

and obtain interesting results on the S-equivalence of negacyclic C-matrices and conference matrices. We do not discuss their results because they do not help us construct Hadamard matrices, except to state one theorem without proof.

THEOREM 3.23. (Delsarte, Goethals and Seidel) *Any Paley matrix is S-equivalent to a negacyclic C-matrix.*

## CHAPTER IV. CONSTRUCTIONS FOR SKEW-HADAMARD MATRICES

Some of the most powerful methods for constructing Hadamard matrices depend on the existence of skew-Hadamard matrices. As we have seen, the only symmetric conference matrices known whose orders are not of the form prime power plus one are those derived from skew-Hadamard matrices.

The properties of these matrices were noticed as long ago as 1933 and 1944 by Paley and Williamson, but it has only been recently when the talents of Szekeres and Whiteman (among others) were directed towards their study that significant understanding of their nature was achieved.

For completeness we will restate results proved earlier. Appendices A and H give lists of the known orders and classes of skew-Hadamard matrices.

THEOREM 4.1. (Williamson) *Let t be a non-negative integer and* $k_i \equiv 0 \pmod 4$ *be a prime power plus one, then there is a skew-Hadamard matrix of order* $2^t \Pi k_i$.

PROOF. In proving in lemma 2.2 that there are amicable Hadamard matrices of orders $2^t \Pi k_i$, we proved that there are skew-Hadamard matrices of these orders. ***

THEOREM 4.2. (Turyn) *Let u be any odd integer and h the order of a skew-Hadamard matrix, then there is a skew-Hadamard matrix of order* $(h-1)^u+1$. ***

See theorem 3.11 and corollary 3.12 for the proof of theorem 4.2.

### 4.1 CONSTRUCTIONS USING SZEKERES DIFFERENCE SETS.

In theorem 2.5 we used two particular supplementary difference sets to obtain amicable Hadamard matrices. Szekeres, himself, and later Whiteman, called these sets "complementary difference sets", but we have noted that in common usage if $D = \{d_1,\dots,d_k\}$, a set of elements from G, is a $(v,k,\lambda)$-difference set and $D' = \{x: x \in G, x \notin D\}$, then D' is a $(v,v-k,v-2k+\lambda)$-difference set and D and D' are called *complementary difference sets* (see Baumert [7; p.3]). Thus we use the definition (recalled from chapter II):

DEFINITION 4.3. Let G be an additive abelian group of order 2m+1, then two subsets $A \subset G$, $B \subset G$, each of size m, will be called *Szekeres difference sets* if

(i) $a \in A \Rightarrow -a \notin A$, and

(ii) for each $d \in G$, $d \neq 0$, the total number of solutions $(a_1,a_2) \in A\times A$, $(b_1,b_2) \in B\times B$ of the equations

$$d = a_1-a_2,\ d = b_1-b_2, \qquad \text{is } m-1.$$

Alternatively 2-$\{2m-1;m;m-1\}$ supplementary difference sets A and B are called *Szekeres difference sets* if $a \in A \Rightarrow -a \notin A$.

THEOREM 4.4. *If A and B are two Szekeres difference sets of size m in an additive abelian group of order 2m+1 then there is a skew-Hadamard matrix of order 4(m+1).*

PROOF. Let A be the set such that $a \in A \Rightarrow -a \notin A$. Let X be the type 1 incidence matrix of A and Y be the type 2 incidence matrix of B. Then

$$XY^T = YX^T, \quad JX^T = JY^T = mJ,$$

and

$$XX^T+YY^T = (m+1)I+(m-1)J.$$

Choose M = J-2X and N = J-2Y. Then if e is the $1\times(2m+1)$ matrix of 1's

$$eM^T = eN^T = e, \quad MN^T = NM^T, \quad N^T = N, \quad (M-I)^T = -M+I$$

and

$$MM^T+NN^T = 4(m+1)I-2J.$$

Thus the required skew-Hadamard matrix is

$$\begin{bmatrix} 1 & 1 & e & e \\ -1 & 1 & e & -e \\ -e^T & -e^T & M & N \\ -e^T & e^T & -N & M \end{bmatrix}.$$

***

THEOREM 4.5. (Szekeres) *If q = 4m+3 is a prime power and G is the cyclic group of order 2m+1, there exist Szekeres difference sets of size m in G.*

PROOF. This is theorem 2.5 restated. ***

This gives no new skew-Hadamard matrices because we know by theorem 4.1 that there is a skew-Hadamard matrix of order q+1 whenever q = 4m+3 is a prime power. However, Blatt and Szekeres [20], Szekeres [129],[130] and Whiteman [172], have constructed Szekeres difference sets which do give new skew-Hadamard matrices.

THEOREM 4.6. (Szekeres) *Let $q = 2m+1 = p^r \equiv 5 \pmod 8$ be a prime power and let G be the elementary abelian group of order $p^k$. Then there exist Szekeres difference sets of size m in G and a skew-Hadamard matrix of order 2(q+1).*

PROOF. We again identify G with the additive group of GF(q). Let x be a primitive root of GF(q) and $G_0$ be the multiplicative group of GF(q), of order q-1, generated by x. We note $m \equiv 2 \pmod 4$ and write

$$H_i = \{x^{4j+i}: j = 0,1,\ldots,\tfrac{1}{2}m-1\} \qquad i = 0,1,2,3.$$

Then we take

$$A = H_0 \cup H_1, \quad B = H_0 \cup H_3.$$

Clearly both contain m elements and since $-1 = x^m \varepsilon H_2$, $a \varepsilon A \Rightarrow -a \notin A$.

We now consider for fixed $d \varepsilon H_0$ the following equations in $a,b \varepsilon A$, $f,g \varepsilon B$:

$$d = a-b \tag{4.60}$$

$$xd = f-g \tag{4.61}$$

$$x^2 d = a-b \tag{4.62}$$

$$x^3 d = f-g\ . \tag{4.63}$$

The number of solutions of these equations is independent of $d \varepsilon H_0$ since

$$z \varepsilon A,\quad w \varepsilon B \Rightarrow x^{4i} z \varepsilon A,\quad x^{4i} w \varepsilon B$$

for every i. Furthermore, the number of solutions of (4.60) and (4.63) are equal to each other because $z \varepsilon A \Rightarrow w = x^3 z \varepsilon B$ and $w \varepsilon B \Rightarrow x^{-3} w = z \varepsilon A$. Similarly the numbers of solutions of (4.61) and (4.62) are equal because

$$w \varepsilon B \Rightarrow xw \varepsilon A.$$

Finally (4.60) and (4.62) have the same number of solutions because

$$z \varepsilon A \Rightarrow -x^2 z \varepsilon A.$$

By the same argument it can be shown that the number of solutions of each of the equations

$$d = f-g$$

$$xd = a-b$$

$$x^2 d = f-g$$

$$x^3 d = a-b$$

is the same. Hence for each $d \neq 0$ the total number of solutions of

$$d = a-b,\quad x = f-g$$

is the same number $\mu$. Therefore $\mu(q-1) = 2\mu m$ is equal to the total number of differences between elements of A and B, i.e. to $2m(m-1)$, giving $\mu = m-1$ as required.

Thus we have shown A and B are Szekeres difference sets. By theorem 5.4 there is a skew-Hadamard matrix of order $4(m+1) = 2(q+1)$. ***

Theorem 4.6 gives many skew-Hadamard matrices. It was theorems 4.5 and 4.6 that showed that *cyclotomy* was likely to be very useful in studying Hadamard matrices.

Blatt and Szekeres used a computer to search the cyclic group of order 17 and found there were no Szekeres difference sets of size 8 in $C_{17}$. They also found there were no Szekeres difference sets of size 12 in $C_{25}$. However, a complete machine search of the elementary abelian group of order 25, $C_5 \times C_5$ produced 480 different pairs of Szekeres difference sets. Examination of these difference sets has shown that the corresponding Hadamard matrices are all equivalent under permutation and multiplication by ±1 of rows and columns.

The following theorem was proved independently by Szekeres [130] and Whiteman [172]:

THEOREM 4.7. (Szekeres-Whiteman) *If* $q = p^t = 8m+1$ *is a prime power such that* $p \equiv 5 \pmod 8$ *and* $t \equiv 2 \pmod 4$, *then there exist Szekeres difference sets of size* $4m$ *and a skew-Hadamard matrix of order* $2(q+1)$.

PROOF. Let x be a primitive root of GF(q). Let $C_0$ be the multiplicative subgroup of index 8 generated by $x^8$, $C_i$ $(i = 1,\ldots,7)$ is given by $x^i C_0$. We wish to show that

$$A = C_0 \cup C_1 \cup C_2 \cup C_3 \quad \text{and} \quad B = C_0 \cup C_1 \cup C_6 \cup C_7$$

are Szekeres difference sets. Now $q \equiv 9 \pmod{16}$ and hence $-1 \in C_4$. So $a \in A \Rightarrow -a \notin A$ and $b \in B \Rightarrow -b \notin B$. We wish to show that for each $d \in G$, $d \neq 0$, the total number of solutions $(a_1,a_2) \in A\times A$, $(b_1,b_2) \in B\times B$ of the equations

$$a_1-a_2 = d, \quad b_1-b_2 = d \tag{4.70}$$

is $4m-1$.

We now find the number $N_k$ of solutions of $y-x = d$ with $y,x \in A$ and $d \in C_k$: we use the notation of §1.5

$$\begin{aligned} N_k = {} & (-k,-k)+(1-k,-k)+(2-k,-k)+(3-k,-k)\\ & +(-k,1-k)+(1-k,1-k)+(2-k,1-k)+(3-k,1-k)\\ & +(-k,2-k)+(1-k,2-k)+(2-k,2-k)+(3-k,2-k)\\ & +(-k,3-k)+(1-k,3-k)+(2-k,3-k)+(3-k,3-k). \end{aligned} \tag{4.71}$$

The corresponding number of $N_k'$ of solutions of $y-x = d$ with $y,x \in B$ and $d \in C_k$ is

$$\begin{aligned} N_k' = {} & (-k,-k)+(1-k,-k)+(6-k,-k)+(7-k,-k)\\ & +(-k,1-k)+(1-k,1-k)+(6-k,1-k)+(7-k,1-k)\\ & +(-k,6-k)+(1-k,6-k)+(6-k,6-k)+(7-k,6-k)\\ & +(-k,7-k)+(1-k,7-k)+(6-k,7-k)+(7-k,7-k). \end{aligned} \tag{4.72}$$

Since every solution of $y-x = d$ with $y,x \in A$ and $d \in C_k$ yields a solution of $x-y = -d$, and since $-1 \in C_4$, it follows that $N_k = N_{k+4}$ $(k = 0,1,2,3)$. Similarly, $N_k' = N_{k+4}'$ $(k = 0,1,2,3)$. Furthermore, since $a \in A \Rightarrow x^6 a \in B$ and $b \in B \Rightarrow x^2 b \in A$, we have also $N_{k+2} = N_k'$ $(k = 0,1,2,3)$. Hence we find that

$$\begin{aligned} N_0+N_0' &= N_2+N_2' = N_4+N_4' = N_6+N_6'\\ N_1+N_1' &= N_3+N_3' = N_5+N_5' = N_7+N_7'. \end{aligned} \tag{4.73}$$

The application of lemma 8 of part 1 to the evaluation of $N_k$ and $N_k'$ depends upon whether or not 2 is a fourth power in G. We now show that 2 is not a fourth power in G when $p \equiv 5 \pmod 8$ and $t \equiv 2 \pmod 4$. It is convenient to put

$r = (p^t-1)/(p-1)$. The number $x$ is a generator of the cyclic group of non-zero elements of $GF(p)$. Since $(2|p) = -1$ the exponent $k$ in the equation $g^k = 2$ is odd. Furthermore, since $r = p^{t-1}+p^{t-2}+\ldots+1 \equiv t \equiv 2 \pmod 4$, the exponent $rk$ in the equation $x^{rk} = 2$ is $\equiv 2 \pmod 4$. Therefore 2 is not a fourth power in $G$.

In view of (4.73) it suffices to evaluate $N_0, N_0'$ and $N_1, N_1'$. From (4.71), (4.72) and the array (2) we get

$$N_0 = AINJBJOOCKNODLMI,$$

$$N_0' = AINJBJMKGLNMHMOI,$$

$$N_1 = JAINKBJOLCKNIHMK,$$

$$N_1' = JAINLFJKMGONIHOO,$$

where, for brevity, we have omitted the plus signs between adjacent letters. Applying case II of lemma 8 we may now derive the following formulae

$$64N_0 = 16q-48-8x+8a+16y-32b,$$

$$64N_0' = 16q-48+8x-8a-16y,$$

$$64N_1 = 16q-48+8x-8a+16y,$$

$$64N_1' = 16q-48-8x+8a-16y+32b.$$

Consequently

$$64(N_0+N_0') = 32q-96-32b,$$

$$64(N_1+N_1') = 32q-96+32b.$$

Thus $N_0+N_0'$ and $N_1+N_1'$ are equal if and only if $b = 0$. Statement (ii) at the end of lemma 8 guarantees that $b = 0$ when $p \equiv 5 \pmod 8$. It follows that for each $d \in G$, $d \neq 0$, the total number of solutions of the equations (4.70) is $(q-3)/2 = 4m-1$. Thus we have shown that $A$ and $B$ are Szekeres difference sets.

By theorem 4.4 there is a skew-Hadamard matrix of order $4(4m+1) = 2(q+1)$.

***

In [130] Szekeres says he has proved theorem 4.7 for $t \equiv 0 \pmod 4$ also, but this is not so because $p^{2r} \not\equiv 9 \pmod{16}$ for $r$ even.

Theorem 4.7 gives many skew-Hadamard matrices including those of orders 52 and 340.

Thus theorem 4.5 shows there are Szekeres difference sets in orders congruent to 3,7,11,15 (mod 16), theorem 4.6 covers the orders congruent to 5,13 (mod 16) and theorem 4.7 covers the orders congruent to 9 (mod 16).

4.2 **THE WILLIAMSON TYPE.** Jennifer Wallis [158] used a computer to obtain skew-Hadamard matrices using the Williamson matrix

$$\begin{bmatrix} A & B & C & D \\ -B & A & D & -C \\ -C & -D & A & B \\ -D & C & -B & A \end{bmatrix}.$$

Those of order < 92 only took at most a few minutes to find, but the matrix of order 92 took many hours on an ICL 1904A. Subsequently Szekeres and Hunt, using a bigger computer, developed indexing techniques that allowed the matrix of order 100 to be found in about one hour.

THEOREM 4.8. *Let A,B,C and D be square matrices of order m. Further let A be skew-type and circulant and B,C,D be back-circulant matrices whose first rows satisfy the following equations:*

$$\left.\begin{aligned} a_{1,j} &= -a_{1,m+2-j} \\ b_{1,j} &= b_{1,m+2-j} \\ c_{1,j} &= c_{1,m+2-j} \\ d_{1,j} &= d_{1,m+2-j} \\ a_{11} &= d_{11} = c_{11} = c_{11} = +1, \end{aligned}\quad 2 \le j \le m \right\} \tag{4.81}$$

*where $A = (a_{ij})$, $B = (b_{ij})$, $C = (c_{ij})$, $D = (d_{ij})$ and every element is +1 or -1.*

*Then if*

$$AA^T + BB^T + CC^T + DD^T = 4mI_m,$$

*H given by*

$$\begin{bmatrix} A & B & C & D \\ -B & A & D & -C \\ -C & -D & A & B \\ -D & C & -B & A \end{bmatrix} \tag{4.82}$$

*is skew-Hadamard of order 4m.*

Further we note:

THEOREM 4.9. *Let A,B,C and D be square matrices of order m. Further let A be derived from a type 1 incidence matrix and be skew-type: let B,C,D be derived from type 2 incidence matrices all of whose generating sets $X_i$ are such that $x_i \in X_i \Rightarrow -x_i \in X_i$.*

*Then if*

$$AA^T + BB^T + CC^T + DD^T = 4mI_m,$$

*H given by (4.82) is a skew-Hadamard matrix of order 4m.* ***

The following first rows generate the matrices of theorem 4.8: the results for m = 21 and 25 were found by D. Hunt and G. Szekeres. We use + to signify +1 and - to signify -1.

| | | |
|---|---|---|
| m = 3: | A = | + -, + |
| | B = | + -, - |
| | C = | + -, - |
| | D = | + +, + |
| m = 5: | A = | + - -, + + |
| | B = | + - -, - - |
| | C = | + - -, - - |
| | D = | + - +, + - |
| m = 7: | A = | + - - -, + + + |
| | B = | + - - -, - - - |
| | C = | + - - +, + - - |
| | D = | + - + -, - + - |
| m = 9: | A = | + - - - +, - + + + |
| | B = | + - - - +, + - - - |
| | C = | + + - + -, - + - + |
| | D = | + + + - +, + - + + |
| m = 11: | A = | + - - - - +, - + + + + |
| | B = | + - - - - +, + - - - - |
| | C = | + + - + - +, + - + - + |
| | D = | + + - + + -, - + + - + |
| m = 13: | A = | + - - - - + -, + - + + + + |
| | B = | + - + + + - +, + - + + + - |
| | C = | + + - - + - +, + - + - - + |
| | D = | + + + + - - +, + - - + + + |
| m = 15: | A = | + - - - - - + -, + - + + + + + |
| | B = | + - + - - + + -, - + + - - + - |
| | C = | + + + - - + - +, + - + - - + + |
| | D = | + + - + + + + -, - + + + + - + |
| m = 17: | A = | + - - - - - + - -, + + - + + + + + |
| | B = | + + - - - + - - +, + - - + - - - + |
| | C = | + + - - - + - + -, - + - + - - - + |
| | D = | + - - + - + - - -, - - - + - + - - |

```
m = 21:  A = + - - - - - + - - + -,  + - + + - + + + + +
         B = + + + - - + - + + + -,  - + + + - + - - + +
         C = + + - + - - - - - + -,  - + - - - - - + - +
         D = + - + - - + + - - - +,  + - - - + + - - + -

m = 23:  A = + - - - - - - - + + - +,  - + - - + + + + + + +
         B = + + - - + - - + + + + -,  - + + + + - - + - - +
         C = + + - - - + - + - + - +,  + - + - + - + - - - +
         D = + - - - - + - - + - - +,  + - - + - - + - - - -

m = 25:  A = + + - - - - - - - + + - +,
                                 - + - - + + + + + + + -
         B = + + + - - + + - + + + + -,
                                 - + + + + - + + - - + +
         C = + - - - - + + + - + - - +,
                                 + - - + - + + + - - - -
         D = + - + - - + + - + - + - -,
                                 - - + - + - + + - - + -
```

Thus using these results and theorem 4.8 we have:

THEOREM 4.10. *There exist skew-Hadamard matrices of order 12,20,28,36,44,52,60,68,76,84,92,100.*

These matrices may be inequivalent to those of theorems 4.5 and 4.6.

We now proceed to analyse the matrices used in theorem 4.8 in the manner of Williamson:

Let T be the m×m matrix and R be the m×m matrix

$$T = \begin{bmatrix} 0 & 1 & 0 & \dots & 0 \\ 0 & 0 & 1 & \dots & 0 \\ \cdot & \cdot & \cdot & \dots & \cdot \\ 0 & 0 & 0 & \dots & 1 \\ 1 & 0 & 0 & \dots & 0 \end{bmatrix}, \quad T^m = I_m, \quad R = \begin{bmatrix} 0 & \dots & 0 & 0 & 1 \\ 0 & \dots & 0 & 1 & 0 \\ \cdot & \dots & \cdot & \cdot & \cdot \\ 0 & \dots & 0 & 0 & 0 \\ 1 & \dots & 0 & 0 & 0 \end{bmatrix} \tag{4.100}$$

Polynomials in T commute with each other. So consider

$$\left.\begin{aligned} A &= a_{11}I + a_{12}T + \dots + a_{1m}T^{m-1}, \\ B &= (b_{11}I + b_{12}T + \dots + b_{1m}T^{m-1})R, \\ C &= (c_{11}I + c_{12}T + \dots + c_{1m}T^{m-1})R, \\ D &= (d_{11}I + d_{12}T + \dots + d_{1m}T^{m-1})R, \end{aligned}\right\} \tag{4.101}$$

where $a_{1i}$, $b_{1i}$, $c_{1i}$, $d_{1i}$ are given by equations (4.81). Then provided

$$AA^T + BB^T + CC^T + DD^T = 4mI_m \tag{4.102}$$

we have all the conditions of theorem 4.8 satisfied and so we would have a skew-Hadamard matrix of order 4m.

Now let us consider equation (4.102) further.

Write

$$A = P_1 - N_1 \tag{4.103}$$

with $P_1$ the sum of the terms with positive coefficient in A and $-N_1$ the sum of the terms with negative coefficient in A, whence

$$P_1 = \sum_j a_{1j}T^{j-1}, \quad a_{1j} = +1, \quad -N_1 = \sum_j a_{1j}T^{i-1}, \quad a_{1j} = -1. \tag{4.104}$$

In the same way write

$$B = P_2 - N_2, \quad C = P_3 - N_3, \quad D = P_4 - N_4. \tag{4.105}$$

Since $a_{11} = +1$ and by (4.81) $a_{1j} = -a_{1,m+2-j}$, $2 \le j \le m$, there are

$$\tfrac{1}{2}(m+1) = p_1 \tag{4.106}$$

positive terms in A, so the number of terms in $P_1$ is an odd number if $m \equiv 1 \pmod 4$ and an even number of $m \equiv 3 \pmod 4$.

Since by (4.81) $b_{11} = +1$ and $b_{1j} = b_{1,m+2-j}$, $2 \le j \le m$, the positive terms occur in pairs, so $p_2$, the number of positive terms in $P_2$, is an odd number. Similarly $p_3$ and $p_4$ are odd numbers.

We now write

$$J = I + T + T^2 + \ldots T^{m-1} = (I + T + T^2 \ldots T^{m-1})R.$$

Then

$$P_i + N_i = J, \quad i = 1,2,3,4,$$

so (4.102) becomes

$$AA^T + B^2 + C^2 + D^2 = 4mI_m$$

and by (4.103), (4.104), (4.105) this becomes

$$(2P_1 - J)(2P_1 - J)^T + (2P_2 - J)^2 + (2P_3 - J)^2 + (2P_4 - J)^2 = 4mI_m,$$

that is,
$$(2P_1 - J)(-2P_1 + J + 2I) + (2P_2 - J)^2 + (2P_3 - J)^2 + (2P_4 - J)^2 = 4mI_m,$$

since A is skew-type. So we have since $P_iJ = p_iJ$ and $J^2 = mJ$

$$4(-P_1^2 + P_2^2 + P_3^2 + P_4^2 + P_1) + 4(p_1 - p_2 - p_3 - p_4)J + (2m-2)J = 4mI_m. \tag{4.107}$$

Now from (4.106)

$$4p_1 + 2m - 2 = 4m$$

so (4.107) becomes

$$(-P_1^2+P_2^2+P_3^2+P_4^2+P_1) = -(m-p_2-p_3-p_4)J+mI_m. \qquad (4.108)$$

If m is odd then since $p_2,p_3$ and $p_4$ are all odd, we have the coefficients of J are all even.

Now we notice B,C,D are polynomials in $T^jR$ so

$$P_i = (\sum_j e_{1j}T^{j-1})R$$

for i = 2,3,4 and $e_{1j} = b_{1j}, c_{1j}, c_{1j}$ respectively; also $P_i = P_i^T$, so

$$P_i^2 = P_iP_i^T = (\sum_j e_{1j}T^{j-1})RR^T(\sum_k e_{1k}T^{k-1}) = \sum_n f_nT^n,$$

that is, $P_1^2,P_2^2,P_3^2,P_4^2$ and $P_1$ may all be regarded as polynomials in T.

For each t = 1,...,m-1, there is a unique s such that $(T^s)^2 = T^t$. In $P_j^2 = (\sum T^k)^2$, k in a subset of 1,2,...,m, we have

$$P_j^2 \equiv \sum T^{2k} \pmod 2.$$

Then since the coefficient of J in (4.108) is always even, we have shown:

THEOREM 4.11. *If m is odd, T is given by (4.100), $P_1,P_2,P_3,P_4$ are the terms with positive coefficients of A,B,C,D as defined by (4.101) respectively, and if*

$$AA^T + BB^T + CC^T + DD^T = 4mI_m$$

*then writing $P_1+P_1^2 = \sum_{i \neq 0} f_iT^i$ and $P_2^2+P_3^2+P_4^2 = \sum_{i\neq 0} g_iT^i$ we have $g_i \equiv f_i \pmod 2$ when $i \neq 0$.*

4.3 THE GOETHALS-SEIDEL TYPE. Goethals and Seidel modified the Williamson matrix so that the matrix entries did not have to be circulant and symmetric. Their matrix has been valuable in constructing many new Hadamard matrices.

THEOREM 4.12. (Goethals and Seidel) *If A,B,C,D are square circulant matrices of order m, and $R = (r_{ij})$ is defined by $r_{i,m-i} = 1$, $i = 1,\ldots,m$, then if A is skew-type, and if*

$$AA^T + BB^T + CC^T + DD^T = 4mI, \qquad (4.121)$$

*then*

$$H = \begin{bmatrix} A & BR & CR & DR \\ -BR & A & -D^TR & C^TR \\ -CR & D^TR & A & -B^TR \\ -DR & -C^TR & B^TR & A \end{bmatrix} \qquad (4.122)$$

*is skew-Hadamard of order 4m.*

PROOF. This is easily verified using lemma 1.8. ***

THEOREM 4.13. (Goethals and Seidel) *There exist skew-Hadamard matrices of orders 36 and 52.*

PROOF. Apply theorem 4.12 with the following circulant matrices of order 9, where the comma denotes the middle of the string:

A = + + + - +, - + - -;  B = + - + + -, - + + -;

C = + + - - -, - - - +;  D = + + + - +, + - + +.

By inspection the skew-type A and the symmetric B,C,D are seen to satisfy the hypotheses. Hence a skew-Hadamard matrix of order 36 is obtained. Next consider the following circulant matrices of order 13:

A = + + + + - + +, - - + - - -;

B = + - + - - + +, + + - - + -;

C = D = + + - + - - -, - - + - - -.

Applying theorem 4.12 to A,B,C,D yields a skew-Hadamard matrix of order 52.

***

The following results of Whiteman use the Goethals-Seidel matrix to obtain a new construction for skew-Hadamard matrices. His method is a modification of his reproof of the result of Turyn that if q is a prime power $\equiv 1 \pmod 4$, then there exists an Hadamard matrix of Williamson type of order $2(q+1)$.

Let $GF(q^2)$ denote the finite field of order $q^2$, where $q = p^t$ is a prime power. Let $\gamma$ denote a generator of the cyclic group of non-zero elements of $GF(q^2)$. Then $\gamma^{q+1} = g$ is a generator of the cyclic group of non-zero elements of the finite field $GF(q)$ of order q. For arbitrary $\xi \in GF(q^2)$ define

$$tr(\xi) = \xi+\xi^q, \tag{4.131}$$

so that $tr(\xi) \in GF(q)$. It follows at once from this definition that

$$tr(\gamma^k) = \gamma^{(q+1)k}tr(\gamma^{-k}), \tag{4.132}$$

where k is any integer.

Suppose henceforth that $q \equiv 3 \pmod 8$. Then the polynomial $P(x) = x^2+1$ is irreducible in the finite field $GF(q)$. Hence the polynomial $ax+b$ $(a,b \in GF(q))$ modulo $P(x)$ form a finite field $GF(q^2)$. We shall employ this concrete representation of $GF(q^2)$ in the rest of this section. For $\xi \in GF(q^2)$, $\xi \neq 0$, let ind $\xi$ be the least non-negative integer t such that $\gamma^t = \xi$. Let $\beta$ denote a primitive eighth root of unity. Then

$$\chi(\xi) = \left\{ \begin{array}{ll} \beta^{\text{ind } \xi} & (\xi \neq 0) \\ 0 & (\xi = 0) \end{array} \right\} \tag{4.133}$$

defines an eighth power character $\chi$ of $GF(q^2)$. For $a \in GF(q)$, $a \neq 0$, put $g^j = a$.

Then (4.133) implies that $\chi(a) = \beta^{(q+1)j} = (-1)^j$. This means that the character $\chi(a)$ reduces to the ordinary Legendre symbol over GF(q). Thus $\chi(a) = +1,-1$ or 0 according as a is a non-zero square, a non-square, or zero in GF(q). Accordingly, we deduce from (4.132) that

$$\chi tr(\gamma^k)\chi tr(\gamma^{-k}) = (-1)^k, \qquad \{tr(\gamma^k) \neq 0\}. \tag{4.134}$$

LEMMA 4.14. *If r is a non-negative integer, then*

$$\sum_{k=0}^{q} \chi tr(\gamma^k)\chi tr(\gamma^{k+r}) = \begin{cases} (-1)^j q & (q+1 \mid r), \\ 0 & (q+1 \nmid r), \end{cases} \tag{4.141}$$

*where, in the first case, $r = j(q+1)$.*

PROOF. For fixed $\eta \in GF(q^2)$ put $\eta = cx+d$ $\{c,d \in GF(q)\}$. Then $\eta \in GF(q)$ if $c = 0$ and $\eta \notin GF(q)$ if $c \neq 0$. We first show that

$$\sum_{\xi} \chi tr(\xi)\chi tr(\eta\xi) = \begin{cases} \chi(d)q(q-1) & (c = 0) \\ 0 & (c \neq 0) \end{cases} \tag{4.142}$$

where the summation extends over all $\xi \in GF(q^2)$. Put $\xi = ax+b$ $\{a,b \in GF(q)\}$. By (4.131) we have $tr(\xi) = 2b$ and $tr(\eta\xi) = 2(bd-ac)$. Therefore

$$\sum_{\xi} \chi tr(\xi)\chi tr(\eta\xi) = \sum_{b} \chi(b) \sum_{a} \chi(bd-ac),$$

and (4.142) follows at once.

For $\eta \neq 0$ we may put $\eta = \gamma^r$ $(0 \leq r \leq q^2-2)$ so that $c = 0$ if $q+1 \mid r$ and $c \neq 0$ if $q+1 \nmid r$. If $c = 0$ put $r = j(q+1)$; then we get $\chi(d) = (-1)^j$.

The sum in (4.142) now becomes

$$\sum_{k=0}^{q^2-2} \chi tr(\gamma^k)\chi tr(\gamma^{k+r}) = \sum_{h=0}^{q-2} \sum_{k=h(q+1)}^{h(q+1)+q} \chi tr(\gamma^k)\chi tr(\gamma^{k+r}).$$

In view of (4.142) the double sum on the right has the value 0 if $q+1 \nmid r$. Since $\chi tr(\gamma^{k+q+1}) = -\chi tr(\gamma^k)$ the value of the inner sum is the same for each h. In particular for h = 0 we obtain the result stated in the lemma. ****

THEOREM 4.15. (Whiteman) *Let q be a prime power $\equiv 3 \pmod 8$ and put $n = (q+1)/4$. Let $\gamma$ be a primitive element of $GF(q^2)$. Put $\gamma^k = ax+b$ $\{a,b \in GF(q)\}$ and define*

$$a_k = \chi(a), \quad b_k = \chi(b). \tag{4.150}$$

*Let A,B,C,D be square circulant matrices of order n whose initial rows are given by $a_0,a_8,a_{16},\ldots,a_{8(n-1)}$; $b_0,b_8,b_{16},\ldots,b_{8(n-1)}$; $a_1,a_9,a_{17},\ldots,a_{8n-7}$; $b_1,b_9,b_{17},\ldots,b_{8n-7}$; respectively. Then the matrix H defined by (4.122) is a skew-Hadamard matrix of order 4n.*

PROOF. We must show that A is skew-type and that the condition (4.121) is satisfied. We first demonstrate that A is skew-type. Since $\gamma$ is a primitive element of $GF(q^2)$ the integer $k = (q+1)/2 = 2n$ is the only value of k in the interval $0 \leq k \leq q$ for which $tr(\gamma^k) = 0$. Put $\gamma^{2n} = wx$ $\left(w \in GF(q)\right)$ so that $-w^2 = g$. Then the numbers $a_k, b_k$ in (4.150) satisfy the relations

$$\left.\begin{aligned} a_k &= -a_{k+4n} = a_{k+8n}; \quad b_k = -b_{k+4n} = b_{k+8n}; \\ a_{k+2n} &= -a_{k+6n} = \chi(w)b_k; \quad b_{k+2n} = -b_{k+6n} = -\chi(w)a_k. \end{aligned}\right\} \tag{4.151}$$

From the relation (4.134) we deduce also the second of the two relations

$$a_k = (-1)^{k+1}a_{8n-k}; \quad b_k = (-1)^k b_{8n-k}. \tag{4.152}$$

The first relation in (4.152) follows from the second by replacing k by k+2n and then applying (4.151). For k = 8i we get in particular

$$a_{8i} = -a_{8(n-i)} \qquad (i = 0,1,\ldots,n-1). \tag{4.153}$$

Relation (4.153) expresses the skew property of the matrix A.

It remains to prove that the matrices A,B,C,D satisfy the orthogonality condition (4.121). We shall employ an alternative formulation of this condition. Put

$$\left.\begin{aligned} F(8r) = &\sum_{j=0}^{n-1} a_{8j}a_{8j+8r} + \sum_{j=0}^{n-1} b_{8j}b_{8j+8r} \\ &+ \sum_{j=0}^{n-1} a_{8j+1}a_{8j+1+8r} + \sum_{j=0}^{n-1} b_{8j+1}b_{8j+1+8r}, \end{aligned}\right\} \tag{4.154}$$

where the subscripts are reduced modulo 8n. Then we may verify directly that condition (4.121) is equivalent to the condition

$$F(8r) = \begin{cases} 4n-1 & (r = 0), \\ 0 & (1 \leq r \leq n-1). \end{cases} \tag{4.155}$$

We also put

$$G(8r) = \sum_{k=0}^{q} b_k b_{k+8r} \, . \tag{4.156}$$

Since n is an odd integer it follows that $q+1 \nmid 8r$ for $1 \leq r \leq n-1$. Hence the lemma 4.14 yields

$$G(8r) = \begin{cases} 4n-1 & (r = 0), \\ 0 & (1 \leq r \leq n-1). \end{cases} \tag{4.157}$$

The right members of (4.155) and (4.156) are identical. Accordingly, we shall complete the proof by transforming the right member of (4.156) into the right member of (4.154). We first rewrite (4.156) in the form

$$G(8r) = \sum_{t=0}^{3} \sum_{i=0}^{n-1} b_{4i+t}b_{4i+t+8r}. \tag{4.158}$$

Put $n = 2f+1$. Then for each fixed t we have

$$\sum_{i=0}^{n-1} b_{4i+t}b_{4i+t+8r} = \sum_{j=0}^{f} b_{8j+t}b_{8j+t+8r} + \sum_{j=0}^{f-1} b_{8j+4+t}b_{8j+4+t+8r}$$

$$= \sum_{j=0}^{f} b_{8j+t}b_{8j+t+8r} + \sum_{j=0}^{f-1} b_{8j+4n+4+t}b_{8j+4n+4+t+8r}$$

$$= \sum_{j=0}^{f} b_{8j+t}b_{8j+t+8r} + \sum_{j=f+1}^{n-1} b_{8j+t}b_{8j+t+8r}$$

$$= \sum_{j=0}^{n-1} b_{8j+t}b_{8j+t+8r} \, ,$$

where we have made use of the periodicity relation $b_k = -b_{k+4n}$ given in (4.151). Hence (4.158) becomes

$$G(8r) = \sum_{t=0}^{3} \sum_{j=0}^{n-1} b_{8j+t}b_{8j+t+8r}. \qquad (4.159)$$

Thus for $t = 0$ we get the second sum in (4.154), and for $t = 1$ we get the fourth.

We proceed to consider the summands $t = 2$ and $t = 3$. From the periodicity relation $b_{k+8n} = b_k$ in (4.151) it is evident that

$$\sum_{j=0}^{n-1} b_{8j+t}b_{8j+t+8r} = \sum_{j=0}^{n-1} b_{8j+u}b_{8j+u+8r}$$

whenever $t \equiv u \pmod 8$. Since $2 \equiv \pm 2n \pmod 8$ according as $n \equiv \pm 1 \pmod 4$ we find in particular for $t = 2$ that

$$\sum_{j=0}^{n-1} b_{8j+2}b_{8j+2+8r} = \sum_{j=0}^{n-1} b_{8j\pm 2n}b_{8j\pm 2n+8r}.$$

In view of the periodicity relation $b_{k+2n} = -b_{k-2n} = -\chi(w)a_k$ in (4.151), we obtain

$$\sum_{j=0}^{n-1} b_{8j+2}b_{8j+2+8r} = \sum_{j=0}^{n-1} a_{8j}a_{8j+8r}.$$

Thus for $t = 2$ the inner sum in (4.159) is transformed into the first sum in (4.154). Finally, we note that $3 \equiv \pm 2n+1 \pmod 8$ according as $n \equiv \pm 1 \pmod 4$. It follows that

$$\sum_{j=0}^{n-1} b_{8j+3}b_{8j+3+8r} = \sum_{j=0}^{n-1} a_{8j+1}a_{8j+1+8r}.$$

Thus for $t = 3$ the inner sum in (4.159) is carried into the third sum in (4.154). This gives the result. ***

4.4 AN ADAPTION OF WALLIS-WHITEMAN. Finally we note the following adaption of the Goethals-Seidel matrix which does not require the matrix entries to be circulant at all.

THEOREM 4.15. (Wallis-Whiteman) *Suppose X,Y and W are type 1 incidence matrices and Z is a type 2 incidence matrix of 4-*$\{v;k_1,k_2,k_3,k_4;\ \sum_{i=1}^{4}k_i-v\}$ *supplementary difference sets, then if*

$$A = 2X-J,\quad B = 2Y-J,\quad C = 2Z-J,\quad D = 2W-J,$$

$$H = \begin{bmatrix} A & B & C & D \\ -B^T & A^T & -D & C \\ -C & D^T & A & -B^T \\ -D^T & -C & B & A^T \end{bmatrix} \tag{4.160}$$

*is an Hadamard matrix of order 4v.*

*Further if A is skew-type (*$C^T = C$ *as Z is of type 2) then H is skew-Hadamard.*

***

This matrix can be used when the sets are from any finite abelian group. We now show how (4.160) may be further modified to obtain useful results.

THEOREM 4.17. (Wallis-Whiteman) *Suppose X,Y and W are type 1 incidence matrices and Z is a type 2 incidence matrix of 4-*$\{2m+1;m;2(m-1)\}$ *supplementary difference sets. Then, if*

$$A = 2X-J$$
$$B = 2Y-J$$
$$C = 2Z-J$$
$$D = 2W-J$$

*and e is the* $1\times(2m+1)$ *matrix with every entry 1,*

$$H = \begin{bmatrix} -1 & -1 & -1 & -1 & e & e & e & e \\ 1 & -1 & 1 & -1 & -e & e & -e & e \\ 1 & -1 & -1 & 1 & -e & e & e & -e \\ 1 & 1 & -1 & -1 & -e & -e & e & e \\ e^T & e^T & e^T & e^T & A & B & C & D \\ -e^T & e^T & -e^T & e^T & -B^T & A^T & -D & C \\ -e^T & e^T & e^T & -e^T & -C & D^T & A & -B^T \\ -e^T & -e^T & e^T & e^T & -D^T & -C & B & A^T \end{bmatrix}$$

*is an Hadamard matrix of order 8(m+1). Further if A is skew-type H is skew-Hadamard.*

***

THEOREM 4.18. (Wallis-Whiteman) *Let* $q = 2m+1 = 8f+1$ *(f odd) be a prime power, then there exist 4-*$\{2m+1;m;2(m-1)\}$ *supplementary difference sets* $X_1,X_2,X_3,X_4$ *for which* $y \in X_i \Rightarrow -y \notin X_i$, $i = 1,2,3,4$.

PROOF. Let x be a primitive root of GF(q) and G the cyclic group with generator x.

Define the sets

$$C_i = \{x^{8s+i} : s = 0,1,\ldots,f-1\} \quad i = 0,1,\ldots,7,$$

and choose

$$X_1 = C_0 \cup C_1 \cup C_2 \cup C_3,$$

$$X_2 = C_0 \cup C_1 \cup C_2 \cup C_7,$$

$$X_3 = C_0 \cup C_1 \cup C_6 \cup C_7,$$

$$X_4 = C_0 \cup C_5 \cup C_6 \cup C_7.$$

Write

$$\overset{7}{\underset{s=0}{\&}} a_s C_s , \qquad \left( \sum_{s=0}^{7} a_s = f-1 \right)$$

where the $a_i$ are non-negative integers, for the differences between elements of $C_0$. Thus with $H_s = C_s \cup C_{s+4}$, since $q = 8f+1$ (f odd), $-1 \in C_4$ and the differences become

$$\overset{3}{\underset{s=0}{\&}} a_s H_s , \qquad \sum_{s=0}^{3} a_s = \tfrac{1}{2}(f-1).$$

Now write

$$\overset{3}{\underset{s=0}{\&}} b_s H_s , \quad \overset{3}{\underset{s=0}{\&}} c_s H_s , \quad \overset{3}{\underset{s=0}{\&}} d_s H_s$$

for the differences between

$$C_0 \text{ and } C_1, \quad C_0 \text{ and } C_2, \quad C_0 \text{ and } C_3,$$

respectively, where

$$\sum_{s=0}^{3} b_s = \sum_{s=0}^{3} c_s = \sum_{s=0}^{3} d_s = f.$$

Then the differences from $X_1$ become

$$\overset{3}{\underset{s=0}{\&}} a_s (H_s \cup H_{s+1} \cup H_{s+2} \cup H_{s+3})$$

$$\& \overset{3}{\underset{s=0}{\&}} b_s (H_s \cup H_{s+1} \cup H_{s+2})$$

$$\& \overset{3}{\underset{s=0}{\&}} c_s (H_s \cup H_{s+1})$$

$$\& \overset{3}{\underset{s=0}{\&}} d_s H_s .$$

The differences from $X_2$ are

$$\mathop{\&}_{s=0}^{3} a_s(H_s \cup H_{s+1} \cup H_{s+2} \cup H_{s+3})$$
$$\& \mathop{\&}_{s=0}^{3} b_s(H_s \cup H_{s+1} \cup H_{s+3})$$
$$\& \mathop{\&}_{s=0}^{3} c_s(H_s \cup H_{s+3})$$
$$\& \mathop{\&}_{s=0}^{3} d_s(H_{s+3}).$$

The differences from $X_3$ are

$$\mathop{\&}_{s=0}^{3} a_s(H_s \cup H_{s+1} \cup H_{s+2} \cup H_{s+3})$$
$$\& \mathop{\&}_{s=0}^{3} b_s(H_s \cup H_{s+2} \cup H_{s+3})$$
$$\& \mathop{\&}_{s=0}^{3} c_s(H_{s+2} \cup H_{s+3})$$
$$\& \mathop{\&}_{s=0}^{3} d_s(H_{s+2}).$$

Finally, the differences from $X_4$ are

$$\mathop{\&}_{s=0}^{3} a_s(H_s \cup H_{s+1} \cup H_{s+2} \cup H_{s+3}) \;\&\; \mathop{\&}_{s=0}^{3} b_s(H_{s+1} \cup H_{s+2} \cup H_{s+3})$$
$$\& \mathop{\&}_{s=0}^{3} c_s(H_{s+1} \cup H_{s+2}) \;\&\; \mathop{\&}_{s=0}^{3} d_s H_{s+1}\,.$$

Now $G = H_s \cup H_{s+1} \cup H_{s+2} \cup H_{s+3}$. So the totality of differences from $X_1, X_2, X_3$ and $X_4$ is

$$4\sum_{s=0}^{3} a_s G \;\&\; 3\sum_{s=0}^{3} b_s G \;\&\; 2\sum_{s=0}^{3} c_s G \;\&\; \sum_{s=0}^{3} d_s G$$
$$= \left[2(f-1)+6f\right]G$$
$$= (8f-2)G.$$

Hence $X_1, X_2, X_3, X_4$ are $4-\{2m+1;m;2(m-1)\}$ supplementary difference sets.

Clearly since $y \in C_s \Rightarrow -y \in C_{s+4}$, $X_1, X_2, X_3, X_4$ all satisfy $y \in X_i \Rightarrow -y \notin X_i$.

***

COROLLARY 4.19. (Wallis-Whiteman). *Let $q = 8f+1$ (f odd) be a prime power, then there exists a skew-Hadamard matrix of order $4(q+1)$.*

PROOF. Use theorems 4.17 and 4.18.

4.5 CONSTRUCTIONS USING AMICABLE HADAMARD MATRICES. We recall that $M = I+S$ and $N$ are amicable Hadamard matrices of order m if

(i) $S^T = -S$, $SS^T = (m-1)I$ and $M$ is a skew-Hadamard matrix;

(ii) $N^T = N$ is a symmetric Hadamard matrix;

(iii) $MN^T = NM^T$.

Then we have

THEOREM 4.20. (Wallis) *Let m and m' be the orders of amicable Hadamard matrices, then if there is a skew-Hadamard matrix of order (m-1)m'/m then there is a skew-Hadamard matrix of order m'(m'-1)(m-1).*

PROOF. Let

$$M = \begin{bmatrix} 1 & e \\ -e^T & P \end{bmatrix} \quad \text{and} \quad N = \begin{bmatrix} 1 & e \\ e^T & D \end{bmatrix}$$

be the amicable Hadamard matrices of order m', where $e = (1,\ldots,1)$ is of order $1\times(m'-1)$. Then

$$JP^T = J, \quad (P-I)^T = -(P-I), \quad D^T = D, \quad JD^T = -J$$

$$PP^T = m'I-J = DD^T.$$

Since

$$MN^T = \begin{bmatrix} 1 & e \\ -e^T & P \end{bmatrix}\begin{bmatrix} 1 & e \\ e^T & D^T \end{bmatrix} = \begin{bmatrix} m' & 0 \\ 0 & PD^T-J \end{bmatrix}$$

$$= \begin{bmatrix} m' & 0 \\ 0 & DP^T-J \end{bmatrix} = \begin{bmatrix} 1 & e \\ e^T & D \end{bmatrix}\begin{bmatrix} 1 & -e \\ e^T & P^T \end{bmatrix} = NM^T,$$

$$PD^T = DP^T.$$

Now write $I+U$ for a skew-Hadamard matrix of size t. Write $A = I+W$ and $B$ for the amicable Hadamard matrices of order m where

$$AB^T = (I+W)B^T = B(I+W^T) = BA^T,$$

$W^T = -W$, $B^T = B$ and hence

$$WW^T = (m-1)I, \quad WB^T = BW^T.$$

Now consider

$$H = U\times B\times D + I\times W\times J + I\times I\times P.$$

Then $(H-I)^T = -(H-I)$. Also

$$HH^T = UU^T\times BB^T\times DD^T + I\times WW^T\times(m'-1)J + I\times I\times PP^T$$

$$= (t-1)I\times mI\times[m'I-J] + (m-1)(m'-1)I\times I\times J + I\times I\times[m'I-J]$$

$$= [mm't-mm'+m']I_{mt(m'-1)} + [-m(t-1)+(m-1)(m'-1)-1]I_{mt}\times J_{m'-1}$$

$$= m'(m'-1)(m-1)I \qquad \text{when } t = (m-1)m'/m.$$

And so H is the required skew-Hadamard matrix. ***

THEOREM 4.21. *Let h be the order of a skew-Hadamard matrix and m the order of amicable Hadamard matrices, then there is a skew-Hadamard matrix of order mh.*

PROOF. Let I+S be the skew-Hadamard matrix, then $S^T = -S$ and $SS^T = (h-1)I_h$. If M and $N^T = N$ are the amicable Hadamard matrices then

$$K = I \times M + S \times N$$

satisfies

$$KK^T = mkI_{mh}$$

and so is the required skew-Hadamard matrix. ***

# CHAPTER V. CONSTRUCTIONS FOR SYMMETRIC HADAMARD MATRICES

Less attention has been paid to the symmetric Hadamard matrices, but nevertheless interesting results are known.

Again we restate results proved earlier. Appendix D gives known classes of symmetric Hadamard matrices.

THEOREM 5.1. (Williamson) *Let $t$ be a non-negative integer and $k_i \equiv 0 \pmod 4$ be a prime power plus one, then there is a symmetric Hadamard matrix of order $2^t \Pi k_i$.*

PROOF. In proving in lemma 2.2 that there are amicable Hadamard matrices of orders $2^t \Pi k_i$ we proved that there are symmetric Hadamard matrices of these orders.

***

## 5.1 CONSTRUCTIONS USING SYMMETRIC CONFERENCE MATRICES.

LEMMA 5.2. (Wallis) *If there is a symmetric conference matrix of order $n$ then there is a symmetric Hadamard matrix of order $2n$.*

PROOF. Let $N = I+P$ be the symmetric conference matrix of order n. Then $P^2 = (n-1)I_n$ and $P^T = P$. Now choose

$$X = \begin{bmatrix} -P-I & P-I \\ P-I & P+I \end{bmatrix}$$

then $X^T = X$ and $XX^T = 2nI_{2n}$. So $X$ is the required symmetric Hadamard matrix.

***

THEOREM 5.3. (Wallis) *If $N = I+R$ is a symmetric conference matrix of order $n$ then there is a symmetric Hadamard matrix of order $2n(n-1)$.*

PROOF. Let $D$ be the core of $N$ above then $D^T = D$ and $D^2 = (n-1)I_{n-1}-J_{n-1}$. Further $R^T = R$, $R^2 = (n-1)I_n$ and $JD = 0$.

So if $J$ is of order n-1 let

$$X = \begin{bmatrix} J & J \\ J & -J \end{bmatrix} \quad \text{and} \quad Y = \begin{bmatrix} D+I & D-I \\ D-I & -D-I \end{bmatrix}$$

Then $X^T = X$, $Y^T = Y$, $XX^T = 2(n-1)J_{n-1}\times I_2$, $YY^T = (2nI_{n-1}-2J_{n-1})\times I_2$, and $XY^T+YX^T = 0$.

Then if

$$H = I_n\times X + R\times Y,$$

H is symmetric and since

$$HH^T = I_n \times XX^T + R\times(YX^T+XY^T) + RR^T\times YY^T$$
$$= I_n\times 2(n-1)J_n\times I_2 + (n-1)I_n\times(2nI_{n-1}-2J_{n-1})\times I_2$$
$$= 2n(n-1)I_{2n(n-1)},$$

H is Hadamard. ***

COROLLARY 5.4. *If $p^r \equiv 1 \pmod 4$ is a prime power there is a symmetric Hadamard matrix of order $2p^r(p^r+1)$.*

5.2 STRONGLY REGULAR GRAPHS. The *adjacency matrix* $A = (a_{ij})$ of a graph is defined by

$$a_{ij} = \begin{cases} 0 & i = j, \\ 1 & \text{node } i \text{ is joined to node } j, \\ -1 & \text{otherwise.} \end{cases}$$

DEFINITION 5.6. A non-void, non-complete graph of order v is said to be a *strong-graph* if its adjacency matrix A satisfies

$$(A-xI)(A-yI) = (v-1+xy)J, \quad x > y.$$

It is known [106] that the real numbers x and y are odd integers unless $x+y = 0$. They are the only eigenvalues of A with the possible exception of one simple eigenvalue w which satisfies

$$(w-x)(w-y) = v(v-1+xy).$$

DEFINITION 5.7. A *strongly regular graph* is a strong graph whose adjacency matrix satisfies $AJ = wJ$, i.e. it is regular.

LEMMA 5.8. (Goethals and Seidel) *Symmetric Hadamard matrices with constant diagonal have order $4s^2$, s an integer. They exist if and only if strong graphs of that order exist with eigenvalues*

$$x = 2s\mp 1, \quad y = -2s\mp 1.$$

PROOF. Let $H = A\pm I$ be the Hadamard matrix and $H^2 = hI$. Then $(A\pm I-h^{\frac{1}{2}}I)(A\pm I+h^{\frac{1}{2}}I) = 0$. The result follows using the Cayley-Hamilton theorem. ***

For the remaining eigenvalue w of A we have $w = x$ or $w = y$. If in addition $AJ = wJ$, then A, and also the Hadamard matrix $H = A\pm I$, is regular with $HJ = 2sJ$ or $HJ = -2sJ$. For the eigenvalues of these strongly regular graphs on $4s^2$ vertices there are the following possibilities:

$w = x = 2s-1,\ \ y = -2s-1$ and $x = 2s+1,\ \ w = y = -2s+1,$

$x = 2s-1,\ \ x = y = -2s-1$ and $w = x = 2s+1,\ \ y = -2s+1.$

The first case corresponds to the pseudo Latin-square graphs $L_s(2s)$ and their complements. The second case corresponds to negative Latin-square graphs and their complements. (See [46],[106],[86].) Thus

THEOREM 5.9. *Regular symmetric Hadamard matrices with constant diagonal of order* $4s^2$ *exist if and only if pseudo Latin-square graphs* $L_s(2s)$ *or negative Latin-square graphs* $NL_s(2s)$ *exist.*

This result has been modified [169] to

THEOREM 5.10. (W.D. Wallis) *If there exist t-1 mutually orthogonal Latin squares of side 2t, then there is a symmetric regular Hadamard matrix of order* $4t^2$ *which has the form*

$$\begin{bmatrix} J & H_{12} & \cdots & H_{1m} \\ H_{21} & J & \cdots & H_{2m} \\ \cdot & & & \cdot \\ \cdot & & & \cdot \\ \cdot & & & \cdot \\ H_{m1} & H_{m2} & \cdots & J \end{bmatrix} \qquad (5.101)$$

*where the blocks are of size 2t×2t and each* $H_{ij}$ *is a (1,-1) matrix satisfying*

$$H_{ij}J = JH_{ij} = 0. \qquad ***$$

We omit the proof except to note that t-1 of the Latin squares are first used to form a Latin square graph and the other Latin square is then used to ensure the J's on the diagonal. This construction leads to cliques in the associated strongly regular graphs.

Bush [27] has shown that the existence of a matrix (5.101) of order 2t is implied by the existence of a finite projective plane of order 2t, and consequently the non-existence of such a matrix of order $4t^2$ would be of great significance (except of order 2t for the case m = 6). Also a projective plane of order 2t is equivalent to 2t-1 mutually orthogonal Latin squares of that order. Bruck has shown that slightly less than 2t-1 squares are sufficient for a plane, but it is not known whether t-1 squares are sufficient.

## 5.3 SYMMETRIC HADAMARD MATRICES WITH CONSTANT DIAGONAL.

Let S±I be a core satisfying

$$SS^T = (v-1)I-J, \quad SJ = JS = 0, \quad S^T = \pm S.$$

We recall these matrices are symmetric for $v \equiv 2 \pmod 4$ and skew-symmetric for $v \equiv 0 \pmod 4$.

We strengthen a theorem of Goethals and Seidel.

THEOREM 5.11. *Let* $S_a \pm I$ *be a symmetric core of order a and* $S_b \pm I$ *be a skew-symmetric core of order b which is symmetric with respect to its back-diagonal. Let* $a = b \pm 2$,

*$\max(a,b) = n+1$. Then there exists a regular symmetric Hadamard matrix of order $n^2$.*

PROOF. As $S_b$ is symmetric with respect to its back-diagonal, writing $R = (r_{ij})$ for the matrix of order b with

$$r_{ij} = \begin{cases} 1 & i+j = b+1, \\ 0 & \text{otherwise,} \end{cases}$$

we have

$$(RS_b)^T = RS_b, \quad R^2 = I_b.$$

Define

$$K = RS_b \times S_a + R \times (J_a - I_a) - J_b \times I_a.$$

Then

$$K^T = K, \quad KJ = JK = J, \quad KK^T = n^2 I - J.$$

Choose $F_a$, the diagonal matrix whose elements are +1 and -1 alternatively and define

$$H = \begin{bmatrix} -1 & e \\ e^T & K \end{bmatrix}, \quad G = \begin{bmatrix} 1 & 0 \\ 0^T & I_b \times F_a \end{bmatrix}, \quad \overline{H} = GHG.$$

Then H and $\overline{H}$ are symmetric Hadamard matrices with constant diagonal and $\overline{H}$ is regular with $\overline{H}J = -nJ$.

***

COROLLARY 5.12. *If $n+1$ and $n-1$ are prime powers there exists a symmetric Hadamard matrix with constant diagonal of order $n^2$.*

PROOF. Use the Paley construction for the cores in the theorem (see lemma 1.19).

***

THEOREM 5.13. (Goethals and Seidel) *If there exists an Hadamard matrix of order n, then there exists a regular symmetric Hadamard matrix with constant diagonal of order $n^2$.*

PROOF. A BIBD with parameters $(n, \frac{1}{2}n(n-1), n-1, 2, 1)$ exists for all n. If an Hadamard matrix H of order n exists, then we proceed as follows: normalize the first column of H so it may be written as

$$H = \begin{bmatrix} 1 & \\ \cdot & \\ \cdot & L \\ \cdot & \\ 1 & \end{bmatrix} = \begin{bmatrix} e^T L \end{bmatrix}.$$

Then $HH^T = [e^T L] \begin{bmatrix} e \\ L^T \end{bmatrix} = [e^T e + LL^T] = nI_n$ so L is n×n-1 and

$$LL^T = nI - J, \quad JL = 0.$$

Write N for the $n \times \frac{1}{2}n(n-1)$ incidence matrix of the BIBD. We construct an $n^2 \times \frac{1}{2}n(n-1)$ matrix P by transforming each row of N into an $n \times \frac{1}{2}n(n-1)$ matrix thus: replace the ith one (numbering from the left) of each row of N by the ith column of L; replace each 0 by the $n \times 1$ zero matrix. Then P satisfies

$$PP^T = (n-1)I+A$$

for some symmetric A of size $n^2$ with zero on the diagonal and $\pm 1$ elsewhere (this follows because $\lambda = 1$ and $LL^T = nI-J$). If the $n^2$ row vectors of P span a vector space of dimension $\frac{1}{2}n(n-1)$, then the smallest eigenvalue y of A has multiplicity $\frac{1}{2}n(n+1)$, and $y = 1-n$ which is negative.

By [46; theorem 2.1] since $y = 1-n$ is negative

$$y \geq -\left(n^2-\tfrac{1}{2}n(n-1)\right)^{-1}[\tfrac{1}{2}n(n-1)(n^2-1)\left(n^2-n(n-1)\right)^{\frac{1}{2}}] = 1-n.$$

But equality only holds when all the other eigenvalues are equal to w say. Now since A has zero diagonal

$$\text{tr}A = w\tfrac{1}{2}n(n-1)+y\left(n^2-\tfrac{1}{2}n(n-1)\right) = 0$$

and $w = n+1$. Thus A satisfies (using the Cayley-Hamilton theorem):

$$\left(A-(n+1)I\right)\left(A-(1-n)I\right) = 0$$

i.e.,

$$(A-I)^2 = n^2I.$$

Hence A-I is the required Hadamard matrix of order $n^2$. Further $JL = 0$, so $JP = 0$ and $JA = -(n-1)J$ so A is regular.

***

The following theorem appears in [46], but we give the proof of Shrikhande and Singh [112].

THEOREM 5.14. *If there exists a BIBD with parameters*

$$k(2k-1),\ 4k^2-1,\ 2k+1,\ k,\ 1 \tag{5.141}$$

*then there exists a symmetric Hadamard matrix with constant diagonal of order* $4k^2$.

PROOF. First the dual of the BIBD is formed. This is shown in Shrikhande [107] to be a PBIBD(2) (see [145]) with parameters

$$v = 4k^2-1,\ b = k(2k-1),\ r = k,2k+1,\ \lambda_1 = 1,\ \lambda_2 = 0,\ n_1 = 2k^2,\ n_2 = 2k^2-2$$

$$p^1 = \begin{bmatrix} k^2 & k^2-1 \\ k^2-1 & k^2-1 \end{bmatrix}, \quad p^2 = \begin{bmatrix} k^2 & k^2 \\ k^2 & k^2-3 \end{bmatrix}.$$

Consider the $v = 4k^2-1$ blocks formed by the first associates of each of the $v = 4k^2-1$ treatments. Since $p_{11}^1 = k^2$, any two of these blocks intersect in $k^2$ treatments. Thus we have $v = 4k^2-1$ blocks of size $n_1 = 2k^2$ involving $v = 4k^2-1$ treatments. Since each treatment has exactly $n_1 = 2k^2$ first associates, each treatment occurs in exactly $n_1 = 2k^2$ of these blocks.

If we now form the incidence matrix B in the normal way, we find it is of order $4k^2-1$ and has $2k^2$ non-zero elements in each row and column. Further, since any two blocks have $k^2$ treatments in common, the inner product of any two columns is $k^2$. Thus

$$B^TB = k^2I+k^2J, \quad JB = 2k^2J.$$

Then by a theorem of Ryser

$$BB^T = k^2I+k^2J, \quad BJ = 2k^2J.$$

Since an element is not a first associate of itself, B has zero diagonal.

Let x be a first associate of y, then x is in $B_y$ (where $B_i$ means the block of first associates of i), but $y \in B_x$. So B is symmetric.

Now consider

$$H = \begin{bmatrix} -1 & e \\ e^T & 2B-J \end{bmatrix}$$

where e is the $1\times 4k^2-1$ matrix of ones.

$$HH^T = 4k^2I$$

and H is the required symmetric Hadamard matrix. ***

The BIBD's (5.141) are known for k = 3,5,6,7 and $2^t$, $t > 0$ from [42], [95], [145; p.10]. This gives the required Hadamard matrices for orders 16,36,64,100,144, 196,256.

The following result is a generalization of the matrix of order 99 in Goethals and Seidel [46].

THEOREM 5.15. (Wallis-Whiteman) *Let X and Y be Szekeres difference sets of size m in an additive abelian group G of order 2m+1 and further suppose* $y \in Y \Rightarrow -y \in Y$. *Suppose there is a symmetric conference matrix D+I of order 4m+2. Then there is a regular symmetric Hadamard matrix of order* $4(2m+1)^2$ *with constant diagonal.*

PROOF. We note $x \in X \Rightarrow -x \notin X$. Let M and N be the type 1 incidence matrices of X and Y. Then using lemmas 1.10, 1.12

$$M+M^T = J-I, \quad MM^T+NN^T = (m+1)I+(m-1)J, \quad MN = NM, \quad MM^T = M^TM, \quad NN^T = N^TN.$$

Choose $A = J-2N$, $B = 2M-J$, so $B^T+B = -2I$, $AJ = J$, $BJ = -J$

$$AA^T+BB^T = 4(m+1)I-2J, \quad A^T = A, \quad AB = BA. \tag{5.151}$$

Also with C the core of D+I,

$$C^T = C, \quad CJ = 0, \quad CC^T = (4m+1)I-J.$$

Write e,f for the matrices of ones of orders $1\times(2m+1)$ and $1\times(4m+1)$ respectively. Then we will show

$$H = \begin{bmatrix} 1 & f & e\times f & -e\times f \\ f^T & J & e\times(C-I) & e\times(C+I) \\ e^T\times f^T & e^T\times(C-I) & A\times C+J\times I & -(B+I)\times C+I\times J+(I-J)\times I \\ -e^T\times f^T & e^T\times(C+I) & -(B+I)^T\times C+I\times J+(I-J)\times I & A^T\times C+J\times I \end{bmatrix}$$

is the required matrix.

(i) It is clearly symmetric with constant diagonal.

(ii) We now show it is regular.

(a) The row sum of $1, f, e\times f, -e\times f$ is $4m+2$.

(b) $CJ = 0$ so $[C^T, J, e\times(C-I), e\times(C+I)]J = (4m+2)J$.

(c) $[e^T\times f^T, e^T\times(C-I), A\times C+J\times I, -(B+I)\times C+I\times J+(I-J)\times I]J\times J$

$= \{1+(-1)+(0)+(2m+1)+(0)+(4m+1)+(-2m)\}J\times J$

$= (4m+2)J\times J.$

(d) $[-e^T\times f^T, e^T\times(C+I), -(B+I)^T\times C+I\times J+(I-J)\times I, A^T\times -C+J\times I]J\times J$

$= \{-1+(1)+(0)+(4m+1)+(-2m)+(0)+(2m+1)\}J\times J$

$= (4m+2)J\times J.$

(iii) We now show it is Hadamard (write (i) for row i):

$(1)(2)^T = (2)(1)^T = f+fJ+ee^T\times f(C-I)-ee^T\times f(C+I)$
$= (4m+2)f-(2m+1)f-(2m+1)f$
$= 0.$

$(1)(3)^T = (3)(1)^T = e\times f+e\times f(C-I)+eA^T\times fC^T+eJ\times fI+e(B+I)\times fC-eI\times fJ-e(I-J)\times fI$
$= \{1-(2m+1)\}e\times f+0+(2m+1)e\times f+0-(4m+1)e\times f-e\times f+(4m+1)e\times f$
$= 0.$

$(1)(4)^T = (4)(1)^T = -e\times f+e\times f(C+I)-e(B+I)\times fC+eI\times fJ+e(I-J)\times fI-eA^T\times -fC-eJ\times fI$
$= -e\times f+e\times f+0+(4m+1)e\times f+-2me\times f+0-(2m+1)e\times f$
$= 0.$

$(2)(3)^T = (3)(2)^T = e\times f^Tf+e\times J(C-I)+eA^T\times(C-I)C^T+eJ\times(C-I)-e(B+I)^T\times(C+I)C$
$+eI\times(C+I)J+e(I-J)\times(C+I)$
$= e\times J-e\times J+e\times[(4m+1)I-J-C]+(2m+1)e\times(C-I)-0+e\times J-(2m)e\times(C+I)$
$= 0.$

$(2)(4)^T = (4)(2)^T = -e\times f^Tf+e\times J(C+I)-e(B+I)\times(C-I)C+e\times(C-I)J+e(I-J)\times(C-I)$
$+eA\times -(C+I)C+eJ\times(C+I)$

$$= -e\times J+e\times J-0-e\times J-(2m)e\times(C-I)+e\times\left(-(4m+1)I+J-C\right)$$
$$+(2m+1)e\times(C+I)$$
$$= 0.$$

$$(3)(4)^T = (4)(3)^T = -e^Te\times f^Tf+e^Te\times(C^2-I)-A(B+I)\times C^2-J(B+I)\times C+A\times CJ+J\times J$$
$$+A(I-J)\times C+J(I-J)\times I$$
$$-(B+I)A\times-C^2+A\times-JC+(I-J)A\times-C-(B+I)J\times C+J\times J+(I-J)J\times I$$
$$= -J\times J+J\times(4mI-J)-A\times C^2-AB\times C^2-0+0+J\times J+A\times C-J\times C-2mJ\times I$$
$$+A\times C^2+BA\times C^2+0-A\times C+J\times C-0+J\times J-2mJ\times I$$
$$= (BA-AB)\times C^2$$
$$= 0.$$

$$(1)(1)^T = 1+(4m+1)+2(2m+1)(4m+1)$$
$$= 4(2m+1)^2.$$

$$(2)(2)^T = J+(4m+1)J+(2m+1)(C-1)^2+(2m+1)(C+I)^2$$
$$= 4(2m+1)^2I.$$

$$(3)(3)^T = J\times J+J\times(C-I)^2+A^2\times C^2+2AJ\times C+(2m+1)J\times I$$
$$(B+I)(B+I)^T\times C^2+(4m+1)I\times J+[I+(2m+1)J]\times I$$
$$+0-(B+I)(I-J)\times C-(I-J)(B+I)^T\times C+2(I-J)\times J$$
$$= 4(2m+1)^2I\times I.$$

$$(4)(4)^T = J\times J+J\times(C+I)^2+(B+I)^T(B+I)\times C^2+(4m+1)I\times J+I\times I$$
$$+(2m-1)J\times I-(B+I)^T(I-J)\times C-(I-J)(B+I)\times C$$
$$+2(I-J)\times J+A^TA\times C^2+(2m+1)J\times I-2A^TJ\times C$$
$$= 4(2m+1)^2I\times I.$$

So H is Hadamard. ***

In chapter 3 we showed that the required Szekeres difference sets exist for

$$m = 2,6,14,26 \quad \text{(lemma 3.8)}$$

$$m = \frac{p-3}{4}, \text{ p a prime power} \quad \text{(theorems 3.5 and 3.6)}.$$

Thus

COROLLARY 5.16. *If p is a prime power and p-1 is the order of a symmetric conference matrix, there is a regular symmetric Hadamard matrix with constant diagonal of order* $(p-1)^2$.

We note that this corollary (barring the constant diagonal) essentially appears in Shrikhande [108].

Finally we note

THEOREM 5.17. *If A and B of orders a and b are (regular) symmetric Hadamard matrices (with constant diagonal) then A×B is a (regular) symmetric Hadamard matrix (with constant diagonal) of order ab.*

# CHAPTER VI. COMPLEX HADAMARD MATRICES

These matrices were first introduced by Richard J. Turyn [141] who showed how they could be used to construct Hadamard matrices. It appears complex Hadamard matrices may be very important for they exist for orders for which symmetric conference matrices cannot exist.

Appendices F,H and I give known orders and classes of complex Hadamard matrices.

THEOREM 6.1. (Turyn) *If C is a complex Hadamard matrix of order c and H is a real Hadamard matrix of order h then there exists a real Hadamard matrix of order hc.*

PROOF. Let $R = I_{\frac{1}{2}h} \times \begin{bmatrix} 0 & 1 \\ -1 & 0 \end{bmatrix}$. Then with $K = HR$, $HK^T = HR^TH^T = -HRH^T = -KH^T$.

Write $C = X + iY$, then since $CC^* = cI$ we have

$$CC^* = (X + iY)(X^T - iY^T) = XX^T + YY^T + i(YX^T - XY^T)$$

so $XX^T + YY^T = cI$ and $YX^T - XY^T = 0$.

Then $D = X\times H + Y\times K$ satisfies

$$DD^T = XX^T\times hI + YY^T\times hI = chI_{ch},$$

and is the required Hadamard matrix. ***

THEOREM 6.2. *If C and D are complex Hadamard matrices of orders r and q then C×D (where × is the Kronecker product) is a complex Hadamard matrix of order rq.*

PROOF. $CC^* = rI$, $DD^* = qI$, so $(C\times D)(C^*\times D^*) = rqI$. ***

THEOREM 6.3. (Turyn) *If I + N is a symmetric conference matrix then iI + N is a (symmetric) complex Hadamard matrix and I + iN is a complex skew-Hadamard matrix.* ***

THEOREM 6.4. (Turyn) *Let A,B,C,D be (1,-1) matrices of order m such that*

$$AA^T + BB^T + CC^T + DD^T = 4mI_m,$$

*and $MN^T = NM^T$ for $N,M \in \{A,B,C,D\}$ (ie A,B,C,D may be used to form a Williamson type Hadamard matrix) then write*

$$X = \tfrac{1}{2}(A+B),\; Y = \tfrac{1}{2}(A-B),\; V = \tfrac{1}{2}(C+D),\; W = \tfrac{1}{2}(C-D)$$

*and*

$$\begin{bmatrix} X + iY & V + iW \\ V^*-iW^* & -X^*+iY^* \end{bmatrix}$$

*is a complex Hadamard matrix of order 2m.* ***

## 6.1 CONSTRUCTIONS USING AMICABLE HADAMARD MATRICES.

THEOREM 6.5. *Let $W = I+W$ be a complex skew-Hadamard matrix of order s. Let $M = I+U$, $U^* = -U$ and $N$, where $N^* = N$ be complex amicable Hadamard matrices of order m. Suppose A,B and C are matrices of order p with elements 1,-1,i,-i satisfying*

$$AB^* = BA^*, \quad AC^* = CA^*, \quad BC^* = CB^*$$
$$AA^* = aI+(p-a)J,$$
$$BB^* = bI+(p-b)J, \quad b = mp-a(m-1)-m(s-1),$$
$$CC^* = (p+1)I-J.$$

*Then $K = I\times I\times B+I\times U\times A+S\times N\times C$ is a complex Hadamard matrix of order mps.*

PROOF.
$$\begin{aligned}KK^* &= I\times I\times BB^*+I\times UU^*\times AA^*+SS^*\times NN^*\times CC^*\\ &\qquad +I\times(U+U^*)\times AB^*+(S+S^*)\times N\times BC^*+(S+S^*)\times UN^*\times AC^*\\ &= I\times I\times[bI+(p-b)J]+I\times(m-1)I\times[aI+(p-a)J]+(s-1)I\times mI\times[(p+1)I-J]\\ &= [b+a(m-1)+m(s-1)(p+1)]I_{mps}+[p-b+(m-1)(p-a)-m(s-1)]I_{ms}\times J_p\\ &= mpsI_{mps}.\end{aligned}$$

So K is a complex Hadamard matrix of order mps. ***

COROLLARY 6.6. *If in theorem 6.5 $B = I+R$ has $R^* = -R$ and A and C are hermitian, then K is a complex skew-Hadamard matrix.*

COROLLARY 6.7. *Suppose S+I is a complex skew-Hadamard matrix of order s and A and C are matrices of order p with elements 1,-1,i,-i satisfying*

$$AC^* = CA^*, \quad AA^* = (p-s+1)I+(s-1)J,$$
$$CC^* = (p+1)I-J.$$

*Then $K = I\times A+S\times C$ is a complex Hadamard matrix of order ps.*

PROOF. Put m = 1, B = A in the theorem. ***

COROLLARY 6.8. *Suppose S+I is a complex skew-Hadamard matrix of order s and there exists a core of order p = s-1. Then there is a complex Hadamard matrix of order s(s-1).*

PROOF. Choose A = J and C = I + core (if s ≡ 3(mod 4)) and C = I + i core (if s ≡ 1 (mod 4)) in corollary 6.6. ***

We may use corollary 6.8 with s = 18 to obtain a complex Hadamard matrix of order 306, and with s = 26, to obtain a complex Hadamard matrix of order 650 = 59 650 = 59×11+1 for which order (by theorem 1.31) a symmetric conference matrix is impossible.

Using corollary 6.8 we get the following orders for complex Hadamard matrices. * signifies that a symmetric conference matrix for this order is not possible by theorem 1.31.

| s | complex Hadamard order | comment |
|---|---|---|
| 18 | 306 | |
| 26 | 650 = 59×11+1 | * |
| 30 | 870 = 79×11+1 | * |
| 38 | 1406 = 281×5+1 | |
| 50 | 2450 = 79×31+1 | * |
| 62 | 3782 = 199×19+1 | * |

COROLLARY 6.9. *Let $p \equiv 3 \pmod 4$ be a prime power or equal to $q(q+2)$ where both $a$ and $q+2$ are prime powers. Suppose there exists a symmetric conference matrix of order*

(i) $p-1$ (ii) $p-2r^{n-1}+1$ *where* $p = (r^{n+1}-1)/(r-1)$ *and* $r$ *is a prime power.*

*Then there exist complex Hadamard matrices of order*

(i) $2p(p-1)$ (ii) $2(r^{n+1}-1)\big((r^{n+1}-1)/(r-1)-2r^{n-1}+1\big)/(r-1)$

*respectively.*

PROOF. Choose for M and N the real amicable Hadamard matrices of order 2. Let C-I be the core of the Hadamard matrix of order p+1.

(i) Let $n = p-1$, $A = J$ and $B = (J-2I)$ in the theorem.

(ii) Let $n = p-2r^{n-1}+1$, $A = J$ and $B = S$ (using the notation of chapter I) in the theorem - S must be obtained from a type 1 incidence matrix.

## 6.2 USING SPECIAL HADAMARD MATRICES.

THEOREM 6.10. *Let $W = I+S$ be a skew-Hadamard matrix (real or complex) of order $s$, and $M$ and $N$ be special (real or complex) Hadamard matrices of order $m$. Suppose $A$ and $B$ are matrices of order $p$ with elements $1,-1,i,-i$ satisfying*

$$AB^* = BA^*,$$
$$AA^* = aI+(p-a)J, \quad a = p-s+1,$$
$$BB^* = (p+1)I-J.$$

*Then $K = I\times iM\times A+S\times N\times B$ is a complex Hadamard matrix of order $mps$.*

PROOF. $KK^* = I\times MM^*\times AA^*+SS^*\times NN^*\times BB^*+S^*\times iMN^*\times AB^*+S\times -iNM^*\times BA^*$

$= I\times mI\times[aI+(p-a)J]+(s-1)I\times mI\times[(p+1)I-J]$

$$= [ma+m(s-1)(p+1)]I_{mps}+[m(p-a)-m(s-1)]I_{ms}\times J_p$$

$$= mpsI_{mps}$$

and so is the required complex Hadamard matrix. ***

COROLLARY 6.11. *Let m be the order of special Hadamard matrices and s be the order of a symmetric conference matrix. Then there exists a complex Hadamard matrix of order ms(s-1).*

PROOF. Choose $p = s-1$, $A = J$ and $B = I+iN$ in the theorem.

This corollary shows that if $m = p+1$ where $p \equiv 1 \pmod 4$ is a prime power, there is a complex Hadamard matrix for all orders 306m, 650m, 870m, 1406m, 2450m, 3782m even though no symmetric conference matrix is known for the orders 306, 650, 870, 1406, 2450, 3782.

## 6.3 OTHER CONSTRUCTIONS

THEOREM 6.12. *If there exists a skew-Hadamard matrix I+U of order h = n-1 and a real symmetric conference matrix of order n+1, then there exists a complex Hadamard matrix of order 2n(n-1).*

PROOF. Let N be the core of the real symmetric conference matrix. Choose N,I,J of order n and

$$A = \begin{bmatrix} J-I+iI & J-I-iI \\ -J+I-iI & J-I-iI \end{bmatrix} \qquad B = \begin{bmatrix} iI+N & -iI+N \\ iI+N & iI-N \end{bmatrix}.$$

Then $AB^* = BA^*$, $AA^* = I_2\times\big(4I+2(n-2)J\big)$, $BB^* = I_2\times\big(2(n+1)-2J\big)$. Hence

$$K = I\times A+U\times B$$

is the required complex Hadamard matrix. ***

We note that if $X^T = X$, $NX^T = XN^T$ and if X-I+iI has elements 1,-1,i,-i, then X could replace J in the above proof.

THEOREM 6.13. *Suppose there exists symmetric conference matrices N and M+I respectively of orders n and m = n-a. Further, suppose L is the core of N and there exists a complex matrix A of order n-1 satisfying*

$$LA^* = -AL^*,\ AA^* = aI+(n-a-1)J,\ A = A^*$$

*then there exists a complex Hadamard matrix of order (n-1)(n-a).*

PROOF. $K = I\times A+M\times(L+iI)$ is the required matrix. ***

COROLLARY 6.14. *If there exists a symmetric conference matrix of order n there exists a complex Hadamard matrix of order n(n-1).*

PROOF. Choose $A = J_n$ in the theorem 6.13. ***

THEOREM 6.15. *Suppose there exists a symmetric conference matrix C+I of order n = m-a+1 and another of order n+a with core N. Further suppose there exists a real matrix A of order n+a-1 satisfying*

$$A^T = A,\ AN = NA \text{ and } AA^* = aI+(n-1)J.$$

*Then there exists a complex Hadamard matrix of order 2n(n+a-1).*

PROOF. Let

$$X = \begin{bmatrix} iA & iA \\ -iA & iA \end{bmatrix} \qquad Y = \begin{bmatrix} I+N & -I+N \\ -I+N & -I-N \end{bmatrix}$$

then

$$XX^* = I_2\times\big(2aI+2(m-a)J\big),$$
$$YY^* = I_2\times\big(2(m+1)I-2J\big),$$
$$XY^* = -YX^*.$$

Now K = I×X+C×Y is the required complex Hadamard matrix. ***

COROLLARY 6.16. *Suppose there exist symmetric conference matrices of orders*

*(i) m+1 and (ii) m+1 and m-3 respectively, then there exist complex Hadamard matrices of orders*

*(i) 2m(m+1) and (ii) 2m(m-3) respectively.*

PROOF. Choose (i) $A = J_m$ and (ii) $A = J_m-2I$ in the theorem. ***

This corollary gives a complex Hadamard matrix of order 3116 for which no Hadamard matrix is yet known.

COROLLARY 6.17. *Suppose there exists a symmetric conference matrix of order n and another with type 1 core of order*

*(i)* $n+4q^{t-1}-1$, *where* $n = (q^{t+1}-1)/(q-1)+1-4q^{t-1}$, *q a prime power;*

*(ii)* $n+3x^2$, *where* $n = x^2+1$, $4x^2+1$ *is prime, x odd;*

*(iii)* $n+3x^2+8$, *where* $n = x^2+1$, $4x^2+9$ *is prime, x odd;*

*(iv)* $n+4(7b^2+1)-1$, *where* $n = 36b^2+6$, $64b^2+9 = 8a^2+1$ *is prime, a,b odd;*

*(v)* $n+4(7b^2+49)-1$, *where* $n = 36b^2+246$, $64b^2+441 = 8a^2+49$ *is prime, a odd, b even;*

*respectively; then there exist complex Hadamard matrices of orders*

(i) $2n(n+4q^{t-1}-1)$;

(ii) $2(x^2+1)(4x^2+1)$;

(iii) $2(x^2+1)(4x^2+9)$;

(iv) $12(6b^2+1)(64b^2+9)$;

(v) $12(6b^2+41)(64b^2+441)$.

PROOF. In theorem 6.16 choose for A the type 2 matrix given by (where we use the notation of chapter I):

(i) S; (ii) B; (iii) BO; (iv) O; (v) OO;

respectively and for N the type 1 core of the appropriate symmetric conference matrix.

THEOREM 6.18. *Suppose there exists a symmetric conference matrix C+I of order $n = n+1-2a$ and another of order $n+2a$ with core N. Further suppose there exists a real matrix A of order $n+2a-1$ satisfying*

$$A^T = A,\ AN = NA \text{ and } AA^* = 4aI+(n-2a-1)J.$$

*Then there exists a complex Hadamard matrix of order $2n(n+2a-1)$.*

PROOF. Let

$$X = \begin{bmatrix} iJ & iA \\ -iA & iJ \end{bmatrix}, \qquad Y = \begin{bmatrix} I+N & -I+N \\ -I+N & -I-N \end{bmatrix}$$

then

$$XX^* = I_2\times aI+(2m-a)J$$
$$YY^* = I_2\times 2(m+1)-2J$$
$$XY^* = -YX^*.$$

Now $K = I\times X+C\times Y$ is the required complex Hadamard matrix.

COROLLARY 6.19. *Suppose there exist symmetric conference matrices of order n and another with type 1 core of order $n+2a-1$*

(i) $(q^{t+1}-a)/(q-1)$ *where* $n = (q^{t+1}-1)/(q-1)-2a+1$,

$4a = 4q^{t-1} \equiv 0 \pmod 8$, *q a prime power,*

(ii) $64b^2+9$ *where* $n = 64b^2+10-2a$, $4a = 4(7b^2+1) \equiv 0 \pmod 8$,

$64b^2+9 = 8d^2+1$ *prime, d,b odd;*

(iii) $64b^2+441$ *where* $n = 64b^2+442-2a$, $4a = 4(7b^2+49) \equiv 0 \pmod 8$,

$64b^2+441 = 8d^2+49$ *prime, d odd, b even;*

*respectively, then there exist complex Hadamard matrices of orders (i)* $2[(q^{t+1}-1)/(q-1)-2q^{t-1}-1](q^{t+1}-1)/(q-1)$;

*(ii)* $4(25b^2+4)(64b^2+9)$;

*(iii)* $4(25b^2+172)(64b^2+441)$;

*respectively.*

PROOF. Use the type 2 matrices formed from the following classes for A (again using the notation of chapter I):

(i) S; (ii) B; (iii) BO; (iv) O; (v) OO;

respectively.

THEOREM 6.20. *Suppose there exists a symmetric conference matrix I+S of order s and two matrices A,W with elements 1,-1,i,-i of order p satisfying*

$$AW^* = WA^*$$
$$AA^* = aI+(p-a)J, \quad a = 2(p-s+1)$$
$$WW^* = (p+1)I-J.$$

*Then there exists a complex Hadamard matrix of order 2ps.*

PROOF. Let

$$X = \begin{bmatrix} iJ & iA \\ -iA & iJ \end{bmatrix}, \qquad Y = \begin{bmatrix} W & W \\ W & -W \end{bmatrix}$$

then

$$XX^* = I_2\times(aI+(2p-a)J),$$
$$YY^* = I_2\times(2(p+1)I-2J),$$
$$XY^* = -YX^*,$$

when A is type 1 and W is type 2. Now $K = I\times X+S\times Y$ is the required complex Hadamard matrix.

COROLLARY 6.21. *Suppose there exists a symmetric conference matrix of order s and a symmetric matrix W with elements 1,-1,i,-i of order s-3 satisfying $WW^* = (s-2)I-J$. Then there exists a complex Hadamard matrix of order 2s(s-3).*

PROOF. Use $p = s-3$ and $A = J-2I$ in the theorem. ***

# CHAPTER VII. HADAMARD ARRAYS

DEFINITION 7.1. An *Hadamard array* $H[h,k,\lambda]$, based on the indeterminates $x_1, x_2, \ldots, x_k$, $k \le h$, is an $h \times h$ array with entries chosen from $x_1, -x_1, x_2, -x_2, \ldots, x_k, -x_k$ in such a way that:

(i) in any row there are $\lambda$ entries $\pm x_1$, $\lambda$ entries $\pm x_2, \ldots,$ $\lambda$ entries $\pm x_k$; and similarly for columns;

(ii) the rows are formally orthogonal, in the sense that if $x_1, x_2, \ldots, x_k$ are realized as any elements of any commutative ring then the rows of the array are pairwise orthogonal; and similarly for columns.

We will distinguish several types: those of Williamson [175] which have $\lambda = 1$; those of Baumert-Hall [9] in which $h = 4t$, $k = 4$, $\lambda = t$, we write $BH[4t]$ for and $H[4t,4,t]$; those of Baumert-Hall-Welch $BHW[4t] = BH[4t]$ which are the same as the Baumert-Hall arrays except the blocks are circulant; and those which have zero elements.

## 7.1 THE HADAMARD ARRAYS OF WILLIAMSON AND BAUMERT-HALL.

The Baumert-Hall array $BH[12]$ is an $H[12,4,3]$ with each indeterminate repeated three times. It is

$$\begin{bmatrix}
A & A & A & B & -B & C & -C & -D & B & C & -D & -D \\
A & -A & B & -A & -B & -D & D & -C & -B & -D & -C & -C \\
A & -B & -A & A & -D & D & -B & B & -C & -D & C & -C \\
B & A & -A & -A & D & D & D & C & C & -B & -B & -C \\
B & -D & D & D & A & A & A & C & -C & B & -C & B \\
B & C & -D & D & A & -A & C & -A & -D & C & B & -B \\
D & -C & B & -B & A & -C & -A & A & B & C & D & -D \\
-C & -D & -C & -D & C & A & -A & -A & -D & B & -B & -B \\
D & -C & -B & -B & -B & C & C & -D & A & A & A & D \\
-D & -B & C & C & C & B & B & -D & A & -A & D & -A \\
C & -B & -C & C & D & -B & -D & -B & A & -D & -A & A \\
-C & -D & -D & -C & -C & -B & B & B & D & A & -A & -A
\end{bmatrix}$$

This design has been used to generate the Hadamard matrix of order 156.

It is a generalization of Williamson's design from [175] which is an $H[4,4,1]$ with each indeterminate repeated once:

$$\begin{bmatrix} A & B & C & D \\ -B & A & D & -C \\ -C & -D & A & B \\ -D & C & -B & A \end{bmatrix}.$$

This design was used to generate the Hadamard matrices of orders 92, 116 and 172.

7.2 THE ARRAY OF GOETHALS-SEIDEL. Assuming that the indeterminates in an $H[h,k,\lambda]$ are going to be replaced by matrices, it is necessary to ensure that $XY^T$ is symmetric for $X,Y \in \{A,B,C,D\}$. The usual method for doing this is to write $x_1,x_2,\ldots,x_h$ for the first row of $X \in \{A,B,C,D\}$ and then restrict the $x_i$ by demanding

$$x_i = x_{h+2-i}. \qquad (7.01)$$

The matrix X is then made circulant. (By theorem 1.15 of chapter I we see it is actually only necessary for X to be type 1.)

Goethals and Seidel [45] removed these restrictions on the $x_i$ by demanding that $X \in \{A,B,C,D\}$ be circulant and with $R = (r_{ij})$ defined by

$$r_{ij} = \begin{cases} 1 & j = h+1-i, \\ 0 & \text{otherwise}, \end{cases}$$

the matrices A,B,C,D should be used in the Goethals-Seidel array

$$GS = \begin{bmatrix} A & BR & CR & DR \\ -BR & A & -D^TR & C^TR \\ -CR & D^TR & A & -B^TR \\ -DR & -C^TR & B^TR & A \end{bmatrix}.$$

This matrix was first used to find skew-Hadamard matrices of orders 36 and 52.

Wallis and Whiteman further altered these conditions by demanding that the elements $x_1,\ldots,x_h$ of an additive abelian group, G, be ordered in some way. Then with W,Y subsets of G, if $X = (x_{ij}) \in \{A,B,D\}$ and $C = (c_{ij})$ define

$$x_{ij} = \begin{cases} 1 & x_j - x_i \in W \text{ (some W)} \\ -1 & \text{otherwise} \end{cases} \qquad c_{ij} = \begin{cases} 1 & x_j + x_i \in Y \\ -1 & \text{otherwise} \end{cases}.$$

The matrices A,B,C,D should be used in the following matrix.

$$H = \begin{bmatrix} A & B & C & D \\ -B^T & A^T & -D & C \\ -C & D^T & A & -B^T \\ -D^T & -C & B & A^T \end{bmatrix}. \qquad (7.02)$$

That is, that A,B,D should be type 1 and C type 2.

7.3 BAUMERT-HALL-WELCH ARRAYS. Five years passed from the publication of the Baumert-Hall array until L.R. Welch found his deceptively simple BHW[20] with each of A,B,C,D repeated five times; viz.

| | | | | | | | | | | | | | | | | | | | |
|---|---|---|---|---|---|---|---|---|---|---|---|---|---|---|---|---|---|---|---|
| -D | B | -C | -C | -B | C | A | -D | -D | -A | -B | -A | C | -C | -A | A | -B | -D | D | -B |
| -B | -D | B | -C | -C | -A | C | A | -D | -D | -A | -B | -A | C | -C | -B | A | -B | -D | D |
| -C | -B | -D | B | -C | -D | -A | C | A | -D | -C | -A | -B | -A | C | D | -B | A | -B | -D |
| -C | -C | -B | -D | B | -D | -D | -A | C | A | C | -C | -A | -B | -A | -D | D | -B | A | -B |
| B | -C | -C | -B | -D | A | -D | -D | -A | C | -A | C | -C | -A | -B | -B | -D | D | -B | A |
| -C | A | D | D | -A | -D | -B | -C | -C | B | -A | B | -D | D | B | -B | -A | -C | C | -A |
| -A | -C | A | D | D | B | -D | -B | -C | -C | B | -A | B | -D | D | -A | -B | -A | -C | C |
| D | -A | -C | A | D | -C | B | -D | -B | -C | D | B | -A | B | -D | C | -A | -B | -A | -C |
| D | D | -A | -C | A | -C | -C | B | -D | -B | -D | D | B | -A | B | -C | C | -A | -B | -A |
| A | D | D | -A | -C | -B | -C | -C | B | -D | B | -D | D | B | -A | -A | -C | C | -A | -B |
| B | -A | -C | C | -A | A | B | -D | D | B | -D | -B | C | C | B | -C | A | -D | -D | -A |
| -A | B | -A | -C | C | B | A | B | -D | D | B | -D | -B | C | C | -A | -C | A | -D | -D |
| C | -A | B | -A | -C | D | B | A | B | -D | C | B | -D | -B | C | -D | -A | -C | A | -D |
| -C | C | -A | B | -A | -D | D | B | A | B | C | C | B | -D | -B | -D | -D | -A | -C | A |
| -A | -C | C | -A | B | B | -D | D | B | A | -B | C | C | B | -D | A | -D | -D | -A | -C |
| -A | -B | -D | D | -B | B | -A | C | -C | -A | C | A | D | D | -A | -D | B | C | C | -B |
| -B | -A | -B | -D | D | -A | B | -A | C | -C | -A | C | A | D | D | -B | -D | B | C | C |
| D | -B | -A | -B | -D | -C | -A | B | -A | C | D | -A | C | A | D | C | -B | -D | B | C |
| -D | D | -B | -A | -B | C | -C | -A | B | -A | D | D | -A | C | A | C | C | -B | -D | B |
| -B | -D | D | -B | -A | -A | C | -C | -A | B | A | D | D | -A | C | B | C | C | -B | -D |

Then Jennifer Wallis [157] found that the Welch idea of circulant blocks could be used with the Goethals-Seidel matrix to obtain H[4t,4,t] = BHW[4t]. First we exhibit BHW[12].

| | | | | | | | | | | | |
|---|---|---|---|---|---|---|---|---|---|---|---|
| A | B | C | B | -C | D | C | D | -A | D | A | -B |
| C | A | B | -C | D | B | D | -A | C | A | -B | D |
| B | C | A | D | B | -C | -A | C | D | -B | D | A |
| -B | C | -D | A | B | C | -D | B | -A | C | -A | D |
| C | -D | -B | C | A | B | B | -A | -D | -A | D | C |
| -D | -B | C | B | C | A | -A | -D | B | D | C | -A |
| -C | -D | A | D | -B | A | A | B | C | -B | -D | C |
| -D | A | -C | -B | A | D | C | A | B | -D | C | -B |
| A | -C | -D | A | D | -B | B | C | A | C | -B | -D |
| -D | -A | B | -C | A | -D | B | D | -C | A | B | C |
| -A | B | -D | A | -D | -C | D | -C | B | C | A | B |
| B | -D | -A | -D | -C | A | -C | B | D | B | C | A |

Here A,B,C,D have defining relations as in (7.01). (Again, while the $x_i = x_{n+2-i}$ is necessary, it is sufficient for the matrices to be type 1 and not necessarily circulant.)

Write $T = T_n = (t_{ij})$ for the n×n matrix defined by

$$t_{i,i+1} = 1, \quad i = 1,2,\ldots,n-1,$$
$$t_{n,1} = 1,$$
$$t_{i,j} = 0, \quad \text{otherwise.}$$

If P is an n×n array of m×m submatrices $P_{ij}$ where $P_{i+1,j+1} = P_{ij}$ (subscripts reduced modulo n), that is

$$P = \sum_{j=1}^{n} T^{j-1} \times P_{1j}$$

(where × denotes Kronecker product), we shall say P is formed by circulating

$$(P_{11}, P_{12}, \ldots, P_{1n}).$$

We denote by R a square back-diagonal matrix whose order shall be determined by context. If $R = (r_{ij})$ is of order n then

$$r_{ij} = 1 \quad \text{when } i+j = n+1,$$
$$r_{ij} = 0 \quad \text{otherwise.}$$

We consider a set of four n×n arrays X,Y,Z and W which are formed by circulating their first rows; the entries shall be m×m matrices chosen from a set of four matrices {A,B,C,D}.

LEMMA 7.2. *If A,B,C and D commute in pairs then X,Y,Z and W commute in pairs.*

In particular, lemma 7.2 is satisfied if A,B,C and D are circulant or type 1 incidence matrices.

LEMMA 7.3. *If S and P are chosen from {X,Y,Z,W} and if A,B,C and D are circulant matrices then*

$$SRP^T = PRS^T. \tag{7.31}$$

PROOF. It is known (see lemma 1.17), that equation (7.31) would hold if S and P were circulant. In particular

$$E_i R F_j^T = F_j R E_i^T$$

when $E_i$ and $F_j$ belong to {A,B,C,D}, and

$$T^i R T^{n-j} = T^j R T^{n-i}.$$

If we write

$$S = \sum_{i=0}^{n-1} T^i \times E_i, \quad P = \sum_{j=0}^{n-1} T^j \times F_j, \quad \text{then}$$

$$\begin{aligned} SRP^T &= \sum_{i=0}^{n-1}\sum_{j=0}^{n-1} (T^i \times E_i)R(T^{n-j} \times F_j^{\ T}) \\ &= \sum\sum (T^i \times E_i)(R \times R)(T^{n-j} \times F_j^{\ T}) \\ &= \sum\sum (T^i R T^{n-j} \times E_i R F_j^{\ T}) \\ &= \sum\sum (T^j R T^{n-i} \times F_j R E_i^{\ T}) \\ &= PRS^T. \end{aligned}$$

***

LEMMA 7.4. (Joan Cooper and Jennifer Wallis) *Suppose there exist four type 1 (0,1,-1) matrices* $X_1, X_2, X_3, X_4$ *of order* $n$, *defined on the same abelian group* $G$, *such that* (i) $X_i * X_j = 0$, $i \neq j$, (* *the Hadamard product),*

(ii) $\sum_{i=1}^{4} X_i$ *is a* $(1,-1)$ *matrix,*

(iii) $\sum_{i=1}^{4} X_i X_i^T = nI_n.$

*Further suppose* $A,B,C,D$ *are indeterminates. Define*

$$X = X_1 \times A + X_2 \times B + X_3 \times C + X_4 \times D,$$

$$Y = X_1 \times -B + X_2 \times A + X_3 \times D + X_4 \times -C,$$

$$Z = X_1 \times -C + X_2 \times -D + X_3 \times A + X_4 \times B,$$

$$W = X_1 \times -D + X_2 \times C + X_3 \times -B + X_4 \times A,$$

*then with* $R = [1]$ *for order 1, and* $R$ *of order* $n$ *defined on* $G$ *given by lemma 1.17,*

$$H = \begin{bmatrix} X & YR & ZR & WR \\ -YR & X & -W^TR & Z^TR \\ -ZR & W^TR & X & -Y^TR \\ -WR & -Z^TR & Y^TR & X \end{bmatrix}$$

*is a* $BHW[4n]$.

PROOF. The verification is straightforward. ***

THEOREM 7.5. *Suppose there exist four (0,1,-1) matrices* $X_1, X_2, X_3, X_4$ *of order* $n$ *which satisfy*

(i) $X_i * X_j = 0$, $i \neq j$, $i,j = 1,2,3,4$,

(ii) $\sum_{i=1}^{4} X_i X_i^T = nI_n.$

*Let* $x_i$ *be the number of positive elements in each row and column of* $X_i$ *and* $y_i$ *be the number of negative elements in each row and column of* $X_i$. *Then*

(a) $\sum_{i=1}^{4} (x_i + y_i) = n,$

(b) $\sum_{i=1}^{4} (x_i - y_i)^2 = n.$

PROOF. (a) follows immediately from (ii). To prove (b) we consider the four (1,-1) matrices

$$Y_1 = -X_1+X_2+X_3+X_4$$

$$Y_2 = X_1-X_2+X_3+X_4$$

$$Y_3 = X_1+X_2-X_3+X_4$$

$$Y_4 = X_1+X_2+X_3-X_4.$$

From lemma 4.15 we know that $4-\{n;k_1,k_2,k_3,k_4;\ \sum_{i=1}^{4} k_i - n\}$ supplementary difference sets may be used to form an Hadamard matrix of order 4n. Now

$$\sum_{i=1}^{4} Y_i Y_i^T = 4nI_n,$$

so $Z_i = \frac{1}{2}(Y_i+J)$, $i = 1,2,3,4$, are the incidence matrices (or permutations of them) of

$$4-\{n;y_1+x_2+x_3+x_4,\ x_1+y_2+x_3+x_4,\ x_1+x_2+y_3+x_4,\ x_1+x_2+x_3+y_4;\ 2\sum_{i=1}^{4} x_i\}$$

supplementary difference sets. Using (1.21) we have

$$2\sum_{i=1}^{4} x_i(n-1) = \sum_{i=1}^{4} (x_1+x_2+x_3+x_4+y_i-x_i)(x_1+x_2+x_3+x_4+y_i-x_i-1)$$

or writing $x_1+x_2+x_3+x_4 = w$, $t = y_1+y_2+y_3+y_4$, $n = w+t$,

$$2w(n-1) = \sum_{i=1}^{4} (x+y_i-x_i)(w+y_i-x_i-1)$$

$$= 4w^2+2w\sum_{i=1}^{4} (y_i-x_i) + \sum_{i=1}^{4} (y_i-x_i)^2 - \sum_{i=1}^{4} (y_i-x_i) - 4w$$

$$= 4w^2+2w(t-w) + \sum_{i=1}^{4} (y_i-x_i)^2 - (t-w) - 4w.$$

So

$$\sum_{i=1}^{4} (y_i-x_i)^2 = n,$$

as required. ***

LEMMA 7.6. (Cooper and Wallis) *The matrices $X_1, X_2, X_3, X_4$ of theorem 7.4 exist for orders 4t where $t \in \{x$: x is an odd integer, $1 \le x \le 19$ <u>or</u> $x = 25\}$.*

PROOF. In the table sets of elements $g_i$ from some abelian group of order v are given, and to some of the $g_i$ a sign - is attached. This sign does <u>not</u> indicate inverse in the additive group. Rather, for each set one forms the type 1 incidence matrix of the subset of elements which are not preceded by -, and subtracts from it the type 1 incidence matrix of the subset of elements which are preceded by minus. The four matrices thus formed should then be used as in lemma 7.4 to obtain the result.

***

| t | | $X_1, X_2, X_3, X_4$ |
|---|---|---|
| 3 | $1^2+1^2+1^2+0^2$ | {1}, {2}, {3} |
| 5 | $2^2+1^2+0^2+0^2$ | {1,2}, {5}, {3,-4} |
| 7 | $2^2+1^2+1^2+1^2$ | {1,2}, {5}, {3,6,-7}, {4} |
| 9 | $2^2+2^2+1^2+0^2$ | {1,6}, {2,8}, {9}, {3,4,-5,-7} |
| | $3^2+0^2+0^2+0^2$ | {1,2,7}, {3,-9}, {4,-8}, {5,-6} |
| 11 | $3^2+1^2+1^2+0^2$ | {1,5,7,8,-9}, {11}, {2,3,-4,-6,10} |
| 13 | $3^2+2^2+0^2+0^2$ | {1,7,9}, {4,5,8,-10}, {-2,-3,6,11,-12,13}<br>or<br>{1,3,9}, {2,5,6,-13}, {4,-7,-8,10,-11,12} |
| | $2^2+2^2+2^2+1^2$ | {1,5}, {3,4,-6,-9,10,12}, {7,13}, {-2,8,11}<br>or<br>{1,2,5,-9}, {3,4,-6,10,-11,12}, {7,13}, {8} |
| 15 | $3^2+2^2+1^2+1^2$ | {1,2,6}, {8,9}, {10,-11,-3}, {-3,-4,5,7,12,14,-15} |
| 17 | $4^2+1^2+0^2+0^2$ | {1,4,8,16}, {2,13,-15}, {9,-17}, {3,5,-6,-7,-10,-11,12,14}<br>or<br>{1,5,10,12}, {3,4,-9}, {8,-15}, {2,-6,-7,11,-13,14,16,-17}<br>or<br>{1,2,-3,-4,-5,-6,-9,-14,15,-16}, {10,11,-17}, {7,-8}, {12,-13} |
| 19 | $3^2+3^2+1^2+0^2$ | {1,2,13}, {7,11,17}, {4,-9,-12,-14,15,16,18}, {3,5,-6,8,-10,-19} |
| 25 | $5^2+0^2+0^2+0^2$ | (using the irreducible equation $x^2=2x+2$)<br>{-0,1,x+1,2x+3,4x,4x+2,4x+3}, {2,-(x+4),-(2x),-(2x+1),2x+2,3x+1},<br>{x,x+3,-(3x),3x+2,-(3x+4),-(4x+1)},<br>{-3,4,x+2,-(2x+4),-(3x+3),4x+4} |

TABLE

COROLLARY 7.7. *There exists a BHW[4t] for* $t \in \{x: x$ *is an odd integer,* $1 \le x \le 19$ *or* $25\}$ *where each indeterminate occurs t times in each row and column.*

The matrices $X_1,X_2,X_3,X_4$ for $n = 13 = 3^2+2^2+0^2+0^2$, were found by listing the multiplicative cyclic group of order 12 generated by 2 to form the subgroup $C_0 = \{2^{4j}: j = 0,1,2\}$ of order 3 and its cosets $C_i = \{2^{4j+i}: j = 0,1,2\}$, $i = 1,2,3$. Then the first rows of $X_1,X_2,X_3,X_4$ may be obtained by using the sets

$$C_0 \cup (-C_1), \quad C_3 \cup \{-13\}, \quad C_2, \quad \phi$$

or

$$C_2 \cup (-C_3), \quad C_1 \cup \{-13\}, \quad C_0, \quad \phi$$

where $-C_i = \{-i: i \in C_i\}$, and the $X_j$ are formed as described in the proof of theorem 7.6.

For $n = 19 = 3^2+3^2+1^2+0^2$, the multiplicative cyclic group of order 18 generated by 2 was used to form the subgroup $C_0 = \{2^{6j}: j = 0,1,2\}$ of order 3 and its cosets $C_i = \{2^{6j+i}: j = 0,1,2\}$, $i = 1,\ldots,5$. Then $X_1,X_2,X_3,X_4$ were found, as above, by using the sets

$$C_1, C_3, C_0 \cup (-C_2), \quad \{0\} \cup C_4 \cup \{-C_5\}.$$

LEMMA 7.8. *Suppose there exist 4-*$\{2m+1; m; 2(m-1)\}$ *supplementary difference sets,* $X_i$, *and e is the* $1\times(2m+1)$ *matrix with every entry 1. Further suppose* $x \in X_i \Rightarrow -x \in X_i$, $i = 1,2,3,4$, *and the type 1 (1,-1) matrix generated by* $X_i$ *is* $Y_i$. *Then*

$$H = \begin{bmatrix} M(-1,-1,-1,-1) & M(e,e,e,e) \\ -M(e^T,e^T,e^T,e^T) & M(Y_1,Y_2,Y_3,Y_4) \end{bmatrix}, \qquad (7.81)$$

*where* $M(Y_1,Y_2,Y_3,Y_4)$ *is an Hadamard array H[4t,4,t] with each variable repeated t times, is an Hadamard matrix of order* $8t(m+1)$.

## 7.4 ARRAYS WITH EACH ELEMENT REPEATED ONCE.

These arrays have been studied by various authors including Olga Taussky [133,134,135], Folkman [43], Spencer [119], Storer [126], Wallis [152] and Spence [118]. The proofs in these papers used varied and interesting results. We give that of Spence.

Clearly, if H is an H[h,h,1] with each letter repeated only once, we may suppose that the sign associated with each element in the first row and column is positive (by changing the sign of each element in a row or column, if necessary) and that of the first row and column are identical (interchange rows and columns, if necessary). Suppose therefore that H satisfies the above conditions on the letters $x_1,x_2,\ldots,x_h$ and write -H for the design obtained by changing the sign of every indeterminate in H. Then we have the following

LEMMA 7.9.

$$\begin{bmatrix} H & -H \\ -H & H \end{bmatrix} \tag{7.91}$$

*is the multiplication table of a loop [see 2] L of order 2h with elements* $x_1, x_2, \ldots, x_h, -x_1, -x_2, \ldots, -x_h$.

PROOF. Since each indeterminate $x_1, x_2, \ldots, x_h$ occurs once and only once in each row and column of H, the array (7.91) is certainly a latin square. The first row and column are identical, so there is an identity and hence is a loop. ***

In the cases h = 1 and 2 it is easy to verify that L is a cyclic group of order 2 or 4 respectively. Henceforth we assume $h > 2$.

Suppose $h_1$ is the identity of L and write $h_1 = 1$. Then from (7.91) we have $(-1)h_i = -h_i = h_i(-1)$ and $(-1)^2 = 1$, hence $\{1,-1\} \subseteq Z$, the centre of L. Also from the orthogonality of the distinct rows of H, we have:

$$h_i h_k = h_j h_\ell \Rightarrow h_j h_k = -(h_i h_\ell) \quad \text{for} \begin{cases} h_i \neq \pm h_j \\ h_k \neq \pm h_\ell \end{cases} \tag{7.92}$$

Now, $1h_i = h_i 1$ and (7.92) means $h_i^2 = -1$ for $h_i \neq \pm 1$. Thus if $h_i x = x h_i$ for all $x \in L$, we must have for some $x \neq \pm 1, \pm h_i$ (such exists since $2h \geq 6$), $h_i^2 = -x^2 = 1$, i.e. $h_i = \pm 1$ which shows $Z = \{1,-1\}$.

Let $x, y \in L$ and $xy \neq yx$. Now suppose

$$x \neq \pm 1, \quad y \neq \pm 1 \quad \text{and} \quad x \neq \pm y. \tag{7.93}$$

Given $x, y \in L$ there exists a unique $t \in L$ such that $xy = tx$, whence, since $x \neq \pm t$ we have from (7.93), $ty = -x^2 = 1$. However $(-y)y = 1$ and cancellation yields $t = -y$, i.e. $xy = -yx$.

We now show any two elements of L generate a subgroup. First we verify the associative laws

(a) $x(xy) = x^2 y$,

(b) $(xy)x = x(yx)$,

(c) $(yx)x = yx^2$.

Since these are trivial when $x = \pm 1$, $y = \pm 1$ or $x = \pm y$, we assume none of these equalities hold. Write $x(xy) = 1.z$ so that, by (7.92), $xy = -xz = x(-z)$. Cancellation gives $y = -z$, i.e. $z = -y$, so that $x(xy) = -y = x^2 y$ which proves (a). Using the result that $xy \neq yx \Rightarrow xy = -yx$ for $x, y \in L$ and (a), we can prove (b) and (c).

$xy \neq yx \Rightarrow xy = -yx$ and then since any two elements form a subgroup, we have that x and y generate a quaternion group.

Suppose that

$$u, v, w \notin Z \text{ and all lie in difference cosets of } Z \text{ in } L. \tag{7.94}$$

Then there exists a unique $t \in L$ such that $uv = tw = v(-u)$, and hence (7.92) yields

$vw = -t(-u) = -ut$, and $u(vw) = u(-ut) = -u^2t = t$ (since $u^2 = -1$ as $u \notin Z$). But $(uv)w = (tw)w = -t$, i.e. $(uv)w = -u(vw)$ if $u,v,w$ satisfy (7.94). Thus if $(xy)z = x(yz)$, then at least one of $x,y,z \in Z$ or $x = \pm y$ or $x = \pm z$ or $y = \pm z$. It follows then since any two elements generate a subgroup, that if $x(yz) = (xy)z$ then $x$, $y$,$z$ generate a subgroup.

THEOREM 7.10. *The loop L defined in the previous lemma ($h > 2$) satisfies the following conditions.*

*(i) The centre Z has order two, and elements 1,-1 where $(-1)^2 = 1$, $-1 \neq 1$.*

*(ii) If $x \notin Z$ then $x^2 = -1$.*

*(iii) If $xy \neq yx$ then $xy = -yx$ and $x,y$ generate a quaternion group.*

*(iv) If $x(yz) = (xy)z$, then $x,y,z$ generate a subgroup.*

However, it was shown by Norton [90] that a loop satisfying the conditions of the theorem must be either a quaternion group, or a Cayley loop. Since a quaternion group has order 8 and a Cayley loop has order 16 it is immediate that $h = 4$ or $8$.

Clearly the arrays obtained from these loops are $H[h,h,1]$. Thus

THEOREM 7.11. *There are Hadamard arrays $H[h,h,1]$ with each letter repeated once and only once for $h = 1,2,4$ or $8$.*

In [152] this theorem is obtained constructively and in the process it is shown that any Hadamard array with each letter repeated once and only once is equivalent under the operations

(a) interchange rows or columns;

(b) multiply rows or columns by -1;

(c) replace any variable by its negative throughout the design;

to the array of appropriate order now given:

$$H[1,1,1] = [X],$$

$$H[2,2,1] = \begin{bmatrix} X & Y \\ Y & -X \end{bmatrix},$$

$$H[4,4,1] = \begin{bmatrix} A & B & C & D \\ -B & A & D & -C \\ -C & -D & A & B \\ -D & C & -B & A \end{bmatrix},$$

$$H[8,8,1] = \left[\begin{array}{cccc|cccc} A & B & C & D & E & F & G & H \\ -B & A & D & -C & F & -E & -H & G \\ -C & -D & A & B & G & H & -E & -F \\ -D & C & -B & A & H & -G & F & -E \\ \hline -E & -F & -G & -H & A & B & C & D \\ -F & E & -H & G & -B & A & -D & C \\ -G & H & E & -F & -C & D & A & -B \\ -H & -G & F & E & -D & -C & B & A \end{array}\right].$$

Storer [126] points out that Hurwitz [63] implies that an H[16;p] where each letter is repeated once and all the other entries are zero must satisfy

$$p < \frac{2 \log 16}{\log 2} + 2 \sim 10.1$$

that is $p \leq 10$.

The following is an H[16,9,1], but H[16,10,1] does not seem to exist:

$$\begin{bmatrix} H[8,8,1] & I_8 \times X \\ I_8 \times -X & H[8,8,1]^T \end{bmatrix}.$$

It would be interesting to know of other such arrays as this H[16,9,1].

## 7.5 ARRAYS OF ORDER DIVISIBLE BY 8.

DEFINITION 7.12. Hadamard matrices (and arrays) constructed using the array

$$\begin{bmatrix} A & B & C & D \\ -B & A & -D & C \\ -C & D & A & -B \\ -D & -C & B & A \end{bmatrix}$$

will be called *quaternion type*. If in addition A,B,C,D are circulant and symmetric, then the Hadamard matrices will be said to be *Williamson type*.

THEOREM 7.13. (Baumert and Hall) *Let H be an Hadamard matrix of the quaternion type of order $n = 4t$, with A,B,C,D symmetric matrices which are commutative in pairs. Then there exist such matrices of orders $2^i n$ for $i = 1,2,3,\ldots$ and Baumert-Hall arrays $BH[2^{i+1}]$.*

PROOF. Let

$$X = \begin{bmatrix} A & B \\ B & A \end{bmatrix}, \quad Y = \begin{bmatrix} A & -B \\ -B & A \end{bmatrix}, \quad Z = \begin{bmatrix} C & D \\ D & C \end{bmatrix}, \quad W = \begin{bmatrix} C & -D \\ -D & C \end{bmatrix}$$

then X,Y,Z,W are symmetric matrices which are commutative in pairs. Further, X,Y,Z, W satisfy

$$XX^T + YY^T + ZZ^T + WW^T = 2nI.$$

Hence

$$H = \begin{bmatrix} X & Y & Z & W \\ -Y & X & -W & Z \\ -Z & W & X & -Y \\ -W & -Z & Y & X \end{bmatrix}$$

is an Hadamard matrix of order 2n. Clearly this process may be iterated to provide matrices of all the indicated orders. If A,B,C,D are left as signed indeterminates, we have the Baumert-Hall array $BH[2^{i+1}]$. ***

This theorem is not significant in giving new Hadamard matrices, but in that quaternion type Hadamard matrices can be constructed for these orders. Similarly

LEMMA 7.14. *Let S+iR be a complex Hadamard matrix of order s which is symmetric. Further let H be an Hadamard matrix of quaternion type of order n = 4t with A,B,C,D symmetric matrices which commute in pairs. Then there exist such matrices of orders $s^i n$ for i = 1,2,3,... and Baumert-Hall arrays $BH[s^{i+1}]$.*

PROOF. Since S+iR is a complex Hadamard matrix, $SS^T+RR^T = sI_s$, $SR^T-RS^T = 0$. Let

$$\begin{aligned} X &= S\times A + R\times B \\ Y &= S\times -B + R\times A \\ Z &= S\times C + R\times D \\ W &= S\times -D + R\times C. \end{aligned}$$

Then X,Y,Z,W are symmetric and are commutative in pairs. Then proceeding as in the previous theorem we have the result. ***

COROLLARY 7.15. *If there exists a Baumert-Hall array BH[t], then there exists a Baumert-Hall array $BH[s^i t]$, i = 1,2,3,...,s the order of a symmetric complex Hadamard matrix.*

PROOF. The matrices X,Y,Z,W of the previous theorem are symmetric and commute in pairs when A,B,C,D are symmetric and commute in pairs. Then, replacing the signed indeterminates A,B,C,D of any Hadamard array BH[t] by X,Y,Z,W we obtain an array BH[st]. Clearly the process can now be repeated to obtain arrays $BH[s^i t]$ for i = 1,2,3,...

Now there are symmetric complex Hadamard matrices of order c when:

(i) there is a symmetric conference matrix of order c;

(ii) there is a symmetric Hadamard matrix of order c;

(iii) c = r(r+1), r ≡ 1 (mod 4) a prime power (corollary 6.7).

Many others exist, but we only list the most powerful here.

COROLLARY 7.16. *If there exists a Baumert-Hall array BH[t], then there exist Baumert-Hall arrays $BH[c^i t]$ and $BH[r^i(r+1)^i t]$ for i = 1,2,3,..., c the order of a symmetric* *conference matrix, r ≡ 1 (mod 4) a prime power.*

7.6 AN ANNOUNCEMENT OF TURYN. Richard J. Turyn [143] has announced that he has proved the following exciting theorem:

THEOREM 7.17. (Turyn) *There exist Baumert-Hall arrays BH[4t] and BH[20t] for* $t \in \{i: i = 1+2^a 10^b 26^c$*, a,b,c non-negative integers,* <u>*or*</u> $i \leq 23$*, where i is odd* <u>*or*</u> $i = 29\}$.

## CHAPTER VIII. CONSTRUCTIONS FOR HADAMARD MATRICES

In this chapter we attempt to present all the constructions for Hadamard matrices not already given.

First we define

$$i = \begin{bmatrix} 0 & 1 & 0 & 0 \\ -1 & 0 & 0 & 0 \\ 0 & 0 & 0 & -1 \\ 0 & 0 & 1 & 0 \end{bmatrix}, \quad j = \begin{bmatrix} 0 & 0 & 1 & 0 \\ 0 & 0 & 0 & 1 \\ -1 & 0 & 0 & 0 \\ 0 & -1 & 0 & 0 \end{bmatrix}, \quad k = ij = \begin{bmatrix} 0 & 0 & 0 & 1 \\ 0 & 0 & -1 & 0 \\ 0 & 1 & 0 & 0 \\ -1 & 0 & 0 & 0 \end{bmatrix},$$

and

$$\left.\begin{aligned} P &= \begin{bmatrix} 0 & -1 \\ 1 & 0 \end{bmatrix} \times I_{\frac{1}{2}m}, \\ K &= i \times I_{\frac{1}{4}m}, \\ L &= j \times I_{\frac{1}{4}m}, \\ M &= k \times I_{\frac{1}{4}m}, \end{aligned}\right\} \qquad (8.01)$$

which are all of order m.

We note

$$PP^T = KK^T = LL^T = MM^T = I, \quad K^T = -K, \quad L^T = -L, \quad M^T = -M,$$

$$KL = M = -LK, \quad LM = K = -ML, \quad MK = L = -KM.$$

The orders of K,L,M and P should be derived from the context.

Now we recall lemma 30 of part 1.

THEOREM 8.1. *If $H_1$ is an Hadamard matrix of order $h_1$ and $H_2$ is an Hadamard matrix of order $h_2$ then $H_1 \times H_2$ is an Hadamard matrix of order $h_1h_2$.*

There is an Hadamard matrix of order 2,

$$\begin{bmatrix} 1 & 1 \\ 1 & -1 \end{bmatrix},$$

so

COROLLARY 8.2. *There is an Hadamard matrix of order $2^t$, t a non-negative integer.*

For completeness we also restate lemma 2.2:

THEOREM 8.3. *If $m = 2^t\Pi(p_i^{r_i}+1)$ where t is a non-negative integer and $p_i^{r_i} \equiv 3 \pmod 4$ is a prime power then there are amicable Hadamard matrices of order m.*

8.1 USING COMPLEX HADAMARD MATRICES. This idea originated with Richard J. Turyn and leads to interesting results.

THEOREM 8.4. (Turyn) *Let $k > 1$ be the order of an Hadamard matrix $H$ and $c$ the order of a complex Hadamard matrix $C$ then there exists an Hadamard matrix of order $kc$.*

PROOF. If $C = X+iY$,

$$CC^* = (X+iY)(X^T-iY^T) = XX^T+YY^T+i(YX^T-XY^T) = cI$$

so equating real and imaginary parts

$$YX^T = XY^T \text{ and } XX^T+YY^T = cI.$$

Now choose

$$R = H\times X + PH\times Y$$

and

$$RR^T = kI\times cI.$$

Clearly R is (1,-1), so R is Hadamard. ***

COROLLARY 8.5. *There exists a complex Hadamard matrix for every order for which there is a symmetric conference matrix. So, if $n$ is the order of a symmetric conference matrix and $k > 1$ is the order of an Hadamard matrix, there is an Hadamard matrix of order $kn$.*

COROLLARY 8.6. *There exists a symmetric conference matrix of order $q^s+1$, where $q^s \equiv 1 \pmod 4$ is a prime power. So if $k > 1$ is the order of an Hadamard matrix, there exist Hadamard matrices of order $k(q^s+1)$ and hence $2(q^s+1)$.*

8.2 CONSTRUCTIONS FOR HADAMARD MATRICES USING SKEW-HADAMARD AND AMICABLE HADAMARD MATRICES. The theorem of this section is due to the author.

THEOREM 8.7. *If there exist*

*(i) a skew-Hadamard matrix $H = U+I$ of order $h$,*

*(ii) amicable Hadamard matrices $M = W+I$ and $N = N^T$ of order $m$, and*

*(iii) three (1,-1) matrices $X,Y,Z$ of order $x \equiv 3 \pmod 4$ satisfying*

*(a) $XY^T, YZ^T$ and $ZX^T$ all symmetric, and*

*(b)* $XX^T = aI+(x-a)J$, *a integer,*
$YY^T = cI+(x-c)J$, $(m-1)c = m(x-h+1)-a$, *c integer,*
$ZZ^T = (x+1)I-J$,

*then*

$$\overline{H} = U\times N\times Z + I_h\times W\times Y + I_h\times I_m\times X$$

*is an Hadamard matrix of order $mxh$.*

PROOF. Since H is skew-Hadamard $U^T = -U$ and $UU^T = (h-1)I_h$. $M = W+I$ and $N$ being amicable means $W^T = -W$, $WW^T = (m-1)I_m$, $MN^T = NM^T$, $MM^T = NN^T = mI_m$ and $N^T = N$. Now lemma 2.1 shows $WN^T = NW^T$,

$$\begin{aligned}
\overline{H}\overline{H}^T &= (U\times N\times Z + I_h\times W\times Y + I_h\times I_m\times X)(U^T\times N^T\times Z^T + I_h\times W^T\times Y^T + I_h\times I_m\times X^T) \\
&= UU^T\times NN^T\times ZZ^T + I_h\times WW^T\times YY^T + I_h\times I_m\times XX^T \\
&\quad + U^T\times WN^T\times YZ^T + U\times NW^T\times ZY^T + U^T\times N^T\times XZ^T + U\times N\times ZX^T \\
&\quad + I_h\times W^T\times XY^T + I_h\times W\times YX^T \\
&= UU^T\times NN^T\times ZZ^T + I_h\times WW^T\times YY^T + I_h\times I_m\times XX^T \\
&\quad + (U+U^T)\times WN^T\times YZ^T + (U^T+U)\times N\times ZX^T + I_h\times (W+W^T)\times XY^T \\
&= (h-1)I_h\times mI_m\times\{(x+1)I_x - J_x\} \\
&\quad + I_h\times I_m\times\{c(m-1)I_x + (x-c)(m-1)J_x\} + I_h\times I_m\times\{aI_x + (x-a)J_x\} \\
&= I_{mh}\times\{[m(h-1)(x+1)+m+mx-mh]I_x - [m(h-1)-mh+m]J_x\} \\
&= mxhI_{mxh};
\end{aligned}$$

which completes the proof. ***

We now develope corollaries to show that those subclasses listed under class IV in the appendix may be obtained from this theorem.

COROLLARY 8.8. *If h is the order of a skew-Hadamard matrix and there exist two (1,-1) matrices, Y and Z, of order $x \equiv 3 \pmod 4$ satisfying*

$$YZ^T = ZY^T$$
$$YY^T = cI + (x-c)J, \quad c = x-h+1,$$
$$ZZ^T = (x+1)I - J,$$

*then there is an Hadamard matrix of order xh.*

PROOF. Put $X = Y$, $a = c$, $m = 1$ in the theorem. ***

COROLLARY 8.9. *If h is the order of a skew-Hadamard matrix, Z is the type 1 (1,-1) matrix of a (4t-1,v+t,v)-group difference set and Y is the type 2 (1,-1) matrix of a (4t-1,u+t-½h,u)-group difference set both defined on the same abelian group, then there exists an Hadamard matrix of order h(4t-1).*

PROOF. Put $x = 4t-1$. If Z is the type 1 (1,-1) matrix of a (4t-1,v+t,v)-group difference set, then from theorem 1.20

$$ZZ^T = 4tI - J = (x+1)I - J.$$

If Y is the type 2 (1,-1) matrix of a (4t-1,u+t-½h,u)-group difference set then, using theorems 1.10 and 1.20, Y satisfies

$$YY^T = (4t-h)I+(h-1)J = cI+(x-c)J, \quad c = 4t-h.$$

From theorem 1.12 $XY^T = YX^T$ and so using corollary 8.8 we have the result. ***

COROLLARY 8.10. *If h is the order of a skew-Hadamard matrix, Z is the circulant (1,-1) incidence matrix of a (4t-1,v+t,v)-difference set and Y is the backcirculant (1,-1) incidence matrix of a (4t-1,u+t-½h,u) difference set, then there exists an Hadamard matrix of order h(4t-1).*

PROOF. This is corollary 8.9 where the abelian group is the integers. ***

COROLLARY 8.11. *Let* $h = \dfrac{q^{n+1}-4q^n+4q^{n-1}+q-2}{q-1}$, *q a prime power, be the order of a skew-Hadamard matrix and 4t-1 be a prime or product of twin primes and suppose Y is the backcirculant (1,-1) incidence matrix of a*

$$\left(4t-1 = \frac{q^{n+1}-1}{q-1}, \frac{q^n-1}{q-1}, \frac{q^{n-1}-1}{q-1}\right)\text{-difference set,}$$

*(see table 1 section 1.2), then there exists an Hadamard matrix of order*

$$\frac{q^{n+1}-1}{q-1}\left(\frac{q^{n+1}-4q^n+4q^{n-1}+q-2}{q-1}\right). \qquad [\textit{note: } h = (q^n+q^{n-1}+\dots+q+1)-4q^{n-1}+1].$$

PROOF. From section 1.2 a circulant (1,-1) matrix Z exists for these 4t-1 and so we have the result from corollary 8.10. ***

COROLLARY 8.12. *Let h = (q-1)(q-2), q a prime power, be the order of a skew-Hadamard matrix and 4t-1 be a prime or product of twin primes and suppose Y is the back-circulant (1,-1) incidence matrix of a*

$$(4t-1 = q^2+q+1, q+1, 1)\text{-difference set}$$

*(see table 1 section 1.2), then there exists an Hadamard matrix of order* $(q-1)(q-2)(q^2+q+1)$.

COROLLARY 8.13. *If h is the order of a skew-Hadamard matrix H then there exist Hadamard matrices of order*

(i) *h(h-1),*

(ii) *h(h+3) when h+4 is the order of a symmetric Hadamard matrix, R.*

PROOF. (i) Let Z-I be the core of H. Put c = 0 and Y = J in corollary 8.8, then

$$YY^T = xJ = (h-1)J, \quad x = h-1,$$

and we have the result.

(ii) Normalize R and then remove the first row and column to obtain Z. Put c = 4 and Y = J-2I in corollary 8.8, then, since $Z^T = Z$

$$YZ^T = (J-2I)Z^T = ZJ-2Z = Z(J-2I) \text{ and}$$

$$YY^T = 4I+(x-4)J, \quad x = h+3,$$

and we have the result. ***

COROLLARY 8.14. *If $p \equiv 3 \pmod 4$ is a prime and $p^r \equiv 3 \pmod 4$ is a prime power, there exist Hadamard matrices of orders*

*(i) $p(p+1)$,*

*(ii) $p^r(p^r+1)$,*

*(iii) $(p^r+1)(p^r+4)$, when $p^r+4$ is also a prime power.*

PROOF. In lemma 2.2 we saw that for p and $p^r \equiv 3 \pmod 4$ there is a skew-Hadamard matrix of order p+1 and $p^r+1$. We also saw in lemma 2.2 that if $p^r+4$ is a prime power there is a symmetric Hadamard matrix of order $p^r+5$. Hence the conditions of corollary 8.13 are satisfied. ***

COROLLARY 8.15. *If m is the order of an amicable Hadamard matrix then there exists an Hadamard matrix for*

*(i) $m(m-1)$,*

*(ii) $m(m+3)$, when m+4 is the order of an amicable Hadamard matrix,*

*(iii) $u(u+3)$, when both u and u+4 are of the form $2^t\Pi(p_i^{r_i}+1)$.*

PROOF. By definition there is a skew-Hadamard matrix of order m so using corollary 8.13 (i) holds. Since m+5 is the order of an amicable Hadamard matrix there exists a symmetric Hadamard matrix of order m+4 and so using corollary 8.13 (ii) holds. ***

COROLLARY 8.16. *There exist Hadamard matrices of orders*

*(i) $u(u-1)$ when u is of the form $2^t\Pi(p_i^{r_i}+1)$,*

*(ii) $u(u+3)$ when u and u+4 are of the form $2^t\Pi(p_i^{r_i}+1)$.*

PROOF. There are amicable Hadamard matrices of every order $2^t\Pi(p_i^{r_i}+1)$ by lemma 2.2 so corollary 8.15 gives the result. ***

Now two more results using amicable Hadamard matrices.

Let H of order h+1 be any Hadamard matrix written in the form

$$H = \begin{bmatrix} 1 & 1 \ \ldots \ 1 \\ 1 & \\ \cdot & \\ \cdot & F \\ \cdot & \\ 1 & \end{bmatrix}. \tag{8.161}$$

Then $FF^T = (h+1)I-J$ and if

$$E = FG \tag{8.162}$$

where G is the back diagonal matrix, then $FE^T = FG^TF^T = FGF^T = EF^T$.

Then using theorem 8.7 we have

COROLLARY 8.17. *If $k+1$ is the order of any Hadamard matrix and $m$ is the order of amicable Hadamard matrices then, if there is a skew-Hadamard matrix of order $(k+1)(m-1)/m$ there is an Hadamard matrix of order $k(k+1)(m-1)$.*

PROOF. The proof follows from theorem 8.7 with F and E as in (8.161) and (8.162) and $X = F$, $Y = J$, $Z = E$. ***

## 8.3 USING COMPLEX HADAMARD MATRICES AND QUATERNION MATRICES.

THEOREM 8.18. (Jennifer Wallis) *Let $k > 2$ be the order of an Hadamard matrix and $c$ be the order of a complex Hadamard matrix with core $A+iB$ satisfying*

$$AA^T+BB^T = (c-1)I-J,\ AB^T = BA^T,\ A^T = A,\ B^T = B,\ BJ = 0 = AJ.$$

*Let $C+I$ be a symmetric conference matrix of order $n$. Further suppose $W$ is the type 2 $(1,-1)$ matrix of a $\{c-1, \tfrac{1}{2}(c-n)+u, u\}$-group difference set and that*

$$WA^T = AW^T, \quad WB^T = BW^T.$$

*Then*

$$Z = C\times A\times MH + C\times B\times LH + C\times I\times KH + I\times W\times H$$

*is an Hadamard matrix of order $kn(c-1)$.*

PROOF. First we show Z is a $(1,-1)$ matrix. On the diagonal there is $W\times H$, a $(1,-1)$ matrix, while off the diagonal we have

$$A\times MH + B\times LH + I\times KH.$$

Now $MH$, $LH$ and $H$ are $(1,-1)$ matrices and $A+B+I$ is a $(1,-1)$ matrix. So Z is a $(1,-1)$ matrix.

If U is a complex Hadamard matrix it may be normalized to

$$\begin{bmatrix} 1 & 1 \ldots\ldots 1 \\ 1 & \\ \cdot & \\ \cdot & A+iB+I \\ \cdot & \\ 1 & \end{bmatrix} \quad \text{where } (A+iB)(A+iB)^*+J = (c-1)I.$$

Then $AA^T+BB^T = (n-1)I-J$, $AJ = BJ = 0$ and $AB^T = BA^T$.

So

$$ZZ^T = CC^T\times AA^T\times MH(MH)^T + CC^T\times BB^T\times LH(LH)^T + CC^T\times I\times(KH)(KH)^T + I\times WW^T\times HH^T$$

$$= I_n\times k\{(n-1)(c-1)+(n-1)+(c-n)\}\times I_k$$
$$+ I_n\times k\{-(n-1)+(n-1)\}J\times I_k$$
$$= kn(c-1)I_{kn(c-1)}.$$

Hence Z is Hadamard. ***

COROLLARY 8.19. *Let $k > 2$ be the order of an Hadamard matrix H, n be the order of a symmetric conference matrix C+I and suppose there exist (1,-1) matrices X+I,W of order $p \equiv 1 \pmod 4$ satisfying*

$$(a)\quad XW^T = WX^T,\quad X^T = X,\quad W^T = W,$$
$$(b)\quad XX^T = pI-J,$$
$$WW^T = (p-n+1)I+(n-1)J$$

*then*

$$Z = C\times H\times X + I\times LH\times W + C\times MH\times I$$

*is an Hadamard matrix of order kpn.*

PROOF. Off the diagonal we have $H\times X + MH\times I$ a (1,-1) matrix and on the diagonal we have $LH\times W$ a (1,-1) matrix. X replaces A and zero replaces B in the theorem. Then

$$ZZ^T = CC^T\times HH^T\times XX^T + I\times LH(LH)^T\times WW^T + CC^T\times MH(MH)^T\times I$$
$$= (n-1)I\times kI\times\{pI-J\} + I\times kI\times\{(p-n+1)I+(n-1)J\} + k(n-1)I$$
$$= [kpn-kp+kp-kn+k+kn-k]I_{nkp}$$
$$= kpnI_{kpn}.$$

Hence Z is Hadamard. ***

COROLLARY 8.20. *Let $k > 2$ be the order of an Hadamard matrix H, n be the order of a symmetric conference matric C+I, then if $p \equiv 1 \pmod 4$ and there exists a $(p,\frac{1}{2}(p-n+1)+u,u)$-group difference set defined on the abelian group of GF(p)\0, then there exists an Hadamard matrix of order kpn.*

PROOF. Form the matrix Q of lemma 1.19 of order p. Then

$$Q^T = Q,\quad QQ^T = pI-J,$$

and is a type 1 matrix. Let W be the type 2 (1,-1) matrix of the $(p,\frac{1}{2}(p-c+1)+u,u)$-group difference set. Then

$$W^T = W,\quad WW^T = (p-n+1)I+(n-1)J.$$

Put X = Q in the above corollary and we have the result. ***

COROLLARY 8.21. *Let $k > 2$ be the order of an Hadamard matrix and n the order of a symmetric conference matrix, then there exists an Hadamard matrix of order* *(i)* $kn(n-1)$,

*(ii)* $kn(n+3)$, $n = p-3, n+4$ *the order of a symmetric conference matrix,*

*(iii)* $k(q-1)(q-2)(q^2+q+1)$ *when* $q$ *is a prime power,* $q^2+q+1$ *is prime and* $n = q^2-3q+2$,

*(iv)* $k(x^2+1)(4x^2+1)$ *when* $x$ *is odd,* $4x^2+1$ *is prime and* $n = x^2+1$,

*(v)* $k(x^2+1)(4x^2+9)$ *when* $x$ *is odd,* $4x^2+9$ *is prime and* $n = x^2+1$.

PROOF. Use the following results in the above corollary:

(i) from table 1 of section 1.2 a $(p,p,p)$-difference set always exists;

(ii) similarly a $(p,p-1,p-2)$-difference set always exists;

(iii) an S-type difference set $(p = q^2+q+1, q+1, 1)$ exists for q a prime power;

(iv) a B-type difference $\{p = 4x^2+1, x^2, \frac{1}{4}(x^2-1)\}$ exists for p a prime, x odd;

(v) a BO-type difference set $\{p = 4x^2+9, x^2+3, \frac{1}{4}(x^2+3)\}$ exists for p a prime, x odd.

Clearly other difference sets from table 1 could be used in the corollary, but these are omitted because they involve high order Hadamard matrices. W-type is omitted because of the difficulty in satisfying the condition $WX^T = XW^T$ of the corollary.

COROLLARY 8.22. *There exists a symmetric conference matrix of order* $q^s+1$ *where* $q^s \equiv 1 \pmod 4$. *If* $k > 2$ *is the order of an Hadamard matrix then there is an Hadamard matrix of order*

*(i).* $kq^s(q^s+1)$,

*(ii)* $k(q^s+1)(q^s+4)$, *when there is a symmetric conference matrix of order* $q^s+5$.

A result of Goethals and Seidel modified the theorem to allow $k = 2$ when $W = J$. This theorem can also be obtained by using theorem 8.4 and corollary 6.13.

THEOREM 8.23. (Goethals and Seidel) *Let* $k > 1$ *be the order of an Hadamard matrix* $H$ *and* $n$ *the order of a symmetric conference matrix* $C+I$ *then, writing* $X$ *for the core of* $C+I$,

$$Z = C\times H\times X + C\times PH\times I + I\times H\times J$$

*is an Hadamard matrix of order* $kn(n-1)$.

## 8.4 USING SYMMETRIC CONFERENCE MATRICES.

THEOREM 8.24. (Jennifer Wallis) *Let $k > 1$ be the order of an Hadamard matrix $H$, $n$ be the order of a symmetric conference matrix $C+I$, and suppose there exist four $(1,-1)$ matrices $X,Y,Z,W$ of order $p$ satisfying*

$$(a)\quad AB^T = BA^T \text{ for any } A,B \in \{X,Y,Z,W\},$$

$$(b)\quad XX^T+YY^T = 2(p+1)I-2J,$$

$$WW^T = bI+(p-b)J,$$

$$ZZ^T = aI+(p-a)J, \quad a = (p-n-b+1)m+b,$$

*then there exists an Hadamard matrix of order $kmnp$ for $m = 2$ or $4$.*

PROOF. If $m = 2$ use

$$M = \begin{bmatrix} Z & W \\ -W & Z \end{bmatrix} \quad \text{and} \quad N = \begin{bmatrix} X & Y \\ Y & -X \end{bmatrix};$$

for $m = 4$ use

$$M = \begin{bmatrix} Z & W & W & W \\ -W & Z & -W & W \\ -W & W & Z & -W \\ -W & -W & W & Z \end{bmatrix} \quad \text{and} \quad N = \begin{bmatrix} X & Y & X & Y \\ Y & -X & Y & -X \\ X & Y & -X & -Y \\ Y & -X & -Y & X \end{bmatrix}.$$

Then since $ZW^T$ and $XY^T$ are symmetric

$$MN^T = NM^T$$

and

$$MM^T = \{m(p+1-n)I+m(n-1)J\}\times I_m$$

$$NN^T = \{m(p+1)I_p-mJ_p\}\times I_m.$$

Now we consider

$$R = C\times H\times N + I\times PH\times M.$$

Clearly $R$ is a $(1,-1)$ matrix as $H\times N$ and $PH\times M$ are $(1,-1)$ matrices. Now

$$RR^T = CC^T\times HH^T\times NN^T + I\times(PH)(PH)^T\times MM^T$$

$$= I_n\times I_k\times k\{(n-1)m(p+1)+m(p+1-n)\}I_p\times I_m$$

$$= knmpI_{knmp}.$$

So $R$ is Hadamard. ***

A similar result holds for $C$ skew-symmetric but is not as powerful as results obtained in other ways.

COROLLARY 8.25. *Let $k > 1$ be the order of an Hadamard matrix. Suppose there exists a* *symmetric conference matrix of order n, a symmetric Hadamard matrix of order n+2. Then there exists an Hadamard matrix of order 2kn(n+1).*

PROOF. Let X-I of order n+1, be the core of the symmetric Hadamard matrix. Let Z = J and W = J-2I. Then if X = Y, X,Y,Z,W commute in pairs.

$$XX^T+YY^T = 2(n+2)I-2J,$$
$$WW^T = 4I+(n-3)J,$$
$$ZZ^T = (n+1)J,$$

so with p = n+1, a = 0, b = 4, m = 2 in the theorem we have the result. ***

COROLLARY 8.26. *Let n be the order of a symmetric conference matrix. If there exists a $\{p,\frac{1}{2}(p-n+1)m+u,u\}$-difference set for p a prime power, then there exists an Hadamard matrix of order mnp for m = 2 or 4.*

PROOF. If p is a prime form the type 1 matrix Q of (1.181), then with X = Q-I and Y = -Q-I, X and Y are the type 1 (1,-1) matrices of the theorem. Choose W = J, so b = 0. The type 2 (1,-1) matrix of the difference set, Z, satisfies

$$ZZ^T = (p-n+1)mI+[p-(p-n+1)m]J.$$

Checking we see conditions (a) and (b) of the theorem are satisfied and we have the result. ***

THEOREM 8.27. (Jennifer Wallis) *Let k > 1 be the order of an Hadamard matrix, H. Further, let n be the order of a symmetric conference matrix, I+R. Suppose there exist four (1,-1) matrices X,Y,Z,W of order q ≡ 1 (mod 4) satisfying*

*(a) $AB^T$ is symmetric for $A,B \in \{X,Y,Z,W\}$,*

*(b)* $$XX^T+YY^T = 2(q+1)I-J,$$
$$ZZ^T = aI+(q-a)J,$$
$$WW^T = bI+(q-b)J, \quad b = 2(q-n+1)-a.$$

*Then*

$$G = I\times\begin{bmatrix}1&0\\0&-1\end{bmatrix}\times H\times Z + I\times\begin{bmatrix}0&1\\1&0\end{bmatrix}\times H\times W + R\times\begin{bmatrix}1&0\\0&1\end{bmatrix}\times PH\times X + R\times\begin{bmatrix}0&-1\\1&0\end{bmatrix}\times PH\times Y$$

*is an Hadamard matrix of order 2kqn.*

PROOF. Write $M = \begin{bmatrix}1&0\\0&-1\end{bmatrix}\times H\times Z + \begin{bmatrix}0&1\\1&0\end{bmatrix}\times H\times W$

$$N = \begin{bmatrix}1&0\\0&1\end{bmatrix}\times PH\times X + \begin{bmatrix}0&-1\\1&0\end{bmatrix}\times PH\times Y.$$

Then

$$MM^T = k(a+b)I_{2kq}+(2q-a-b)kH_{2k}\times J_q.$$

$$NN^T = 2(q+1)kI_{2kq}-2kI_{2k}\times J_q.$$

$$MN^T+NM^T = 0.$$

Writing

$$G = I\times M + R\times N$$

we have

$$GG^T = I\times MM^T + RR^T\times NN^T$$

$$= [(a+b)k+2(c-1)(q+1)k]I_{2kqc} + [(2q-a-b)k-2k(c-1)]I_{2kc}\times J_q$$

$$= 2kqcI_{2kqc}.$$

So G is Hadamard. ***

COROLLARY 8.28. *Suppose $2t+3 \equiv 3 \pmod 4$ is prime. Suppose there exists a symmetric conference matrix of order*

*(i) $t+2$ (ii) $t-2$*

*then there exists an Hadamard matrix of order*

*(i) $2k(t+1)(t+2)$ (ii) $2k(t+1)(t-2)$*

*respectively, where $k > 1$ is the order of an Hadamard matrix.*

PROOF. For $2t+3$ prime there exist Szekeres difference sets A and B satisfying $x \in A \Rightarrow -x \notin A$, $y \in B \Rightarrow -y \in B$. Let X be the type 1 (1,-1) matrix of A and R the type 2 (1,-1) matrix of Y. The result follows by choosing Y = W the type 2 (1,-1) matrix of

(i) $(t+1,t+1,t+1)$ (ii) $(t+1,t,t-1)$

difference sets respectively. ***

COROLLARY 8.29. *If there is a symmetric conference matrix of order*

*(i) $q^n-q^{n-1}+q^{n-2}+\dots+q^2+q+2$, where $p^r = q^n+q^{n-1}+\dots+q+1$, +n integer,*

*(ii) $q^n-q^{n-1}+q^{n-2}+\dots+q^2+q$, $p^r$ as in (i),*

*(iii) $\frac{5x^2-1}{2}$, where $p = 4x^2+1$,*

*(iv) $\frac{5x^2+7}{2}$, where $p = 4x^2+9$,*

*(v) $50b^2+8$, where $p = 8a^2+1 = 64b^2+9$, b odd,*

*(vi) $50b^2+346$, where $p = 8a^2+49 = 64b^2+441$, b even;*

*with $p^r \equiv 1 \pmod 4$, $p^r$ and q prime powers, p prime and x and a odd, then there are Hadamard matrices of orders*

*(i) $4(q^n+q^{n-1}+\dots+1)(q^n-q^{n-1}+q^{n-2}+\dots+q^2+q+2)$,*

*(ii) $4(q^n+q^{n-1}+\dots+1)(q^n-q^{n-1}+q^{n-2}+\dots+q^2+q)$,*

*(iii)* $2(5x^2-1)(4x^2+1)$,

*(iv)* $2(5x^2+7)(4x^2+9)$,

*(v)* $8(25b^2+4)(64b^2+9)$,

*(vi)* $8(25b^2+173)(64b^2+441)$.

PROOF. In each case W of the theorem is a type 2 (1,-1) matrix generated by a difference set and $X = Q+I$, $Y = Q-I$ where Q is from (1.181). We use the notation of section 1.2. The proof follows with

(i) $Z = J$, $W = S$, q powers of 2,

(ii) $Z = J-2I$, $W = S$, $q \equiv 3 \pmod 4$,

(iii) $Z = J-2I$, $W = B$,

(iv) $Z = J-2I$, $W = BO$,

(v) $Z = J$, $W = O$,

(vi) $Z = J$, $W = OO$.

8.5 USING SKEW-HADAMARD MATRICES. Let F be the type 1 matrix of (1.181) of order $p = q^s \equiv 1 \pmod 4$ a prime power. F satisfies

$$FF^T = q^sI-J, \quad F^T = F.$$

Write

$$\left.\begin{aligned} X &= F+I \\ Y &= F-I \end{aligned}\right\} \qquad (8.291)$$

$$\text{Then } XY^T = (F+I_p)(F^T-I_p) = FF^T-I_p = (F-I_p)(F^T+I_p) = YX^T \text{ and}$$
$$XX^T = YY^T = (FF^T-2F+I_p) + (FF^T-2F+I_p) = 2(p+1)I_p-2J_p.$$

THEOREM 8.30. (Jennifer Wallis) *If there exist*

*(i) a skew-Hadamard matrix of order h,*

*(ii) four (1,-1) matrices X,Y,Z,W of order $p \equiv 1 \pmod 4$*

*satisfying*

*(a)* $XY^T$, $XZ^T$, $XW^T$, $YZ^T$, $YW^T$, $ZW^T$ *all symmetric, and*

*(b)* $XX^T+YY^T = 2(p+1)I_p-2J_p$

$WW^T = aI_p+(p-a)J_p$

$ZZ^T = bI_p+(p-b)J_p$, $\quad b = m(p+1-h-a)+a$

*where m = 2 or 4, then there is an Hadamard matrix of order mph.*

PROOF. If m = 2 use

$$M = \begin{bmatrix} Z & W \\ -W & Z \end{bmatrix} \qquad \text{and} \qquad N = \begin{bmatrix} X & Y \\ Y & -X \end{bmatrix}$$

for m = 4 use

$$M = \begin{bmatrix} Z & W & W & W \\ -W & Z & -W & W \\ -W & W & Z & -W \\ -W & -W & W & Z \end{bmatrix} \qquad \text{and} \qquad N = \begin{bmatrix} X & Y & X & Y \\ Y & -X & Y & -X \\ X & Y & -X & -Y \\ Y & -X & -Y & X \end{bmatrix}.$$

Then since $ZW^T$ and $XY^T$ are symmetric

$$MN^T = NM^T$$

and

$$MM^T = \{m(p+1-h)I_p + m(h-1)J_p\} \times I_m$$

$$NN^T = \{m(p+1)I_p - mJ_p\} \times I_m.$$

Now H is Hadamard so $HH^T = hI_h = UU^T + I_h$ and

$$\overline{H} = U \times N + I_h \times M$$

is the required Hadamard matrix of order mhp since

$$\begin{aligned} \overline{H}\overline{H}^T &= (U \times N + I_h \times M)(U^T \times N^T + I_h \times M^T) \\ &= UU^T \times NN^T + U^T \times MN^T + U \times NM^T + I_h \times MM^T \\ &= (h-1)I_h \times \{m(p+1)I_p - mJ_p\} \times I_m \\ &\quad + I_h \times \{m(p+1-h)I_p + m(h-1)J_p\} \times I_m \\ &= mphI_{mph}; \end{aligned}$$

which completes the proof. ***

COROLLARY 8.31. *Suppose there exist Szekeres difference sets A and B of size m in an abelian group of order $2n+1 \equiv 1 \pmod 4$. Further suppose $a \in A \Rightarrow -a \notin A$ and $b \in B \Rightarrow -b \in B$. Then if there is a skew-Hadamard matrix of order h = 2n*

*(i) $h = 2n$ (ii) $h = 2n+4$*

*then there is an Hadamard matrix of order*

*(i) $4n(2n+1)$ (ii) $4(n+2)(2n+1)$.*

PROOF. Let X be the type 2 (1,-1) matrix of A and Y the type 1 (1,-1) matrix of B. Then with m = 2 and

(i) $W = Z-2I$, $Z = J$ (ii) $Z = J-2I$, $W = J$

in the theorem we have the result. ***

Now we have shown in chapter 2 that such Szekeres difference sets exist for $2n+1 = \frac{1}{2}(p-1)$ when $p \equiv 3 \pmod 4$ is a prime power. So

COROLLARY 8.32. *Suppose* $2h+3 \equiv 3 \pmod 4$ *is a prime power and there exists a skew-Hadamard matrix of order*

(i) $h$ (ii) $h+4$

*then there is an Hadamard matrix of order*

(i) $2h(h+1)$ (ii) $2(h+4)(h+1)$.

More corollaries in the style of the next corollary are available if $2n+1$ is a prime or prime power. But they will not be as powerful as those we now give.

COROLLARY 8.33. *Let* $p^r$ *be a prime power* $\equiv 1 \pmod 4$, $q$ *be a prime power (may be a power of 2),* $x$ *and* $a$ *odd, and* $m = 2$ *or* $4$, *then if there is a skew-Hadamard matrix of order*

(i) $p^r-1$,

(ii) $p^r+1-\frac{4q^{n-1}}{m}$, *where* $p^r = q^n+q^{n-1}+\ldots+q+1$, $n$ *a positive integer,*

(iii) $p^r+1-4q^{n-1}+\frac{4q^{n-1}}{m}$, *with* $p^r$ *as in (ii),*

(iv) $p^r+1-4q^{n-1}+\frac{4(q^{n-1}-1)}{m}$, *with* $p^r$ *as in (ii),*

(v) $p^r-3-\frac{4(q^{n-1}-1)}{m}$, *with* $p^r$ *as in (ii),*

(vi) $\frac{5x^2+3}{2}$, *where* $p = 4x^2+1$,

(vii) $\frac{13x^2-5}{4}$, *where* $p = 4x^2+1$,

(viii) $\frac{7x^2+1}{4}$, *where* $p = 4x^2+1$,

(ix) $\frac{5x^2+11}{2}$, *where* $p = 4x^2+9$,

(x) $2(25b^2+3)$, *where* $p = 8a^2+1 = 64b^2+9$, $b$ *odd,*

(xi) $2(25b^2+173)$, *where* $p = 8a^2+49 = 64b^2+441$, $b$ *even,*

(xii) $57b^2+391$, *where* $p$ *and* $b$ *are as in (x),*

(xiii) $43b^2+297$, *where* $p$ *and* $b$ *are as in (x),*

*then there is an Hadamard matrix of order*

(i) $2p^r(p^r-1)$,

(ii) $[m(p^r+1)-4q^{n-1}]p^r$,

(iii) $[m(p^r+1-4q^{n-1}]p^r$,

(iv) $[m(p^r+1-4q^{n-1})+4(q^{n-1}-1)]p^r$,

(v) $[m(p^r-3)-4(q^{n-1}-1)]p^r$,

(vi) $(5x^2+3)(4x^2+1)$,

(vii) $(13x^2-5)(4x^2+1)$,

*(viii)* $(7x^2+1)(4x^2+1)$,

*(ix)* $(5x^2+11)(4x^2+9)$,

*(x)* $4(25b^2+3)(64b^2+9)$,

*(xi)* $4(25b^2+173)(64b^2+441)$,

*(xii)* $4(57b^2+391)(64b^2+441)$,

*(xiii)* $4(43b^2+297)(64b^2+441)$,

*respectively.*

PROOF. We use the notation of table 1 of section 1.2 and each matrix for Z and W if it is not J or J-2I is type 2. In cases (ii), (iii), (iv) and (v) m is not evaluated as q may be a power of 2. We use X and Y defined in (8.291). The corollary follows with the following substitutions in theorem 8.30:

(i) m = 2, Z = J-2I, W = J,
(ii) Z = S, W = J,
(iii) Z = J, W = S,
(iv) Z = J-2I, W = S,
(v) Z = S, W = J-2I,
(vi) m = 2, Z = B, W = J,
(vii) m = 4, Z = B, W = J-2I,
(viii) m = 4, Z = J-2I, W = B,
(ix) m = 2, Z = BO, W = J,
(x) m = 2, W = J-2I, Z = O,
(xi) m = 2, Z = OO, W = J,
(xii) m = 4, Z = OO, W = J-2I,
(xiii) m = 4, Z = J-2I, W = OO.

## 8.6 THE WILLIAMSON TYPE HADAMARD MATRICES.

Let

$$e = \begin{bmatrix} 1 & 0 & 0 & 0 \\ 0 & 1 & 0 & 0 \\ 0 & 0 & 1 & 0 \\ 0 & 0 & 0 & 1 \end{bmatrix}, \quad i = \begin{bmatrix} 0 & 1 & 0 & 0 \\ -1 & 0 & 0 & 0 \\ 0 & 0 & 0 & -1 \\ 0 & 0 & 1 & 0 \end{bmatrix}, \quad j = \begin{bmatrix} 0 & 0 & 1 & 0 \\ 0 & 0 & 0 & 1 \\ -1 & 0 & 0 & 0 \\ 0 & -1 & 0 & 0 \end{bmatrix},$$

$$k = ij = \begin{bmatrix} 0 & 0 & 0 & 1 \\ 0 & 0 & -1 & 0 \\ 0 & 1 & 0 & 0 \\ -1 & 0 & 0 & 0 \end{bmatrix}$$

as before. The four matrices e,i,j,k are isomorphic to the quaternions and satisfy the usual equations. Further, if A,B,C,D are four Hermitian matrices of order n satisfying the equation

$$AA^*+BB^*+CC^*+DD^* = 4nI \tag{8.331}$$

where $A^*$ is the hermitian conjugate of A. Since A,B,C,D are hermitian (8.331) is equivalent to

$$A^2+B^2+C^2+D^2 = 4nI. \tag{8.332}$$

If the four matrices commute in pairs and if G is the matrix defined by

$$B = A\times e + B\times i + C\times j + D\times k = \begin{bmatrix} A & B & C & D \\ -B & A & -D & C \\ -C & D & A & -B \\ -D & -C & B & A \end{bmatrix},$$

$GG^* = 4nI\times e$. Further if the elements of A,B,C and D are all plus one or minus one, the matrix G is an Hadamard matrix of order 4n. So we have

THEOREM 8.34. (Williamson) *Suppose there exist four symmetric (1,-1) matrices of order n which commute in pairs. Further, suppose*

$$A^2+B^2+C^2+D^2 = 4nI_n.$$

*Then*

$$H = \begin{bmatrix} A & B & C & D \\ -B & A & -D & C \\ -C & D & A & -B \\ -D & -C & B & A \end{bmatrix} \tag{8.340}$$

*is an Hadamard matrix of order 4n of* Williamson type *or* quaternion type.

The *Williamson type Hadamard matrices* have A,B,C,D circulant. When A,B,C and D are all polynomials in the same matrix they certainly commute in pairs. Let

$$T = \begin{bmatrix} 0 & 1 & 0 & \dots & 0 \\ 0 & 0 & 1 & \dots & 0 \\ & \dots & & \dots & . \\ & \dots & & \dots & . \\ & \dots & & \dots & . \\ 0 & 0 & 0 & \dots & 1 \\ 1 & 0 & 0 & \dots & 0 \end{bmatrix}$$

and note

$$T^n = I,\ (T^i)^* = (T^i)^T = T^{n-i}, \qquad i = 0,1,\dots,n-1. \tag{8.341}$$

Let

$$\left.\begin{aligned} A &= \sum_{i=0}^{n-1} a_i T^i \qquad a_i = +1 \text{ or } -1,\ a_{n-i} = a_i \\ B &= \sum_{i=0}^{n-1} b_i T^i \qquad b_i = +1 \text{ or } -1,\ b_{n-i} = b_i \\ C &= \sum_{i=0}^{n-1} c_i T^i \qquad c_i = +1 \text{ or } -1,\ c_{n-i} = c_i \\ D &= \sum_{i=0}^{n-1} d_i T^i \qquad d_i = +1 \text{ or } -1,\ d_{n-i} = d_i \end{aligned}\right\} \tag{8.342}$$

Because of (8.341) the four matrices A,B,C,D are hermitian and in fact real symmetric.

Since $TT^{n-1} = I$ and $T^T = T^{n-1}$, the matrix T is orthogonal and, since it is real, unitary. Hence, there exists unitary matrix U, such that $UTU^* = D$, where D is the diagonal matrix $D = [\omega_1,\omega_2,\ldots,\omega_n]$ and $\omega_1,\omega_2,\ldots,\omega_n$ are the n distinct nth roots of unity. If $UAU^* = A_1$,

$$A_1 = \sum_{i=0}^{n-1} a_i D^i$$

and with similar definitions of $B_1,C_1,D_1$ it follows from (8.332) that

$$A_1^2+B_1^2+C_1^2+D_1^2 = 4nI. \qquad (8.343)$$

Conversely, if (8.343) is true, (8.332) is true and G is an Hadamard matrix. Since the matrices $A_1,B_1,C_1$ and $D_1$ are all diagonal matrices (8.343) is equivalent to the n ordinary equations

$$\left(\sum_{i=0}^{n-1} a_i\omega_j^i\right)^2 + \left(\sum_{i=0}^{n-1} b_i\omega_j^i\right)^2 + \left(\sum_{i=0}^{n-1} c_i\omega_j^i\right)^2 + \left(\sum_{i=0}^{n-1} d_i\omega_j^i\right)^2 = 4n \qquad (8.344)$$

for all n nth roots of unity $\omega_j$. If (8.336) is to be satisfied for all nth roots $\omega_j$ it must be satisfied when $\omega_j = 1$. Hence (8.344) includes the equation

$$\left(\sum_{i=0}^{n-1} a_i\right)^2 + \left(\sum_{i=0}^{n-1} b_i\right)^2 + \left(\sum_{i=0}^{n-1} c_i\right)^2 + \left(\sum_{i=0}^{n-1} d_i\right)^2 = 4n.$$

If $p_1,n_1$; $p_2,n_2$; $p_3,n_3$ and $p_4,n_4$ denote the number of positive and negative $a_i,b_i,c_i$ and $d_i$ respectively, this is equivalent to

$$\sum_{i=1}^{4} (p_i-n_i)^2 = 4n. \qquad (8.345)$$

Since every integer is the sum of the squares of four positive or zero integers we know (8.345) has in all cases integer solutions for the differences $p_i-n_i$. From each such solution we obtain a possible choice for the number of positive $a_i,b_i,c_i$ and $d_i$.

EXAMPLE (due to Williamson). If n = 5, $20 = 3^2+3^2+1^2+1^2$. Therefore if

$$20 = [1-a_1(\omega+\omega^4)+a_2(\omega^2+\omega^3)]^2 + [1+b_1(\omega+\omega^4)+b_2(\omega^2+\omega^3)]^2$$
$$+ [1+c_1(\omega+\omega^4)+c_2(\omega^2+\omega^3)]^2 + [1+d_1(\omega+\omega^4)+d_2(\omega^2+\omega^3)]^2$$

for every fifth root of unity, we must have

$$1+2a_1+2a_2 = 1+2b_1+2b_2 = \pm 3$$

and

$$1+2c_1+2c_2 = 1+2d_1+2d_2 = \pm 1.$$

Hence $a_1 = a_2 = b_1 = b_2 = -1$ and $c_1 = -c_2$, $d_1 = -d_2$. We now try the possibilities

$$c_1 = -c_2 = d_1 = -d_2 = \pm 1$$
$$c_1 = -c_2 = -d_1 = d_2 = \pm 1$$

and find that

$$c_1 = -c_2 = -d_1 = d_2 = 1$$

gives the solution. Thus

$$20 = (1-\omega-\omega^2-\omega^3-\omega^4)^2 + (1-\omega-\omega^2-\omega^3-\omega^4)^2 + (1+\omega-\omega^2-\omega^3+\omega^4)^2 + (1-\omega+\omega^2+\omega^3-\omega^4)^2.$$

THEOREM 8.35. (Williamson) *For $a_i, b_i, c_i, d_i$ satisfying (8.342) and (8.332), if $a_0 = b_0 = c_0 = d_0$ exactly three of $a_j, b_j, c_j, d_j$, $j \neq 0$, have the same sign.*

PROOF. Write $P_1, N_1$; $P_2, N_2$; $P_3, N_3$ and $P_4, N_4$ for the positive and negative elements of A,B,C,D. Then (8.332) becomes, using $P_i + N_i = J$,

$$(2P_1-J)^2 + (2P_2-J)^2 + (2P_3-J)^2 + (2P_4-J)^2 = 4nI. \qquad (8.351)$$

Since $T^i J = J$ for all J we have $P_i J = p_i J$ and $J^2 = nJ$, so (8.351) becomes

$$4(P_1{}^2+P_2{}^2+P_3{}^2+P_4{}^2) - 4(p_1+p_2+p_3+p_4)J + 4nJ = 4nI.$$

So

$$P_1{}^2+P_2{}^2+P_3{}^2+P_4{}^2 = nI + (p_1+p_2+p_3+p_4-n)J.$$

Now $p_1, p_2, p_3, p_4$ and n are all odd so the coefficient of every off diagonal term on the right hand side is odd.

Since the $P_i$ are polynomials in T with integer coefficients so is $P_i{}^2$. Now if

$$P_i = \sum f_j T^j, \; f = 0 \text{ or } 1$$

then

$$P_i{}^2 = \sum (T^j)^2 + 2T^jT^\ell, \quad j \neq \ell$$
$$\equiv \sum (T^j)^2 (\text{mod } 2).$$

Hence

$$P_1{}^2+P_2{}^2+P_3{}^2+P_4{}^2 \equiv \{\sum_{a\in\alpha} (T^a)^2 + \sum_{b\in\beta} (T^b)^2 + \sum_{c\in\gamma} (T^c)^2 + \sum_{d\in\delta} (T^d)^2\} (\text{mod } 2)$$
$$\equiv (J-I)(\text{mod } 2), \qquad (8.352)$$

where $\alpha = \{i: i \neq 1 \text{ and } a_i = 1\}$, $\beta = \{i: i \neq 1 \text{ and } b_i = 1\}$, $\gamma = \{i: i \neq 1 \text{ and } c_i = 1\}$, $\delta = \{i: i \neq 1 \text{ and } d_i = 1\}$ ($a_i, b_i, c_i, d_i$ as in 8.342). Then in (8.352) each $T^i$ occurs 1 or 3 times. Since there is a unique $T^j$ giving $(T^j)^2 = T^i$ we have in $P_1, P_2, P_3, P_4$ each $T^i$ occurs 1 or 3 times. ***

Hence

$$a_i + b_i + c_i + d_i = \pm 2.$$

Now consider the expressions

$$\left.\begin{aligned}\mu_1 &= 1+2t_{11}(\omega+\omega^{n-1})+2t_{12}(\omega^2+\omega^{n-2})+\ldots+2t_{1s}(\omega^s+\omega^{n-s})\\ \mu_2 &= 1+2t_{21}(\omega+\omega^{n-1})+2t_{22}(\omega^2+\omega^{n-2})+\ldots+2t_{2s}(\omega^s+\omega^{n-s})\\ \mu_3 &= 1+2t_{31}(\omega+\omega^{n-1})+2t_{32}(\omega^2+\omega^{n-2})+\ldots+2t_{3s}(\omega^s+\omega^{n-s})\\ \mu_4 &= 1+2t_{41}(\omega+\omega^{n-1})+2t_{42}(\omega^2+\omega^{n-2})+\ldots+2t_{4s}(\omega^s+\omega^{n-s})\end{aligned}\right\} \quad (8.353)$$

where $s = \frac{1}{2}(n-1)$, and exactly one of $t_{ij}$, $i = 1,2,3,4$ is non-zero and equal to +1 or -1. Then

$$\left.\begin{aligned}x_1 &= -\mu_1+\mu_2+\mu_3+\mu_4 = 2\{1+\sum_{j=1}^{s}(-t_{1j}+t_{2j}+t_{3j}+t_{4j})(\omega^j+\omega^{n-j})\}\\ x_2 &= \mu_1-\mu_2+\mu_3+\mu_4 = 2\{1+\sum_{j=1}^{s}(t_{1j}-t_{2j}+t_{3j}+t_{4j})(\omega^j+\omega^{n-j})\}\\ x_3 &= \mu_1+\mu_2-\mu_3+\mu_4 = 2\{1+\sum_{j=1}^{s}(t_{1j}+t_{2j}-t_{3j}+t_{4j})(\omega^j+\omega^{n-j})\}\\ x_4 &= \mu_1+\mu_2+\mu_3-\mu_4 = 2\{1+\sum_{j=1}^{s}(t_{1j}+t_{2j}+t_{3j}-t_{4j})(\omega^j+\omega^{n-j})\}\end{aligned}\right\} \quad (8.354)$$

and if $t_{ij} \neq 0$ then the coefficient of $\omega^j+\omega^{n-j}$ in $x_i$ is different from the coefficient of $\omega^j+\omega^{n-j}$ in the other equations.

Thus (8.353) with

$$\begin{aligned}a_j &= -t_{1j}+t_{2j}+t_{3j}+t_{4j}\\ b_j &= t_{1j}-t_{2j}+t_{3j}+t_{4j}\\ c_j &= t_{1j}+t_{2j}-t_{3j}+t_{4j}\\ d_j &= t_{1j}+t_{2j}+t_{3j}-t_{4j}\end{aligned}$$

gives from (8.344)

$$x_1^2+x_2^2+x_3^2+x_4^2 = 16n$$

or

$$4\mu_1^2+4\mu_2^2+4\mu_3^2+\mu_4^2 = 16n.$$

Thus we have shown

THEOREM 8.36. (Williamson) *If there exist solutions to the equations*

$$\mu_i = 1+2\{\sum_{j=1}^{s} t_{ij}(\omega^j+\omega^{n-j})\} \qquad i = 1,2,3,4$$

*where $s = \frac{1}{2}(n-1)$, $\omega$ is an nth root of unity, exactly one of $t_{1j}, t_{2j}, t_{3j}, t_{4j}$ is non-zero and equals $\pm 1$ for each $j = 1,2,\ldots,s$, and*

$$\mu_1^2+\mu_2^2+\mu_3^2+\mu_4^2 = 4n,$$

*then there exist solutions to the equations*

$$A = \sum_{i=0}^{n-1} a_i T^i, \quad a_0 = 1, \; a_i = a_{n-i} = \pm 1$$

$$B = \sum_{i=0}^{n-1} b_i T^i, \quad b_0 = 1, \; b_i = b_{n-i} = \pm 1$$

$$C = \sum_{i=0}^{n-1} c_i T^i, \quad c_0 = 1, \; c_i = c_{n-i} = \pm 1$$

$$D = \sum_{i=0}^{n-1} d_i T^i, \quad d_0 = 1, \; d_i = d_{n-i} = \pm 1.$$

*That is, there exists an Hadamard matrix of order 4n.* ***

The table shows the $\mu_i$ found by Williamson, Baumert and Hall. We write $\omega_j$ for $\omega^j + \omega^{n-j}$ and $\omega_{2^j}$ for $\omega^{2^j} + \omega^{n-2^j}$. Williamson found the results for 148, 172, Baumert and Hall for 92 and Baumert for 116.

EXAMPLE. From the table we find for $36 = 4\times 9 = 1^2+1^2+3^2+5^2$

$$\mu_1 = 1$$
$$\mu_2 = 1$$
$$\mu_3 = 1-2\omega_2$$
$$\mu_4 = 1+2\omega_1+2\omega_3-2\omega_4.$$

Now $\overline{\mu_i} = \mu_i$ and

$$\begin{aligned}\mu_1^2+\mu_2^2+\mu_3^2+\mu_4^2 &= 1^2+1^2+(1-2\omega^2-2\omega^7)^2+(1+2\omega+2\omega^8+2\omega^3+2\omega^6-2\omega^4-2\omega^5)^2 \\ &= 1^2+1^2+(9+4\omega^4+4\omega^5-4\omega^2-4\omega^7) + (25+4\omega^2+4\omega^7-4\omega^4-4\omega^5) \\ &= 1^2+1^2+3^2+5^2.\end{aligned}$$

Note if $\omega = 1$

$$\mu_1 = 1, \; \mu_2 = 1, \; \mu_3 = -1, \; \mu_4 = 5.$$

We form the $x_i$ of equation (8.354)

$$x_1 = 2(1+\omega_1-\omega_2+\omega_3-\omega_4)$$
$$x_2 = 2(1+\omega_1-\omega_2+\omega_3-\omega_4)$$
$$x_3 = 2(1+\omega_1+\omega_2+\omega_3-\omega_4)$$
$$x_4 = 2(1-\omega_1-\omega_2-\omega_3+\omega_4).$$

Then writing $S_i = T^i+T^{n-i}$

$$A = I+S_1-S_2+S_3-S_4$$
$$B = I+S_1-S_2+S_3-S_4$$
$$C = I+S_1+S_2+S_3-S_4$$
$$D = I-S_1-S_2-S_3+S_4$$

and using them in

$$\begin{bmatrix} A & B & C & D \\ -B & A & -D & C \\ -C & D & A & -B \\ -D & -C & B & A \end{bmatrix}$$

gives an Hadamard matrix of order 36.

Recently Richard Turyn has found the first infinite class of Hadamard matrices of Williamson type, a very important result. We give here Whiteman's proof of this theorem which uses the notation and results of section 4.3 except that $q \equiv 1 \pmod 4$.

We write v for a non-square element in GF(q) and note that the polynomial $P(x) = x^2 - v$ is irreducible in GF(q).

If $\beta$ denotes a primitive fourth root of unity then (4.133) defines a fourth power character $\chi$ of $GF(q^2)$. Thus $\chi(a)$ is the Legendre symbol in GF(q) and lemma 4.14 applies.

THEOREM 8.37. (Whiteman) *Let q be a prime power* $\equiv 1 \pmod 4$ *and put* $n = (q+1)/2$. *Let* $\gamma$ *be a primitive root of* $GF(q^2)$. *Put* $\gamma^k = ax+b$ $\{a, b \in GF(q)\}$ *and define*

$$a_k = \chi(a), \quad b_k = \chi(b). \tag{8.371}$$

*Then the sums*

$$f(\zeta) = \sum_{i=0}^{n-1} a_{4i}\zeta^i, \quad g(\zeta) = \sum_{i=0}^{n-1} b_{4i}\zeta^i \tag{8.372}$$

*satisfy the identity*

$$f^2(\zeta) + g^2(\zeta) = q \tag{8.373}$$

*for each nth root of unity* $\zeta$ *including* $\zeta = 1$.

Note that when q is a prime $\equiv 1 \pmod 4$ and $\zeta = 1$ the identity (8.373) reduces to the classical result that every prime $\equiv 1 \pmod 4$ is representable as the sum of two squares of integers.

PROOF. Since $\gamma$ is a primitive element of $GF(q^2)$ the integer $k = (q+1)/2 = n$ is the only value of k in the interval $0 \leq k \leq q$ for which $tr(\gamma^k) = 0$. Put $\gamma^n = wx$ $\{w \in GF(q)\}$. Then the numbers $a_k, b_k$ in (8.317) satisfy the relations

$$\left.\begin{aligned} b_{k+n} &= -\chi(w)a_k \\ b_{k+2n} &= -b_k \end{aligned}\right\} \tag{8.374}$$

$$b_{k+4n} = b_k. \tag{8.375}$$

Using (8.374) and (8.375) in conjunction with (4.134) we get

$$a_k = (-1)^k a_{4n-k}, \quad b_k = (-1)^k b_{4n-k} \quad (0 \leq k \leq 4n). \tag{8.376}$$

Hadamard Matrices of the Williamson Type

| t | n | $\mu_1^2+\mu_2^2+\mu_3^2+\mu_4^2$ | $\mu_1$ | $\mu_2$ | $\mu_3$ | $\mu_4$ |
|---|---|---|---|---|---|---|
| 3 | 12 | $1^2+1^2+1^2+3^2$ | $1$ | $1$ | $1$ | $1-2\omega_1$ |
| 5 | 20 | $1^2+1^2+3^2+3^2$ | $1$ | $1$ | $1-2\omega_1$ | $1-2\omega_2$ |
| 7 | 28 | $1^2+3^2+3^2+3^2$ | $1$ | $1-2\omega_1$ | $1-2\omega_2$ | $1-2\omega_3$ |
| 7 | 28 | $1^2+1^2+1^2+5^2$ | $1$ | $1$ | $1+2\omega_1-2\omega_2$ | $1+2\omega_3$ |
| 9 | 36 | $3^2+3^2+3^2+3^2$ | $1-2\omega_1$ | $1-2\omega_2$ | $1-2\omega_3$ | $1-2\omega_4$ |
| 9 | 36 | $1^2+1^2+3^2+5^2$ | $1$ | $1+2\omega_1-2\omega_2$ | $1-2\omega_4$ | $1+2\omega_3$ |
| | | | $1$ | $1$ | $1-2\omega_2$ | $1+2\omega_1+2\omega_3+2\omega_4$ |
| 11 | 44 | $1^2+3^2+3^2+5^2$ | $1+2\omega_1-2\omega_2$ | $1-2\omega_4$ | $1-2\omega_5$ | $1+2\omega_3$ |
| 13 | 52 | $1^2+1^2+1^2+7^2$ | $1$ | $1$ | $1+2\omega_1-2\omega_4+2\omega_5$ $-2\omega_6$ | $1-2\omega_2-2\omega_3$ |
| | | | $1$ | $1+2\omega_4-2\omega_5$ | $1-2\omega_1-2\omega_6$ | $1-2\omega_2-2\omega_3$ |
| 13 | 52 | $3^2+3^2+3^2+5^2$ | $1-2\omega_2$ | $1-2\omega_4$ | $1-2\omega_1-2\omega_3+2\omega_5$ | $1+2\omega_6$ |
| 13 | 52 | $1^2+1^2+5^2+5^2$ | $1-2\omega_3+2\omega_4$ | $1-2\omega_2+2\omega_6$ | $1+2\omega_1$ | $1+2\omega_5$ |
| 15 | 60 | $1^2+3^2+5^2+5^2$ | $1$ | $1-2\omega_5$ | $1+2\omega_6$ | $1+2\omega_1-2\omega_2+2\omega_3+2\omega_4$ $-2\omega_7$ |
| | | | $1-2\omega_1+2\omega_7$ | $1-2\omega_3$ | $1+2\omega_2$ | $1+2\omega_4+2\omega_5-2\omega_6$ |
| | | | $1-2\omega_4+2\omega_6$ | $1-2\omega_1-2\omega_3+2\omega_5$ | $1+2\omega_7$ | $1+2\omega_2$ |
| 15 | 60 | $1^2+1^2+3^2+7^2$ | $1$ | $1$ | $1-2\omega_1-2\omega_5+2\omega_7$ | $1+2\omega_2-2\omega_3-2\omega_4-2\omega_6$ |
| 17 | 68 | $3^2+3^2+5^2+5^2$ | $1-2\omega_2$ | $1-2\omega_8$ | $1-2\omega_1+2\omega_5+2\omega_6$ | $1+2\omega_3-2\omega_4+2\omega_7$ |
| 17 | 68 | $1^2+3^2+3^2+7^2$ | $1-2\omega_3-2\omega_5+2\omega_6$ $+2\omega_7$ | $1-2\omega_2$ | $1-2\omega_8$ | $1-2\omega_1-2\omega_4$ |
| | | | $1$ | $1-2\omega_4-2\omega_5+2\omega_6$ | $1-2\omega_1-2\omega_3+2\omega_7$ | $1-2\omega_2-2\omega_8$ |
| 17 | 68 | $1^2+3^2+3^2+7^2$ | $1$ | $1-2\omega_2-2\omega_4+2\omega_5$ | $1-2\omega_1+2\omega_3-2\omega_8$ | $1-2\omega_6-2\omega_7$ |
| 19 | 76 | $1^2+5^2+5^2+5^2$ | $1$ | $1+2\omega_1-2\omega_2+2\omega_4$ | $1-2\omega_3+2\omega_6+2\omega_8$ | $1-2\omega_5+2\omega_7+2\omega_9$ |
| | | | $1-2\omega_3-2\omega_4+2\omega_5$ $+2\omega_9$ | $1+2\omega_2-2\omega_7+2\omega_2$ | $1+2\omega_6$ | $1+2\omega_1$ |
| | | | $1$ | $1-2\omega_3+2\omega_8+2\omega_9$ | $1+2\omega_4-2\omega_5+2\omega_7$ | $1+2\omega_1-2\omega_2+2\omega_6$ |
| 19 | 76 | $3^2+3^2+3^2+7^2$ | None | | | |

| t | n | $\mu_1^2+\mu_2^2+\mu_3^2+\mu_4^2$ | $\mu_1$ | $\mu_2$ | $\mu_3$ | $\mu_4$ |
|---|---|---|---|---|---|---|
| 19 | 76 | $1^2+1^2+5^2+7^2$ | $1$ | $1$ | $1+2\omega_1-2\omega_3+2\omega_8$ | $1+2\omega_2-2\omega_4-2\omega_5+2\omega_6-2\omega_7-2\omega_9$ |
| | | | $1-2\omega_2+2\omega_8$ | $1-2\omega_4+2\omega_7$ | $1+2\omega_3+2\omega_6-2\omega_9$ | $1-2\omega_1-2\omega_5$ |
| | | | $1+2\omega_4-2\omega_8$ | $1+2\omega_2-2\omega_5$ | $1+2\omega_1$ | $1-2\omega_3-2\omega_6+2\omega_7-2\omega_9$ |
| 21 | 84 | $3^2+5^2+5^2+5^2$ | $1-2\omega_7$ | $1+2\omega_3+2\omega_5-2\omega_8$ | $1-2\omega_2+2\omega_4+2\omega_6$ | $1+2\omega_1+2\omega_9-2\omega_{10}$ |
| 21 | 84 | $1^2+1^2+1^2+9^2$ | $1+2\omega_2+2\omega_3$ | $1-2\omega_6+2\omega_{10}$ | $1+2\omega_8-2\omega_9$ | $1+2\omega_1+2\omega_4+2\omega_5-2\omega_7$ |
| | | | $1$ | $1$ | $1-2\omega_5-2\omega_6+2\omega_7+2\omega_9$ | $1+2\omega_1+2\omega_2-2\omega_3+2\omega_4+2\omega_8-2\omega_{10}$ |
| | | | $1-2\omega_3+2\omega_9$ | $1+2\omega_8-2\omega_{10}$ | $1+2\omega_4-2\omega_5$ | $1+2\omega_1+2\omega_2-2\omega_6+2\omega_7$ |
| 21 | 84 | $1^2+3^2+5^2+7^2$ | $1-2\omega_4+2\omega_5$ | $1+2\omega_2-2\omega_6-2\omega_8-2\omega_9+2\omega_{10}$ | $1+2\omega_1$ | $1-2\omega_3-2\omega_7$ |
| | | | $1-2\omega_5+2\omega_9$ | $1+2\omega_2-2\omega_4-2\omega_{10}$ | $1+2\omega_6+2\omega_7-2\omega_8$ | $1-2\omega_1-2\omega_3$ |
| | | | $1-2\omega_6+2\omega_8$ | $1+2\omega_2-2\omega_4-2\omega_{10}$ | $1+2\omega_5+2\omega_7-2\omega_9$ | $1-2\omega_1-2\omega_3$ |
| 23 | 92 | $1^2+1^2+3^2+9^2$ | $1-2\omega_4-2\omega_8+2\omega_9+2\omega_{11}$ | $1+2\omega_5-2\omega_7$ | $1+2\omega_1-2\omega_3-2\omega_{10}$ | $1+2\omega_2+2\omega_6$ |
| 23 | 92 | $3^2+3^2+5^2+7^2$ | None | | | |
| 25 | 100 | $1^2+3^2+3^2+9^2$ | $1+2\omega_6-2\omega_{11}$ | $1-2\omega_1+2\omega_3-2\omega_{12}$ | $1+2\omega_4-2\omega_7-2\omega_9$ | $1+2\omega_2+2\omega_5-2\omega_8+2\omega_{10}$ |
| 25 | 100 | $5^2+5^2+5^2+5^2$ | $1+2\omega_1-2\omega_6+2\omega_9$ | $1+2\omega_7-2\omega_8+2\omega_{12}$ | $1+2\omega_2-2\omega_4+2\omega_5$ | $1-2\omega_3+2\omega_{10}+2\omega_{11}$ |
| 25 | 100 | $1^2+1^2+7^2+7^2$ | $1$ | $1$ | $1-2\omega_2-2\omega_3-2\omega_5+2\omega_6-2\omega_7+2\omega_{12}$ | $1-2\omega_1-2\omega_4+2\omega_8+2\omega_9-2\omega_{10}-2\omega_{11}$ |
| | | | $1+2\omega_3-2\omega_7$ | $1-2\omega_1+2\omega_4$ | $1+2\omega_8-2\omega_9-2\omega_{10}-2\omega_{11}$ | $1-2\omega_2-2\omega_5+2\omega_6-2\omega_{12}$ |
| 27 | 108 | $1^2+1^2+9^2+5^2$ | $1$ | $1$ | $1-2\omega_3+2\omega_4+2\omega_5+2\omega_7-2\omega_9+2\omega_{12}$ | $1-2\omega_1-2\omega_2+2\omega_6+2\omega_8+2\omega_{10}-2\omega_{11}+2\omega_{13}$ |
| 29 | 116 | $1^2+3^2+5^2+9^2$ | $1+2\omega_2-2\omega_4+2\omega_6-2\omega_9-2\omega_{11}+2\omega_{12}$ | $1-2\omega_3-2\omega_5+2\omega_7-2\omega_8+2\omega_{10}$ | $1+2\omega_1$ | $1+2\omega_{13}+2\omega_{14}$ |
| 37† | 148 | $1^2+7^2+7^2+7^2$ | $1$ | $1-2\alpha_0-2\alpha_1+2\alpha_5$ | $1-2\alpha_3-2\alpha_4+2\alpha_8$ | $1+2\alpha_2-2\alpha_6-2\alpha_7$ |
| 43‡ | 172 | $1^2+1^2+1^2+13^2$ | $1+2\alpha_0-2\alpha_2$ | $1-2\alpha_1+2\alpha_3$ | $1+2\alpha_4-2\alpha_6$ | $1+2\alpha_5$ |

† $\alpha_j = \omega_{2^j} + \omega_{2^{9+j}}$

‡ $\alpha_j = \omega_{3^j} + \omega_{3^{7+j}} + \omega_{3^{14+j}}$

TABLE 3

We note also that the periodicity property (8.375) implies

$$\sum_{i=0}^{n-1} b_{4i+r} = \sum_{i=0}^{n-1} b_{4i+s} \quad \{r \equiv s \pmod 4\}. \tag{8.377}$$

For $k = 4i$ (8.376) becomes

$$a_{4(n-i)} = a_{4i},\ b_{4(n-i)} = b_{4i} \quad (0 \leq i \leq n). \tag{8.378}$$

The sums $f(\zeta)$ and $g(\zeta)$ in (8.372) are therefore real. Applying the finite Parseval relation

$$\sum_{i=0}^{n-1} a_i\bar{a}_{i+r} = \frac{1}{n}\sum_{j=0}^{n-1} |f(\zeta^j)|^2\zeta^{jr}$$

and

$$\sum_{i=0}^{n-1} b_i\bar{b}_{i+r} = \frac{1}{n}\sum_{j=0}^{n-1} |g(\zeta^j)|^2\zeta^{jr}$$

where $r$ is some fixed integer, $\bar{a}$ is the complex conjugate of $a$. We now obtain

$$\sum_{i=0}^{n-1} (a_{4i}a_{4i+4r} + b_{4i}b_{4i+4r}) = \frac{1}{n}\sum_{j=0}^{n-1} \{f^2(\zeta^j) + g^2(\zeta^j)\}\zeta^{jr}, \tag{8.379}$$

where $\zeta = \exp(2\pi i/n)$. Denote the sum in (4.141) by $F(r)$. The assumption $q \equiv 1 \pmod 4$ implies that $r = 0$ is the only value of $r$ in the interval $0 \leq r \leq n-1$ for which $4r$ is divisible by $q+1$. Thus it follows from (4.141) that

$$F(4r) = \sum_{k=0}^{q} b_k b_{k+4r} = \begin{cases} q & (r = 0) \\ 0 & (1 \leq r \leq n-1) \end{cases}. \tag{8.3710}$$

On the other hand we have by (8.374) and (8.377)

$$\begin{aligned} F(4r) &= \sum_{i=0}^{n-1} b_{4i}b_{4i+4r} + \sum_{i=0}^{n-1} b_{4i+n}b_{4i+n+4r} \\ &= \sum_{i=0}^{n-1} b_{4i}b_{4i+4r} + \sum_{i=0}^{n-1} a_{4i}a_{4i+4r}, \end{aligned} \tag{8.3711}$$

where we note that $n \equiv 1$ or $3 \pmod 4$ according as $q \equiv 1$ or $5 \pmod 8$.

Combining (8.379) and (8.3711) we get

$$F(4r) = \frac{1}{n}\sum_{j=0}^{n-1} \{f^2(\zeta^j) + g^2(\zeta^j)\}\zeta^{jr}. \tag{8.3712}$$

The inverted form of (8.3712) is given by

$$f^2(\zeta^j) + g^2(\zeta^j) = \sum_{r=0}^{n-1} F(4r)\zeta^{-rj} \quad (j = 0,1,\ldots,n-1).$$

By (8.3710) we have $F(0) = q$ and $F(4r) = 0$ for $1 \leq r \leq n-1$. Hence the last sum reduces to $q$. This completes the proof of the theorem. ***

COROLLARY 8.38. *Let $q = 2n-1$ be a prime power $\equiv 1 \pmod 4$. Put*

$$\psi_1(\zeta) = 1+f(\zeta), \quad \psi_2(\zeta) = 1-f(\zeta), \quad \psi_3(\zeta) = \psi_4(\zeta) = g(\zeta),$$

*where $f(\zeta)$ and $g(\zeta)$ are the polynomials defined by*

$$f(\zeta) = \sum_{i=0}^{n-1} a_{4i}\zeta^i, \quad g(\zeta) = \sum_{i=0}^{n-1} b_{4i}\zeta^i.$$

*Then the identity*

$$\psi_1^2(\zeta) + \psi_2^2(\zeta) + \psi_3^2(\zeta) + \psi_4^2(\zeta) = 4n$$

*is satisfied for each nth root of unity $\zeta$ including $\zeta = 1$.* ***

Returning to the Williamson matrix in (8.340) we may now derive the following theorem of Turyn.

THEOREM 8.39. (Turyn's theorem, proof by Whiteman) *Let $q = 2n-1$ be a prime power $\equiv 1 \pmod 4$. Then there exists a Williamson matrix of order $4n$ in which A and B agree only on the main diagonal and, moreover, $C = D$.*

We employ the construction described in this section. By (8.371) we have $a_0 = 0$, $b_0 = 1$. The successive elements in the first row of A are $1, a_4, a_8, \ldots, a_{4(n-1)}$; the successive elements in the first row of B are $1, -a_4, -a_8, \ldots, -a_{4(n-1)}$; the successive elements in the first rows of C and D are $1, b_4, b_8, \ldots, b_{4(n-1)}$. The matrices A,B,C,D are circulants. Furthermore, in view of (8.378), A,B,C,D are symmetric. Theorem 8.39 now follows readily from the corollary. ***

Turyn has reported the following result [143].

THEOREM 8.40. (Turyn) *There exist Williamson matrices of order $4.9^t$ which are symmetric with constant diagonal.*

8.7 THE GOETHALS-SEIDEL TYPE HADAMARD MATRIX AND ADAPTIONS. The Williamson type Hadamard matrices use four circulant symmetric (1,-1) matrices A,B,C,D in the array

$$G = \begin{bmatrix} A & B & C & D \\ -B & A & -D & C \\ -C & D & A & -B \\ -D & -C & B & A \end{bmatrix}.$$

The A,B,C,D commute in pairs because they are polynomials in T,

$$T = \begin{bmatrix} 0 & 1 & 0 & \ldots & 0 & 0 \\ 0 & 0 & 1 & \ldots & 0 & 0 \\ & \cdot & & \ldots & & \cdot \\ & \cdot & & \ldots & & \cdot \\ & \cdot & & \ldots & & \cdot \\ 0 & 0 & 0 & \ldots & 1 & 0 \\ 1 & 0 & 0 & \ldots & 0 & 0 \end{bmatrix}.$$

But A symmetric is no use in constructing skew-Hadamard matrices so, as we saw in chapter IV, Goethals and Seidel used the array:

$$H = \begin{bmatrix} A & BR & CR & DR \\ -BR & A & -D^TR & C^TR \\ -CR & D^TR & A & -B^TR \\ -DR & -C^TR & B^TR & A \end{bmatrix}, \qquad (8.401)$$

where A is skew-type and

$$R = \begin{bmatrix} 0\,0\,0 & \cdots & 0\,1 \\ 0\,0\,0 & \cdots & 1\,0 \\ \cdot & \cdots & \cdot \\ \cdot & \cdots & \cdot \\ \cdot & \cdots & \cdot \\ 0\,1\,0 & \cdots & 0\,0 \\ 1\,0\,0 & \cdots & 0\,0 \end{bmatrix}.$$

This, of course, may be used with other than skew-type A and so we have:

THEOREM 8.41. (Goethals and Seidel) *Suppose there exist four circulant (1,-1) matrices A,B,C,D of order n satisfying*

$$AA^T+BB^T+CC^T+DD^T = 4nI_n.$$

*Then, using (8.401) there is an Hadamard matrix of order 4n of Goethals-Seidel type.*

**

Now these four matrices are circulant and so are specified by their first rows and in particular by the positive elements of the first row. This leads us to consider supplementary difference sets to specify these first rows. So

COROLLARY 8.42. *Suppose there exist 4-$\{v;k_1,k_2,k_3,k_4;\Sigma k_i-v\}$ supplementary difference sets with circulant incidence matrices. Then there exists an Hadamard matrix of Goethals-Seidel type of order 4v.*

PROOF. Using lemma 1.20, the type 1 (circulant) (1,-1) matrices A,B,C,D of the supplementary difference sets satisfy

$$AA^T+BB^T+CC^T+DD^T = 4\Big(\sum_{j=1}^{4} k_j - \sum_{j=1}^{4} k_j+v\Big)I + [4v-4v]J = 4vI_v.$$

Hence, using the theorem we have the result.

***

Goethals and Seidel used their matrix to form the first skew-Hadamard matrix of order 36 and the second of order 52.

We note that from (1.21)

$$\Big(\sum_{i=1}^{4} k_i-v\Big)(v-1) = \sum_{i=1}^{4} k_i(k_i-1).$$

So

$$-4v(v-1) = 4\sum_{i=1}^{4} (k_i^2-vk_i)$$

and

$$4v = \sum_{i=1}^{4} (2k_i-v)^2.$$

Thus, once again we are led to a dependence on writing 4v as a sum of four squares.

In order to use supplementary difference sets defined on abelian groups rather than just cyclic groups, Wallis and Whiteman slightly modified the Goethals-Seidel array to

$$H = \begin{bmatrix} A & B & C & D \\ -B^T & A^T & -D & C \\ -C & D^T & A & -B^T \\ -D^T & -C & B & A^T \end{bmatrix} \qquad (8.421)$$

Then we have

THEOREM 8.43. *Suppose there exist 4-$\{v;k_1,k_2,k_3,k_4;\Sigma k_i-v\}$ supplementary difference sets $S_1,S_2,S_3,S_4$. Suppose A,B,D are the type 1 (1,-1) matrices of $S_1,S_2$ and $S_4$, respectively, and C is the type 2 (1,-1) matrix of $S_3$. Then (8.421) is an Hadamard matrix of order 4v.*

Wallis and Whiteman further modified (8.421) so that 4-$\{v;k_1,k_2,k_3,k_4;\Sigma k_i-v-1\}$ supplementary difference sets could be used. Using (1.21) we see

$$\left(\sum_{i=1}^{4} k_i-v-1\right)(v-1) = \sum_{i=1}^{4} k_i(k_i-1)$$

and so

$$-4(v^2-1) = 4\sum_{i=1}^{4} (k_i{}^2-vk_i)$$

$$4 = \sum_{i=1}^{4} (2k_i-v)^2.$$

Hence if v is odd $k_i = \frac{1}{2}(v\pm1)$. Clearly if A is the type 1 (1,-1) matrix of a set with $\frac{1}{2}(v+1)$ elements then -A is the type 1 (1,-1) matrix of a set with $\frac{1}{2}(v-1)$ elements. Thus we have

THEOREM 8.44. (Wallis-Whiteman) *Suppose there exist 4-$\{v;k_1,k_2,k_3,k_4;\sum_{i=1}^{4}k_i-v-1\}$ supplementary difference sets $S_1,S_2,S_3,S_4$. Let $R_1,R_2,R_4$ be the type 1 (1,-1) matrices of $S_1,S_2$ and $S_4$ respectively, and $R_3$ be the type 1 (1,-1) matrix of $S_3$. Choose $A = \pm S_1$ according as $S_1J = \pm J$. Similarly choose $B = \pm S_2$, $C = \pm S_3$, $D = \pm S_4$ according as $S_2J = \pm J$, $S_3J = \pm J$, $S_4 = \pm J$ respectively. Then*

$$H = \left[\begin{array}{cccc|cccc} -1 & -1 & -1 & -1 & e & e & e & e \\ 1 & -1 & 1 & -1 & -e & e & -e & e \\ 1 & -1 & -1 & 1 & -e & e & e & -e \\ 1 & 1 & -1 & -1 & -e & -e & e & e \\ \hline e^T & e^T & e^T & e^T & A & B & C & D \\ -e^T & e^T & -e^T & e^T & -B^T & A^T & -D & C \\ -e^T & e^T & e^T & -e^T & -C & D^T & A & -B^T \\ -e^T & -e^T & e^T & e^T & -D^T & -C & B & A^T \end{array}\right]$$

*which may be written as*

$$\begin{bmatrix} -H(1,1,1,1) & H(e,e,e,e) \\ H(e^T,e^T,e^T,e^T) & H(A,B,C,D) \end{bmatrix}$$

*is an Hadamard matrix of order $4(v+1)$.*

8.8 USING HADAMARD ARRAYS. The idea of using Hadamard arrays to construct Hadamard matrices is due to Baumert and Hall who used the Hadamard array $H[12,4,3] = BH[12]$, to construct an Hadamard matrix of order 156.

THEOREM 8.45. (Baumert and Hall) *Suppose there exists an Hadamard array $H[4t,4,t] = BH[4t]$ and four $(1,-1)$ matrices $A,B,C,D$ of order $n$ which are symmetric, commute in pairs and satisfy*

$$AA^T+BB^T+CC^T+DD^T = 4nI_n.$$

*Then there exists an Hadamard matrix of order $4nt$.*

Write $S = \{x\colon x = 1+2^i10^j26^k$ <u>or</u> $x$ (odd) $\leq 23$ <u>or</u> $x = 29\}$, and

$T = \{b\colon$ Baumert-Hall array $BH[4b]$ exists$\}$

$= \{b\colon b$ is an odd integer between 3 and 19 inclusive or 25 (from theorem 7.6)$\}$

$\cup \{b\colon b \in S$ (from [143])$\} \cup \{5b\colon b \in S$ (from [143])$\}$.

Then we can say

COROLLARY 8.46. *Let $t \in T$ and $\omega \in T \cup \{25\} \cup \{31\} \cup \{37\} \cup \{43\} \cup \{x\colon x = \frac{1}{2}(q^s+1)$ where $q^s$ is a prime power $\equiv 1 \pmod 4\}$. Then there is an Hadamard matrix of order $4t\omega$.*

PROOF. $\omega \in T$ since if there exists a Baumert-Hall array and we put $A = B = C = D = [1]$ we get an Hadamard matrix. $\omega = 31$ is quoted from [143]. The other values for $\omega$ come from table 3 and theorem 8.39. Then using the theorem we have the result.

Clearly to be able to remove the restriction that the $(1,-1)$ matrices $A,B,C,D$ must be symmetric and commute in pairs would be very important. So far this restriction has only been eased for BH[4t] when t = 1.

8.9 THE RESULTS OF YANG. Suppose C and D are the circulant incidence matrices of $2-\{m;k_1,k_2;k_1+k_2+\frac{1}{2}(1-m)\}$ supplementary difference sets X and Y. Then

$$CC^T+DD^T = \tfrac{1}{2}(m-1)I + \big(k_1+k_2+\tfrac{1}{2}(1-m)\big)J. \tag{8.461}$$

With

$$A = J-2C \quad \text{and} \quad B = J-2D$$

$$AA^T+BB^T = 2(m-1)I+2J.$$

THEOREM 8.47. *Suppose there exist $2-\{m;k_1,k_2;k_1+k_2+\frac{1}{2}(1-m)\}$ supplementary difference sets for m a prime, then there exists an Hadamard matrix of order 4m.*

PROOF. In the Goethals-Seidel array use A,B as above and (i) for $m \equiv 3 \pmod 4$ C = D, use the incidence matrix of the $\big(m,\frac{1}{2}(m-1),\frac{1}{4}(m-3)\big)$ difference set, (ii) for $m \equiv 1 \pmod 4$ use the incidence matrices of the supplementary difference sets of lemma 1.8. ***

Let ω be an mth root of unity. Further let X and Y be as above. Form two polynomials C(ω) and D(ω) by choosing

$$C(\omega) = \sum_{i\in X} \omega^i,$$

$$D(\omega) = \sum_{j\in Y} \omega^j,$$

then (8.461) becomes

$$|C(\omega)|^2 + |D(\omega)|^2 = \tfrac{1}{2}(m-1).$$

COROLLARY 8.48. *If there exist polynomials C(ω) and D(ω) of degree m in ω, where ω is an mth root of unity and m is prime satisfying*

$$|C(\omega)|^2 + |D(\omega)|^2 = \tfrac{1}{2}(m-1),$$

*then there exists an Hadamard matrix of order 4m.*

Yang has found such polynomials for various orders (see table 4) and so

COROLLARY 8.49. *There exist Hadamard matrices of Goethals-Seidel type of order 4 for $\omega \in \{3,5,7,13,19,23,31\}$.*

If P,Q are circulant (0,1) matrices then P and Q are polynomials in T (see 8.340). Denote this P(T) and Q(T).

THEOREM 8.50. (Yang) *Suppose there exist $2-\{m;k_1,k_2;k_1+k_2-\frac{1}{2}m\}$ supplementary difference sets X and Y with circulant incidence matrices P and Q. Write $Q' = J-Q$. Then with $S = T_{2m}$*

| m | $C(\omega)$ | $D(\omega)$ |
|---|---|---|
| 3 | 0 | 1 |
| 5 | 1 | 1 |
| 7 | 1 | $1+\omega+\omega^3$ |
| 9 | $1+\omega$ | $1+\omega^2+\omega^5$ |
| 13 | $1+\omega+\omega^3+\omega^9$ | $1+\omega_0+\omega_0{}^3+\omega_0{}^9$ |
| | $1+\omega+\omega^4$ | $1+\omega+\omega^2+\omega^4+\omega^7+\omega^9$ or $1+\omega+\omega^3+\omega^5+\omega^8+\omega^9$ |
| 15 | $1+\omega+\omega^4+\omega^6$ | $1+\omega+\omega^3+\omega^4+\omega^8+\omega^{10}$ or $1+\omega+\omega^3+\omega^5+\omega^8+\omega^9$ |
| | $1+\omega+\omega^4+\omega^{10}$ | $1+\omega+\omega^3+\omega^5+\omega^7+\omega^8$ |
| 19 | $1+\omega+\omega^3+\omega^5+\omega^8+\omega^9$ | $1+\omega+\omega^3+\omega^7+\omega^9+\omega^{10}+\omega^{14}$ |
| | $1+\omega+\omega^2+\omega^4+\omega^7+\omega^{12}$ or $1+\omega+\omega^3+\omega^{12}+\omega^{14}+\omega^{15}$ | $1+\omega+\omega^3+\omega^4+\omega^8+\omega^{10}+\omega^{14}$ or $1+\omega+\omega^3+\omega^5+\omega^9+\omega^{10}+\omega^{16}$ |
| | $1+\omega+\omega^2+\omega^3+\omega^6+\omega^{12}$ | $1+\omega+\omega^5+\omega^7+\omega^9+\omega^{12}+\omega^{15}$ |
| | $1+\omega+\omega^2+\omega^3+\omega^7+\omega^{12}$ | $1+\omega+\omega^3+\omega^6+\omega^8+\omega^{12}+\omega^{16}$ or $1+\omega+\omega^3+\omega^7+\omega^{11}+\omega^{14}+\omega^{16}$ |
| 21 | $1+\omega+\omega^5+\omega^7+\omega^9+\omega^{10}$ | $1+\omega+\omega^2+\omega^3+\omega^6+\omega^8+\omega^{10}+\omega^{11}+\omega^{14}+\omega^{17}$ |
| 23 | $1+\omega+\omega^2+\omega^5+\omega^7+\omega^{11}+\omega^{14}$ | $1+\omega+\omega^2+\omega^3+\omega^6+\omega^8+\omega^{10}+\omega^{11}+\omega^{14}+\omega^{18}$ |
| 25 | $1+\omega+\omega^2+\omega^6+\omega^8+\omega^{10}+\omega^{11}+\omega^{14}+\omega^{15}$ | $1+\omega^2+\omega^5+\omega^6+\omega^7+\omega^{14}+\omega^{17}+\omega^{20}+\omega^{23}$ |
| 27 | $1+\omega+\omega^2+\omega^3+\omega^6+\omega^{10}+\omega^{12}+\omega^{15}+\omega^{23}$ | $1+\omega+\omega^2+\omega^5+\omega^7+\omega^9+\omega^{10}+\omega^{17}+\omega^{20}+\omega^{21}+\omega^{23}$ |
| | $1+\omega+\omega^2+\omega^6+\omega^9+\omega^{18}+\omega^{21}+\omega^{22}$ | $1+\omega+\omega^2+\omega^3+\omega^9+\omega^{11}+\omega^{13}+\omega^{16}+\omega^{19}+\omega^{23}+\omega^{24}$ |
| 31 | $1+\omega+\omega^2+\omega^3+\omega^4+\omega^8+\omega^{13}+\omega^{19}+\omega^{23}+\omega^{26}$ | $1+\omega+\omega^2+\omega^3+\omega^6+\omega^7+\omega^{10}+\omega^{12}+\omega^{14}+\omega^{15}+\omega^{17}+\omega^{18}+\omega^{24}+\omega^{26}+\omega^{28}$ |

TABLE 4

$$P_2 = P(S^2) + S^k Q(S^2)$$

*and*

$$Q_2 = P(S^2) + S^k Q'(S^2),$$

*with k any odd integers, are* $2\text{-}\{2m;k_1+k_2,k_1-k_2+m;2k_1\}$ *supplementary difference sets with circulant incidence matrices.*

PROOF. $S^T = S^{-1}$, so

$$\begin{aligned} P_2P_2^T + Q_2Q_2^T &= [P(S^2)P(S^{-2}) + Q(S^2)Q(S^{-2})] + [P(S^2)P(S^{-2}) + Q'(S^2)Q'(S^{-2})] \\ &\quad + S^kP(S^2)[Q(S^{-2}) + Q'(S^{-2})] \\ &\quad + S^{-k}P(S^{-2})[Q(S^2) + Q'(S^2)] \\ &= [\, mI + (k_1+k_2-m)\sum_{i=0}^{m-1} S^{2i}] + [\tfrac{1}{2}mI + (k_1-k_2+\tfrac{1}{2}m)\sum_{i=0}^{m-1} S^{2i}] \\ &\quad + 2k_1\sum_{i=0}^{m-1} S^{2i+1} \\ &= mI+2k_1J. \end{aligned}$$

Hence $P_2$ and $Q_2$ are $2\text{-}\{2m;k_1+k_2,k_1-k_2+m;2k_1\}$ supplementary difference sets as required. ***

Now the (1,-1) incidence matrices A and B of $2\text{-}\{m;k_1,k_2;k_1+k_2-\tfrac{1}{2}m\}$ supplementary difference sets satisfy

$$AA^T+BB^T = 2mI_m$$

(from lemma 1.20). So the existence of A and B is equivalent to saying

$$\begin{bmatrix} A & B \\ -B^T & A^T \end{bmatrix}$$

is an Hadamard matrix and we have

COROLLARY 8.51 (Yang) *Suppose there exists an Hadamard matrix of order m and of the form*

$$\begin{bmatrix} A & B \\ -B^T & A^T \end{bmatrix}$$

*where A and B are circulant incidence matrices, then there exists an Hadamard matrix of order 2m and of the same form.*

THEOREM 8.52. (Yang) *Let* $P_1,P_2,P_3,P_4$ *be the circulant incidence matrices of* $4\text{-}\{m;k_1,k_2,k_3,k_4;\sum_{i=1}^{4}k_i-m\}$ *supplementary difference sets and write* $P_i' = J-P_i$. *Then*

$$P = P_1(S^2) + S^mP_2(S^2), \quad Q = P_1(S^2) + S^mP_2'(S^2)$$

$$U = P_3(S^2) + S^mP_4(S^2), \quad W = P_3(S^2) + S^mP_4'(S^2)$$

*are* $4\text{-}\{2m;k_1+k_2,k_1-k_2+m,k_3+k_4,k_3-k_4+m;2k_1+2k_2+m\}$ *supplementary difference sets with circulant incidence matrices.*

PROOF. The proof is similar to that of the previous theorem. ***

COROLLARY 8.53. (Yang) *Suppose there exists a Williamson type Hadamard matrix of order 4m comprising circulant matrices of order m. Then there exists a Williamson type Hadamard matrix of order 8m comprising circulant incidence matrices.*

8.10 A COMPUTER SEARCH. Because it is very easy to consume computer looking for combinatorial matrices, we outline the ideas that led to one computer search and the techniques used to handle the problem.

We know that if there exist 4-$\{2q;q,q+1,q+1,q+1;2q+2\}$ supplementary difference sets then there exists an Hadamard matrix of order $4(2q+1)$.

Whiteman has suggested that if four supplementary difference sets

$$S_1,\ S_2,\ \{0\}\cup Q\cup\{q+Q\},\ \{0\}\cup Q\cup\{q+Q\},$$

exist where Q is the set of quadratic residues of a prime q, then an Hadamard matrix of order 188 would be obtained when q = 23. In fact

$q = 3$ gives $H_{28}$
$q = 7$ gives $H_{60}$
$q = 11$ gives $H_{92}$
$q = 19$ gives $H_{156}$
$q = 23$ gives $H_{188}$

where $H_n$ denotes the Hadamard matrix of order n. Now

LEMMA 8.54. (Whiteman) *The differences from $\{0\}\cup Q\cup\{q+Q\}$, where Q is the set of quadratic elements of GF(q) and G is the non-zero element of GF(q), q a prime, is*

$$\tfrac{1}{2}(q+1)G\ \&\ \tfrac{1}{2}(q-1)\{q\}.$$

PROOF. Write $r = \frac{1}{2}(q-1)$, $s = \frac{1}{2}(q-3)$, $P = \{1,2,\ldots,q-1\}$. Then the totality of differences from $0\cup Q\cup\{q+Q\}$ is

$$\begin{aligned}
&[\text{differences between elements of } Q]\\
&\quad \&\ [\text{differences between elements of } \{q+Q\}]\\
&\qquad \&\ Q\ \&\ -Q\ \&\ \{q+Q\}\ \&\ -\{q+Q\}\ \&\ Q-\{q+Q\}\ \&\ \{q+Q\}-Q\\
&= sP\ \&\ sP\ \&\ Q\ \&\ -\{q+Q\}\ \&\ -Q\ \&\ \{q+Q\}\ \&\ r\{q\}\ \&\ r\{-q\}\ \&\ 2s\{q+P\}\\
&= 2sG\ \&\ P\ \&\ \{q+P\}\ \&\ (2r-2s)\{q\}\\
&= (2s+1)G\ \&\ (2r-2s-1)\{q\}\\
&= \tfrac{1}{2}(q-1)G\ \&\ \tfrac{1}{2}(q-1)\{q\};
\end{aligned}$$

which is the required result. ***

This leads to the result

LEMMA 8.55. (Whiteman) *If there exist two sets* $S_1$ *and* $S_2$ *of size* $q$ *and* $q+1$ *in the cyclic group* $C_{2q}$, *where* $q$ *is a prime, whose totality of differences* $T_1$ *and* $T_2$ *satisfy*

$$T_1 \& T_2 = (q-1)(G \setminus q)$$

*then there is an Hadamard matrix of order* $4(2q+1)$.

PROOF. Let $S_3 = S_4 = \{0\} \cup Q \cup \{q+Q\}$, where $Q$ is the set of quadratic elements of $GF(q)$. Then using the previous lemma $S_1,S_2,S_3,S_4$ are $4-\{2q;q,q+1,q+1,q+1;2q+2\}$ supplementary difference sets. Suppose their $(1,-1)$ incidence matrices are $X,Y,Z,W$ where $Z$ is type 2 and $X,Y,W$ are type 1. Then with $A = X$, $B = -Y$, $C = -Z$, $D$-$W$ in (8.441) we have the required Hadamard matrix. ***

Whiteman found the first two of the following sets which satisfy this lemma and H.C. Rumsey Jr. studied the others.

| | | |
|---|---|---|
| $v = 6$, | $q = 3$, | $S_1 = \{0,4\}$, $S_2 = \{0,1,2\}$, |
| $v = 14$, | $q = 7$, | $S_1 = \{0,1,4,6,9,10\}$, $S_2 = \{0,1,2,4,10,12,13\}$, |
| $v = 22$, | $q = 11$, | impossible - see next lemma, |
| $v = 38$, | $q = 19$, | $S_1 = \{3,5,6,7,10,11,15,17,18,20,21,23,27,28,31,32,33,35\}$, |
| | | $S_2 = \{1,3,4,5,6,7,8,10,12,13,14,15,18,19,21,28,30,35,36\}$, |
| $v = 46$, | $q = 23$, | not found by a complete computer search. |

COROLLARY 8.56. (Rumsey) *There exist* $4-\{38;19,20,20,20;40\}$ *supplementary difference sets and an Hadamard matrix of order 156.*

LEMMA 8.57. (Rumsey) *If there exist two sets* $S_1$ *and* $S_2$ *of size* $q$ *and* $q+1$ *in the cyclic group* $C_{2q}$, *where* $q$ *is prime, whose totality of differences* $T_1$ *and* $T_2$ *satisfy*

$$T_1 \& T_2 = (q-1)(G \setminus q)$$

*then*

$$2q-1 = a^2 + b^2$$

*for some integers* $a$ *and* $b$.

PROOF. Consider the *Hall polynomials* (see [7,p.8]) of $S_1$ and $S_2$, call them $\theta(x)$ and $\psi(x)$ respectively. Then we need

$$\theta(x)\theta(x^{-1}) + \psi(x)\psi(x^{-1}) \equiv q + (q-1)(1+x+\ldots+x^{2q-1})-(q-1)x^q(\text{mod } x^{2q}-1),$$

with $\theta(1) = q-1$, $\psi(1) = q$. Reducing this modulo $x^q+1$ we get

$$\theta(x)\theta(x^{-1}) + \psi(x)\psi(x^{-1}) \equiv (2q-1) \text{ mod } (x^q+1),$$

when $\theta,\psi$ have $\pm 1$ coefficients with one coefficient of $\theta$ being zero. Make the transformation $-x \to x$. Then the "new" $\theta,\psi$ have the same structure but the equation becomes

$$\theta(x)\theta(x^{-1}) + \psi(x)\psi(x^{-1}) \equiv (2q-1) \text{ mod } (x^q-1). \qquad (8.571)$$

Let x = 1 then

$$[\theta(1)]^2 + [\psi(1)]^2 = 2q-1 = a^2+b^2.$$

***

If q = 11, 2q-1 = 21 which is not the sum of two squares and so the construction is not possible for v = 22.

We now briefly outline the technique used by Rumsey to programme this search for q = 23. From the last lemma we have

$$[\theta(1)]^2 + [\psi(1)]^2 = 2q-1 = 45 = 6^2+3^2.$$

Now $\theta$ has 22 terms so $\theta(1) = \pm 6$ and without loss of generality we may assume $\theta$ has 11 +1's and 8 -1's while $\psi$ has 13 +1's and 10 -1's.

We consider $\psi$'s with the property that

$$\psi(\zeta)\psi(\zeta^{-1}) \leq 45 \qquad (8.572)$$

for all 23rd roots of unity $\zeta \neq 1$. The arithmetic is carried out in the complex plane, so if $\psi(x)$ satisfies (8.572) then so does

$$x^j\psi(x^k) \quad \text{for all } j,k.$$

Rumsey generated all binary 23-tuples and tested

(1) $\psi(1) = 3$ (if not reject it),

(2) reject the 23-tuple if it is not the lexicographically least of all cyclic shifts and also if it is not the lexicographically least of all the sequences run backward (i.e., $x \to x^{-1}$),

(3) fix $\zeta = \exp(2\pi i/23)$ and insist that

$$|\psi(x^k)|^2$$

evaluated at $x = \zeta$ is maximum over k.

This effectively reduces out the automorphs. (8.572) was then tested using all 23rd roots and some 60 sequences passed all the tests.

Now all possible $\theta$'s (i.e., 23-tuples with 8 -1's) were generated and an attempt made to match up exactly by equation (8.571). Roundoff error has to be allowed for in the complex plane and this was kept in mind. No $\theta$ and $\psi$ were found for q = 23.

## CHAPTER IX. GENERALIZED HADAMARD MATRICES

### 9.1 THE RESULTS OF BUTSON.

DEFINITION 9.1. A square matrix H of order h all of whose elements are complex pth roots of unity is called a *generalized Hadamard matrix* H(p,h) if $HH^* = hI$ or equivalently if the matrix $\sqrt{h}H$ is unitary.

LEMMA 9.2. (Butson) *If H is a generalized Hadamard matrix H(p,h), then:*

*(i)* $HH^* = hI$ *implies* $H^*H = hI$;

*(ii) a permutation of rows or columns, or multiplying any row or column by a fixed pth root of unity leaves the Hadamard property unchanged;*

*(iii) H may be normalized so that its first row and column contain only ones;*

*(iv) if* $H = (h_{ij})$, *then*

$$\sum_{j=1}^{h} h_{ij} = \sum_{j=1}^{h} h^*_{ij} = 0, \qquad i = 2,3,\ldots,h,$$

$$\sum_{i=1}^{h} h_{ij} = \sum_{i=1}^{h} h^*_{ij} = 0, \qquad j = 2,3,\ldots,h;$$

*(v) if* $H_1$ *is an* $H(p_1,h_1)$ *and* $H_2$ *is an* $H(p_2,h_2)$, $h = h_1h_2$, $p = \ell.c.m(p_1,p_2)$, *then* $H_1 \times H_2$ *is an H(p,h) matrix;*

*(vi) if* $H_1$ *is an* $H(p_1,h)$ *and x is a primitive* $p_2$ *root of unity, and* $p = \ell.c.m(p_1,p_2)$ *then* $xH_1$ *is an H(p,h) matrix.*

PROOF.

(i) $HH^* = hI$, so $H^* = hH^{-1}$ and $H^*H = hI$.

(ii) Clearly permutation of rows or columns does not alter the Hadamard property. Now let $H = (h_{ij})$ then $\sum_{j=1}^{h} h_{ij}h^*_{kj} = 0$ and $\sum_{j=1}^{h} h_{ji}h^*_{jk} = 0$, $i \neq k$. If we multiply row i or column i by x a pth root of unity, we just get $x\sum_{j=1}^{h} h_{ij}h^*_{kj} = 0$ and $x\sum_{j=1}^{h} h_{ji}h^*_{jk} = 0$, $i \neq k$, i.e. the Hadamard property is unchanged.

(iii) Using property (ii) each row or column is multiplied by the appropriate pth root of unity until the first element in each row or column is one.

(iv) In the normalized H(p,h) by taking the product of the jth row and column with the first row and column gives the required equations. But every matrix $\overline{H}(p,h)$ is obtained from some normalized $H(p,h)$ by multiplying rows or columns by pth roots of unity.

(v) $(H_1 \times H_2)(H_1^* \times H_2^*) = p_1 I_{h_1} \times p_2 I_{h_2} = p_1 p_2 I_h = hI_h$ and so $H_1 \times H_2$ is an $H(p,h)$.

(vi) $(xH_1)(x^* H_1^*) = xx^* hI_h = hI_h$ and so $xH_1$ is an $H(p,h)$. ***

LEMMA 9.3. (Butson) *When p is a prime, an H(p,h) matrix can exist only for the values h = pt, where t is a positive integer.*

PROOF. Let x be a primitive pth root of unity then $x^i$ and $x^j$ are independent for $i \neq j$ and $\sum_{i=0}^{p-1} x^i = 0$ since $(x-1)\sum_{i=0}^{p-1} x^i = x^p - 1 = 0$ and $x \neq 1$.

Then in any normalized $H(p,h)$ the first and second rows may be written as (after suitable rearranging of columns):

$$\begin{matrix} 1 & 1 & \cdots & 1 & 1 & 1 & \cdots & 1 & \cdots & 1 & 1 & \cdots & 1 \\ 1 & x & \cdots & x^{p-1} & 1 & x & \cdots & x^{p-1} & \cdots & 1 & x & \cdots & x^{p-1} \end{matrix}$$

and so each row must have a multiple of p elements. ***

But when p is not a prime lemma 9.3 need not apply.

LEMMA 9.4. (Butson) *If there exists a real Hadamard matrix H(2,h) and p any integer, then there exists an H(2p,h).*

PROOF. Use lemma 9.2. ***

The following result is well known: we quote from Butson [31].

THEOREM 9.5. *Let x be a primitive pth root of unity, then the Vandermonde matrix* $V = (v_{ij}) = (x^{ij})$ *of order p is a symmetric generalized Hadamard matrix H(p,p).*

PROOF. $\sum_{i=1}^{p} v_{ji} v_{ki}^* = \sum_{i=1}^{p} x^{ij} x^{p-ik} = \sum_{i=1}^{p} x^{i(j-k)} = y.$

When $j = k$, $y = p$. Suppose $j \neq k$. If $(j-k,p) = 1$, then $x^{j-k}$ is a primitive pth root of unity and $y = 0$. If $(j-k,p) = d > 1$, let $p = qd$ and $j-k = \ell d$. Then $(j-k)q = \ell dq = \ell p \equiv 0 \pmod p$, so that $x^{j-k}$ is a qth root of unity. Here $y = d\sum_{i=1}^{q} x^{(j-k)i} = 0$. Thus we have the result. ***

The matrix V of this theorem for $p = q^r$, q prime, is not equivalent to that obtained by finding $V_q = H(q,q)$ and then forming $V_q \times V_q \times \ldots \times V_q$.

THEOREM 9.6. (Butson) *When p is a prime a generalized Hadamard matrix H(p,2p) exists.*

PROOF. Since the result is known for $p = 2$ (see chapter I) we assume $p = 2q+1$. Let n be the smallest quadratic non-residue of p and x be a primitive root of p. Denote by U that permutation matrix such that W = VU has elements $w_{ij} = x^{nij}$, $i,j = 0,1,\dots,p-1$ where V is defined in theorem 9.5. Define the matrix Q to be the diagonal matrix $(x^q, x^{4q}, x^{9q}, \dots, x^{i^2q}, \dots)$. Then $C = QVQ$ and $B = Q^nWQ^n$ are, by lemma 9.2 part (ii), generalized Hadamard matrices H(p,p). Using $-2q \equiv 1 \pmod p$, we have $C = (c_{ij})$, $B = (b_{ij})$ satisfy $c_{ij} = x^{i^2q}\cdot x^{ij}\cdot x^{j^2q} = x^{q(i^2-2ij+j^2)} = x^{q(i-j)^2}$ and $b_{ij} = x^{ni^2q}\cdot x^{nij}\cdot x^{nj^2q} = x^{nq(i-j)^2}$. Now clearly C and B are circulant and symmetric. Furthermore, because of $(i-j)^2$ occurring in the definition of C and B they both contain at most q+1 distinct pth roots of unity.

If $v_i$ and $v_j$ are two rows of V we note $v_i \cdot v_j \pmod p = v_{i+j}$ (· is the dot product). So the rows of V (under dot product) form a cyclic group with generator $v_1$. Similarly the columns of V, the rows of W and the columns of W all form cyclic groups with generators $v_1^T$, $w_1 = v_n$, and $w_1^T = v_n^T$, respectively. Let D be the diagonal matrix $(1, x, x^2, x^3, \dots, x^{p-1})$ and $T = (t_{ij})$ be given by $t_{i+1,i} = 1$, $i \in \{0,1,\dots,p-1\}$, $t_{ij} = 0$ otherwise. Then $D^kV = VT^k$ and $D^{nk}W = WT^k$.

Let Y and Z be the 1×p matrix of ones and zeros respectively. Then the kth column of B can be written as $T^kQ^nY^T$ and the kth column of CP as $T^{kn}QY^T$ (where P is an appropriate permutation matrix.

Now consider

$$K = \left[\begin{array}{c|c} QV & Q^nW \\ \hline (CP)^* & B^* \end{array}\right].$$

Now $CY^T = (YQY^T)Y^T$ and $BY^T = (YQ^nY^T)Y^T$ since as k takes all the values 0 to p-1, $k^2$ and $(k-i)^2$ cover the same values, the quadratic residues. When $p = 2q+1$ is a prime, the number of quadratic residues and non-residues are both equal to q. So

$$YQY^T = YQ^nY^T = \sum_{i=0}^{p-1} x^{qi^2} + \sum_{i=0}^{p-1} x^{nqi^2}$$
$$= 2\sum_{j=0}^{p-1} x^{qj} = 0.$$

Thus $CY^T + BY^T = Z^T$. Now

$$KK^* = \left[\begin{array}{c|c} QVV^*Q^* + Q^nWW^*Q^{n*} & (QV)(CP)+Q^nWB \\ \hline (CP)^*(QV)^* + B^*(Q^nW)^* & (CP)^*CP+B^*B \end{array}\right].$$

The kth column of $(QV)(CP)+(Q^nW)B$ is given by

$$QVT^{kn}QY^T + Q^nWT^kQ^nY^T = D^{kn}QVQY^T + D^{kn}Q^nWQ^nY^T$$
$$= D^{kn}(CY^T + BY^T)$$
$$= Z^T,$$

since D and Q are both diagonal. Thus $(QV)(CP)+Q^nWB$ and its Hermitian conjugate are both zero. Now using lemma 9.2 part (ii) all of $QV, Q^nW, CP$ and B are generalized Hadamard matrices $H(p,p)$. So $KK^* = 2pI_p \times I^2$. ***

COROLLARY 9.7. (Butson) *When p is a prime, $H(p, 2^m p^k)$ matrices can be constructed for any non-negative integers $m \leq k$.*

PROOF. Use theorem 9.6 and lemma 9.2 part (v). ***

We note the following result from Butson [32].

THEOREM 9.8. (Butson) *There exists a generalized Hadamard matrix $H(p,p^u)$ whose matrix forms a group and whose core is cyclic.*

## 9.2 DELSARTE AND GOETHALS RESULTS.

Theorem 9.8 indicated that generalized Hadamard matrices might have other interesting properties. The following results are due to Delsarte and Goethals [37].

Let T be any linear mapping from GF(q) onto GF(p), for example the trace

$$T(\alpha) = \sum_{k=0}^{m-1} \alpha^{p^k}. \tag{9.81}$$

DEFINITION 9.9. By the *T-character* on GF(q) is meant the complex-valued function defined for the primitive root $\alpha$ of GF(q), by

$$e(\alpha) = \exp(2\pi i T_\alpha / p) \tag{9.91}$$

where $T_\alpha$ is any integer whose residue class modulo p is $T(\alpha)$.

Clearly

$$e(\alpha+\beta) = e(\alpha)e(\beta)$$

and so e is a character on the additive group GF(q), that is a homomorphism of the additive group of GF(q) onto the multiplicative group of complex pth roots of unity.

LEMMA 9.10. *The polynomial $z^{q+1}-\alpha$ has exactly (q+1) distinct roots in $GF(q^2)$ when is any non-zero element in GF(q).*

PROOF. $y^{q-1}-1$ may be factored into linear factors over GF(q), thus

$$y^{q-1}-1 = (y-\alpha_1)(y-\alpha_2)\dots(y-\alpha_{q-1}),$$

where the $\alpha_i$ are the q-1 non-zero elements of GF(q). This is also valid for $z^{q^2-1}-1$ in $GF(q^2)$, and since

$$z^{q^2-1}-1 = (z^{q+1})^{q-1}-1$$
$$= (z^{q+1}-\alpha_1)(z^{q+1}-\alpha_2)\dots(z^{q+1}-\alpha_{q-1}).$$

Now since $z^{q^2-1}-1$ factors into linear factors over $GF(q^2)$, the same is true for each factor $(z^{q+1}-\alpha_i)$. ***

Let $\alpha$ be any non-zero element of GF(q), and let us define a square matrix $M_\alpha$ of order $q^2$, whose rows and columns are numbered with the elements of $GF(q^2)$. Define the (x,y) element of $M_\alpha$ by

$$M_\alpha(x,y) = T[\alpha^{-1}(y-x)^{q+1}] \qquad (9.101)$$

where T is defined by (9.81).

Now $(y-x)^{q+1}$ belongs to GF(q) when x and y are any elements of $GF(q^2)$. Thus the elements (9.101) of $M_\alpha$ belong to GF(p), since T maps GF(q) onto GF(p). With each non-zero element $\alpha$ of GF(q), associate a square matrix $H_\alpha$ of order $q^2$, whose entries are the complex pth roots of unity defined by

$$H_\alpha(x,y) = e[\alpha^{-1}(y-x)^{q+1}], \qquad (9.102)$$

where e is the T-character defined in (9.91).

Now define a matrix S of order $q^2$ whose entries are

$$S(x,y) = e(xy+x^q y^q) \qquad (9.103)$$

where e is the T-character (9.91).

THEOREM 9.11. (Delsarte and Goethals) *The matrix S is a symmetric generalized Hadamard matrix* $H(p,p^{2m})$.

PROOF. The entries of S are pth roots of unity, since $xy+(xy)^q$ belongs to GF(q) for $xy \in GF(q^2)$, and e maps GF(q) onto the pth roots of unity. Since $S(x,y) = e(xy+x^q y^q) = S(y,x)$, S is symmetric.

The (i,j) element of SS* is

$$(SS^*)_{ij} = \sum_{y \in GF(q^2)} e(iy+i^q y^q)e(-jy-j^q y^q)$$

$$= \sum_{y \in GF(q^2)} e[(i-j)y+(i-j)^q y^q]$$

$$= \begin{cases} q^2 e(0) & \text{for } i-j = 0 \\ q \sum_{\alpha \in GF(q)} e(\alpha) & \text{for } i-j \neq 0 \end{cases}$$

as each element of GF(q) appears exactly q times in the expression under brackets, when y runs through $GF(q^2)$, thus

$$(SS^*)_{ij} = \begin{cases} q^2 & i = j \\ 0 & i \neq j \end{cases}$$

since e(0) = 1 and e is a non-trivial character. ***

Now associate with each non-zero element $\alpha$ of GF(q) a diagonal matrix $D_\alpha$ of order $q^2$, whose diagonal entries are

THEOREM 9.12. (Delsarte and Goethals) *The matrix S transforms any matrix $H_\alpha$ into the corresponding diagonal matrix $D_\alpha$, that is*

$$S^{-1}H_\alpha S = D_\alpha.$$

PROOF. The (i,j) element $h_\alpha(i,j)$ of $H_\alpha S$ is

$$h_\alpha(i,j) = \sum_{y\in GF(q^2)} e[\alpha^{-1}(y-i)^{q+1}+yj+y^qj^q]$$

$$= \sum_{y\in GF(q^2)} e[\alpha^{-1}z^{q+1}+(ij+i^qj^q)-\alpha j^{q+1}]$$

where $z = y-i+\alpha j^q$

$$= e(ij+i^qj^q)e(-\alpha j^{q+1}) \sum_{z\in GF(q^2)} e(\alpha^{-1}z^{q+1}).$$

Using lemma 9.10, $z^{q+1}$ takes (q+1) times each non-zero value in GF(q) when z runs through the $q^2$-1 non-zero elements of GF($q^2$), and obviously is zero for z = 0. Hence

$$h_\alpha(i,j) = e(ij+i^qj^q)e(-\alpha j^{q+1})\Big[(q+1) \sum_{\alpha\in GF(q)} e(\alpha)-qe(0)\Big] \tag{9.121}$$

$$= -qe(ij+i^qj^q)e(-\alpha j^{q+1})$$

$$= -qS(i,j)D_\alpha(j,j) \quad \text{using (9.103) and (9.111)}$$

and

$$H_\alpha S = SD_\alpha.$$

***

THEOREM 9.13. (Delsarte and Goethals) *(i) The q-1 matrices $H_\alpha$ are symmetric and regular generalized Hadamard matrices. (ii) The (q-1) matrices $(-\frac{1}{q}H_\alpha)$, together with the unit matrix, form a multiplicative group of order q, simply isomorphic to the additive group of GF(q).*

PROOF. From (9.102), we have $H_\alpha(y,x) = H_\alpha(x,y)$, hence proving that $H_\alpha$ is symmetric. The sum

$$\sum_{y\in GF(q^2)} e[\alpha^{-1}(y-x)^{q+1}]$$

of the elements of any row in $H_\alpha$ can be shown to be -q using (9.121), hence proving that $H_\alpha$ is regular.

Now, from theorem 9.12, and using the result $H_\alpha{}^* = H_{-\alpha}$, we get

$$H_\alpha H_\alpha{}^* = SD_\alpha D_{-\alpha}S^{-1} = q^2I.$$

Thus (i) is proved.

Now (ii) follows from

$$D_{\alpha_1}D_{\alpha_2} = -qD_{\alpha_1+\alpha_2},$$

which can be obtained from (9.111). Thus from theorem 9.12

$$(-\frac{1}{q}H_{\alpha_1})(-\frac{1}{q}H_{\alpha_2}) = (-\frac{1}{q}H_{\alpha_1+\alpha_2})$$

**and thus we have (ii).** ***

# CHAPTER X. EQUIVALENCE OF HADAMARD MATRICES

This problem has mainly been studied for small orders, though Stiffler, Baumert and W.D. Wallis have obtained other results.

DEFINITION 10.1. If A and B are matrices over the ring Z of integers, A and B are called *integral equivalent* or *Z-equivalent* ($A \underset{\sim}{Z} B$) if there are Z-matrices P and Q with determinant ±1, such that

$$B = PAQ.$$

If A and B are over an Euclidean domain E then A and B are *E-equivalent* if P and Q are E-matrices.

Then Z-equivalent is the same as saying B can be obtained from A by performing some sequence of the following operations:

(a) add an integer multiple of one row to another,

(b) negate some row,

(c) reorder the rows,

and the corresponding column operations.

DEFINITION 10.2. Two Hadamard matrices, $H_1$ and $H_2$, will be said to be *Hadamard equivalent* or *H-equivalent* ($H_1 \underset{\sim}{H} H_2$) if $H_2$ can be obtained from $H_1$ by a sequence of the following operations:

(a) negate some row,

(b) reorder the rows,

and the corresponding column operations.

DEFINITION 10.3. Two Hadamard matrices, $H_1$ and $H_2$, will be said to be *Seidel equivalent* or *S-equivalent* ($H_1 \underset{\sim}{S} H_2$) if $H_2$ can be obtained from $H_1$ by a sequence of the following operations:

(a) negate some row j and negate column j,

(b) interchange row i and row j and also interchange column i and column j.

We will establish the results of the table on the equivalence of Hadamard matrices.

We note:

| Our Reference Number | Order of Hadamard matrix | $N_S$ | Reference | $N_H$ | Reference | $N_Z$ | Reference |
|---|---|---|---|---|---|---|---|
| 1 | 2 | | | 1 | §10.4 | 1 | theorem 10.7 |
| 2 | 4 | | | 1 | §10.4 | 1 | theorem 10.7 |
| 3 | 8 | | | 1 | §10.4 | 1 | theorem 10.7 |
| 4 | 12 | | | 1 | §10.4 | 1 | theorem 10.7 |
| 5 | 16 | | | 5 | §10.5 | 4 | §10.5 |
| 6 | 20 | | | 3 | §10.4 | 1 | theorem 10.7 |
| 7 | 24 | | | | | 1 | theorem 10.7 |
| 8 | 28 | | | | | 1 | theorem 10.7 |
| 9 | 32 | | | ≥11 | §10.7 | 11 | §10.7 |
| 10 | 36 | ≥91 | §10.8 | ≥4 | §10.8 | ≥4, ≤12 | §10.8 |
| 11 | $2^n$ | | | $\geq 10[\frac{n}{5}]+1$ | §10.9 | $\geq 10[\frac{n}{5}]+1$ | §10.9 |
| 12 | 4m* | | | | | 1 | theorem 10.7 |
| 13 | 16t** | | | | | ≥2 | §10.3 |

$N_A$ is the number of A-inequivalent Hadamard matrices.

* m square-free integer.

** 8t is the order of a skew-Hadamard matrix.

TABLE

LEMMA 10.4.

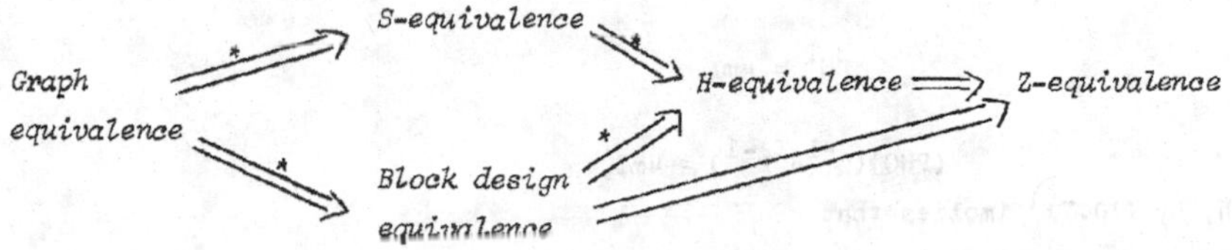


* *where applicable.*

## 10.1 Z-EQUIVALENCE: THE SMITH NORMAL FORM

The following theorem is due to Smith [1] and has been reworded from the theorems and proofs in MacDuffee [1, p.41] and Marcus and Minc [1, p.44].

THEOREM 10.5. *If* $A = (a_{ij})$ *is any Z-matrix of order n and rank r, then there is a unique Z-matrix*

$$D = diag(a_1,a_2,\ldots,a_r,0,\ldots,0),$$

*such that* $A \overset{Z}{\sim} D$ *and* $a_i|a_{i+1}$ *where the* $a_i$ *are non-negative. The greatest common divisor of the* $i \times i$ *sub-determinants of A is*

$$a_1a_2a_3\ldots a_i.$$

*If* $A \overset{Z}{\sim} E$ *where*

$$E = diag(a_1,a_2,\ldots,a_i)\oplus F$$

*then* $a_{i+1}$ *is the greatest common divisor of the non-zero elements of F.* ***

DEFINITION 10.6. The $a_i$ of theorem 10.5 are called the *invariants* of A, and the diagonal matrix D is called the *Smith normal form.*

The following result of Jennifer Wallis and W.D. Wallis for odd m has been reproved and strengthened by Newman, Spence and W.D. Wallis.

THEOREM 10.7. *Any Hadamard matrix of order 4m, where m is square-free, is Z-equivalent to*

$$diag(1,\underbrace{2,2,\ldots,2}_{2m-1\text{ times}},\underbrace{2m,2m,\ldots,2m}_{2m-1\text{ times}},4m).$$

PROOF. By theorem 10.5 an Hadamard matrix, H of order 4m, is Z-equivalent to $diag(a_1,a_2,\ldots,a_{4m})$ where $a_i|a_{i+1}$. Write

$$PHQ = diag(a_1,a_2,\ldots,a_{4m}). \tag{10.71}$$

Now

$$HH^T = 4mI$$

so

$$(PHQ)(Q^{-1}H^TP^{-1}) = 4mI,$$

which, by (10.71) implies that

$$Q^{-1}H^TP^{-1} = diag(4m/a_1,4m/a_2,\ldots,4m/a_{4m}). \tag{10.72}$$

However, it is clear that H and $H^T$ have the same invariant factors. Thus, since $4m/a_{i+1}|4m/a_i$ $(1 \le i \le 4m)$, it follows at once that the invariant factors of $H^T$ are $(4m/a_{4m},4m/a_{4m-1},\ldots,4m/a_1)$ which can be identified with $(a_1,a_2,\ldots,a_{4m})$.

Consequently

$$a_{4m+1-i}a_i = 4m. \tag{10.73}$$

Using the fact that $a_i | a_{i+1}$ and $d_{2m}d_{2m+1} = 4m$ from (10.73), it is seen that

$$4m \equiv 0 \pmod{d_{2m}^2}.$$

Thus if m is square-free, $d_{2m} = 1$ or 2. However, since $d_2$ is the greatest common divisor of the two rowed minors of H, clearly $d_2 = 2$, for any 2×2 matrix with entries ±1 has determinant 0 or ±2. It follows that $d_{2m} = 2$ and the invariant factors of H are

$$1,\underbrace{2,2,\ldots,2}_{2m-1 \text{ times}},\underbrace{2m,2m,\ldots,2m}_{2m-1 \text{ times}},4m.$$

***

COROLLARY 10.8. *Every Hadamard matrix of order 4m, m square-free, is Z-equivalent to*

$$diag(a_1,a_2,\ldots,a_{4m})$$

*where* $a_1 = 1$, $a_2 = 2$, $a_{4m-1} = 2m$, $a_{4m} = 4m$, $a_i | a_{i+1}$ *and* $a_i a_{4m+1-i} = 4m$.

DEFINITION 10.9. The diagonal matrix of order 4m

$$(1,\underbrace{2,2,\ldots,2}_{2m-1 \text{ times}},\underbrace{2m,\ldots,2m}_{2m-1 \text{ times}},4m)$$

will be called the *standard form.*

The following result can be obtained by noting that the rank of the incidence matrix of a (v,k,λ)-configuration is v (see Marcus and Minc p.46):

THEOREM 10.10. *The incidence matrix of a (v,k,λ)-configuration is equivalent to a diagonal matrix with entries*

$$\begin{cases} 1 & \frac{v+1}{2} \text{ times}, \\ (k-\lambda) & \frac{v-3}{2} \text{ times}, \\ k(k-\lambda) & \text{once}, \end{cases}$$

*when k-λ is square-free and (k-λ,k) = 1.*

***

## 10.2 Z-EQUIVALENCE: THE NUMBER OF SMALL INVARIANTS

We know from corollary 10.8 that an Hadamard matrix has exactly one invariant equal to 1, and that the next invariant is 2. In this section a lower limit for the number of invariants equal to 2 is found as a consequence of a general result of independent interest. We write [x] for the largest integer not exceeding x.

THEOREM 10.11. (W.D. Wallis) *Suppose B is an n×n matrix with non-zero determinant whose entries are all 0 and 1, the zero identity elements of a Euclidean domain E. Then the number of invariants of B under E-equivalence which equal 1 is at least*

$[\log_2 n]+1$.

PROOF. Write t for $[\log_2 n]$; that is, t is the unique integer such that $2^t \leq n \leq 2^{t+1}$. We shall use a sequence of E-equivalence operations to transform B to

$$I_t \oplus D,$$

where 1 is a greatest common divisor of the entries in D.

The first part of the process is to reorder the rows and columns of B so that, in the reordered matrix, columns 1 and 2 have different entries in the first row, columns 3 and 4 are identical in the first row but have different entries in the second row, and in general columns 2i-1 and 2i are identical in the first i-1 rows but differ in row i for i = 1,2,...,t. This is done using the following algorithm.

STEP 1. Select two columns of B which have different entries in the first row. (If the first row of B has every entry 1, it will first be necessary to subtract some other row from row 1.) Reorder columns so that the two chosen columns become columns 1 and 2 (in either order).

STEP 2. Select two columns of the matrix just formed, neither of them being columns 1 or 2, which have identical entries in the first row. Reorder the rows of the matrix other than row 1 so that the two columns chosen have different entries in the new second row. Reorder columns after column 2 so that the new pair become columns 3 and 4.

STEP k. In the matrix resulting from step k-1, select two columns to the right of column 2k-2 which are identical in rows 1 to k-1. Reorder the rows after row k-1 and the columns after column 2k-2 so that the chosen columns become columns 2k-1 and 2k and differ in their kth row.

It is always possible in step k to find a row in which the two chosen columns differ, since the matrix cannot have two identical columns. Therefore, step k only requires that we can choose two columns from the n-2k+2 available ones which are identical in their first k-1 places. Since there are only $2^{k-1}$ different (0,1)-vectors of length k-1, this will be possible provided

$$2^{k-1} < n-2k+2,$$

and this is always true for $1 \leq k \leq t$ except when k = t and n = 4 or n = 8. In the case n = 4, it is easy to check by hand that every (0,1) matrix of non-zero determinant is Z-equivalent to a matrix on which t steps can be carried out. If n = 8, step 3 will be impossible if the first two rows of the last four columns contain all (0,1)-vectors of length 2; typically the first two rows are

$$\begin{matrix} 1 & 0 & 1 & 1 & 1 & 1 & 0 & 0 \\ * & * & 1 & 0 & 1 & 0 & 1 & 0 \end{matrix}$$

and in this case we can proceed with step 3 if we first apply the column permutation (37)(48). Consequently t steps can always be carried out.

In the second stage, select if possible two columns (columns a and b say) which are identical in rows 1 to t; reorder the later rows so that columns a and b differ in row t+1. If the selection was impossible then $n = 2^t$ and the first t rows of the matrix constitute the $2^t$ different column vectors, so one column (column a say) will start with t zeros; reorder the rows from t+1 on so that there is a 1 in the (a,t+1) position. In either case, if column a were in a pair chosen in stage 1, reorder the pair if necessary so that a is even; and similarly for b if two were chosen.

The third stage isolates certain entries ±1 by carrying out t steps:

STEP 1. Subtract column 2 from column 1, so that the (1,1) entry becomes ±1. Then add a suitable multiple of the first column to every other column to ensure that row 1 has every entry 0 except the first, and similarly eliminate all entries except the first from column 1 by adding suitable multiples of row 1 to the other rows.

After step 1 the matrix has first row and column (±1,0,0,...,0); whatever multiple of column 1 was added to column 2k-1 ($2 \leq k \leq t$), the same multiple was added to column 2k, so that the (k,2k-1) and (k,2k) entries still differ by 1. It will be seen from the description of the general step that, after k-1 steps, the matrix will have its first k-1 rows zero except for entries ±1 in the (i,2i-1) positions, $1 \leq i \leq k-1$; the first k-1 odd-numbered columns will be zero except at those positions; and for $k \leq i \leq t$ the (i,2i-1) and (i,2i) entries differ by 1 and the (j,2i-1) and (j,2i) entries are equal when $j < i$. After step k this description can be extended by replacing k-1 by k.

STEP k. Subtract column 2k from column 2k-1, so that the (k,2k-1) entry becomes ±1. Add a suitable multiple of column 2k-1 to every subsequent column, and then add suitable multiples of row k to the later rows, so that row k and column 2k-1 become zero except at their intersection.

Observe that if $k < i \leq t$ the (k,2i-1) and (k,2i) entries were equal before step k, so the same multiple of column 2k-1 was added to both columns 2i-1 and 2i and the differences between these columns is unchanged. If two columns, a and b, were chosen at stage two, then the difference between those columns is unaltered in the t steps; if only one column was chosen then that column is unaltered in the t steps since it has never had a non-zero entry in its kth row to be eliminated.

Finally, reorder the columns so that the former columns 1,3,...,2t-1 become the first t columns. We obtain

$$\begin{bmatrix} I_t & 0 \\ 0 & D \end{bmatrix};$$

D contains either two entries which differ by 1 (corresponding to the former (a,t+1) and (b,t+1) entries) or has an entry 1 (the former (a,t+1) entry) depending on the course followed at stage 2, and in either case the greatest common divisor of entries of D is 1.

Therefore, B has at least t+1 invariants equal to 1. ***

COROLLARY 10.12. *An Hadamard matrix of order 4m has at least*

$$[\log_2 (4m-1)]+1$$

*invariants equal to 2, and by theorem 10.11 it has at least this number of invariants equal to 2m.*

PROOF. Let A be an Hadamard matrix of order 4m; assume A to be normalized. Subtract row 1 from every other row and them column 1 from every other column; we obtain

$$A = \begin{bmatrix} 1 & 0 \\ 0 & -2B \end{bmatrix}$$

where B is an (0,1)-matrix of size 4m-1 with non-zero determinant. The first invariant of A is 1; the others are double the invariants of B. The result follows from theorem 10.11. ***

THEOREM 10.13. (W.D. Wallis) *Suppose E is a Euclidean domain with characteristic not equal to 2, and suppose f and g are monotonic non-decreasing functions which satisfy:*

*(i) any Hadamard matrix of order N has at least f(N) invariants equal to 2;*

*(ii) any (0,1)-matrix over E which has non-zero determinant and is of size r×r has at least g(r) invariants equal to 1.*

*Then*

$$f(N) \leq [\log_2 (N-1)]+1 \qquad (10.131)$$

*and, if 2 is a non-unit of E,*

$$g(r) \leq [\log_2 r]+1.$$

PROOF. The function on the right hand side of (10.131) is a step-function which increases in value just after N takes as its value a power of 2. So, if (10.131) is false, we must have

$$f(2^t) > [\log_2 (2^t-1)]+1 = t$$

for some t. However, for every t, there is an Hadamard matrix A of order $2^t$ which has precisely t invariants 2. Let

$$H = \begin{bmatrix} 1 & 1 \\ 1 & -1 \end{bmatrix}$$

which has invariants {1,2}, and define A as the direct product of t copies of H.

Then $A \sim D$, where D is the direct product of t copies of diag(1,2); D is a diagonal matrix whose entries are powers of 2, and $2^a$ occurs $\binom{t}{a}$ times. These must be the invariants of A. A has precisely t invariants equal to 2.

If we consider A as a matrix over E, rather than an integer matrix, and pass to B as in the proof of corollary 10.12, then B is a (0,1)-matrix over E of size $r = 2^t-1$, and has non-zero determinant (as the characteristic of E is not 2). The invariants of B are $2^a$, $\binom{t}{a+1}$ times each, for a = 0,1,...,t-1. $2^a$ and $2^b$ are the same invariant if and only if $2^{b-a}$ is a unit of E, and this cannot occur when $a \neq b$ provided 2 is a non-unit. So the matrix B can be used to prove the part of the theorem involving g. ***

Theorem 10.13 shows that the results of theorem 10.11 and corollary 10.12 are best possible in a certain sense unless E has characteristic 2 or 2 is a unit of E. If 2 were a unit then the matrix B has every invariant 2 (or 1, which is the same thing), and if E had characteristic 2 then we could not divide by 2 to get B.

## 10.3 Z-EQUIVALENCE: MATRICES DERIVED FROM SKEW-HADAMARD MATRICES AND SYMMETRIC CONFERENCE MATRICES

THEOREM 10.14. (W.D. Wallis) *If there is a skew-Hadamard matrix of order n then there is a skew-Hadamard matrix of order 2n Z-equivalent to the standard form.*

PROOF. The theorem is easily proven when n = 1 or 2, so put n = 4m. Suppose A is a skew-Hadamard matrix with canonical diagonal matrix D; suppose P and Q are unimodular integral matrices such that

$$D = PAQ.$$

Then

$$Q^{-1}A^TP^{-1} = nD^{-1},$$

and $nD^{-1}$ is the matrix D with the order of its entries reversed (10.72). For convenience write

$$D = (1) \oplus 2C \oplus (4m);$$

C is a diagonal integral matrix of order n-2.

Consider the matrix

$$K = \begin{bmatrix} A & A \\ -A^T & A^T \end{bmatrix}$$

which is skew-Hadamard of order 2n. K is equivalent to

$$\begin{bmatrix} P & 0 \\ Q^{-1} & -Q^{-1} \end{bmatrix} \begin{bmatrix} A & A \\ -A^T & A^T \end{bmatrix} \begin{bmatrix} Q & P^{-1} \\ 0 & P^{-1} \end{bmatrix}$$

$$= \begin{bmatrix} PAQ & 0 \\ Q^{-1}(A+A^T)Q & 2Q^{-1}A^TP^{-1} \end{bmatrix}$$

$$= \begin{bmatrix} D & 0 \\ 2I & 2nD^{-1} \end{bmatrix}$$

using the fact that $A+A^T = 2I$. This last matrix is

$$\left[\begin{array}{c|c|c|c|c|c} 1 & & & & & \\ \hline & 2C & & & & \\ \hline & & n & & & \\ \hline 2 & & & 2n & & \\ \hline & I & & & nC^{-1} & \\ \hline & & 2 & & & 2 \end{array}\right].$$

Subtract twice row 1 from row n+1; subtract column 2n from column n; then we can reorder the columns and rows to obtain

$$\begin{bmatrix} 1 & & & \\ & 2 & & \\ & & n & \\ & & & 2n \end{bmatrix} \oplus \left[\begin{array}{c|c} 2C & 0 \\ \hline 2I & nC^{-1} \end{array}\right].$$

Every entry of C divides $\frac{1}{2}n$, so $\frac{1}{2}nC^{-1}$ is integral. So the second direct summand is integrally equivalent to

$$\begin{bmatrix} -I & C \\ 0 & I \end{bmatrix} \begin{bmatrix} 2C & 0 \\ 2I & nC^{-1} \end{bmatrix} \begin{bmatrix} -\frac{1}{2}nC^{-1} & I \\ I & 0 \end{bmatrix} = \begin{bmatrix} nI & 0 \\ 0 & 2I \end{bmatrix}$$

and the invariants of K are as required.

LEMMA 10.15. *If there is an Hadamard matrix of order $n = 8m$, then there is an Hadamard matrix of order $2n$ with at least $12m-1$ invariants divisible by 4.*

PROOF. If A is Hadamard of order n and has canonical diagonal matrix D, then

$$H = \begin{bmatrix} A & A \\ -A & A \end{bmatrix}$$

is Hadamard of order 2n and is equivalent to the diagonal matrix

$$D \oplus 2D.$$

Since $n = 8m$, the last $4m$ invariants of A are divisible by 4. Every entry of 2D except the first is divisible by 4. So $D \oplus 2D$ has at least $12m-1$ entries divisible by 4. Even if $D \oplus 2D$ is not in canonical form, it is easy to deduce that $D \oplus 2D$ (and consequently H) has at least 12m-1 invariants divisible by 4.

COROLLARY 10.16. *If there is a skew-Hadamard matrix of order 8m then there exists a pair of inequivalent Hadamard matrices of order 16m.*

PROOF. Lemmas 10.14 and 10.15. ***

Recall that if C+I is a symmetric conference matrix of order n, C+I is a (1,-1)-matrix and we may subtract or add multiples of row 1 from other rows and then multiples of column 1 to other columns till we have

$$C+I \overset{Z}{\sim} \begin{bmatrix} 1 & 0 \\ 0 & E \end{bmatrix} \overset{Z}{\sim} D \quad \text{(the Smith normal form).}$$

Writing

$$P(C+I)Q = D,$$

where P and Q are unimodular matrices, D has first entry 1 and all other entries divisible by 2.

Now C satisfies

$$C = C^T, \quad C^2 = (n-1)I,$$

$$(C+I)(C-I) = (n-2)I,$$

and we have

$$(C+I)^{-1} = (n-2)^{-1}(C-I).$$

Also

$$Q^{-1}(C+I)^{-1}P^{-1} = D^{-1}$$

$$Q^{-1}(C-I)P^{-1} = (n-2)D^{-1}.$$

THEOREM 10.17. (W.D. Wallis) *If there is a symmetric conference matrix of order n then there exists an Hadamard matrix of order 2n which is Z-equivalent to the standard form.*

PROOF. Write H for the Hadamard matrix of order 2n given by

$$H = \begin{bmatrix} -C-I & C-I \\ C-I & C+I \end{bmatrix},$$

then noting that

$$\begin{bmatrix} -Q^{-1} & Q^{-1} \\ 0 & P \end{bmatrix} \text{ and } \begin{bmatrix} -P^{-1} & 0 \\ P^{-1}+(n-2)QD^{-1} & Q \end{bmatrix}$$

are also unimodular matrices, we have

$$\begin{bmatrix} -Q^{-1} & Q^{-1} \\ 0 & P \end{bmatrix} \begin{bmatrix} -C-I & C-I \\ C-I & C+I \end{bmatrix} \begin{bmatrix} -P^{-1} & 0 \\ P^{-1}+(n-2)QD^{-1} & Q \end{bmatrix}$$

$$\begin{bmatrix} 0 & 2I \\ nI & D \end{bmatrix} = \begin{bmatrix} 0 & 2I \\ & 10 \\ nI & \\ & 0E \end{bmatrix}$$

$$\overset{Z}{\sim} 2I_{n-1} \oplus nI_{n-1} \oplus \begin{bmatrix} 0 & 2 \\ n & 1 \end{bmatrix}$$

$\overset{Z}{\sim}$ standard form. ***

10.4 H-EQUIVALENCE: ORDERS 1,2,4,8,12,20. We have noted before that every Hadamard matrix is H-equivalent to an Hadamard matrix in normalized form and further, that the columns of a matrix of order 4m can be rearranged into the form

```
1 1 ... 1 1 1 1 ... 1 1 1 1 ... 1 1 1 1 ... 1 1
1 1 ... 1 1 1 1 ... 1 1 - - ... - - - - ... - -
1 1 ... 1 1 - - ... - - 1 1 ... 1 1 - - ... - -
```

where - is written for -1.

Clearly, the only possibilities with the above first three rows for orders 1,2 and 4 are

$$[1], \quad \begin{bmatrix} 1 & 1 \\ 1 & - \end{bmatrix} \quad \text{and} \quad \begin{bmatrix} 1 & 1 & 1 & 1 \\ 1 & 1 & - & - \\ 1 & - & 1 & - \\ 1 & - & - & 1 \end{bmatrix} .$$

Now every Hadamard matrix of order 4m has 2m positive elements in each column. So the rows of the matrix of order 8 may be rearranged until the element (2,4) is 1 and (2,5) is -1. Then with the above first three rows we have

$$\begin{bmatrix} 1 & 1 & 1 & 1 & 1 & 1 & 1 & 1 \\ 1 & 1 & 1 & 1 & - & - & - & - \\ 1 & 1 & - & - & 1 & 1 & - & - \\ 1 & 1 & & & & & & \\ 1 & - & & & & & & \\ 1 & & & & & & & \\ 1 & & & & & & & \\ 1 & & & & & & & \end{bmatrix}$$

which can only be completed in one way to make an Hadamard matrix after we have noted that the columns may be rearranged to make row 5

1 - 1 - 1 - 1 - .

It can also be shown that all (7,3,1)-configurations are equivalent.

By a similar kind of argument Hussain [45] shows that all (11,5,2)-configurations are equivalent and hence

LEMMA 10.18. *All Hadamard matrices of order m, where m = 1,2,4,8,12, are H-equivalent.*

Clearly the Hadamard matrices of orders 1 and 2 have invariants 1 and 1,2 respectively and hence using theorem 10.7:

LEMMA 10.19. *All Hadamard matrices of order m, where m = 1,2,4,8,12, are Z-equivalent.*

The result - that integral equivalence and Hadamard-equivalence-plus-transposition come to the same thing - is trivially true for Hadamard matrices of order less than 16 (there is only one matrix of each order under Hadamard equivalence); however, it is false for order 20. Hall [39] finds three inequivalent Hadamard matrices under Hadamard equivalence, and these remain inequivalent if transposition is allowed; however, all Hadamard matrices of order 20 are integrally equivalent by theorem 10.7.

THEOREM 10.20. (Marshall Hall Jr.) *There are three H-inequivalent matrices of order 20.*

10.5 H-EQUIVALENCE: ORDER 16. The results of the preceding sections place restrictions on the possible invariants of Hadamard matrices, even when the order is divisible by 16 or by an odd square. For example, the invariants of an Hadamard matrix of order 16 must be of the form

$$\begin{array}{rl} 1 & \text{once,} \\ 2 & \alpha \text{ times,} \\ 4 & \beta \text{ times,} \\ 8 & \gamma \text{ times,} \\ 16 & \text{once;} \end{array}$$

since the number of invariants is 16 and their product is $16^8$, we must have

$$\alpha+\beta+\gamma = 14,$$
$$\alpha+2\beta+3\gamma = 28;$$

$\gamma \geq 1$ and $\alpha \geq 4$ ($4 = [\log_2 16]+1$). So the invariants must be of the form

$$\begin{array}{rl} 1 & \text{once,} \\ 2 & \alpha \text{ times,} \\ 4 & 14-2\alpha \text{ times,} \\ 8 & \alpha \text{ times,} \\ 16 & \text{once;} \end{array}$$

$4 \leq \alpha \leq 7$ and

$$(\alpha,\beta,\gamma) = (4,6,4),\ (5,4,5),\ (6,2,6) \text{ or } (7,0,7).$$

TABLE

A =

| | | | | | | | | | | | | | | | |
|---|---|---|---|---|---|---|---|---|---|---|---|---|---|---|---|
| 1 | 1 | 1 | 1 | 1 | 1 | 1 | 1 | 1 | 1 | 1 | 1 | 1 | 1 | 1 | 1 |
| 1 | 1 | 1 | 1 | 1 | 1 | 1 | 1 | - | - | - | - | - | - | - | - |
| 1 | 1 | 1 | 1 | - | - | - | - | 1 | 1 | 1 | 1 | - | - | - | - |
| 1 | 1 | 1 | 1 | - | - | - | - | - | - | - | - | 1 | 1 | 1 | 1 |
| 1 | 1 | - | - | 1 | 1 | - | - | 1 | 1 | - | - | 1 | 1 | - | - |
| 1 | 1 | - | - | 1 | 1 | - | - | - | - | 1 | 1 | - | - | 1 | 1 |
| 1 | 1 | - | - | - | - | 1 | 1 | 1 | 1 | - | - | - | - | 1 | 1 |
| 1 | 1 | - | - | - | - | 1 | 1 | - | - | 1 | 1 | 1 | 1 | - | - |
| 1 | - | 1 | - | 1 | - | 1 | - | 1 | - | 1 | - | 1 | - | 1 | - |
| 1 | - | 1 | - | 1 | - | 1 | - | - | 1 | - | 1 | - | 1 | - | 1 |
| 1 | - | 1 | - | - | 1 | - | 1 | 1 | - | 1 | - | - | 1 | - | 1 |
| 1 | - | 1 | - | - | 1 | - | 1 | - | 1 | - | 1 | 1 | - | 1 | - |
| 1 | - | - | 1 | 1 | - | - | 1 | 1 | - | - | 1 | 1 | - | - | 1 |
| 1 | - | - | 1 | 1 | - | - | 1 | - | 1 | 1 | - | - | 1 | 1 | - |
| 1 | - | - | 1 | - | 1 | 1 | - | 1 | - | - | 1 | - | 1 | 1 | - |
| 1 | - | - | 1 | - | 1 | 1 | - | - | 1 | 1 | - | 1 | - | - | 1 |

Hall's Class I

$(\alpha,\beta,\gamma) = (4,6,4)$

B =

| | | | | | | | | | | | | | | | |
|---|---|---|---|---|---|---|---|---|---|---|---|---|---|---|---|
| 1 | 1 | 1 | 1 | 1 | 1 | 1 | 1 | 1 | 1 | 1 | 1 | 1 | 1 | 1 | 1 |
| 1 | 1 | 1 | 1 | 1 | 1 | 1 | 1 | - | - | - | - | - | - | - | - |
| 1 | 1 | 1 | 1 | - | - | - | - | 1 | 1 | 1 | 1 | - | - | - | - |
| 1 | 1 | 1 | 1 | - | - | - | - | - | - | - | - | 1 | 1 | 1 | 1 |
| 1 | 1 | - | - | 1 | 1 | - | - | 1 | 1 | - | - | 1 | 1 | - | - |
| 1 | 1 | - | - | 1 | 1 | - | - | - | - | 1 | 1 | - | - | 1 | 1 |
| 1 | 1 | - | - | - | - | 1 | 1 | 1 | 1 | - | - | - | - | 1 | 1 |
| 1 | 1 | - | - | - | - | 1 | 1 | - | - | 1 | 1 | 1 | 1 | - | - |
| 1 | - | 1 | - | 1 | - | 1 | - | 1 | - | 1 | - | 1 | - | 1 | - |
| 1 | - | 1 | - | 1 | - | 1 | - | - | 1 | - | 1 | - | 1 | - | 1 |
| 1 | - | 1 | - | - | 1 | - | 1 | 1 | - | 1 | - | - | 1 | - | 1 |
| 1 | - | 1 | - | - | 1 | - | 1 | - | 1 | - | 1 | 1 | - | 1 | - |
| 1 | - | - | 1 | 1 | - | - | 1 | 1 | - | - | 1 | - | 1 | 1 | - |
| 1 | - | - | 1 | 1 | - | - | 1 | - | 1 | 1 | - | 1 | - | - | 1 |
| 1 | - | - | 1 | - | 1 | 1 | - | 1 | - | - | 1 | 1 | - | - | 1 |
| 1 | - | - | 1 | - | 1 | 1 | - | - | 1 | 1 | - | - | 1 | 1 | - |

Hall's Class II

$(\alpha,\beta,\gamma) = (5,4,5)$

C =

| | | | | | | | | | | | | | | | |
|---|---|---|---|---|---|---|---|---|---|---|---|---|---|---|---|
| 1 | 1 | 1 | 1 | 1 | 1 | 1 | 1 | 1 | 1 | 1 | 1 | 1 | 1 | 1 | 1 |
| 1 | 1 | 1 | 1 | 1 | 1 | 1 | 1 | - | - | - | - | - | - | - | - |
| 1 | 1 | 1 | 1 | - | - | - | - | 1 | 1 | 1 | 1 | - | - | - | - |
| 1 | 1 | 1 | 1 | - | - | - | - | - | - | - | - | 1 | 1 | 1 | 1 |
| 1 | 1 | - | - | 1 | 1 | - | - | 1 | 1 | - | - | 1 | 1 | - | - |
| 1 | 1 | - | - | 1 | 1 | - | - | - | - | 1 | 1 | - | - | 1 | 1 |
| 1 | 1 | - | - | - | - | 1 | 1 | 1 | 1 | - | - | - | - | 1 | 1 |
| 1 | 1 | - | - | - | - | 1 | 1 | - | - | 1 | 1 | 1 | 1 | - | - |
| 1 | - | 1 | - | 1 | - | 1 | - | 1 | - | 1 | - | 1 | - | 1 | - |
| 1 | - | 1 | - | 1 | - | 1 | - | - | 1 | - | 1 | - | 1 | - | 1 |
| 1 | - | 1 | - | - | 1 | - | 1 | 1 | - | - | 1 | 1 | - | - | 1 |
| 1 | - | 1 | - | - | 1 | - | 1 | - | 1 | 1 | - | - | 1 | 1 | - |
| 1 | - | - | 1 | 1 | - | - | 1 | 1 | - | - | 1 | - | 1 | 1 | - |
| 1 | - | - | 1 | 1 | - | - | 1 | - | 1 | 1 | - | 1 | - | - | 1 |
| 1 | - | - | 1 | - | 1 | 1 | - | 1 | - | 1 | - | - | 1 | - | 1 |
| 1 | - | - | 1 | - | 1 | 1 | - | - | 1 | - | 1 | 1 | - | 1 | - |

Hall's Class III

$(\alpha,\beta,\gamma) = (6,2,6)$

D =

| | | | | | | | | | | | | | | | |
|---|---|---|---|---|---|---|---|---|---|---|---|---|---|---|---|
| 1 | 1 | 1 | 1 | 1 | 1 | 1 | 1 | 1 | 1 | 1 | 1 | 1 | 1 | 1 | 1 |
| 1 | 1 | 1 | 1 | 1 | 1 | 1 | 1 | - | - | - | - | - | - | - | - |
| 1 | 1 | 1 | 1 | - | - | - | - | 1 | 1 | 1 | 1 | - | - | - | - |
| 1 | 1 | 1 | 1 | - | - | - | - | - | - | - | - | 1 | 1 | 1 | 1 |
| 1 | 1 | - | - | 1 | 1 | - | - | 1 | 1 | - | - | 1 | 1 | - | - |
| 1 | 1 | - | - | 1 | 1 | - | - | - | - | 1 | 1 | - | - | 1 | 1 |
| 1 | 1 | - | - | - | - | 1 | 1 | 1 | 1 | - | - | - | - | 1 | 1 |
| 1 | 1 | - | - | - | - | 1 | 1 | - | - | 1 | 1 | 1 | 1 | - | - |
| 1 | - | 1 | - | 1 | - | 1 | - | 1 | - | 1 | - | 1 | - | 1 | - |
| 1 | - | 1 | - | 1 | - | - | 1 | 1 | - | - | 1 | - | 1 | - | 1 |
| 1 | - | 1 | - | - | 1 | 1 | - | - | 1 | - | 1 | 1 | - | - | 1 |
| 1 | - | 1 | - | - | 1 | - | 1 | - | 1 | 1 | - | - | 1 | 1 | - |
| 1 | - | - | 1 | 1 | - | 1 | - | - | 1 | - | 1 | - | 1 | 1 | - |
| 1 | - | - | 1 | 1 | - | - | 1 | - | 1 | 1 | - | 1 | - | - | 1 |
| 1 | - | - | 1 | - | 1 | 1 | - | 1 | - | 1 | - | - | 1 | - | 1 |
| 1 | - | - | 1 | - | 1 | - | 1 | 1 | - | - | 1 | 1 | - | 1 | - |

Hall's Class IV

$(\alpha,\beta,\gamma) = (7,0,7)$

Nandi [61] found five non-isomorphic (15,7,3)-configurations which put an upper bound on the number of H-inequivalent Hadamard matrices of order 16. In [38] Marshall Hall Jr. studied the triples (3-tuples) of the Hadamard matrices of order 16 and found that the number of possible triples is 35,19,11 or 7. For 7 he shows there are two types (actually the matrix and its transpose), one of which has a common element in all triples. Representative matrices A,B,C,D of classes one to four are exhibited in the table.

THEOREM 10.21. (Marshall Hall Jr.) *There are exactly five H-inequivalent Hadamard matrices of order 16.*

A calculation shows that the matrices A,B,C and D have four, five, six and seven invariants respectively equal to 2. So that

A corresponds to $(\alpha,\beta,\gamma) = (4,6,4)$

B corresponds to $(\alpha,\beta,\gamma) = (5,4,5)$

C corresponds to $(\alpha,\beta,\gamma) = (6,2,6)$

D corresponds to $(\alpha,\beta,\gamma) = (7,0,7)$

and Hall's class V which has representative matrix $D^T$ has the same invariants as class IV. So all the theoretically possible Z-inequivalent Hadamard matrices of order 16 actually occur.

10.6 H-EQUIVALENCE: ORDER 24. By theorem 10.7 all Hadamard matrices of order 24 are Z-equivalent.

Bhat and Shrikhande have studied the number of inequivalent (23,11,5)-configurations, but the theoretical tools available are not adequate to find the number of H-inequivalent Hadamard matrices of order 24.

10.7 Z-EQUIVALENCE: ORDER 32. The restrictions in corollary 10.8 and 10.12 imply that the invariants of an Hadamard matrix of order 32 are

1 once,

2 $\alpha$ times,

4 $(15-\alpha)$ times,

8 $(15-\alpha)$ times,

16 $\alpha$ times,

32 once;

and that $5 \leq \alpha \leq 15$. We shall say a matrix with the invariants shown is "of class $\alpha$"; W.D. Wallis constructed Hadamard matrices of order 32 in all eleven possible classes, thus proving

THEOREM 10.22. (W.D. Wallis) *There are precisely eleven Z-inequivalent Hadamard matrices of order 32.*

In section 10.10 the invariants of some Hadamard matrices of order a power of 2 will be calculated by generating functions. In particular, if A is Hadamard of order 16 with $\omega$ invariants equal to 2 and if $H_2$ is Hadamard of order 2, the direct product

$$H_2 \times A = \begin{bmatrix} A & A \\ -A & A \end{bmatrix}$$

has exactly $\omega+1$ invariants equal to 2. So the existence of 16×16 Hadamard matrices with 4,5,6 and 7 invariants equal to 2 implies the existence of 32×32 Hadamard matrices of classes 5,6,7 and 8. There is a skew-Hadamard matrix of order 16, so by theorem 10.14 class 15 exists.

An Hadamard matrix of order 16 can be constructed from a symmetric balanced incomplete block design with parameters (15,7,3): first construct a matrix with (i,j) entries 1 if treatment j belongs to block i and -1 elsewhere; then add on a first row and column with every entry 1. The (15,7,3)-designs have been found by Nandi [61], and are also listed in [13]. Write A for the 16×16 Hadamard matrix constructed from Nandi's design $(\alpha_2\alpha_2')$, and B for the matrix constructed from Nandi's $(\alpha_1\alpha_1')_1$ after applying a permutation $\pi$ to the blocks. Then consider the matrix

$$H = \begin{bmatrix} A & B \\ -A & B \end{bmatrix}.$$

It is found that

when $\pi = (1)$ H is of class 9,
when $\pi = (1,3)$ H is of class 10,
when $\pi = (1,3,4)$ H is of class 11,
when $\pi = (2,7,12,13)$ H is of class 12,
when $\pi = (3,4,5,6,7)$ H is of class 13,
when $\pi = (2,3,4,5,6,7)$ H is of class 14.

Therefore, examples of all classes can be found. (These results were found in a computer test of various 32×32 Hadamard matrices. Details of the computation may be found in [127].

## 10.8 ORDER 36 AND S-EQUIVALENCE.

For order 36 the invariants are

1 once,
2 $\alpha$ times,
6 $34-2\alpha$ times,
18 $\alpha$ times,
36 once;

by corollary 10.12, $6 \leq \alpha \leq 17$, and so there are 12 theoretically possible Z-inequivalent Hadamard matrices of order 36.

W.D. Wallis has found four Z-inequivalent matrices.

Bussemaker and Seidel [29], [30], have studied the S-equivalence of symmetric Hadamard matrices of order 36 with constant diagonal. They define S-equivalence by "switching" in graphs which multiplies the rows and columns of the adjacency matrix by -1. Of course, the graph is not altered by relabelling the nodes so our definition of S-equivalence includes interchanging rows or columns. See definition 10.3.

Bussemaker and Seidel obtained symmetric Hadamard matrices with constant diagonal of order 36 from the 12 Latin squares on 6 symbols, and from the 80 Steiner triple systems on 15 symbols. They used a computer to study the S-equivalence of the associated graphs.

All these Hadamard matrices except for one pair of Latin square type are S-inequivalent. Among the 80 matrices of Steiner type exactly 23 are S-equivalent to regular Hadamard matrices H satisfying $HJ = -6J$. All 80 are S-equivalent to regular Hadamard matrices H satisfying $HJ = 6J$, some in several ways.

10.9 HIGHER POWERS OF 2. Baumert [4] found that for all orders $2^n$, $n \geq 5$, there were at least six inequivalent Hadamard matrices. This result is improved here.

To discuss higher powers of 2 we use a generating function for the numbers of invariants. Suppose H, of order $2^a$, has invariants $2^i$ occurring $a_i$ times. Put

$$f(H,t) = 1+\alpha_1 t+\alpha_2 t^2+\ldots+t^a.$$

Then if

$$f(K,t) = 1+\beta_1 t+\beta_2 t^2+\ldots+t^b,$$

the direct product $H\times K$ will be equivalent to the direct product of the two diagonal matrices, which has $2^k$ as an entry $\alpha_k+\alpha_{k-1}\beta_1+\ldots+\beta_k$ times; so

$$f(H\times K,t) = f(H,t)f(K,t).$$

Suppose H and K are Hadamard matrices of order $2^a$ with

$$f(H,t) = \sum\alpha_i t^i, \quad f(K,t) = \sum\beta_i t^i.$$

Write $H_2$ for an Hadamard matrix of order 2;

$$f(H_2,t) = (1+t),$$

so

$$f(H_2\times H,t) = \sum_{i=0}^{a+1} (\alpha_i+\alpha_{i-1})t^i$$

$$f(H_2\times K,t) = \sum_{i=0}^{a+1} (\beta_i+\beta_{i-1})t^i$$

(with the conventions $\alpha_{-1} = \alpha_{a+1} = \beta_{-1} = \beta_{a+1} = 0$). Clearly these functions cannot be identical unless $f(H,t) = f(K,t)$. Therefore

LEMMA 10.23. (W.D. Wallis) *$H_2 \times H$ and $H_2 \times K$ are integrally equivalent if and only if H and K are integrally equivalent.*

Let us write $g(H)$ for the coefficient of t in $f(H,t)$. Then

$$f(H \times K,t) = \big(1+g(H)t+\ldots\big)\big(1+g(K)t+\ldots\big)$$
$$= 1+[g(H)+g(K)]t+\ldots$$

so

$$g(H \times K) = g(H)+g(K).$$

THEOREM 10.24. (W.D. Wallis) *There are at least 10x+1 Z-inequivalent Hadamard matrices of order $2^{5x}$. Consequently the number of Z-inequivalent (and therefore H-inequivalent) Hadamard matrices of order $2^m$ is $\geq 10[\frac{m}{5}]+1$.*

PROOF. We show by induction that there is an Hadamard matrix H of order $2^{5x}$ with $g(H) = y$ whenever $5x \leq y \leq 15x$.

The case $x = 1$ is covered by the result on Hadamard matrices of order 32. Assume that $w > 1$, and that the result holds for $x = w$. We shall prove it holds for $x = w+1$. We write $H_n$ for an Hadamard matrix of order 32 and class m (that is, $g(H_m) = m$), and $K_n$ for an Hadamard matrix of order $2^{5w}$ satisfying $g(K_n) = n$. Then

$$g(H_m \times K_n) = m+n,$$

and $H_m \times K_n$ is Hadamard of order $2^{5(w+1)}$.

There is an $H_m$ whenever $5 \leq m \leq 15$; and by the induction hypothesis there is a $K_n$ whenever $5w \leq n \leq 15w$. So we can construct an Hadamard matrix H of order $2^{5(w+1)}$ with $g(H) = y$ whenever

$$y \in \{m+n: 5 \leq m \leq 15,\ 5w \leq n \leq 15w\}$$
$$= \{5(w+1), 5(w+1)+1, \ldots, 15(w+1)\}.$$

From lemma 10.24 the existence of 10x+1 inequivalent Hadamard matrices of order $2^{5x}$ implies the existence of 10x+1 inequivalent Hadamard matrices of order $2^{5x+1}$, $2^{5x+2}$, $2^{5x+3}$, $2^{5x+4}$. So the number of integrally inequivalent Hadamard matrices of order $2^m$ is at least

$$10[\tfrac{m}{5}]+1.$$

***

COROLLARY 10.25. *Given any positive integer N, there are infinitely many orders of which there are at least N integrally inequivalent (and therefore Hadamard inequivalent) Hadamard matrices of that order. The smallest such order is at most $32^n$ where $n = [\frac{N+9}{10}]$.*

These results are certainly not the best possible. For example, there is a skew-Hadamard matrix of order 2, so by theorem 10.14 there will be an Hadamard matrix with invariants in standard form of every order which is a power of 2, and this has not been taken into account. Of the eleven inequivalent matrices of order 32, only four can be found as direct products of smaller matrices.

# CHAPTER XI. USES OF HADAMARD MATRICES

This chapter is intended to be a brief survey rather than a detailed description of the use of Hadamard matrices.

Marshall Hall Jr. [57] mentions that the Mariner '69 telemetry system was based on an Hadamard matrix of order 32. Belevitch uses skew-Hadamard matrices and symmetric conference matrices in constructing most efficient telephone conference networks.

Hadamard matrices are important in Information Theory in constructing sequences with small correlation which may be used to simulate white noise, maximal codes and in the generation of Walsh functions.

Statistically they are important as they yield many block designs and give the most efficient weighing designs.

Round-robin tournaments and Bridge (yes! the card game) tournaments have their links with Hadamard matrices and so even do simple groups.

Johnsen [67] points out they even have an application in the behavioural sciences in studying the dominance relations and structure of animal societies.

Paley originally studied Hadamard matrices because of a problem with polytopes.

## 11.1 BALANCED INCOMPLETE BLOCK DESIGNS.

The existence of an Hadamard matrix of order $4t$ implies the existence of a $(4t-1,2t-1,t-1)$ SBIBD or its complement a $(4t-1,2t,t)$ SBIBD. A regular Hadamard matrix of order $4s^2$ gives a $(4s^2,2s^2\pm s,s^2\pm s)$ SBIBD.

In addition if $C$ is the incidence matrix of a $(4t-1,2t-1,t-1)$ SBIBD then

$$\left[\begin{array}{c|c} C & J-C \\ \hline e & 0 \end{array}\right]$$

where $e$ is the $1\times(4t-1)$ matrix of ones, is a $\{4t,2(4t-1),4t-1,2t,2t-1\}$ BIBD.

Van Lint and Seidel [146] noticed that if $D$ is the core of a skew-Hadamard matrix or symmetric conference matrix of order $2n$, then if $e$ is the $1\times(2n-1)$ matrix of ones

$$\left[\begin{array}{c|c} \frac{1}{2}(J-D-I) & \frac{1}{2}(J-D+I) \\ \hline e & 0 \end{array}\right]$$

is a $(2n,4n-2,2n-1,n,n-1)$ BIBD.

Since the Kronecker product of two regular Hadamard matrices of orders $4s^2$ and $4t^2$ is a regular Hadamard matrix of order $16s^2t^2$ we have similar results to Shrikhande [108] purely by considering Hadamard matrices.

More recent studies of Hadamard 3-designs have been made by Marion E. Kimberley [71].

## 11.2 INTEGRAL SOLUTIONS TO THE INCIDENCE EQUATION FOR FINITE PROJECTIVE PLANES.

Ryser [103] gave the following example of an integer matrix, where H is a normalized Hadamard matrix of order h, and H is repeated h+1 times

$$K = \begin{array}{c|c|c|c|c|c}
 & 1\,0\,\ldots\,0 & 1\,0\,\ldots\,0 & & 1\,0\,\ldots\,0 & 1\,0\,\ldots\,0 \\ \hline
\begin{matrix}1\\ \cdot\\ \cdot\\ \cdot\\ 1\end{matrix} & H & O & \cdots\cdots & O & O \\ \hline
\begin{matrix}1\\ \cdot\\ \cdot\\ \cdot\\ 1\end{matrix} & O & H & & O & O \\ \hline
 & \vdots & \vdots & & \vdots & \vdots \\ \hline
\begin{matrix}1\\ \cdot\\ \cdot\\ \cdot\\ 1\end{matrix} & O & O & & H & O \\ \hline
\begin{matrix}1\\ \cdot\\ \cdot\\ \cdot\\ 1\end{matrix} & O & O & \cdots\cdots & O & H
\end{array}$$

which satisfies

$$KK^T = hI+J$$

the incidence equation for a projective plane of order h.

Johnsen [65] studies integer solutions of these equations for $n \equiv 2 \pmod 4$. He gives

$$
K = \left[\begin{array}{c|cccc|cccc|cccc|cccc|cccc}
0 & 1 & 0 & 0 & 0 & 1 & 0 & 0 & 0 & 1 & 0 & 0 & 0 & 1 & 0 & 0 & 0 & 1 & 0 & 0 & 0 \\
\hline
1 & 1 & 1 & 1 & -1 & & & & & & & & & & & & & & & & \\
1 & 1 & 1 & -1 & 1 & & & & & & & & & & & & & & & & \\
1 & 1 & -1 & & & & & 1 & -1 & & & & & & & & & & & & \\
1 & 1 & -1 & & & & & -1 & 1 & & & & & & & & & & & & \\
\hline
1 & & & 1 & 1 & 1 & 1 & & & & & & & & & & & & & & \\
1 & & & -1 & -1 & 1 & 1 & & & & & & & & & & & & & & \\
1 & & & & & 1 & -1 & & & & & 1 & -1 & & & & & & & & \\
1 & & & & & 1 & -1 & & & & & -1 & 1 & & & & & & & & \\
\hline
1 & & & & & & & 1 & 1 & 1 & 1 & & & & & & & & & & \\
1 & & & & & & & -1 & -1 & 1 & 1 & & & & & & & & & & \\
1 & & & & & & & & & 1 & -1 & & & & & 1 & -1 & & & & \\
1 & & & & & & & & & 1 & -1 & & & & & -1 & 1 & & & & \\
\hline
1 & & & & & & & & & & & 1 & 1 & 1 & 1 & & & & & & \\
1 & & & & & & & & & & & -1 & -1 & 1 & 1 & & & & & & \\
1 & & & & & & & & & & & & & 1 & -1 & & & & & 1 & -1 \\
1 & & & & & & & & & & & & & 1 & -1 & & & & & -1 & 1 \\
\hline
1 & & & & & & & & & & & & & & & 1 & 1 & 1 & 1 & & \\
1 & & & & & & & & & & & & & & & -1 & -1 & 1 & 1 & & \\
1 & & & & & & & & & & & & & & & & & 1 & -1 & 1 & 1 \\
1 & & & & & & & & & & & & & & & & & 1 & -1 & -1 & -1
\end{array}\right]
$$

which satisfies

$$KK^T = 4I+J$$

and shows

THEOREM 11.1. (Johnsen) *Let $t_1, t_2, u_1, u_2, x_1, x_2$ be integers which satisfy*

$$w = [t_1+\tfrac{1}{2}(q-1)(x_1+u_1)]^2 + [t_2+\tfrac{1}{2}(q-1)(x_2+u_2)]^2$$

*where $q \equiv 3 \pmod 4$ and $q+1$ is the order a skew-Hadamard matrix. Further suppose $w \equiv t_1{}^2+t_2{}^2+\tfrac{1}{2}(q-1)(x_1{}^2+u_1{}^2+x_2{}^2+u_2{}^2$ , $x_1+u_1 = 2$, $t = (r+2)/2$ and $w = 2rq+2$ for the positive even integer r. Then there exists a matrix $K$ of order $w^2+w^2+1$ satisfying*

$$KK^T = wI+J.$$

THEOREM 11.2. (Johnsen) *Let $t_1, t_2, u_1, u_2, x_1, x_2, w$, and $q$ satisfy the same conditions as theorem 11.1 except that $w = 2rq+1$. Then there exists a matrix $K$ of order $w^2+w+1$ satisfying*

$$KK^T = wI+J.$$

11.3 STRONGLY REGULAR GRAPHS. As we observed in chapter V symmetric Hadamard matrices with constant diagonal of order $4s^2$ exist if and only if a strongly regular graph of the same order exists.

These have been studied in relation to Hadamard matrices by Bussemaker, Delsarte, Goethals and Seidel [44], [46], [36], [37]; other authors who have considered these graphs are Bose, Shrikhande and Bhagawandas [23], [24], [11].

Goethals and Seidel [47] point out that the matrix of a quasisymmetric block design is the adjacency matrix of a strongly regular graph. They say: "The notion of quasisymmetric block design essentially goes back to Shrikhande, who investigated the dual of block designs with $\lambda = 1$; cf also Shrikhande and Bhagawandas, [11], concluding remarks. Stanton and Kalbfleisch [121], introduced the name quasisymmetric for a more restricted class of such block designs." See also W.D. Wallis [164].

11.4 DESIGN GRAPHS, $(v,k,\lambda)$-GRAPHS, $G_2(\lambda)$ GRAPHS. All these names apply to the same graphs: those with v vertices which are regular of valency k and such that every pair of vertices is adjacent to exactly $\lambda$ further vertices.

Infinitely many of the known design graphs are related to Hadamard matrices, see W.D. Wallis [161], [162], [163], Bose and Shrikhande [23], [24], and other papers by these authors. These graphs are also studied by Ahrens and Szekeres [1].

An example of such a (16,6,2) graph is

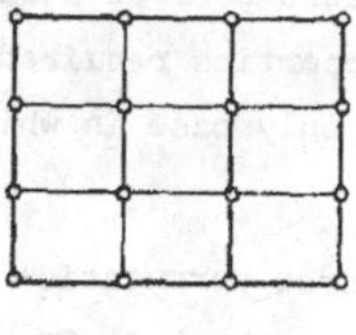

where all the vertices in any row or column are connected. The adjacency matrix of this graph -I is a regular symmetric Hadamard matrix of order 16.

11.5 TOURNAMENTS. Szekeres [129], and later Reid and Brown [100], showed that doubly regular tournaments are equivalent to skew-Hadamard matrices. These tournaments are studied further by W.D. Wallis [165], Esther and George Szekeres [128], and Delsarte, Goethals and Seidel [36].

Berlekamp and Hwang [17], construct balanced Howell rotations (used for Bridge tournaments) whenever $n \equiv 0 \pmod 4$ and n-1 is a prime power. They show the existence of such a rotation on n partnerships implies the existence of an Hadamard matrix of order n.

11.6 ORTHOGONAL ARRAYS AND PROJECTIVE PLANES. Shrikhande [107] shows how generalized Hadamard matrices may be used to construct orthogonal arrays.

Bush [27], [28], points out that the existence of an Hadamard matrix of the form

$$H = \begin{bmatrix} J & H_{12} & \cdots & H_{1m} \\ H_{21} & J & \cdots & H_{2m} \\ \cdot & & & \\ \cdot & & & \\ \cdot & & & \\ H_{m1} & H_{m2} & \cdots & J \end{bmatrix}$$

where the blocks are of size m×m, m even, and each $H_{ij}$ is a (1,-1) matrix with every row-sum and column-sum zero is implied by the existence of a finite projective plane of order m. Consequently the non-existence of such a matrix of order $m^2$ would be of great significance (except for the case m = 6).

W.D. Wallis [169] gives two constructions for such matrices.

11.7 AUTOMORPHISM AND SIMPLE GROUPS. Dade and Goldberg [34], give a construction which yields an Hadamard matrix from any transitive permutation group with certain properties.

Consider the finite field $GF(p^r)$, and the group of all linear mappings on $GF(p^r)$, $x \to a^2x+b$, for which $a^2$ is a non-zero perfect square in $GF(p^r)$. When $q \equiv 3 \pmod 4$ is a prime this group has the properties required by Dade and Goldberg. But Marshall Hall Jr. [57] shows this is the only case in which the properties can be satisfied.

Marshall Hall Jr. calls a rank three permutation group G on 4m+1 letters in which the orbits of a stabilizer are of lengths 1,2m,2m and in which $\lambda = m-1$, $\mu = m$ in Higman's notation, an *Hadamard group.* He proves

THEOREM 11.3. (Marshall Hall Jr.) *If 4m+1 is a prime power, then an Hadamard group on 4m+1 letters exist. If an Hadamard group on 4m+1 letters exist, then the following three conditions all hold:*

*(i)* $4m+1 = u^2+v^2$;

*(ii) 4m+1 is either prime or else is divisible by a perfect square (that is, if 4m+1 has more than one distinct prime factor, then it must have some repeated prime factor);*

*(iii)* $4m+1 \neq 45$.

It would be interesting if the conditions of this theorem apply to the construction of symmetric conference matrices.

An automorphism of an Hadamard matrix H of size n is a pair (P,Q) of n×n monomial matrices such that PHQ = H. The automorphisms form a group Γ. 1 = (I,I) and σ = (-I,-I) are the centre of Γ. $\overline{\Gamma}/\langle\sigma\rangle$ acts faithfully as a permutation group on the union of the sets of rows and columns of H.

Let q > 3 be a prime power ≡ 3 (mod 4). Write Π for a subgroup of Γ containing σ such that $\overline{\Pi} = \Pi/\langle\sigma\rangle$ acts faithfully on both the rows and columns of H as the group of all permutations of $GF(q)\cup\{\infty\}$ of the form $x \to (ax^\theta+b)/(cx^\theta+d)$, $a,b,c,d \in GF(q)$, $ad-bc = 1$, and $\theta \in \mathrm{Aut}\ GF(q)$.

Marshall Hall Jr. [53] has shown that the Mathieu group $M_{12}$ is the automorphism group of the Hadamard matrix of order 12; no similar representation is possible for $M_{24}$.

Kantor has studied automorphism groups of Hadamard designs and Hadamard matrices and shown

THEOREM 11.4. (Kantor) *If H is the Hadamard matrix obtained by bordering a Paley matrix of prime power order q ≡ 3 (mod 4), q > 11, then Γ = Π.*

THEOREM 11.5. (Kantor) *Paley designs are the only Hadamard designs admitting automorphism groups which are transitive on point-block pairs but which are not 2-transitive on points.*

COROLLARY 11.6. (Kantor) *Paley designs and their complementary designs are the only symmetric designs admitting automorphism groups which are 2-homogeneous but not 2-transitive on points.*

Goethals and Seidel in talking about strongly regular graphs [46], point out that Witt has proved the existence and uniqueness of the Steiner system (1; 24,8,5) which has the Mathieu group $M_{24}$ as its automorphism group. They also mention that the existence and uniqueness of the strongly regular graph on 56 vertices was first proved by Gewirtz who discovered its complement as a graph of diameter 2, girth 4, valency 10, order 56. Gewirtz also proved the uniqueness of the graph of order 100 which, together with its subgraph of order 77, was first constructed by Higman and Sims, while discovering the simple group which carries their names.

It is interesting to speculate on whether such combinatorial structures will yield more simple groups.

11.8 WEIGHING DESIGNS. Suppose that a chemical balance has an error of weighing, independent of the loads on the pans. This is quite realistic if the objects to be weighed have masses small compared with the mass of the moving parts of the balance. It is further assumed that the errors at different weighings are mutually independent.

For distribution theory it is assumed that the errors are normally distributed. Let us assume that the weights are always put on the right hand pan but that there is a choice at each weighing, between putting the object on the left hand pan (+1), in the right hand pan (-1), or in neither (0). At any weighing the masses of the objects will act as a sum of the form

$$m_1+m_2+\ldots-m_1'-m_2'\ldots$$

where m'are the masses in the left pan, m' is the mass in the right pan and the $m_i$ and $m_j'$ are just the masses in some order. Let $y_i$ be the result of the ith weighing. Then we have for each weighing

$$x_i^T m_i = y_i = z_i + e_i$$

where $x_i$ is a vector of 0,1, and -1's, $z_i$ is the true weight and $e_i$ is the error. Thus we write

$$X\underset{\sim}{m} = \underset{\sim}{y} = \underset{\sim}{z} + \underset{\sim}{e}$$

where X is a (0,1,-1) matrix, $\underset{\sim}{m}$ is the masses to be weighed, $\underset{\sim}{z}$ is the true weights and $\underset{\sim}{e}$ is the errors. Now we have assumed the $e_i$ are mutually independent normal variable with expectation zero (we are assuming no bias in the balance) and variance, $\sigma^2$.

The least squares theory states the best estimates for a given design with matrix D is given by

$$\hat{\underset{\sim}{m}} = (D^TD)^{-1}D^T\underset{\sim}{y}.$$

$\sigma^2 D^T D$ is then the variance-covariance matrix of the estimates, $\hat{m}_i$, so that the variance and covariance estimates are given by the elements of $\sigma^2(D^TD)^{-1}$.

We have not yet chosen D, but we should like to minimize the diagonal elements of $(D^TD)^{-1}$. A further desirable feature of the design is that the errors in estimating the weights should be independent, which is equivalent, under the assumption of normality of the $e_i$, to their being uncorrelated.

We now note some results from Raghavarao [98], to whom we refer the reader for a more detailed study.

THEOREM 11.7. *For any weighing design X,*

$$var(\hat{m}_i) \geq \frac{\sigma^2}{n}, \qquad i = 1,2,\ldots,p.$$

We showed above that the variance of the estimates is given by the elements of $\sigma^2(D^TD)^{-1}$ which is $\frac{\sigma^2}{n}I$ if D is an Hadamard matrix and $\frac{\sigma^2}{n-1}I$ if D is a symmetric conference matrix. This leads us to decide that Hadamard matrices give optimal weighing designs for $n \equiv 0 \pmod 4$ and symmetric conference matrices give the optimal answer for $n \equiv 2 \pmod 4$. These two types of matrices may also be used for odd n. See [98].

More generally we might assume that our weighing matrices X has more zeros. This lead M. Bhaskar Rao [13] to consider *balanced orthogonal matrices*, B, of order v×b which have elements (±1,0) and satisfy

(i) $BB^T$ = diagonal $(\lambda_1,\lambda_2,\ldots,\lambda_v)$;

(ii) when the -1's are changed to 1's the resulting matrix is a BIBD.

Bhaskar Rao shows these designs may be used in constructing BIBD's and group divisible designs and he gives some constructions.

In constructing Hadamard arrays we used four subsets of signed integers from the integers modulo v with incidence matrices $X_i$, i = 1,2,3,4 which satisfied

(i) $\sum_{i=1}^{4} X_i$ is a (1,-1) matrix; and

(ii) $\sum_{i=1}^{4} X_i X_i^T = vI_v$.

This is one possible generalization of balanced orthogonal designs (BOD's), another is to consider orthogonal (0,1,-1) matrices which have exactly k non-zero elements in each row and column.

**LEMMA 11.8.** (Wallis) *Let n = 4,8,12,16,20,24,28,32,40. Then there exists an orthogonal (0,1,-1) matrix with k non-zero elements per row and column for every k = 0,1,2,...,n.*

**CONJECTURE.** For n ≡ 0 (mod 4) there exists a (0,1,-1) matrix A satisfying

$$A^T A = AA^T = kI_n \quad \text{for every } k = 0,1,\ldots,n.$$

## 11.9 MODULAR HADAMARD MATRICES.

O. Marrero and A.T. Butson [82] have studied (1,-1) matrices H of order h which satisfy

$$HH^T = nI_h.$$

Write H(n,h) for such a matrix. An ordinary Hadamard matrix is written H(0,h). They show how these matrices are related to other combinatorial designs.

Marrero and Butson prove:

**THEOREM 11.9.** *Each of the following is a necessary condition for the existence of an H(n,h) matrix:*

*(i) if h ⩾ 3, then h ≡ 4t (mod n) for some integral t,*

*(ii) if h is odd, then n is odd and $h^h$ is a quadratic residue (mod n).*

They give the following (among other) constructions for modular Hadamard matrices.

THEOREM 11.10. *Let M be an $H(n_1,h_1)$ matrix and let N be an $H(n_2,h_2)$ matrix. Then $M{\times}N$ is an*

(i) *$H(n,h_1h_2)$ matrix, when $n = gcd(h_1n_2,n_1n_2,h_2n_1)$,*

(ii) *$H(n_2h_1,h_1h_2)$ matrix when $n_1 = 0$.*

THEOREM 11.11. *If $n|h$ or $n|(h-4)$ then $H(n,h)$ matrices exist.*

THEOREM 11.12. *If $H(0,4u)$ and $H(0,4t)$ exist, then an $H(n,4u+4t-2)$ exists where $n = gcd(4u-2,4t-2)$, in particular if $u = t$ there exists an $H(4t-2,8t-2)$.*

THEOREM 11.13. *If a $(v,k,\lambda)$-configuration exists then $H\{v-4(k-\lambda),v\}$ and $H\{2v-4(k-\lambda),2v\}$ matrices exist.*

They show

THEOREM 11.14. *A necessary and sufficient condition for the existence of an $H(n,h)$ is that for*

(i) *$n = 2$, $h \equiv 0 \pmod 2$,*

(ii) *$n = 3$, $h \equiv 0,2,3,4,6,7,8,9,10 \pmod{12}$,*

(iii) *$n = 6$, $h \equiv 0 \pmod 2$.*

11.10 SEQUENCES. In his paper on "Sequences with small correlation" [140], R.J. Turyn writes of the problem of finding complex functions $f(t)$ such that $c(s) = \int f(t)\overline{f}(t+s)dt$ is small except near $s = 0$. He then considers $f(t)$ as a step function which can be expressed as a finite sequence of roots of 1. Thus the problem becomes to find a sequence of complex numbers $x_1,\dots,x_n$ with **aperiodic correlation** function

$$c_j = \sum_1^{n-j} x_i\overline{x}_{i+j}, \quad i+j \text{ reduced mod } n$$

and $a_j$ the periodic correlation function,

$$a_j = c_j + \overline{c}_{n-j} = \sum_1^n x_i\overline{x}_{i+j}$$

Such that for $i \neq j$, $c_j$ is small.

In 1953 R.H. Barker, in connection with a problem in digital communications, considered the case where the finite sequences of ones and minus ones had their a-periodic correlation coefficients $c_j$ as small as possible. That is, he asked that

$$c_j = \sum_{i=1}^{v-j} x_i x_{i+j} = 0 \text{ or } -1, \quad \text{for all } j = 1,\dots,n-1.$$

Sequences with these properties may be used to simulate white noise. Barker found such sequences for $n = 3,7,11$. It has become customary to refer to *Barker sequences* as those finite sequences of $\pm 1$, whose aperiodic correlations $c_j$ are restricted to $-1,0,1$. Only the following Barker sequences are known: (+ denotes +1 and - denotes -1) n = 2  + +

n = 3  + + -

n = 4  + + + -; + + - +

n = 5  + + + - +

n = 7  + + + - - + -

n = 11  + + + - - - + - - + -

n = 13  + + + + + - - + + - + - +

together with the sequences which may be derived from them by the following transformations:

$$b_i = (-1)^i x_i$$

$$b_i = (-1)^{i+1} x_i$$

$$b_i = -x_i.$$

In fact, Storer and Turyn [124] have shown that any further Barker sequences which may exist must be of even length, indeed they show that $n \equiv 0 \pmod 4$ is necessary.

All these sequences correspond to difference sets. For n = 2,3,4,5 the difference sets are trivial; for n = 7,11,13 the sets have parameters 7,4,2; 11,5,2 and 13,9,6 respectively.

Any further Barker sequences exist if and only if there exist difference sets with parameters $4s^2, 2s^2-s, s^2-s$ for $s > 1$. In fact it may be shown that if further Barker sequences exist they must have $s \geq 55$, i.e. $n \geq 12{,}100$.

The matrix

$$\begin{matrix} + & + & + & - \\ - & + & + & + \\ + & - & + & + \\ + & + & - & + \end{matrix}$$

is the only known circulant Hadamard matrix. There is a one-to-one correspondence between Barker sequences of even length $n \geq 14$ and circulant Hadamard matrices. Thus from the results on Barker sequences, if there are any further circulant Hadamard matrices they have order $n \geq 12{,}100$. It has been conjectured that there are no further Barker sequences or circulant Hadamard matrices.

The related problem of finding 1,-1 sequences of length n for which the maximum aperiodic correlation coefficient is of least magnitude (i.e. for which $\max_j |c_j|$ is minimized) and indeed the problem of finding this minimum, at least asymptotically as a function of n, is unsolved.

For the non-binary case we already know there are cyclic generalized Hadamard matrices (see Butson [31,32]) for any odd order. The connection between generalized Hadamard matrices was also noted by Shrikhande [109], Turyn [140] and Delsarte [35].

THEOREM 11.15. (Turyn) *If q is an odd prime power, there exists a sequence of length q, with terms qth roots of 1, with* $a_j = 0$. *For any n, there exists a sequence of length* $n^2$ *with terms nth roots of 1,* $a_j = 0$.

We recall that the quadratic character $\chi$ satisfies

$$\sum_{g\in GF(q)}\chi(g) = 0$$

and

$$\sum_{g\in GF(q)}\chi(g)\ (g+s) = -1, \quad s \neq 0.$$

This was used extensively in constructing Hadamard matrices and it clearly is related to sequences with small correlation.

See Turyn [140] for more details.

Delsarte and Goethals [38] consider the problem: for an odd integer v, and an abelian group G of order v, with elements $g_1, g_2, \ldots, g_v$, where $g_1 = 1$ is the identity, does there exist an ordered sequence $s(g_1), s(g_2), \ldots, s(g_v)$, with

$$s(g_1) = 0,\ s(g_i) = 1, \quad \text{for } i = 2,3,\ldots,v$$

such that the element

$$s_G = \sum_{i=1}^{v} g_i s(g_i)$$

in the group algebra RG of G over R (the field of rationals), satisfies

$$s_G \sum_{i=1}^{v} g_i = 0, \quad s_G^{\,2} = (-1)^{\frac{1}{2}(v-1)}\left(v - \sum_{i=1}^{v} g_i\right).$$

They call any solution of this problem a *(0,1,-1) G-sequence.* They point out that the equivalent formulation for cyclic groups was considered by Kelly [70], who showed that for p prime there are two solutions, but there are none for p non-prime. For a (0,1,-1) G-sequence in an abelian group of order $v \equiv 3 \pmod 4$, it can be shown that the set D of elements g of G such that $s(g) = +1$, forms a skew-Hadamard abelian group difference set. The existence of these difference sets was considered by Camion [33] and Johnsen [67], who proved that they do not exist, unless G is a p-group.

They show:

THEOREM 11.16. (Delsarte-Goethals-Turyn) *There exists no (0,1,-1) G-sequence for an abelian group G of order* $v = p^2q^2$, *where p and q are distinct odd primes.*

THEOREM 11.17. (Delsarte and Goethals) *There exists no (0,1,-1) G-sequence in the abelian p-groups* $G = p^2 \times p^2$, *of type (1,1,2).*

THEOREM 11.18. (Delsarte and Goethals) *There exists a (0,1,-1) G-sequence if and only if there exists an orthogonal matrix with zero diagonal (see chapter 3) and other elements ±1 of order* $v = |G|$, *having in its automorphism group a regular subgroup isomorphic to G.*

THEOREM 11.19. (Delsarte and Goethals) *Suppose there exists a (0,1,-1) $G_i$-sequence in the groups $G_1, G_2, \ldots, G_n$, each of the same order v. Then there exists a (0,1,-1) G-sequence in the group $G = G_1 \times G_2 \times \ldots G_n$, of order $v^n$.*

We note in passing that each of the row vectors in the matrices $X_1$ of theorem 7.5 have inner product r, $|r| < 1$.

## 11.11 BELEVITCH ON 2N-TERMINAL NETWORKS.

Belevitch [14,15] constructed matched non-dissipative networks interconnecting n telephone circuits and giving a loss of $10 \log_{10}(n-1)$ decibels between all their terminals for certain values of n. He says that these are the most efficient possible. In particular, the networks form symmetric conference matrices (hence their name, for Belevitch studied them in connection with conference telephony), are non-dissipative and the loss, $10 \log_{10}(n-1)$ decibels, merely results from the division of power; they are composed only of ideal transformers.

## 11.12 WALSH FUNCTIONS.

Harmuth [61] says of Walsh functions: "The acid test of any theory in engineering is its practical applications. Several such applications are known [for Walsh functions] and they are all intimately tied to semi-conductor technology. The little known system of Walsh functions appears to be ideal for linear, time-variable circuits, if based on binary digital components, as the system of sine and cosine functions is for linear, time-invariant circuits, based on resistors, capacitors and coils. Very simple sequency [a generalization of frequency] filters based on these Walsh functions have been developed. Furthermore, an experimental sequency multiplex system using Walsh functions as carriers has been developed that has advantages over frequency or time multiplex systems in certain applications. Digital filters and digital multiplex equipment are among the most promising applications for the years ahead. They are simpler and faster when based on Walsh functions rather than on sine and cosine functions. Their practical application, however, will require considerable progress in the development of large scale integrated circuits."

The Walsh functions are wal(0,θ), sal(i,θ) and cal(i,θ). There is a close connection between sal and sine functions (both are even functions), as well as between sal and cosine functions (both are odd functions). The letters s and c in cal and sal were chosen to indicate this connection, while the letters "al" are derived from the name Walsh.

For computations sometimes sine and cosine functions are used, while at other times exponential functions are more convenient. A similar duality of notation exists for Walsh functions. A single function wal(j,θ) may be defined:

$$\begin{aligned} \mathrm{wal}(2i,\theta) &= \mathrm{cal}(i,\theta), \\ \mathrm{wal}(2i-1,\theta) &= \mathrm{sal}(i,\theta), \end{aligned} \qquad i = 1,2,\ldots$$

The functions wal(j,θ) may be defined by the following difference equation:

$$\mathrm{wal}(2j+p,\theta) = (-1)^{(j/2)+p}\{\mathrm{wal}[j,2(\theta+\tfrac{1}{4})] + (-1)^{j+p}\mathrm{wal}[j,2(\theta-\tfrac{1}{4})]\}$$

$$p = 0 \text{ or } 1, \quad j = 0,1,2,\ldots$$

$$\mathrm{wal}(0,\theta) = 1 \qquad \text{for } -\tfrac{1}{2} \leq \theta < \tfrac{1}{2}$$

$$\mathrm{wal}(0,\theta) = 0 \qquad \text{for } \theta < -\tfrac{1}{2},\ \theta > \tfrac{1}{2}.$$

The first few Walsh and trigonometric functions are:

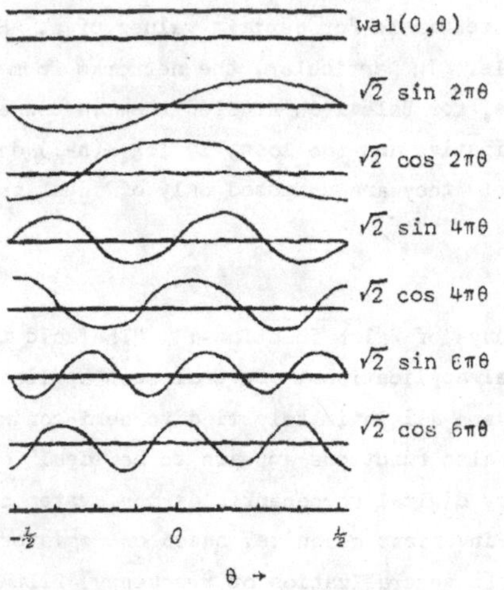


Fig. 1 Orthogonal sine and cosine elements.

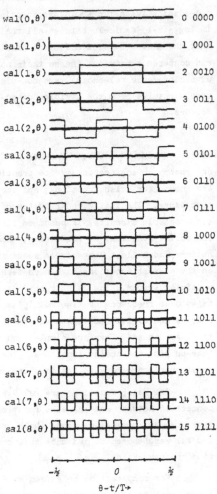

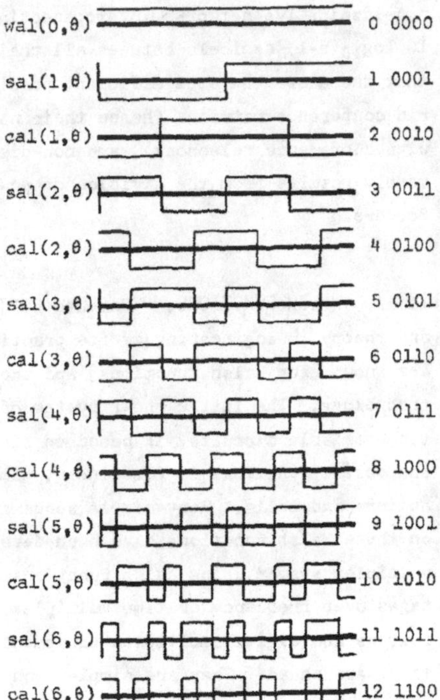


Fig. 2 Orthogonal Walsh elements. The numbers on the right give j in decimal and binary form, if the notation wal(j,θ) is used.

$$\mathrm{wal}(2i,\theta) = \mathrm{cal}(i,\theta)$$

$$\mathrm{wal}(2i-1,\theta) = \mathrm{sal}(i,\theta)$$

Clearly Figure 2 is equivalent to the following Hadamard matrix of order 16:

| | | | | | | | | | | | | | | | |
|---|---|---|---|---|---|---|---|---|---|---|---|---|---|---|---|
| 1 | 1 | 1 | 1 | 1 | 1 | 1 | 1 | 1 | 1 | 1 | 1 | 1 | 1 | 1 | 1 |
| -1 | -1 | -1 | -1 | -1 | -1 | -1 | -1 | 1 | 1 | 1 | 1 | 1 | 1 | 1 | 1 |
| -1 | -1 | -1 | -1 | 1 | 1 | 1 | 1 | 1 | 1 | 1 | 1 | -1 | -1 | -1 | -1 |
| 1 | 1 | 1 | 1 | -1 | -1 | -1 | -1 | 1 | 1 | 1 | 1 | -1 | -1 | -1 | -1 |
| 1 | 1 | -1 | -1 | -1 | -1 | 1 | 1 | 1 | 1 | -1 | -1 | -1 | -1 | 1 | 1 |
| -1 | -1 | 1 | 1 | 1 | 1 | -1 | -1 | 1 | 1 | -1 | -1 | -1 | -1 | 1 | 1 |
| -1 | -1 | 1 | 1 | -1 | -1 | 1 | 1 | 1 | 1 | -1 | -1 | 1 | 1 | -1 | -1 |
| 1 | 1 | -1 | -1 | 1 | 1 | -1 | -1 | 1 | 1 | -1 | -1 | 1 | 1 | -1 | -1 |
| 1 | -1 | -1 | 1 | 1 | -1 | -1 | 1 | 1 | -1 | -1 | 1 | 1 | -1 | -1 | 1 |
| -1 | 1 | 1 | -1 | -1 | 1 | 1 | -1 | 1 | -1 | -1 | 1 | 1 | -1 | -1 | 1 |
| -1 | 1 | 1 | -1 | 1 | -1 | -1 | 1 | 1 | -1 | -1 | 1 | -1 | 1 | 1 | -1 |
| 1 | -1 | -1 | 1 | -1 | 1 | 1 | -1 | 1 | -1 | -1 | 1 | -1 | 1 | 1 | -1 |
| 1 | -1 | 1 | -1 | -1 | 1 | -1 | 1 | 1 | -1 | 1 | -1 | -1 | 1 | -1 | 1 |
| -1 | 1 | -1 | 1 | 1 | -1 | 1 | -1 | 1 | -1 | 1 | -1 | -1 | 1 | -1 | 1 |
| -1 | 1 | -1 | 1 | -1 | 1 | -1 | 1 | 1 | -1 | 1 | -1 | 1 | -1 | 1 | -1 |
| 1 | -1 | 1 | -1 | 1 | -1 | 1 | -1 | 1 | -1 | 1 | -1 | 1 | -1 | 1 | -1 |

Those that are symmetric about the middle (even) are $\mathrm{cal}(i,\theta)$ and those that are skew-symmetric (odd) about the middle are $\mathrm{sal}(i,\theta)$. In all cases $i$ = half the number of times $\mathrm{cal}(i,\theta)$ is zero which corresponds with a definition of frequency for trigonometric functions.

The function $\mathrm{wal}(j,2\theta)$ has the same shape as $\mathrm{wal}(j,\theta)$ but is defined over the interval $-\frac{1}{4} \leq \theta \leq \frac{1}{4}$ rather than $-\frac{1}{2} \leq \theta \leq \frac{1}{2}$.

$\mathrm{wal}\{j,2(\theta+\frac{1}{4})\}$ is obtained by shifting $\mathrm{wal}(j,2\theta)$ to the left into the interval $-\frac{1}{2} \leq \theta < 0$, and similarly $\mathrm{wal}\{j,2(\theta-\frac{1}{4})\}$ is defined on $0 \leq \theta < \frac{1}{2}$.

The product of two Walsh functions is a Walsh function:

$$\mathrm{wal}(h,\theta)\mathrm{wal}(k,\theta) = \mathrm{wal}\{(h+k)\bmod 2,\theta\}.$$

So
$$\mathrm{wal}(h,\theta)\mathrm{wal}(h,\theta) = \mathrm{wal}(0,\theta),$$

and
$$\mathrm{wal}(h,\theta)\mathrm{wal}(0,\theta) = \mathrm{wal}(h,\theta).$$

We can easily show that Walsh functions form an abelian group with respect to multiplication.

The Walsh functions $\mathrm{wal}(i,\theta)$ for integral $i$ are equivalent to Hadamard matrices whose orders are powers of 2. They have been extended to real $i$ and there seems no reason why other Hadamard matrices cannot be used since the functions they give would still be orthogonal and complete.

11.13 CODING THEORY. Although more than twenty years have passed since the appearance of Shannon's papers, a still unsolved problem of coding theory is to construct block codes which attain a low probability of error at rates close to capacity. However, for moderate block lengths many good codes are known, the best known being the BCH codes discovered in 1959 (these essentially have the rows of generalized Hadamard matrices as codewords.

An *(n,k) linear code* over the field GF(q) consists of $q^k$ vectors (called *codewords*) of length n with components from GF(q) such that

(a) the vector sum of two codewords is a codeword;

(b) the multiplication of any codeword by a scalar which is any element of GF(q) yields a codeword.

The *redundancy* of the code is r = n-k and the *rate* is R = k/n.

A *cyclic code* is a linear code with the property that a cyclic shift of any codeword is also a codeword. BCH codes (and others) are of this type.

Coding theorists have concentrated on linear codes, cyclic codes and binary codes as these are simpler to implement and simpler to analyse (and that means cheaper). They aim to minimize the redundancy and maximize the *distance*, which is the minimum number of places in which two codewords differ.

We write S for the orthogonal (0,1,-1) matrix with zero diagonal of chapter 3 and $C = \frac{1}{2}(S+J)$ for the binary equivalent. Let S' and C' be their respective cores. Then the (23,12) Golay code can be written as

$$\left[ I_{12} \quad \begin{matrix} C' \\ 1 \ldots\ldots 1 \end{matrix} \right]$$

where C' is the circulant core obtained from the Paley Hadamard matrix of order 12. Clearly

$$\left[ I_{n+1} \quad \begin{matrix} C_n' \\ 1 \ldots\ldots 1 \end{matrix} \right]$$

is also a binary code for $n+1 \equiv 0 \pmod 4$ but is most useful for $C_n'$ circulant.

Vera Pless observed [93],[94], that if the code is taken over GF(3), and I and S are of order q+1

$$[I \; S]$$

is a linear self-orthogonal (2q+2,q+1) code for $q \equiv 2 \pmod 3$. Clearly other combinations of I and S, for example [I,I,S], should be studied, as should the circulant complex Hadamard cores we saw earlier.

But the greatest use of Hadamard matrices is in forming non-linear binary codes (some of these wermention actually have smaller redundancy over GF(3), and while codes over GF(p), $p > 2$, can be implemented, binary codes seem to be more practical).

An *(n,M,d) code* is a set of M codewords of length n, with symbols from GF(q) and minimum distance d. The dimension of this code is $k = \log_q M$, the redundancy is $r = n - \log_q M$, and the rate is $R = k/n$. Now k and r need not be integers.

An (n,M,d) code is said to be *optimal* if it has the largest possible number of codewords for given values of n and d, because as M increases the redundancy, r, decreases. This use of optimal is naive as it omits the consideration of encoding and decoding. But it can be argued that once good codes have been found, the techniques for their implementation will be developed later, as happened with BCH codes.

Plotkin (see [16; p.316]) found that max(M) for an (n,M,d) code is given by

$$M \leq 2\left[\frac{d}{2d-n}\right] \quad \text{where } 2d > n \geq d,$$

$$M \leq 2n \quad \text{where } 2d = n,$$

this later is reached for n the order of an Hadamard matrix, H. Codes for which M attains its upper bound are called *optimal*. Thus

$$\begin{bmatrix} \frac{1}{2}(H-J) \\ \frac{1}{2}(J-H) \end{bmatrix}$$

is an optimal $(n,2n,\frac{1}{2}n)$ binary code.

Bose and Shrikhande [22] pointed out that if H is normalized and the first column removed to form H', then

$$\tfrac{1}{2}(J+H)$$

is a (4t-1,4t,2t) optimal binary code. Further if we take the rows of $\frac{1}{2}(J+H')$ with first element 1 and form a new matrix from these rows with the 1 removed, we have a (4t-2,2t,2t) optimal binary code.

Sloan and Seidel [113] noticed that

$$\begin{bmatrix} \frac{1}{2}(S-I+J) \\ \frac{1}{2}(J-S+I) \end{bmatrix},$$

with S of order n, has $d = \frac{1}{2}(n-2)$ for $n \equiv 2 \pmod 4$ and $d = \frac{1}{2}n$ for $n \equiv 0 \pmod 4$; thus if the first column is removed an $(n-1,2n,\frac{1}{2}(n-2))$ binary code is obtained. If this is left over GF(3) as

$$\begin{bmatrix} 0 \;\cdots\; 0 \\ S \\ -S \end{bmatrix}$$

then we have an $(n,2n+1,\frac{1}{2}(n+2))$ ternary code but $n-\log_3 M > n-\log_2 M$. Further possibilities arise if complex Hadamard matrices (with or without zero diagonal) are used; or other combinations such as

$$\begin{bmatrix} -S & I \\ I & S \end{bmatrix} \text{ or } \begin{bmatrix} S & S \\ S & -S \end{bmatrix} \quad \text{which are orthogonal,}$$

$$\text{or}\quad\begin{bmatrix} & I & S & \\ & I & -S & \\ & -S & -I & \\ & S & I & \\ 1 & 1..1 & & 1 \\ 0 & 0..0 & & 0 \end{bmatrix}\quad\text{or}\quad\begin{bmatrix} 1 & \cdots\cdots & & 1 \\ 0 & \cdots\cdots & & 0 \\ I & S & S & S \\ -S & I & S & -S \\ -S & -S & I & S \\ -S & S & -S & I \end{bmatrix}.$$

Of course these may also be altered in the manner of Bose and Shrikhande.

Levenshtein [75] showed how various combinations of binary Hadamard matrices and those codes derived from them by Bose and Shrikhande may be combined to give large distance (as the expense of the number of codewords and hence redundancy).

For example, let $H_1$ be a $(4t-1,4t-1,2t)$ code and $H_2$ be a $(4s-2,2s,2s)$ code, then

$$\underbrace{H_1,H_1,\ldots,H_1}_{a\text{ times}},\ \underbrace{H_2,H_2,\ldots,H_2}_{b\text{ times}}$$

is a $\{a(4s-2)+b(4t-1),\min(2s,4t-1),2(as+bt)\}$ maximal binary non-linear code.

Coding theorists do not need the orthogonality of Hadamard matrices so the constructions used for Hadamard matrices can be used to give $(n,M,d)$ codes of length $n = 4s$ where no Hadamard matrix is known, with reasonable $M$ and $d$.

For example if $L$ is a $(4n-1,4n-1,2n)$ code and $L' = \frac{1}{2}(J-L)$, then if we form a matrix by putting $L$ in place of every 1 and $L'$ in place of every $-1$ in $\begin{bmatrix}-H\\H\end{bmatrix}$ where $H$ is an Hadamard matrix of order $4t$, we get a $\{(4n-1)4t,(4n-1)8t,(4n-1)2t\}$ code. e.g. if $t = 1$, $4n-1 = 47$, we get a $(188,376,94)$ code even though an Hadamard matrix of order 188 is not yet known.

This is actually a direct product of two codes (see [16] p.338). But if we put one code in the positions of the diagonal of an orthogonal matrix with zero diagonal and another code in the other positions, or put different codes in place of the A,B,C,D in Hadamard arrays, the results may be more interesting.

**11.14 PAIRWISE STATISTICAL INDEPENDENCE.** Hadamard matrices give simple examples of pairwise independence. In fact, for any pair of random variables taking distinct values zero correlation is equivalent to mutual independence. In the Hadamard case the columns can be taken as the elements of the constant function and $(n-1)$ random variables, $X_1, X_2, \ldots, X_{n-1}$. The X's are pairwise independent. If the values $\pm 1$ are taken as the coordinates, a space of dimension $(n-1)$ with $2^{n-1}$ points is defined. Only $n$ of these points are associated with positive probability. In [182] Lancaster proves the following theorem

THEOREM 11.20. *For any probability measure on a space of n distinct points, a set of at most (n-1) pairwise independent random variables can be defined. A maximal set can be obtained only if each random variable takes precisely two distinct values with positive measure. A maximal set can be obtained for each* $n \zeta > 3$. *If the measure,* $n^{-1}$, *is assigned to each point of the space the solution is equivalent to determining an Hadamard matrix of order* $n$.

## CHAPTER XII. UNANSWERED QUESTIONS

1. Recently M. Bhaskar Rao [13] introduced B.O.D.s (balanced orthogonal arrays). How may Hadamard arrays with zero elements be used in constructing B.O.D.s?

2. It is known that symmetric conference matrices must be of order $p+1 \equiv 2 \pmod 4$ where p is the sum of two integer squares. 46,66,86,118,146 are the first few orders for which symmetric conference matrices are unknown. How may these orders be constructed?

3. Szekeres difference sets X,Y with $y \in Y \Rightarrow -y \in Y$ are used in theorem 5.15 to make regular symmetric Hadamard matrices with constant diagonal and in constructing amicable Hadamard matrices. In lemma 2.8 a construction is given for orders 5,13,29,53. Can this construction be generalized? How may the sets be constructed for other orders?

4. Find constructions for Szekeres difference sets for orders congruent to 1 or 17 (mod 32).

5. Hadamard arrays H[4n;n] where each letter is repeated n times have been found for $n = 2t+1$, $3 \leq 2t+1 \leq 17$, and other n by Turyn. Find more!

6. If two subsets X and Y of size m from the set of 2m+1 integers with circulant incidence matrices A and B exist such that $X \cap Y = \phi$ and

$$AA^T + BB^T = 2mI_m$$

then M = A+B and N = A-B satisfy

$$MM^T + NN^T = 4mI_m.$$

These have been used in constructing the Hadamard arrays H[4n;n] for n = 3,11. Can more be found?

7. $(k(2k-1), 4k^2-1, 2k+1, k, 1)$ BIBD are known to give symmetric Hadamard matrices with constant diagonal of order $4k^2$. These BIBD are known for k = 2,3,4,5,6,7,8 and $2^t$, $t \geq 4$. Are there others?

8. Do the different constructions for skew-Hadamard matrices of the same order (see chapter V) give H-inequivalent or Z-inequivalent matrices?

9. If there is a symmetric and skew-symmetric Hadamard matrix of the same order, are they equivalent?

10. Do different constructions for Hadamard matrices of the same order give H-inequivalent matrices (part of the answer is given by the fact that different doubling operations give Z-inequivalent matrices)?

11. Compare the diagonalizations over Z of N+I, N-I, N when N+I is a symmetric conference matrix.

12. Do complex Hadamard matrices exist for all orders ≡ 2 (mod 4)?

13. What happens to complex Hadamard matrices when we diagonalize over the Gaussian integers?

14. No Z-inequivalent skew-Hadamard matrices are known. Are all skew-Hadamard (symmetric conference) matrices of the same order Z-equivalent? Can this be used to construct skew-Hadamard matrices of size n from skew-Hadamard matrices of order 2n?

15. Study the relationship between inequivalence of BIBD's with parameters $(4t+3,2t+1,t)$ or $(4t^2,2t^2+t,t^2+t)$ and the H-inequivalence of their associated Hadamard matrices.

16. Theorem 4.8 used $4$-$\{v;k_1,k_2,k_3,k_4;\Sigma k_i-v\}$ supplementary difference sets to construct (skew-) Hadamard matrices of order 4v. $4$-$\{v;k_1,k_2,k_3,k_4;\Sigma k_i-v-1\}$ supplementary difference sets may be used (lemma 4.17) to form (skew-) Hadamard matrices of order 4(v+1). Find constructions for these sets.

17. Do skew-Hadamard matrices, symmetric Hadamard matrices and amicable Hadamard matrices exist for every order ≡ 0 (mod 4)?

18. Fill the gaps in the appendices.

19. Investigate the suggestions in chapter XI, section 11.12 on codes.

20. Many of these problems have been tackled on a computer but they take large amounts of time and/or storage. There are several problems:

(a) study indexing techniques such as hash addressing, and other efficient algorithms;

(b) develope software to handle discrete structures, matrices and sets which are not graph theoretic;

(c) Hadamard matrices and incidence matrices use at most two *bits* per element; autocodes are frustratingly inefficient for handling these matrices. For example, compare the coding for taking the dot product of two rows of an incidence matrix with each element in one word with the hardware needed if each row is put into one word and the bit by bit "logical and" of the words is obtained and then summed. Also compare the time. Do these subjects really need special hardware?

21. A *regular 2-graph* $(\Omega,\tau)$, based on a set $\Omega$ of objects and a set $\tau$ of 3-element subsets of $\Omega$, satisfies the following conditions:

(i) $\tau \neq \phi$, $\tau$ does not contain every 3-element subset of $\Omega$;

(ii) every 2-set of $\Omega$ is contained in the same number of elements of $\tau$;

(iii) every 4-set of $\Omega$ contains an even number of elements of $\tau$.

A subset of a regular 2-graph is *coherent* if all of its 3-element subsets belong to $\tau$.

Given a regular 2-graph let $\rho_1, \rho_2$, $\rho_1 > \rho_2$, be the eigenvalues (with multiplicities $\mu_1$ and $\mu_2$, respectively) of any associated adjacency matrix. Then (theorem told to me by Dr. D.E. Taylor):

*if $m$ is the size of a coherent set, then*

$$m \leq 1-\rho_2,$$
$$m \leq \mu_1.$$

For the regular 2-graph on 26 vertices a coherent set has $\leq 6$ points. In fact, in all the known examples there exists a coherent set of 6 points.

Prove (question told me by Taylor) that in any regular 2-graph on 26 points there *must* exist either a coherent or incoherent set of 6 points, i.e. a generalized Ramsey number problem.

## APPENDIX A: KNOWN CLASSES OF HADAMARD MATRICES

In this section we use the following notations:

| | |
|---|---|
| $p^r$, $p_i^{r_i}$ | prime powers ≡ 3 (mod 4); |
| $q^s$ | prime power ≡ 1 (mod 4); |
| g | prime power ≡ 5 (mod 8); |
| $x^u$ | any prime power; |
| r | an odd positive integer; |
| t | a non-negative integer; |
| $k,k_1,k_2$ | $k > 1, k_1 > 1, k_2 > 1$ the orders of Hadamard matrices; |
| a | the order of an Hadamard array H[a,4] with each letter occurring ¼a times; |
| c | the order of a complex Hadamard matrix; |
| h | the order of a skew-Hadamard matrix; |
| m | the order of amicable Hadamard matrices; |
| n | the order of symmetric conference matrix; |
| w | $w \in W = \{i: i$ an odd integer, $3 \leq i \leq 29$ or $i = 37$ or $i = 43$ or $i = \frac{1}{2}(q^s+1)$ or $i = 9^t\}$. |

$H\{X,Y,Z,W,q,m,t,a,b,$ with $b = f(q,m,t,a)\}$:

If there exist four (1,-1) matrices X,Y,Z,W of order q satisfying

(a) $AB^T = BA^T$ symmetric for any $A,B \in \{X,Y,Z,W\}$

(b) $XX^T+YY^T = 2(q+1)I-J$,

$ZZ^T = aI+(q-a)J$,

$WW^T = bI+(q-b)J$, $\quad b = m(q-t-a+1)+a$

where t = n if a symmetric conference matrix of order n is required or t = h if a skew-Hadamard matrix of order h is required: then we will say *condition* $H\{X,Y,Z,W,q,m,t,a,b,$ with $b = f(m,q,t,a)\}$ *holds*.

In all cases the class indexed with an asterisk, *, is stronger than the others classified under the same number and includes the subclasses. Thus HII* which is $h(p^r+1)$ includes the other member of the class HII which is $p^r+1$. If we say as in HIII* that the class is proved in theorem 5.4 then it means theorem 5.4 of this part (i.e., part IV). The subclasses are obtained from corollaries to the theorem that gave the class.

The following classes of Hadamard matrices are known:

| | | |
|---|---|---|
| HI | $2^t$* | Sylvester; corollary 8.2. |
| HII | $2^t(p_i^{r_i}+1)$* | Paley, Williamson; theorem 8.3. |
| | $p^r+1$ | Paley; theorem 8.3. |
| HIII | kc* | Turyn; theorem 8.4. |
| | kn | Goethals and Seidel; theorem 8.5. |
| | $2(q^s+1)$ | Paley; corollary 8.6. |
| | $k(q^s+1)$ | Williamson; corollary 8.6. |
| | $kn(n-1)$ | Goethals and Seidel; theorem 8.4 and corollary 6.14. |
| | $2n(n-1)$ | Goethals and Seidel; theorem 8.4 and corollary 6.14. |
| HIV | $h^2$ or $n^2$ | $n = h-2$ or $h = n-2$ : Goethals and Seidel; from theorems 5.11 and 3.7. |
| | $x^\mu(x^\mu+2)+1$ | $x^\mu+2$ a prime power: Brauer, Whiteman, Stanton and Sprott; from corollary 5.12. |
| | $(h-2)^2$ | h-3 a prime power: Ehlich; from corollary 5.12. |
| HV | $(h-1)^c+1$* | $c > 0$ an odd integer: Turyn, from corollary 3.12. |
| | $(h-1)^d+1$ | $d = 3^a5^b7^c$ a,b,c non-negative integers: Wallis, from corollary 3.12. |
| | $(h-1)^3+1$ | Goldberg; from corollary 3.12. |
| HVI | 4w | Williamson; theorem 8.34. |
| | 92 | Baumert, Golomb, Hall; §8.6. |
| | 116 | Hall; §8.6. |
| | 148,172 | Williamson; §8.6. |
| | $2(q^s+1)$ | Turyn; theorem 8.39. |
| | $4\cdot 9^t$ | Turyn; from [143]. |
| HVII | aw* | Baumert and Hall; §7.1. |
| | 156 | Baumert and Hall; §7.1. |
| | 12w | Baumert and Hall; §7.1. |
| | 20w | Welch; §7.3. |
| | 28w,36w,44w,100w | Wallis; §7.3. |
| | 52w,60w,68w,76w | Cooper and Wallis; §7.3. |
| | 4bw,20bw | $b \in \{i$: i an odd integer $3 \le i \le 23$ or $i = 29$ or $i = 1+2^x10^y26^z$, where x,y,z are non- |

| | | |
|---|---|---|
| | | negative integers} : Turyn [143]. |
| HVIII | mrh* | $r \equiv 3 \pmod 4$, condition $H(X,Y,Z,Z,r,m,h,a,b,$ with $(m-1)b = m(r-h+1)-a)$ holds: from theorem 8.7. |
| | $p(p+1)$ | Scarpis; from corollary 8.14. |
| | $p^r(p^r+1)$ | Williamson; from corollary 8.14. |
| | $h(h-1)$ | Williamson; from corollary 8.13. |
| | $m(m-1)$ | from corollary 8.15. |
| | $h(h+3)$ | $h+4$ a symmetric Hadamard matrix: Hall, from corollary 8.13. |
| | $(p^r+1)(p^r+4)$ | $p^r+4$ a prime power; from corollary 8.14. |
| | $u(u+3)$ | $u$ and $u+4$ both of the form $2^t\Pi(p_i^{r_i}+1)$: Williamson from corollary 8.16. |
| | $m(m+3)$ | $m+4$ the order of an amicable Hadamard matrix; from corollary 8.15. |
| | $k(4t-1)$ | whenever $(4t-1,u+t-\frac{1}{2}h,u)$ and $(4t-1,v+t,v)$ cyclic difference sets exist: Wallis; from corollary 8.9. |
| | $h\left(\frac{q^{n+1}-1}{q-1}\right)$ | $h = (q^n+q^{n-1}+\ldots+q+1)-4q^{n-1}+1$, $q$ a prime power, $(q^{n+1}-1)/(q-1)$ a prime or product of twin primes: Spence; from corollary 8.11. |
| | $(q-1)(q-2)(q^2+q+1)$ | $h = (q-1)(q-2)$, $q$ a prime power, $q^2+q+1 \equiv 3 \pmod 4$ a prime or product of twin primes: Wallis; from corollary 8.12. |
| | $k(k+1)(m-1)$ | $(k+1)(m-1)/m$ the order of a skew-Hadamard matrix: from corollary 8.17. |
| HIX | $kn(c-1)$* | $k > 2$. The core of the complex Hadamard matrix, $A+iB$, satisfies $A^T = A$, $B^T = B$. $W$ the type 2 $(1,-1)$ matrix of a $\{c-1,\frac{1}{4}(c-n)+u,u\}$ difference set and $WA^T, WB^T$ symmetric: from theorem 8.18. |
| | $knp$ | $k > 2$. There exist $X$ and $W$ of order $p$ satisfying $XW^T = WX^T$, $X^T = X$, $W^T = W$, $XX^T = pI-J$, $WW^T = (p-n+1)I+(n-1)J$: from corollary 8.19. |
| | $kn(n-1)$ | $k > 2$: Goethals and Seidel; from corollary 8.21. |
| | $kn(n+3)$ | $k > 2$, $n+4$ the order of a symmetric conference matrix: Goethals and Seidel; from |

| | | |
|---|---|---|
| | | corollary 8.21. |
| | $k(q-1)(q-2)(q^2+q+1)$ | $k > 2$, $q$ a prime power, $q^2+q+1$ prime, $n = q^2-3q+2$: from corollary 8.21. |
| | $k(x^2+1)(4x^2+1)$ | $k > 2$, $x$ odd, $4x^2+1$ a prime, $n = x^2+1$; from corollary 8.21. |
| | $k(x^2+1)(4x^2+9)$ | $k > 2$, $x$ odd, $4x +9$ a prime, $n = x^2+1$; from corollary 8.21. |
| | $kq^s(q^s+1)$ | $k > 2$: from corollary 8.22. |
| | $k(q^s+1)(q^s+4)$ | $k > 2$, $n = q^s+4$: from corollary 8.22. |
| | $k_1k_2q^s(q^s+1)$ | Williamson; from corollary 8.22. |
| | $k_1k_2(q^s+1)(q^s+4)$ | $n = q^s+4$, Williamson; from corollary 8.22. |
| HX | $kmnr$* | $m = 2$ or $4$, condition $H(X,Y,Z,W,r,m,n,a,b$, with $b = (p-n-a+1)m+a)$ holds; from theorem 8.24. Corollaries 8.25 and 8.26 give cases where the conditions are satisfied. |
| | $2kn(n+1)$ | $k > 1$, $n+2$ the order of a symmetric Hadamard matrix; from corollary 8.25. |
| HXI | $2krn$* | $k > 1$, condition $H(X,Y,Z,W,r,2,n,a,b$, with $b = 2(q-n+1)-a)$ holds; from theorem 8.27. Corollary 8.29 gives cases where the conditions are satisfied. |
| | $2(5x^2-1)(4x^2+1)$ | $n = \frac{1}{2}(5x^2-1)$, $q = 4x^2+1$ a prime; from corollary 8.29. |
| | $2(5x^2+7)(4x^2+9)$ | $n = \frac{1}{2}(5x^2+7)$, $q = 4x^2+9$ a prime; from corollary 8.29. |
| | $2k(t+1)(t+2)$ | $n = t+2$, $2t+3 = p$ a prime; from corollary 8.28. |
| | $2k(t+1)(t-2)$ | $n = t-2$, $2t+3 = p$ a prime; from corollary 8.28. |
| HXII | $mrh$* | $m = 2$ or $4$, condition $H(X,Y,Z,W,r,m,h,a,b$, with $b = m(p+1-h-a)+a)$ holds: from theorem 8.30. Corollary 8.33 gives cases where the conditions are satisfied. |
| | $2h(h+1)$ | if there exist Szekeres difference sets A,B of order $h+1$, with $a \in A \Rightarrow -a \notin A$ and $b \in B \Rightarrow -b \in B$; from corollary 8.31. |
| | $2h(h-3)$ | as above |
| | $2h(h+1)$ | $2h+3$ a prime power; from corollary 8.32. |
| | $2h(h-3)$ | $2h-5$ a prime power; from corollary 8.32. |
| | $2q^s(q^s-1)$ | Wallis: $h = q^s-1$; from corollary 8.33. |
| | $(5x^2+3)(4x^2+1)$ | $x$ odd, $q = 4x^2+1$, $h = \frac{1}{2}(5x^2+3)$; from corollary 8.33. |
| HXIII | $k_1k_2$ | the product of orders from any class. |

## APPENDIX B: KNOWN CLASSES OF SKEW-HADAMARD MATRICES

| | | |
|---|---|---|
| SI | $2^t \Pi k_i$ | $t, r_i$ all positive integers, $k_i = p_i^{r_i}+1 \equiv 0 \pmod 4$, $p_i$ a prime; from theorem 4.1. |
| SII | $(p-1)^u+1$ | p the order of a skew-Hadamard matrix, $u > 0$ an odd integer; from theorem 4.2. |
| SIII | $2(q+1)$ | $q \equiv 5 \pmod 8$ a prime power; from theorem 4.5. |
| SIV | $2^s(q+1)$ | $q = p^t$ is a prime power such that $p \equiv 5 \pmod 8$, $t \equiv 2 \pmod 4$, $s > 1$ an integer; from theorem 4.7. |
| SV | 4m | $m \in$ {odd integers between 3 and 25 inclusive}; from theorem 4.8, theorem 4.10 and unpublished results of D. Hunt and G. Szekeres. |
| SVI | $m^1(m^1-1)(m-1)$ | m and $m^1$ the order of amicable Hadamard matrices, where a skew-Hadamard matrix of order $(m-1)m^1/m$ exists; from theorem 4.20. |
| SVII | $4(q+1)$ | $q = 8f+1$ (f odd) is a prime power; from corollary 4.19. |
| SVIII | hm | h the order of a skew-Hadamard matrix, m the order of amicable Hadamard matrices; from theorem 4.21. |

## APPENDIX C: KNOWN CLASSES OF SYMMETRIC CONFERENCE MATRICES

| | | |
|---|---|---|
| NI | $p^r+1$ | r a positive integer, $p^r$ (prime power) $\equiv 1 \pmod 4$; from theorem 3.13. |
| NII | $(h-1)^2+1$ | h the order of a skew-Hadamard matrix; from theorem 3.6. |
| NIII | $(n-1)^\mu+1$ | n the order of a symmetric conference matrix, $\mu$ an odd integer; from corollary 3.12. |

## APPENDIX D: KNOWN CLASSES OF SYMMETRIC HADAMARD MATRICES

The following are orders of symmetric Hadamard matrices.

| | | |
|---|---|---|
| SHI | $2^t$ | t a non-negative integer; theorem 5.1. |
| SHII | $p^r+1$ | $p^r \equiv 3 \pmod 4$ is a prime power; theorem 5.1. |
| SHIII | $2(q^s+1)$ | $q^s \equiv 1 \pmod 4$ is a prime power. |
| SHIV | 2n | n the order of a symmetric conference matrix; from lemma 5.2. |
| SHV | 2n(n-1) | n as above; from theorem 5.3. |
| SHVI | $4s^2$ | if strong graphs of order $4s^2$ exist with eigenvalues $\rho_1 = 2s\mp 1$, $\rho_2 = -2s\mp 1$; or if pseudo Latin-square graphs $L_s(2s)$ or negative Latin-square graphs $NL_s(2s)$ exist; from Goethals and Seidel [46]. |
| SHVII | $(a-1)^2$ | a and a-2 both the orders of cores, one of which is symmetric and the other skew-symmetric with respect to its back-diagonal; from theorem 5.10. |
| | $(a-1)^2$ | a and a-2 both prime powers; from corollary 5.11. |
| SHVIII | $h_1{}^2$ | $h_1$ the order of an Hadamard matrix; from theorem 5.12. |
| SHIX | $(2k)^2$ | if there exists a $(k(2k-1), 4k^2-1, 2k+1, k, 1)$-configuration; from theorem 5.13. |
| SHX | $4(2m+1)^2$ | if there exist Szekeres difference sets in an additive abelian group of order 2m+1 and a symmetric conference matrix of order 2(2m+1); from theorem 5.14. |
| | $(p-1)^2$ | p a prime power and p-1 the order of a symmetric conference matrix; from corollary 5.15. |
| SHXI | k | k is a product of orders from the other classes SHI - SHX. |

## APPENDIX E: KNOWN CLASSES OF SYMMETRIC HADAMARD MATRICES WITH CONSTANT DIAGONAL

We note that the existence of an Hadamard matrix of order 4t always implies that a

$$4t-1,\ 2t-1,\ t-1$$

configuration exists, but if a regular Hadamard matrix of order $4s^2$ exists then there exists a

$$4s^2,\ 2s^2 \pm s,\ s^2 \pm s$$

configuration.

We indicate with an asterisk those matrices that are regular.

| | | |
|---|---|---|
| GI | $4s^2$ | if strong graphs of order $4s^2$ exist with eigenvalues $\rho_1 = 2s+1$, $\rho_2 = -2s+1$; from lemma 5.8. |
| GII* | $4s^2$ | if pseudo Latin-square graphs $L_s(2s)$ or negative Latin-square graphs $NL_s(2s)$ exist; from Goethals and Seidel [46]. |
| GIII* | $(a-1)^2$ | a and a-2 the orders of cores, one of which is symmetric and the other skew-symmetric, but symmetric with respect to its back-diagonal; from theorem 5.10. |
| | $(a-1)^2$ | a and a-2 both prime powers; from corollary 5.11. |
| GIV* | $h^2$ | h the order of an Hadamard matrix; from theorem 5.12. |
| GV* | $4k^2$ | if there exists a $(k(2k-1),4k^2-1,2k+1,k,1)$-configuration; from theorem 5.13. |
| GVI* | $4(2m+1)^2$ | if there exist Szekeres difference sets in an additive abelian group of order 2m+1 and a conference matrix of order 2(2m+1); from theorem 5.14. |
| | $(p-1)^2$ | p a prime power and p-1 the order of a symmetric conference matrix; from corollary 5.15. |
| GVII | k | k the product of any of the above orders. The product is regular if both the factors are regular. |

## APPENDIX F: KNOWN CLASSES OF COMPLEX HADAMARD MATRICES

Where the class is marked with an asterisk, *, that class is stronger than the others classified under the same number and includes the subclasses; we use the following notation

| | |
|---|---|
| s | the order of a complex skew-Hadamard matrix, |
| m | the order of complex amicable Hadamard matrices, |
| n | the order of a symmetric conference matrix, |
| t | the order of special Hadamard matrices. |

| | | |
|---|---|---|
| CI | n | from theorem 6.3. |
| CII | 2w | 4w the order of a Williamson type Hadamard matrix; from theorem 6.4. |
| CIII | mps* | p the order of three matrices A,B,C with elements 1,-1,i,-i satisfying $AB^* = BA^*$, $AC^* = CA^*$, $BC^* = CB^*$, $AA^* = aI + (p-a)J$, $BB^* = bI + (p-b)J$, $b = mp-a(m-1)-m(s-1)$, $CC^* = (p+1)I - J$; from theorem 6.5. |
| | s(s-1) | p = s-1 as above is the order of a core, m = 1; from corollary 6.8. |
| | 2p(p-1) | n = p-1, m = 2, p ≡ 3 (mod 4), p a prime power or equal to q(q+2) where q and q+2 are both prime powers. |
| CIV | mwt* | w the order of two matrices A,B with elements 1,-1,i,-i satisfying $AB^* = BA^*$, $AA^* = aI + (p-a)J$, $a = w-t+1$, $BB^* = (w+1)I - J$; from theorem 6.10. |
| CV | 2n(n-1) | s = n-2; from theorem 6.12. |
| CVI | (n-1)(n-a)* | n, n-a the order of symmetric conference matrices N and M, if L is the core of N, and there exists a complex matrix A of order n-1 satisfying $LA^* = -AL^*$, $A = A^*$, $AA^* = aI + (n-a-1)J$; from theorem 6.13. |
| | n(n-1) | from corollary 6.14. |

| | | |
|---|---|---|
| CVII | 2n(n+a-1) | n+a, n the order of symmetric conference matrices N and M, if L is the core of N and there exists a real matrix A of order n+a-1 satisfying $A^T = A$, $AL = LA$, $AA^* = aI + (n-1)J$; from theorem 6.15, (see corollary 6.17 for applications). |
| | 2n(n-1) | a = 0; from corollary 6.16. |
| | 2n(n+3) | a = 4, n+4 the order of a symmetric conference matrix; from corollary 6.16. |
| CVIII | 2n(n+2a-1)* | n+2a, n the order of symmetric conference matrices N and M, if L is the core of N and there exists a real matrix A of order n-2a-1 satisfying $A^T = A$, $AL = LA$, $AA^* = 4aI + (n-2a-1)J$; from theorem 6.18, (see corollary 6.19 for applications). |
| CIX | 2wn* | if there exist two complex matrices A and W of order w satisfying $AW^* = WA^*$, $AA^* = aI + (w-a)J$, $a = 2(w-n+1)$, $WW^* = (w+1)I - J$; from theorem 6.20. |
| | 2n(n-3) | w = n-3 the order of a complex matrix W satisfying $WW^* = (w-2)I - J$; from corollary 6.21. |
| CX | ab | where a and b are the orders of complex Hadamard matrices from theorem 6.2. |

## APPENDIX G: KNOWN CLASSES OF AMICABLE HADAMARD MATRICES

| | | |
|---|---|---|
| AI | $2^t$ | t a non-negative integer. |
| AII | $p^r+1$ | $p^r$ (prime power) $\equiv$ 3 (mod 4). |
| AIII | 2(q+1) | 2q+1 is prime power, q (prime power) $\equiv$ 1 (mod 4). |
| AIV | S | where S is a product of the above orders. |

## APPENDIX H: A TABLE OF ORDERS ≡ 0 (mod 4) < 4,000 FOR WHICH HADAMARD, SKEW-HADAMARD AND COMPLEX HADAMARD MATRICES ARE KNOWN

2 is omitted from the table, but there exists both an Hadamard and a skew-Hadamard matrix of order 2.

The table shows one class from appendices A,B or F for each matrix: the order of the matrix may also belong to other classes. The numbers down the left hand side of the page give the last two digits of the order while the number at the top of each column gives the hundreds and thousands. The heading "type" gives a class of Hadamard matrices to which the order belongs while "skew" gives a class of skew-Hadamard matrices to which the order belongs. If a complex Hadamard matrix is known but no real Hadamard matrix, then the class of complex Hadamard matrices to which the order belongs is indicated: in all cases the real matrix is preferred.

A blank in the table indicates that no matrix is known.

| last two digits | type 00 | skew 00 | type 100 | skew 100 | type 200 | skew 200 | type 300 | skew 300 | type 400 | skew 400 |
|---|---|---|---|---|---|---|---|---|---|---|
| | | | | | hundreds | | | | | |
| 0 | | | HIII | SV | HII | SI | HIII | SIII | HII | SI |
| 4 | HI | SI | HII | SI | HIII | SIII | HII | SI | | |
| 8 | HI | SI | HII | SI | HII | SI | HII | SI | HIII | SIII |
| 12 | HII | SI | HII | SI | HII | SI | HII | SI | | |
| 16 | HI | SI | HVI | | HII | SI | HIII | SIII | HII | SI |
| 20 | HII | SI | HII | SI | HIII | SIII | HII | SI | HII | SI |
| 24 | HII | SI | HIII | SIII | HII | SI | HIV | | HII | SI |
| 28 | HII | SI | HI | SI | HII | SI | HII | SI | | |
| 32 | HI | SI | HII | SI | HXIII | | HII | SI | HII | SI |
| 36 | HIII | SV | HII | SI | | | HII | SI | | |
| 40 | HII | SI | HII | SI | HII | SI | HIII | SIV | HII | SI |
| 44 | HII | SI | HII | SI | HII | SI | HII | SI | HII | SI |
| 48 | HII | SI | HIII | | HIII | SIII | HII | SI | HII | SI |
| 52 | HIII | SIV | HII | SI | HII | SI | HII | SI | HIII | |
| 56 | HII | SI | HVII | | HI | SI | | | HII | SI |
| 60 | HII | SI | HII | SI | HVII | | HII | SI | HIII | SIII |
| 64 | HI | SI | HII | SI | HII | SI | HIII | SIII | HII | SI |
| 68 | HII | SI | HII | SI | | | HII | SI | HII | SI |
| 72 | HII | SI | HVI | | HII | SI | HVIII | | | |
| 76 | HIII | SIII | HII | SI | HIII | | | | HVII | |
| 80 | HII | SI | HII | SI | HII | SI | HII | SI | HII | SI |
| 84 | HII | SI | HVI | SVIII | HII | SI | HII | SI | HIII | |
| 88 | HII | SI | | | HII | SI | HIII | | HII | SI |
| 92 | HVI | SV | HII | SI | | | HIII | | HII | SI |
| 96 | HII | SI | HIII | | HIII | SVII | HIII | SIII | HIII | SIII |

| | hundreds | | | | | | | | | |
|---|---|---|---|---|---|---|---|---|---|---|
| last two digits | type 500 | skew 500 | type 600 | skew 600 | type 700 | skew 700 | type 800 | skew 800 | type 900 | skew 900 |
| 0 | HII | SI | HII | SI | HIII | SIII | HII | SI | HIII | |
| 4 | HII | SI | | | HII | SI | HIII | | HIII | |
| 8 | | | HII | SI | HIII | | | | HII | SI |
| 12 | HI | SI | HIII | | | | HII | SI | HII | SI |
| 16 | HIII | | HII | SI | | | HIII | SIII | HIII | |
| 20 | HIX | | HII | SI | HII | SI | HIII | | HII | SI |
| 24 | HII | SI | HII | SI | HIII | | HII | SI | HIII | SIII |
| 28 | HII | SI | HIII | | HII | SI | HII | SI | HII | SI |
| 32 | HVII | | HII | SI | HVI | | HII | SI | | |
| 36 | | | HIII | SIII | HII | SI | HVII | | HII | SI |
| 40 | HIII | SIII | HII | SI | HII | SI | HII | SI | | |
| 44 | HII | SI | HII | SI | HII | SI | HIII | SIII | | |
| 48 | HII | SI | HII | SI | HIII | SIII | HII | SI | HII | SI |
| 52 | HIII | SVI | | | HII | SI | | | HIX | |
| 56 | HIII | SIII | HII | SI | HVIII | | | | | |
| 60 | HII | SI | HII | SI | HII | SI | HII | SI | HII | SI |
| 64 | HII | SI | HII | SI | | | HII | SI | | |
| 68 | HII | SI | | | HII | SI | HIII | | HII | SI |
| 72 | HII | SI | HII | SI | | | | | HII | SI |
| 76 | HII | SI | HIII | | HIII | | | | HII | SI |
| 80 | HIII | | HIII | SVIII | HIII | SIII | HII | SI | HVII | |
| 84 | | | HII | SI | HIII | | HII | SI | HII | SI |
| 88 | HII | SI | HII | SI | HII | SI | HII | SI | HVII | |
| 92 | HIII | SVIII | HII | SI | HIII | SIII | | | HII | SI |
| 96 | | | HII | SI | HIII | SIII | HII | SI | | |

hundreds

| last two digits | type 1000 | skew 1000 | type 1100 | skew 1100 | type 1200 | skew 1200 | type 1300 | skew 1300 | type 1400 | skew 1400 |
|---|---|---|---|---|---|---|---|---|---|---|
| 0 | HII | SI | HVII | | HII | SI | HIII | | HII | SI |
| 4 | | | HII | SI | HIII | | HII | SI | HIII | SIII |
| 8 | HII | SI | | | | | HII | SI | HII | SI |
| 12 | HVII | | HIII | SIII | | | HII | SI | | |
| 16 | | | HIII | SIII | HII | SI | | | HIII | |
| 20 | HII | SI | HII | SI | HVII | | HII | SI | HIII | SIII |
| 24 | HI | SI | HII | SI | HII | SI | HIII | SIII | HII | SI |
| 28 | | | HII | SI | HIII | SIII | HII | SI | HII | SI |
| 32 | HII | SI | | | HII | SI | HII | SI | | |
| 36 | HVII | | HII | SI | HIII | | | | | |
| 40 | HII | SI | HIII | | HII | SI | | | HII | SI |
| 44 | HIII | | HII | SI | | | HII | SI | HVII | |
| 48 | HII | SI | HVII | | HII | SI | HIII | | HII | SI |
| 52 | HII | SI | HII | SI | HIII | | HIII | | HII | SI |
| 56 | HII | SI | HIII | | HIII | | HIII | SIII | HII | SI |
| 60 | HIII | | HIII | | HII | SI | HIII | SVIII | HII | SI |
| 64 | HII | SI | HII | SI | HII | SI | HVII | | HIII | |
| 68 | | | | | | | HII | SI | HIII | SIII |
| 72 | | | HII | SI | HIII | SIII | HVII | | HII | SI |
| 76 | | | HII | SI | HVII | | HII | SI | HVII | |
| 80 | HIII | SIII | | | HII | SI | HVI | | HII | SI |
| 84 | HIII | SIII | HIII | SVIII | HII | SI | HII | SI | HII | SI |
| 88 | HII | SI | HII | SI | HII | SI | | | HII | SI |
| 92 | HII | SI | | | HII | SI | HII | SI | | |
| 96 | HII | SI | HVII | | HII | SI | | | HIII | SIII |

| | hundreds | | | | | | | | | |
|---|---|---|---|---|---|---|---|---|---|---|
| last two digits | type 1500 | skew 1500 | type 1600 | skew 1600 | type 1700 | skew 1700 | type 1800 | skew 1800 | type 1900 | skew 1900 |
| 0 | HII | SI | HII | SI | HII | SI | HIII | | HVII | |
| 4 | HII | SI | | | HIII | | | | HIX | |
| 8 | HVII | | HII | SI | HIII | SIII | HIII | | HII | SI |
| 12 | HII | SI | HVII | | | | HII | SI | | |
| 16 | HIII | SIII | | | HIII | | HII | SI | | |
| 20 | HII | SI | HII | SI | HII | SI | HVII | | HII | SI |
| 24 | HII | SI | HII | SI | HII | SI | HII | SI | HIII | |
| 28 | | | HII | SI | HII | SI | | | | |
| 32 | HII | SI | HIII | SIII | | | HII | SI | HII | SI |
| 36 | HII | SI | | | HIII | | HVII | | HII | SI |
| 40 | HIII | | HIII | SVII | HIII | | HII | SI | HVII | |
| 44 | HII | SI | HIII | SIII | | | | | HII | SI |
| 48 | HIII | SIII | HII | SI | HII | SI | HII | SI | | |
| 52 | HIII | | | | | | | | HII | SI |
| 56 | | | HII | SI | HIII | SIII | HII | SI | HIII | |
| 60 | HII | SI | HIII | SIII | HII | SI | HIII | | HIX | |
| 64 | HVII | | HII | SI | HIII | | | | | |
| 68 | HII | SI | HII | SI | HII | SI | HII | SI | HII | SI |
| 72 | HII | SI | HXIII | | | | HII | SI | HVII | |
| 76 | HII | SI | | | HII | SI | HIII | | HIII | |
| 80 | HII | SI | HII | SI | HVII | | HII | SI | HII | SI |
| 84 | HII | SI | HIII | SIV | HII | SI | HIII | SIII | HII | SI |
| 88 | | | HIII | SIII | HII | SI | | | HII | SI |
| 92 | HIII | SIII | HVI | | HII | SI | HVIII | | HIII | |
| 96 | HIII | SIII | HII | SI | | | HII | SI | HIII | SIII |

hundreds

| last two digits | type 2000 | skew 2000 | type 2100 | skew 2100 | type 2200 | skew 2200 | type 2300 | skew 2300 | type 2400 | skew 2400 |
|---|---|---|---|---|---|---|---|---|---|---|
| 0 | HII | SI | HII | SI | HIX | | HVII | | HII | SI |
| 4 | HII | SI | HII | SI | HII | SI | HII | SI | HIII | |
| 8 | | | HVII | | HII | SI | HIII | | HIII | SVII |
| 12 | HII | SI | HII | SI | HVII | | HII | SI | HII | SI |
| 16 | HII | SI | | | | | | | | |
| 20 | HIII | | HIII | | HIII | SIII | HIII | | HVII | |
| 24 | HXIII | | HIII | SIII | HIII | SIII | | | HII | SI |
| 28 | HII | SI | HII | SI | | | HII | SI | HIII | SIII |
| 32 | | | HII | SI | HIII | SIII | | | HII | SI |
| 36 | | | | | HIII | SIII | | | HIII | |
| 40 | HII | SI | HIII | SIII | HII | SI | HII | SI | HIX | |
| 44 | HIII | SIII | HII | SI | HII | SI | HII | SI | | |
| 48 | HI | SI | | | HII | SI | HII | SI | HII | SI |
| 52 | HVII | | | | HII | SI | HII | SI | HIII | |
| 56 | | | HVII | | HII | SI | HVII | | HIII | SIII |
| 60 | HIX | | HIII | SIII | HIII | | HIX | | HII | SI |
| 64 | HII | SI | | | | | HIII | SIII | HII | SI |
| 68 | HIII | | HIII | SIII | HII | SI | HIII | SVIII | HII | SI |
| 72 | HIII | | HVI | | HII | SI | HII | SI | HIII | SVII |
| 76 | | | HII | SI | | | HII | SI | HIII | SIII |
| 80 | HII | SI | HII | SI | HIII | SVII | HVII | | HII | SI |
| 84 | HII | SI | HII | SI | | | HII | SI | HVII | |
| 88 | HII | SI | HII | SI | HII | SI | HIII | | | |
| 92 | | | HII | SI | | | HIII | | | |
| 96 | HII | SI | HIII | | HIII | | | | HII | SI |

hundreds

| last two digits | type 2500 | skew 2500 | type 2600 | skew 2600 | type 2700 | skew 2700 | type 2800 | skew 2800 | type 2900 | skew 2900 |
|---|---|---|---|---|---|---|---|---|---|---|
| 0 | HIII | | HIII | | HII | SI | HII | SI | HVII | |
| 4 | HII | SI | HIII | SIII | HIII | | HII | SI | HII | SI |
| 8 | HVII | | HII | SI | HII | SI | HIII | SIII | HIII | SIII |
| 12 | HIII | | | | HII | SI | HIII | | HII | SI |
| 16 | HVII | | HII | SI | HVII | | HII | SI | HIV | |
| 20 | HII | SI | | | HII | SI | HII | SI | HII | SI |
| 24 | | | HII | SI | HIII | | | | HVII | |
| 28 | HII | SI | | | HIII | | | | HII | SI |
| 32 | HII | SI | | | HII | SI | HIII | | | |
| 36 | | | | | HII | SI | | | HIII | SIII |
| 40 | HII | SI | HII | SI | HIII | SIV | HIII | SIII | HII | SI |
| 44 | HII | SI | HIII | | HIII | | HII | SI | HII | SI |
| 48 | HVII | | HII | SI | HIII | SIII | HII | SI | | |
| 52 | HII | SI | HVIII | | HII | SI | HII | SI | HIII | |
| 56 | HIII | SIII | HII | SI | | | HII | SI | | |
| 60 | HII | SI | HII | SI | HIII | | HIII | SIII | HII | SI |
| 64 | | | HII | SI | HII | SI | | | HII | SI |
| 68 | HII | SI | | | HII | SI | HIII | | HII | SI |
| 72 | | | HII | SI | HVI | | | | HII | SI |
| 76 | HII | SI | | | | | | | HII | SI |
| 80 | HII | SI | HIX | | HVII | | HII | SI | HIII | |
| 84 | HII | SI | HII | SI | HII | SI | | | | |
| 88 | | | HII | SI | HVII | | HII | SI | HIII | SIII |
| 92 | HII | SI | | | HII | SI | | | HIII | SIII |
| 96 | HIII | | HIII | | | | HII | SI | | |

| last two digits | type 3000 | skew 3000 | type 3100 | skew 3100 | type 3200 | skew 3200 | type 3300 | skew 3300 | type 3400 | skew 3400 |
|---|---|---|---|---|---|---|---|---|---|---|
| | hundreds | | | | | | | | | |
| 0 | HII | SI | HIII | SIII | HII | SI | HII | SI | HII | SI |
| 4 | | | HIII | | HII | SI | HVIII | | | |
| 8 | HII | SI | HIII | | | | HII | SI | HII | SI |
| 12 | HII | SI | | | | | HII | SI | | |
| 16 | HIX | | HVII | | HII | SI | HIII | | HIII | SIII |
| 20 | HII | SI | HII | SI | HIII | | HII | SI | HIII | SIII |
| 24 | HII | SI | | | HIII | | HII | SI | | |
| 28 | | | HVII | | HIII | SIII | HII | SI | | |
| 32 | HIII | SIII | HVI | | | | HII | SI | HIII | SVII |
| 36 | HVII | | HII | SI | | | HII | SI | | |
| 40 | HII | SI | | | HII | SI | HIII | SIII | HII | SI |
| 44 | HIII | | HII | SI | HIII | SIII | HII | SI | HIII | |
| 48 | HII | SI | | | HII | SI | HII | SI | HII | SI |
| 52 | | | HII | SI | HII | SI | | | | |
| 56 | | | | | HII | SI | | | HII | SI |
| 60 | HIX | | HII | SI | HII | SI | HII | SI | | |
| 64 | HII | SI | HII | SI | HIII | SIII | HIII | SIV | HII | SI |
| 68 | HII | SI | HII | SI | | | HIII | SIX | HII | SI |
| 72 | HII | SI | HVII | | HII | SI | HII | SI | HIII | SVIII |
| 76 | | | | | HIII | SIII | HIII | SII | HVII | |
| 80 | HII | SI | HVII | | HIII | SI | HVII | | HIII | |
| 84 | HII | SI | HIII | SIII | | | HIII | | HIII | SIII |
| 88 | HII | SI | HII | SI | HIII | SIII | HIII | SIII | | |
| 92 | | | HII | SI | | | HII | SI | HII | SI |
| 96 | HIII | SIII | HIII | SIII | HXIII | SI | HIII | | HII | SI |

| last two digits | hundreds<br>type 3500 | skew 3500 | type 3600 | skew 3600 | type 3700 | skew 3700 | type 3800 | skew 3800 | type 3900 | skew 3900 |
|---|---|---|---|---|---|---|---|---|---|---|
| 0 | HII | SI | HIII | SVIII | HIII | | HIX | | HIII | SIII |
| 4 | | | HIII | | | | HII | SI | HII | SI |
| 8 | HIII | | HII | SI | HVI | | HIX | | HII | SI |
| 12 | HII | SI | HVI | | HII | SI | | | HII | SI |
| 16 | | | HIII | | | | HII | SI | | |
| 20 | HII | SI | HVII | | HII | SI | | | HII | SI |
| 24 | | | HII | SI | HIII | SIII | HII | SI | HII | SI |
| 28 | HII | SI | | | HII | SI | HIII | | | |
| 32 | | | HII | SI | | | | | HII | SI |
| 36 | HII | SI | | | HII | SI | | | HII | SI |
| 40 | HII | SI | HIII | | HII | SI | HII | SI | | |
| 44 | | | HII | SI | HII | SI | HIV | | HII | SI |
| 48 | HII | SI | HII | SI | HIII | | HII | SI | HII | SI |
| 52 | HII | SI | | | HIII | SVII | HII | SI | HIII | |
| 56 | HIII | | | | HIII | SIII | | | | |
| 60 | HII | SI | HII | SI | HII | SI | | | HII | SI |
| 64 | HVI | | HII | SI | | | HII | SI | | |
| 68 | HII | SI | | | HII | SI | HIII | SIII | HII | SI |
| 72 | HII | SI | HII | SI | | | HII | SI | HVI | |
| 76 | HII | SI | | | | | HVII | | HII | SI |
| 80 | HIII | SIII | HII | SI | HII | SI | HIX | | HVII | |
| 84 | HII | SI | HVI | | HXIII | | | | HXIII | SI |
| 88 | | | | | | | HII | SI | HIII | |
| 92 | | | HII | SI | HII | SI | HVII | | HIII | SIII |
| 96 | | | HII | SI | | | | | HIII | SIII |

## APPENDIX I: A TABLE OF ORDERS ≡ 2 (MOD 4) < 4,000 FOR WHICH SYMMETRIC CONFERENCE AND COMPLEX HADAMARD MATRICES ARE KNOWN

We recall that complex Hadamard matrices always exist when symmetric conference matrices exist.

The first page of this appendix uses the following notation:

'NI, NII' - indicates that a symmetric conference matrix is known and it belongs to the class NI or NII respectively. See appendix C.

'non' - theorem 1.31 shows that a symmetric conference matrix cannot exist for this order.

blank - a symmetric conference matrix is not known but this order is not excluded by theorem 1.31.

As in appendix H the numbers on the left of the page are the last two digits of the order and the top of the column gives the hundreds.

On the second page of this appendix we have listed the orders between 1,000 and 4,000 for which symmetric conference matrices are known. An asterisk after a number indicates it is from class NII, otherwise all orders belong to class NI.

We then list those orders between 1,000 and 4,000 for which a complex Hadamard matrix exists, but a symmetric conference matrix is not known.

| last two digits | hundreds 00 | 100 | 200 | 300 | 400 | 500 | 600 | 700 | 800 | 900 |
|---|---|---|---|---|---|---|---|---|---|---|
| 2 | NI | NI | non | non | NI | non | NI | NI | | |
| 6 | NI | non | | † | | | | non | non | |
| 10 | NI | NI | non | non | NI | NI | non | NI | NI | |
| 14 | NI | NI | non | NI | non | non | NI | non | non | non |
| 18 | NI | | non | NI | non | non | NI | non | non | non |
| 22 | *non | NI | | non | NI | NI | non | non | NI | non |
| 26 | NI | NI | NII | | | non | NI | | non | |
| 30 | NI | non | NI | non | non | NI | | NI | NI | NI |
| 34 | *non | non | NI | | NI | | non | NI | | non |
| 38 | NI | NI | non | NI | non | non | | non | non | NI |
| 42 | NI | non | NI | non | | NI | NI | non | NI | NI |
| 46 | * | | | non | | | non | | | non |
| 50 | NI | NI | non | NI | NI | †non | non | non | | |
| 54 | NI | | non | NI | non | non | NI | non | NI | NI |
| 58 | *non | NI | NI | non | NI | NI | | NI | NI | non |
| 62 | NI | non | | NI | NI | non | NI | NI | non | NI |
| 66 | | non | | | non | | non | | | |
| 70 | non | NI | NI | | non | NI | non | NI | †non | non |
| 74 | NI | NI | non | NI | non | non | NI | NI | | non |
| 78 | non | non | NI | | non | NI | NI | non | NI | NI |
| 82 | NI | NI | NI | non | | non | non | non | NI | |
| 86 | * | | non | non | | | | | non | |
| 90 | NI | non | NI | NI | non | non | | non | non | non |
| 94 | non | NI | NI | non | | NI | non | | non | non |
| 98 | NI | NI | non | NI | non | non | | NI | non | NI |

* a complex Hadamard matrix exists for these orders from CII,

† a complex Hadamard matrix exists for these orders from CIII.

| | | | | | |
|---|---|---|---|---|---|
| 1010 | 1482 | 1994 | 2442 | 3002 | 3542 |
| 1014 | 1490 | 1998 | 2474 | 3026* | 3558 |
| 1022 | 1494 | 2018 | 2478 | 3038 | 3582 |
| 1034 | 1522* | 2030 | 2522 | 3042 | 3594 |
| 1050 | 1550 | 2054 | 2550 | 3050 | 3614 |
| 1062 | 1554 | 2070 | 2558 | 3062 | 3618 |
| 1070 | 1598 | 2082 | 2594 | 3090 | 3638 |
| 1094 | 1602 | 2090 | 2602* | 3110 | 3674 |
| 1098 | 1610 | 2114 | 2610 | 3122 | 3678 |
| 1110 | 1614 | 2130 | 2618 | 3126 | 3698 |
| 1118 | 1622 | 2138 | 2622 | 3138 | 3702 |
| 1130 | 1638 | 2142 | 2634 | 3170 | 3710 |
| 1154 | 1658 | 2154 | 2658 | 3182 | 3722 |
| 1182 | 1670 | 2162 | 2678 | 3210 | 3734 |
| 1194 | 1682 | 2198 | 2690 | 3218 | 3762 |
| 1202 | 1694 | 2210 | 2694 | 3222 | 3770 |
| 1214 | 1698 | 2214 | 2714 | 3230 | 3794 |
| 1218 | 1710 | 2222 | 2730 | 3254 | 3798 |
| 1226* | 1722 | 2238 | 2742 | 3258 | 3822 |
| 1230 | 1734 | 2270 | 2750 | 3302 | 3834 |
| 1238 | 1742 | 2274 | 2778 | 3314 | 3854 |
| 1250 | 1754 | 2282 | 2790 | 3330 | 3878 |
| 1278 | 1778 | 2294 | 2798 | 3362 | 3882 |
| 1290 | 1790 | 2298 | 2802 | 3374 | 3890 |
| 1298 | 1802 | 2310 | 2810 | 3390 | 3918 |
| 1302 | 1850 | 2334 | 2834 | 3414 | 3930 |
| 1322 | 1862 | 2342 | 2838 | 3434 | 3970* |
| 1362 | 1874 | 2358 | 2858 | 3450 | 3990 |
| 1370 | 1878 | 2378 | 2862 | 3458 | |
| 1374 | 1890 | 2382 | 2898 | 3462 | |
| 1382 | 1902 | 2390 | 2910 | 3470 | |
| 1410 | 1914 | 2394 | 2918 | 3482 | |
| 1430 | 1934 | 2402 | 2954 | 3518 | |
| 1434 | 1950 | 2418 | 2958 | 3530 | |
| 1454 | 1974 | 2438 | 2970 | 3534 | |

Symmetric conference matrices are known to exist for the above orders. Complex Hadamard matrices also exist for the following orders:

2450, 3782 (for which orders a symmetric conference matrix cannot exist), and 1046 from class CIII.

* Indicates the matrix is from class NII.

## APPENDIX J: A TABLE OF ORDERS FOR WHICH AMICABLE HADAMARD MATRICES EXIST

These matrices exist for order 2. The number in the table below gives the class from appendix G from which the order is obtained.

| last two digits | hundreds 0 | 100 | 200 | 300 | 400 | 500 | 600 | 700 | 800 | 900 |
|---|---|---|---|---|---|---|---|---|---|---|
| 0 | | | AII | | AIV | AII | AII | | AIV | |
| 4 | AI | AII | | AIV | | AII | | AIV | | |
| 8 | AI | AII | AIV | AII | | | AII | | | AII |
| 12 | AII | AIV | AII | AII | | AI | | | AII | AII |
| 16 | AI | | AIV | | AIV | | AIV | | AIV | |
| 20 | AII | AIV | | AIV | AII | | AII | AII | | AII |
| 24 | AII | | AII | | AIV | AII | AIV | | AII | |
| 28 | AII | AI | AII | AIV | | AIV | | AII | AII | AIV |
| 32 | AI | AII | | AII | AII | | AII | | AIV | |
| 36 | | AIV | | AIV | | | | AIV | | AIV |
| 40 | AIV | AII | AII | | AII | | AIV | AII | AII | |
| 44 | AII | AIV | AII | AII | AII | AIV | AII | AII | | |
| 48 | AII | | | AII | AIV | AII | AII | | AIV | AII |
| 52 | | AII | AII | AIV | | | | AII | | |
| 56 | AIV | | AI | | AIV | | AIV | | | |
| 60 | AII | AIV | | AII | | AIV | AII | AIV | AII | AIV |
| 64 | AI | AII | AII | | AII | AII | AIV | | AII | |
| 68 | AII | AII | | AII | AII | AIV | | AIV | | AII |
| 72 | AII | | AII | | | AII | AIV | | | AII |
| 76 | | AIV | | | | AIV | | | | AIV |
| 80 | AII | AII | AIV | AII | AII | | | | AIV | |
| 84 | AII | | AII | AII | | | AII | AIV | AII | AII |
| 88 | AIV | | AIV | | AII | AII | AIV | AII | AII | |
| 92 | | AII | | | AII | | AII | | | AII |
| 96 | AIV | | | | | | AIV | | AIV | |

## APPENDIX K: SKEW-HADAMARD MATRICES OF ORDER ≤ 100

Using the construction of theorem 4.8 Dr. David Hunt has constructed skew-Hadamard matrices for orders ≤ 100. We give a table of the number of different matrices of each order that he found. The results for orders less than 48 were found by hand and the remainder by computer.

| 4n | Decomposition of 4n into squares | | | number of apparently different solutions |
|---|---|---|---|---|
| 100 | $1^2+5^2+5^2+7^2$ | $1^2+1^2+7^2+7^2$ | $1^2+3^2+3^2+9^2$ | ≥ 1, ≥ 1, * |
| 92 | $1^2+1^2+3^2+9^2$ | | | ≥ 1 |
| 84 | $1^2+3^2+5^2+7^2$ | $1^2+1^2+1^2+9^2$ | | 6, 4 |
| 76 | $1^2+1^2+5^2+7^2$ | $1^2+5^2+5^2+5^2$ | | 5, 3 |
| 68 | $1^2+3^2+3^2+7^2$ | | | 2 |
| 60 | $1^2+1^2+3^2+7^2$ | $1^2+3^2+5^2+5^2$ | | 7, 4 |
| 52 | $1^2+1^2+1^2+7^2$ | $1^2+1^2+5^2+5^2$ | | 2, 4 |
| 44 | $1^2+3^2+3^2+5^2$ | | | 2 |
| 36 | $1^2+1^2+3^2+5^2$ | | | 1 |
| 28 | $1^2+3^2+3^2+3^2$ | $1^2+1^2+1^2+5^2$ | | 2, 1 |
| 20 | $1^2+1^2+3^2+3^2$ | | | 1 |
| 12 | $1^2+1^2+1^2+3^2$ | | | 1 |
| 4 | $1^2+1^2+1^2+1^2$ | | | 1 |

* zero with probability .9999.

We recall that the construction of theorem 4.8 requires four circulant (1,-1) matrices A,B,C,D, where A is skew-type and B,C,D are symmetric. We now give Hunt's results by giving the first rows of A,B,C,D. We recall that the sum of the elements in the first rows of A,B,C,D of order n, a,b,c,d respectively, in this construction satisfy

$$a^2+b^2+c^2+d^2 = 4n.$$

The solutions for orders 4n, n = 1,3,5,...,21 are:

$4n = 4 = 1^2+1^2+1^2+1^2$ one solution:

$4n = 12 = 1^2+1^2+1^2+3^2$ one solution:

1 1, -1
1 -1, -1
1 -1, -1
1 1, 1

$4n = 20 = 1^2+1^2+3^2+3^2$ one solution:

1 1 1, -1 -1
1 -1 1, 1 -1
1 -1 -1, -1 -1
1 -1 -1, -1 -1

$4n = 28 = 1^2+3^2+3^2+3^2$ two solutions:

1 1 1 1, -1 -1 -1
1 1 1 -1, -1 1 1
1 1 -1 1, 1 -1 1
1 -1 1 1, 1 1 -1

and

1 1 1 1, -1 -1 -1
1 1 -1 1, 1 -1 1
1 1 -1 1, 1 -1 1
1 -1 1 1, 1 1 -1

$1^2+1^2+1^2+5^2$ one solution:

1 1 1 1, -1 -1 -1
1 -1 1 -1, -1 1 -1
1 -1 -1 1, 1 -1 -1
1 -1 -1 -1, -1 -1 -1

$4n = 36 = 1^2+1^2+3^2+5^2$ one solution:

1 1 1 1 -1, 1 -1 -1 -1
1 1 -1 1 -1, -1 1 -1 1
1 -1 -1 -1 1, 1 -1 -1 -1
1 1 1 -1 1, 1 -1 1 1

$4n = 44 = 1^2+3^2+3^2+5^2$ two solutions:

1 1 -1 1 -1 -1, 1 1 -1 1 -1
1 1 1 1 -1 -1, -1 -1 1 1 1
1 -1 1 1 1 -1, -1 1 1 1 -1
1 -1 -1 -1 1 -1, -1 1 -1 -1 -1

and

1 1 -1 -1 1 -1, 1 -1 1 1 -1
1 1 1 -1 -1 1, 1 -1 -1 1 1
1 1 -1 1 -1 1, 1 -1 1 -1 1
1 1 -1 -1 -1 -1, -1 -1 -1 -1 1

$4n = 52 = 1^2+1^2+1^2+7^2$

two solutions:

1 1 -1 1 -1 -1 -1, 1 1 1 -1 1 -1
1 -1 -1 -1 1 1 1, 1 1 1 -1 -1 -1
1 1 -1 1 -1 -1 1, 1 -1 -1 1 -1 1
1 -1 -1 -1 -1 1 -1, -1 1 -1 -1 -1 -1

and

1 1 -1 -1 1 -1 1, -1 1 -1 1 1 -1
1 -1 -1 1 1 1 -1, -1 1 1 1 -1 -1
1 1 1 -1 1 -1 -1, -1 -1 1 -1 1 1
1 -1 -1 -1 -1 -1 1, 1 -1 -1 -1 -1 -1

$1^2+1^2+5^2+5^2$

four solutions:

1 1 -1 -1 -1 -1 1, -1 1 1 1 1 -1
1 1 -1 1 1 -1 -1, -1 -1 1 1 -1 1
1 -1 1 1 1 1 -1, -1 1 1 1 1 -1
1 -1 1 1 1 -1 1, 1 -1 1 1 1 -1

and

1 1 -1 -1 -1 -1 1, -1 1 1 1 1 -1
1 1 -1 1 -1 1 -1, -1 1 -1 1 -1 1
1 1 1 -1 -1 1 1, 1 1 -1 -1 1 1
1 1 1 -1 1 1 -1, -1 1 1 -1 1 1

and

1 1 1 1 1 -1 -1, 1 1 -1 -1 -1 -1
1 -1 1 1 -1 -1 1, 1 -1 -1 1 1 -1
1 1 1 1 -1 1 -1, -1 1 -1 1 1 1
1 1 -1 1 -1 1 1, 1 1 -1 1 -1 1

and

1 1 -1 -1 -1 -1 -1, 1 1 1 1 1 -1
1 -1 -1 1 1 1 -1, -1 1 1 1 -1 -1
1 1 1 -1 1 -1 1, 1 -1 1 -1 1 1
1 -1 1 1 1 -1 1, 1 -1 1 1 1 -1

$4n = 60 = 1^2+1^2+3^2+7^2$

seven solutions:

1 1 1 1 1 -1 1 -1, 1 -1 1 -1 -1 -1 -1
1 -1 -1 -1 1 1 -1 1. 1 -1 1 1 -1 -1 1
1 1 1 -1 -1 1 -1 1, 1 -1 1 -1 -1 1 1
1 1 1 -1 1 1 1 -1, -1 1 1 1 -1 1 1

and

1 1 1 1 1 1 -1 1, -1 1 -1 -1 -1 -1 -1
1 1 -1 -1 -1 1 -1 1, 1 -1 1 -1 -1 -1 1
1 -1 1 1 -1 1 1 -1, -1 1 1 -1 1 1 -1
1 1 1 -1 1 1 1 -1, -1 1 1 1 -1 1 1

and

1 1 1 1 1 1 -1 1, -1 1 -1 -1 -1 -1 -1
1 -1 1 -1 -1 1 1 -1, -1 1 1 -1 -1 1 -1
1 1 1 -1 -1 1 -1 1, 1 -1 1 -1 -1 1 1
1 1 -1 1 1 1 1 -1, -1 1 1 1 1 -1 1

and

1 1 -1 1 1 1 -1 1, -1 1 -1 -1 -1 1 -1
1 -1 -1 -1 1 1 -1 1, 1 -1 1 1 -1 -1 -1
1 1 -1 -1 -1 1 1 1, 1 1 1 -1 -1 -1 1
1 -1 1 1 -1 1 1 1, 1 1 1 -1 1 1 -1

and

1 1 -1 1 1 1 1 -1, 1 -1 -1 -1 -1 1 -1
1 -1 1 -1 -1 -1 1 1, 1 1 -1 -1 -1 1 -1
1 -1 -1 1 1 -1 1 1, 1 1 -1 1 1 -1 -1
1 1 -1 1 1 1 1 -1, -1 1 1 1 1 -1 1

and

1 1 -1 1 1 1 1 -1, 1 -1 -1 -1 -1 1 -1
1 -1 1 -1 -1 -1 1 1, 1 1 -1 -1 -1 1 -1
1 -1 1 1 -1 -1 1 1, 1 1 -1 -1 1 1 -1
1 1 1 1 -1 1 1 -1, -1 1 1 -1 1 1 1

and

1 1 -1 1 1 1 1 -1, 1 -1 -1 -1 -1 1 -1
1 1 -1 1 -1 -1 -1 1, 1 -1 -1 -1 1 -1 1
1 1 1 -1 1 -1 -1 1, 1 -1 -1 1 -1 1 1
1 1 -1 1 1 1 1 -1, -1 1 1 1 1 -1 1

$1^2+3^2+5^2+5^2$ four solutions:

1 1 1 1 1 1 -1 -1, 1 1 -1 -1 -1 -1 -1
1 -1 1 1 -1 -1 1 1, 1 1 -1 -1 1 1 -1
1 -1 -1 -1 -1 1 -1 1, 1 -1 1 -1 -1 -1 1
1 -1 1 -1 -1 -1 1 -1, -1 1 -1 -1 -1 1 -1

and

1 1 -1 1 1 1 1 1, -1 -1 -1 -1 -1 1 -1
1 1 -1 -1 1 -1 1 1, 1 1 -1 1 -1 -1 1
1 -1 -1 -1 -1 1 1 -1, -1 1 1 -1 -1 -1 -1
1 -1 1 -1 -1 -1 1 -1, -1 1 -1 -1 -1 1 -1

and

1 1 1 1 -1 1 1 1, -1 -1 -1 1 -1 -1 -1
1 -1 1 1 1 -1 1 -1, -1 1 -1 1 1 1 -1
1 -1 -1 1 1 -1 -1 -1, -1 -1 -1 1 1 -1 -1
1 -1 -1 -1 -1 1 -1 1, 1 -1 1 -1 -1 -1 -1

and

1 1 -1 -1 -1 1 1 1, -1 -1 -1 1 1 1 -1
1 1 1 1 -1 -1 1 -1, -1 1 -1 -1 1 1 1
1 -1 -1 -1 -1 1 -1 1, 1 -1 1 -1 -1 -1 -1
1 -1 1 -1 -1 -1 1 -1, -1 1 -1 -1 -1 1 -1

$4n = 68 = 1^2+3^2+3^2+7^2$ two solutions:

1 1 -1 -1 1 1 -1 1 -1, 1 -1 1 -1 -1 1 1 -1
1 1 -1 -1 1 -1 -1 -1 -1, -1 -1 -1 -1 1 -1 -1 1
1 1 1 -1 -1 -1 -1 1 -1, -1 1 -1 -1 -1 -1 1 1
1 1 -1 -1 -1 1 -1 1 -1, -1 1 -1 1 -1 -1 -1 1

and

1 1 1 -1 -1 -1 -1 -1 -1, 1 1 1 1 1 1 -1 -1
1 -1 -1 -1 -1 -1 1 1 -1, -1 1 1 -1 -1 -1 -1 -1
1 -1 -1 -1 1 1 -1 1 -1, -1 1 -1 1 1 -1 -1 -1
1 -1 1 -1 1 -1 -1 1 -1, -1 1 -1 -1 1 -1 1 -1

$4n = 76 = 1^2+7^2+5^2+1^2$ five solutions:

1 -1 1 -1 -1 -1 -1 -1 1 1, -1 -1 1 1 1 1 1 -1 1
1 1 -1 1 1 1 1 1 -1 -1, -1 -1 1 1 1 1 1 -1 1
1 -1 1 -1 -1 -1 -1 1 1 -1, -1 1 1 -1 -1 -1 -1 1 -1
1 -1 -1 1 -1 1 -1 1 1 -1, -1 1 1 -1 1 -1 1 -1 -1

and

1 -1 1 -1 -1 -1 -1 1 -1 -1, 1 1 -1 1 1 1 1 -1 1
1 1 -1 1 1 1 -1 1 -1 1, 1 -1 1 -1 1 1 1 -1 1
1 -1 -1 1 1 -1 -1 -1 -1 1, 1 -1 -1 -1 -1 1 1 -1 -1
1 -1 1 -1 -1 -1 -1 1 1 1, 1 1 1 -1 -1 -1 -1 1 -1

and

1 -1 1 -1 -1 -1 -1 1 1 1, -1 -1 -1 1 1 1 1 -1 1
1 1 -1 1 1 1 1 -1 1 -1, -1 1 -1 1 1 1 1 -1 1
1 -1 -1 -1 1 1 -1 -1 1 -1, -1 1 -1 -1 1 1 -1 -1 -1
1 -1 1 1 1 -1 1 -1 -1 -1, -1 -1 -1 1 -1 1 1 1 -1

and

1 -1 1 -1 -1 -1 1 1 1 1, -1 -1 -1 -1 1 1 1 -1 1
1 -1 1 1 1 1 -1 1 1 1, 1 1 1 -1 1 1 1 1 -1
1 1 -1 1 -1 1 -1 -1 -1 -1, -1 -1 -1 -1 1 -1 1 -1 1
1 -1 -1 1 -1 -1 1 1 1 -1, -1 1 1 1 -1 -1 1 -1 -1

and

1 -1 1 -1 -1 -1 1 1 1 1, -1 -1 -1 -1 1 1 1 -1 1
1 -1 1 1 -1 1 1 1 1 -1, -1 1 1 1 1 -1 1 1 -1
1 1 1 -1 -1 1 -1 -1 -1 -1, -1 -1 -1 -1 1 -1 -1 1 1
1 -1 1 -1 1 -1 -1 1 1 -1, -1 1 1 -1 -1 1 -1 1 -1

$1^2+5^2+5^2+5^2$ three solutions:

1 1 1 -1 -1 -1 -1 -1 -1 1, -1 1 1 1 1 1 1 -1 -1
1 -1 -1 -1 1 -1 -1 1 -1 1, 1 -1 1 -1 -1 1 -1 -1 -1
1 -1 1 -1 -1 -1 -1 1 1 -1, -1 1 1 -1 -1 -1 -1 1 -1
1 -1 -1 -1 1 -1 1 1 -1 -1, -1 -1 1 1 -1 1 -1 -1 -1

and

1 1 1 -1 1 -1 1 -1 1 1, -1 -1 1 -1 1 -1 1 -1 -1
1 -1 1 1 -1 -1 1 -1 -1 -1, -1 -1 -1 1 -1 -1 1 1 -1
1 1 -1 -1 -1 -1 1 -1 -1 1, 1 -1 -1 1 -1 -1 -1 -1 1
1 1 1 -1 -1 -1 -1 1 -1 -1, -1 -1 1 -1 -1 -1 -1 1 1

and

1 1 -1 -1 1 1 1 1 -1 1, -1 1 -1 -1 -1 -1 1 1 -1
1 -1 -1 -1 1 -1 1 -1 -1 1, 1 -1 -1 1 -1 1 -1 -1 -1
1 -1 1 1 -1 1 -1 -1 -1 -1, -1 -1 -1 -1 1 -1 1 1 -1
1 1 -1 -1 -1 -1 -1 1 1 -1, -1 1 1 -1 -1 -1 -1 -1 1

$4n = 84 = 1^2+3^2+5^2+7^2$ six solutions:

1 -1 -1 -1 -1 -1 1 -1 -1 1 -1, 1 -1 1 1 -1 1 1 1 1 1
1 1 1 -1 -1 1 -1 1 1 1 -1, -1 1 1 1 -1 1 -1 -1 1 1
1 1 -1 1 -1 -1 -1 -1 -1 1 -1, -1 1 -1 -1 -1 -1 -1 1 -1 1
1 -1 1 -1 -1 1 1 -1 -1 -1 1, 1 -1 -1 -1 1 1 -1 -1 1 -1

and

1 -1 -1 -1 -1 -1 1 1 -1 1 -1, 1 -1 1 -1 -1 1 1 1 1 1
1 1 1 -1 -1 1 1 1 -1 -1 1, 1 -1 -1 1 1 1 -1 -1 1 1
1 1 -1 1 -1 -1 1 -1 -1 -1 -1, -1 -1 -1 -1 1 -1 -1 1 -1 1
1 -1 1 -1 -1 1 1 -1 1 -1 -1, -1 -1 1 -1 1 1 -1 -1 1 -1

and

1 -1 -1 -1 -1 1 -1 -1 1 1 -1, 1 -1 -1 1 1 -1 1 1 1 1
1 -1 1 1 1 1 1 -1 -1 -1 1, 1 -1 -1 -1 1 1 1 1 1 -1
1 -1 -1 -1 -1 -1 1 -1 1 -1 1, 1 -1 1 -1 1 -1 -1 -1 -1 -1
1 -1 1 -1 1 1 -1 -1 1 -1 -1, -1 -1 1 -1 -1 1 1 -1 1 -1

and

1 -1 -1 -1 1 -1 -1 -1 1 1 -1, 1 -1 -1 1 1 1 -1 1 1 1
1 1 -1 -1 1 -1 -1 1 1 1 1, 1 1 1 1 -1 -1 1 -1 -1 1
1 -1 -1 -1 1 -1 1 -1 -1 1 -1, -1 1 -1 -1 1 -1 1 -1 -1 -1
1 1 1 -1 -1 -1 1 -1 1 -1 -1, -1 -1 1 -1 1 -1 -1 -1 1 1

and

1 -1 -1 -1 1 -1 -1 1 -1 -1 1, -1 1 1 -1 1 1 -1 1 1 1
1 1 -1 1 1 1 1 -1 1 -1 -1, -1 -1 1 -1 1 1 1 1 -1 1
1 -1 1 -1 -1 -1 -1 -1 1 1 -1, -1 1 1 -1 -1 -1 -1 -1 1 -1
1 1 -1 1 -1 -1 -1 1 1 -1 -1, -1 -1 1 1 -1 -1 -1 1 -1 1

and

1 -1 -1 -1 1 -1 1 -1 -1 -1 1, -1 1 1 1 -1 1 -1 1 1 1
1 1 1 1 1 -1 -1 -1 1 -1 1, 1 -1 1 -1 -1 -1 1 1 1 1
1 1 -1 -1 1 -1 -1 -1 -1 1 -1, -1 1 -1 -1 -1 -1 1 -1 -1 1
1 -1 -1 -1 1 1 -1 1 -1 -1 1, 1 -1 -1 1 -1 1 1 -1 -1 -1

$1^2+9^2+1^2+1^2$ four solutions:

1 1 -1 -1 1 1 1 1 1 1 1, -1 -1 -1 -1 -1 -1 -1 1 1 -1
1 1 1 -1 -1 1 1 1 1 -1 1, 1 -1 1 1 1 1 -1 -1 1 1
1 -1 -1 1 1 -1 1 -1 1 -1 1, 1 -1 1 -1 1 -1 1 1 -1 -1
1 -1 -1 -1 1 1 1 -1 1 -1 1, 1 -1 1 -1 1 1 1 -1 -1 -1

and

1 1 -1 -1 -1 -1 -1 -1 1 1 1, -1 -1 -1 1 1 1 1 1 1 -1
1 -1 1 1 1 -1 -1 1 1 1 1, 1 1 1 1 -1 -1 1 1 1 -1
1 1 -1 1 1 -1 -1 1 -1 1 -1, -1 1 -1 1 -1 -1 1 1 -1 1
1 -1 1 -1 1 1 -1 1 1 -1 -1, -1 -1 1 1 -1 1 1 -1 1 -1

and

1 1 1 -1 -1 1 -1 -1 1 -1 1, -1 1 -1 1 1 -1 1 1 -1 -1
1 1 1 1 1 1 -1 1 -1 -1 1, 1 -1 -1 1 -1 1 1 1 1 1
1 -1 -1 -1 1 1 1 -1 1 -1 1, 1 -1 1 -1 1 1 1 -1 -1 -1
1 1 -1 1 1 -1 -1 -1 -1 1 1, 1 1 -1 -1 -1 -1 1 1 -1 1

and

1 1 -1 1 -1 -1 1 -1 -1 -1 1, -1 1 1 1 -1 1 1 -1 1 -1
1 1 -1 1 -1 -1 1 1 1 1 1, 1 1 1 1 -1 -1 1 -1 1
1 -1 -1 -1 1 1 1 -1 1 -1 1, 1 -1 1 -1 1 1 1 -1 -1 -1
1 -1 -1 1 1 -1 -1 -1 1 1 1, 1 1 1 -1 -1 -1 1 1 -1 -1

## REFERENCES

[1] R. AHRENS and G. SZEKERES, On a combinatorial generalization of 27 lines associated with a cubic surface. *J. Aust. Math. Soc.* 10 (1969), 485-492.

[2] A.A. ALBERT, *Studies in Modern Algebra.* (M.A.A. Studies in Mathematics, Vol. 2), Prentice-Hall, Englewood Cliffs, N.J., 1963.

[3] E.R. ASSMUS JR. and H.F. MATTSON JR., On automorphism groups of Paley-Hadamard matrices. *Combinatorial mathematics and its applications,* R.C. Bose and T.A. Dowling (ed.), (Univ. of North Carolina Press, Chapel Hill, 1969), 98-103.

[4] L.D. BAUMERT, Six inequivalent Hadamard matrices of order $2^n$, $n \geq 5$. J.P.L. Research Summary No. 36-12, 1 (1962), 74-76.

[5] L.D. BAUMERT, Hadamard matrices of the Williamson type. *Math. of Comp.* 19 (1965), 442-447.

[6] L.D. BAUMERT, Hadamard matrices of orders 116 and 232. *Bull. Amer. Math. Soc.* 72 (1966), 237.

[7] L.D. BAUMERT, *Cyclic Difference Sets.* Springer-Verlag, Berlin-Heidelberg-New York, 1971.

[8] L.D. BAUMERT, S.W. GOLOMB and MARSHALL HALL JR., Discovery of an Hadamard matrix of order 92. *Bull. Amer. Math. Soc.* 68 (1962), 237-238.

[9] L.D. BAUMERT and MARSHALL HALL JR., A new construction for Hadamard matrices. *Bull. Amer. Math. Soc.* 71 (1965), 169-170.

[10] L.D. BAUMERT and MARSHALL HALL JR., Hadamard matrices of Williamson type. *Math. of Comp.* 19 (1965), 442-447.

[11] BHAGAWANDAS and S.S. SHRIKHANDE, Seidel-equivalence of strongly regular graphs. *Sankhyā Ser. A* 30 (1968), 359-368.

[12] M. BHASKAR RAO, On two-parameter family of BIB designs. *Sankhyā Ser. A* 32 (1970), 259-264.

[13] M. BHASKAR RAO, Balanced orthogonal designs and their application in the construction of some BIB and group divisible designs. *Sankhyā Ser. A* 32 (1970), 439-448.

[14] V. BELEVITCH, Conference networks and Hadamard matrices. *Ann. Soc. Scientifique Brux.* T. 82 (1968), 13-32.

[15] V. BELEVITCH, Theory of 2n-terminal networks with applications to conference telephony. *Electr. Commun.* 27, 3 (1950), 231-244.

[16] E.R. BERLEKAMP, *Algebraic Coding Theory.* McGraw-Hill, New York, 1968.

[17] E.R. BERLEKAMP and F.K. HWANG, Constructions for balanced Howell rotations for bridge tournaments. *J. Combinatorial Th.* 12(1972),159-166.

[18] E.R. BERLEKAMP, J.H. VAN LINT and J.J. SEIDEL, A strongly regular graph derived from the perfect ternary Golay code. (to appear)

[19] VASANTI N. BHAT and S.S. SHRIKHANDE, Non-isomorphic solutions of some balanced incomplete block designs, I. *J. Combinatorial Theory* 9 (1970), 174-191.

[20] D. BLATT and G. SZEKERES, A skew Hadamard matrix of order 52. *Can. J. Math.* 22 (1970), 1319-1322.

[21] R.C. BOSE, On the construction of balanced incomplete block designs. *Ann. of Eugenics* 9 (1939), 353-399.

[22] R.C. BOSE and S.S. SHRIKHANDE, A note on a result in the theory of code construction. *Information and Control* 2 (1959), 183-194.

[23] R.C. BOSE and S.S. SHRIKHANDE, Graphs in which each pair of vertices is adjacent to the same number d of other vertices. *Studia Sci. Math. Hungar.* 5 (1970), 181-195.

[24] R.C. BOSE and S.S. SHRIKHANDE, Some further constructions for $G_2(d)$ graphs. (to appear)

[25] A. BRAUER, On a new class of Hadamard determinants. *Math. Z.* 58 (1953), 219-225.

[26] W.G. BRIDGES and H.J. RYSER, Combinatorial designs and related systems. *J. of Algebra* 13 (1969), 432-446.

[27] K.A. BUSH, Unbalanced Hadamard matrices and finite projective planes of even order. *J. Combinatorial Theory* 11 (1971), 38-44.

[28] K.A. BUSH, An inner orthogonality of Hadamard matrices. *J. Austral. Math. Soc.* 12 (1971), 242-248.

[29] F.C. BUSSEMAKER and J.J. SEIDEL, Symmetric Hadamard matrices of order 36. Dept. Mathematics, Technological University, Eindhoven, T.H. Report 70-WSK-02, 1970.

[30] F.C. BUSSEMAKER and J.J. SEIDEL, Symmetric Hadamard matrices of order 36. Ann. N.Y. Acad. Sci. 175 (1970).

[31] A.T. BUTSON, Generalized Hadamard matrices. *Proc. Amer. Math. Soc.* 13 (1962), 894-898.

[32] A.T. BUTSON, Relations among generalized Hadamard matrices, relative difference sets and maximal length recurring sequences. *Canad. J. Math.* 15 (1963), 42-48.

[33] P. CAMION, Antisymmetric Hadamard difference sets. Univ. of North Carolina, Institute of Statistics mimeo series, Chapel Hill, N.C., 1970.

[34] E.C. DADE and K. GOLDBERG, The construction of Hadamard matrices. *Michigan Math. J.* 6 (1959), 247-250.

[35] P. DELSARTE, Orthogonal matrices over a group and related tactical configurations. Scientific Report R90, MBLE Research Laboratory, Brussels, 1968.

[36] P. DELSARTE, J.M. GOETHALS and J.J. SEIDEL, Orthogonal matrices with zero diagonal. II. *Canad. J. Math.* 23 (1971), 816-832.

[37] P. DELSARTE and J.M. GOETHALS, Tri-weight codes and generalized Hadamard matrices. *Information and Control*, 15 (1969), 196-206.

[38] P. DELSARTE and J.M. GOETHALS, On quadratic residue like sequences in Abelian groups. Report R168, MBLE Research Laboratory, Brussels, 1971.

[39] P. DEMBOWSKI, *Finite Geometries*. Springer-Verlag, New York, (1968).

[40] H. EHLICH, Neue Hadamard-matrizen. *Arch. Math.* 16 (1965), 34-36.

[41] J.E.H. ELLIOTT and A.T. BUTSON, Relative difference sets. *Illinois J. Math.* 10 (1966), 517-531.

[42] R.A. FISHER and F. YATES, *Statistical Tables for Biological, Agricultural and Medical Research.* 2nd ed. London, Oliver & Boyd Ltd., 1943.

[43] JON FOLKMAN, A nonexistence theorem for Hadamard matrices in many variables. Rand Document, D-16326-PR (1967).

[44] J.M. GOETHALS and J.J. SEIDEL, Orthogonal matrices with zero diagonal. *Canad. J. Math.* 19 (1967), 1001-1010.

[45] J.M. GOETHALS and J.J. SEIDEL, A skew-Hadamard matrix of order 36. *J. Aust. Math. Soc.* 11 (1970), 343-344.

[46] J.M. GOETHALS and J.J. SEIDEL, Strongly regular graphs derived from combinatorial designs. *Canad. J. Math.* 22 (1970), 597-614.

[47] J.M. GOETHALS and J.J. SEIDEL, Quasisymmetric block designs. *Combinatorial Structures and their Applications. Proceedings Calgary International Conference,* Gordon and Breach, New York, 1970.

[48] K. GOLDBERG, Hadamard matrices of order cube plus one. *Proc. Amer. Math. Soc.* 17 (1966), 744-746.

[49] S.W. GOLOMB, et al. *Digital communications with space applications.* Prentice-Hall, Englewood Cliffs, New Jersey, 1964.

[50] HARRIET GRIFFIN, *Elementary Theory of Numbers.* McGraw-Hill, New York, 1954.

[51] MARSHALL HALL JR., A survey of difference sets. *Proc. Amer. Math. Soc.* 7 (1956), 975-986.

[52] MARSHALL HALL JR., Hadamard matrices of order 16. J.P.L. Research Summary No. 36-10, 1 (1961), 21-26.

[53] MARSHALL HALL JR., Note on the Mathieu group $M_{12}$. *Arch. Math.* 13 (1962), 334-340.

[54] MARSHALL HALL JR., Hadamard matrices of order 20. *J.P.L. Technical Report* No. 32-761.

[55] MARSHALL HALL JR., On Hadamard matrices. SIAM Meeting, Santa Barbara, 1967.

[56] MARSHALL HALL JR., *Combinatorial Theory.* Blaisdell, Waltham, Mass., 1967.

[57] MARSHALL HALL JR., Automorphisms of Hadamard matrices. *SIAM J. Appl. Math.* 7 (1969), 1094-1101.

[58] MARSHALL HALL JR. and H.J. RYSER, Cyclic incidence matrices. *Canad. J. Math.* 3 (1951), 495-502.

[59] MARSHALL HALL JR. and H.J. RYSER, Normal completions of incidence matrices. *Amer. J. Math.* 76 (1954), 581-589.

[60] G.H. HARDY and E.M. WRIGHT, *An introduction to the Theory of Numbers.* Oxford, Clarendon Press, 1938.

[61] HENNING F. HARMUTH, *Transmission of Information by Orthogonal Functions.* 2nd printing corrected, Springer-Verlag, New York-Heidelberg-Berlin, 1970.

[62] Q.M. HUSAIN, On the totality of the solutions for the symmetrical incomplete block designs: $\lambda = 2$, $k = 5$ or 6. *Sankhyā* 7 (1945), 204-208.

[63] ADOLF HURWITZ, Uber die Komposition der quadratischen, Formen von beliebig vielen Variabeln. *Nachr. Konigl. Ges. Wiss. Göttingen, Math-phys. Klasse* (1898), 309-316; *Math. Werke,* Band 2, 565-571, Birkhäuser, Basel, 1933.

[64] A. HURWITZ, Uber die komposition der quadratischen Formen. *Math. Ann.* 88 (1923), 1-25; *Math. Werke,* Band 2, 641-666, Birkhäuser, Basel, 1933.

[65] E.C. JOHNSEN, The inverse multiplier for abelian group difference sets. *Can. J. Math.* 16 (1964), 787-796.

[66] E.C. JOHNSEN, Integral solutions to the incidence equation for finite projective plane cases of orders $n \equiv 2 \pmod 4$. *Pacific J. of Math.* 17 (1) (1966), 97-120.

[67] E.C. JOHNSEN, Skew-Hadamard abelian group difference sets. *J. Algebra* 4 (1966), 388-402.

[68] W.M. KANTOR, Automorphism groups of Hadamard matrices. *J. Combinatorial Theory* 6 (1969), 279-281.

[69] W.M. KANTOR, 2-Transitive symmetric designs. *Trans. Amer. Math. Soc.* 146 (1970), 1-28.

[70] J.B. KELLY, A characteristic property of quadratic residues. *Proc. Amer. Math. Soc.* 5 (1954), 38-46.

[71] MARION E. KIMBERLEY, On the construction of certain Hadamard designs. *Math. Z.* 119 (1971), 41-59.

[72] MARION E. KIMBERLEY, On collineations of Hadamard designs. (to appear)

[73] D.H. LEHMER, *List of prime numbers from 1 to 10,006,721.* Hafner, New York, 1956.

[74] E. LEHMER, On the number of solutions of $u^k + D = w^2 \pmod p$. *Pacific J. Math.* 5 (1955), 103-118.

[75] V.I. LEVENSHTEIN, The application of Hadamard matrices to a problem in coding. *Problems of Cybernetics* 5 (1964), 166-184.

[76] C.C. MACDUFFEE, *The Theory of Matrices.* Reprint of First Ed., Chelsea, N.Y., 1964.

[77] F.J. MACWILLIAMS, Orthogonal circulant matrices over finite fields and how to find them. *J. Combinatorial Th. Ser. A* 10 (1971), 1-18.

[78] K.N. MAJINDAR, On integer matrices and incidence matrices of certain combinatorial configurations, I: square matrices. *Can. J. Math.* 18 (1966), 1-5.

[79] H.B. MANN, *Addition theorems.* Wiley, New York, 1965.

[80] M. MARCUS and H. MINC, *A survey of Matrix Theory and Matrix Inequalities.* Allyn and Bacon, Boston, 1964.

[81] O. MARRERO, Modular Hadamard matrices and related combinatorial designs. Dissertation, University of Miami, 1970.

[82] O. MARRERO and A.T. BUTSON, Modular Hadamard matrices and related designs. (preprint) to appear in *J. Combinatorial Th.* and *Canad. J. Math.*

[83] P. KESAVA MENON, Difference sets in abelian groups. *Proc. Amer. Math. Soc.* 11 (1960), 368-376.

[84] P. KESAVA MENON, Certain Hadamard designs. *Proc. Amer. Math. Soc.* 13 (1962), 524-531.

[85] DALE M. MESNER, Negative latin square designs. Institute of Statistics, University of North Carolina, Mimeo Series No. 410, 1964.

[86] DALE M. MESNER, A new family of partially balanced incomplete block designs with some latin square design properties. *Ann. Math. Stat.* 38 (1967), 571-581.

[87] H.K. NANDI, A further note on non-isomorphic solutions of incomplete block designs. *Sankhyā Ser. A* 7 (1945-46), 313-316.

[88] H.K. NANDI, Enumeration of non-isomorphic solutions of balanced incomplete block designs. *Sankhyā Ser. A* 7 (1945-46), 305-312.

[89] MORRIS NEWMAN, Invariant factors of combinatorial matrices. *Israel J. Math.* 10 (1971), 126-130.

[90] D.A. NORTON, Hamiltonian loops. *Proc. Amer. Math. Soc.* 3 (1952), 56-65.

[91] R.E.A.C. PALEY, On orthogonal matrices. *J. Math. Phys.* 12 (1933), 311-320.

[92] R.L. PLACKETT and J.P. BURMAN, Designs of optimum multifactorial experiments. *Biometrika* 33 (1946), 305-325.

[93] V.S. PLESS, On a new family of symmetry codes and related new five designs. *Bull. Amer. Math. Soc.* 75 (1969), 1339-1342.

[94] V.S. PLESS, Symmetry codes over GF(3) and new five designs. *J. Combinatorial Th. Ser. A* 12 (1972), 119-142.

[95] E.C. POSNER, Combinatorial structure in planetary reconnaissance. Proc. Symposium on Error Correcting Codes, Mathematics Research Center, University of Wisconsin Press, Madison, 1968.

[96] D. RAGHAVARAO, Some optimum weighing designs. *Ann. Math. Stat.* 30 (1959), 295-303.

[97] D. RAGHAVARAO, Some aspects of weighing designs. *Ann. Math. Stat.* 31 (1960), 878-884.

[98] D. RAGHAVARAO, *Constructions and Combinatorial Problems in Design of Experiments.* Wiley Series in Probability and Mathematical Statistics, Wiley, New York-Sydney, 1971.

[99] C.R. RAO, A study of BIB designs with replications 11 to 15. *Sankhyā Ser. A* 23 (1961), 117-127.

[100] K.B. REID and E. BROWN, Doubly regular tournaments are equivalent to skew Hadamard matrices. (to appear)

[101] H.J. RYSER, A note on a combinatorial problem. *Proc. Amer. Math. Soc.* 1 (1950), 422-424.

[102] H.J. RYSER, Matrices with integer elements in combinatorial investigations. *Amer. J. Math.* 74 (1952), 769-773.

[103] H.J. RYSER, *Combinatorial Mathematics.* (Carus Monograph No. 14), Wiley, New York, 1963.

[104] H.J. RYSER, An extension of a theorem of de Bruijn and Erdös on combinatorial designs. *J. of Algebra* 10 (1968), 246-261.

[105] V. SCARPIS, Sui determinanti di valore massimo. *Rend. R. Inst. Lombardo Sci. e Lett.*, (2) 31 (1898), 1441-1446.

[106] J.J. SEIDEL, Strongly regular graphs with (-1,1,0) adjacency matrix having eigenvalue 3. *Linear Algebra and Appl.* 1 (1968), 281-298.

[107] S.S. SHRIKHANDE, On the dual of some balanced incomplete block designs. *Biometrics* 8 (1952), 66-72.

[108] S.S. SHRIKHANDE, On a two parameter family of balanced incomplete block designs. *Sankhyā, Ser. A* 24 (1962), 33-40.

[109] S.S. SHRIKHANDE, Generalized Hadamard matrices and orthogonal arrays of strength two. *Can. J. Math.* 16 (1964), 736-740.

[110] S.S. SHRIKHANDE and BHAGWANDAS, Duals of incomplete block designs. *Journal Indian Statist. Assoc.* 3 (1965), 30-37.

[111] S.S. SHRIKHANDE, D. RAGHAVARAO and S.K. THARTHARE, Non-existence of some partially balanced incomplete block designs. *Can. J. Math.* 15 (1963), 686-701.

[112] S.S. SHRIKHANDE and N.K. SINGH, On a method of constructing symmetrical balanced incomplete block designs. *Sankhyā Ser. A* 24 (1962), 25-32.

[113] N.J.A. SLOAN, A survey of recent results in constructive coding theory. *NTC '71 RECORD* (1971), 218-226.

[114] H.J.S. SMITH, *Proc. London Math. Soc.* 4 (1873), 236-253.

[115] E. SPENCE, A new class of Hadamard matrices. *Glasgow J.* 8 (1967), 59-62.

[116] EDWARD SPENCE, Some new symmetric block designs. *J. Combinatorial Th.* 11 (1971), 299-302.

[117] EDWARD SPENCE, A note on equivalence of Hadamard matrices, Abstract 71T-A99. *Notices Amer. Math. Soc.* 18 (1971) 624 (to appear).

[118] EDWARD SPENCE, Hadamard designs. (to appear)

[119] JOEL SPENCER, Hadamard matrices in many variables. Rand Document D-15925-PR(1967).

[120] D.A. SPROTT, A list of BIB designs with r = 16 to 20. *Sankhyā Ser. A* 24 (1962), 203-204.

[121] R.G. STANTON and J.G. KALBFLEISCH, Quasi-symmetric balanced incomplete block designs. *J. Combinatorial Th.* 4 (1968), 391-396.

[122] R.G. STANTON and D.A. SPROTT, A family of difference sets. *Can. J. Math.* 10 (1958), 73-77.

[123] J.J. STIFFLER and L.D. BAUMERT, Inequivalent Hadamard matrices. J.P.L. Research Summary No. 36-9, 1 (1961), 28-30.

[124] J. STORER and R. TURYN, On binary sequences. *Proc. Amer. Math. Soc.* 12 (1961), 394-399.

[125] T. STORER, *Cyclotomy and difference sets.* Markham Publishing Company, Chicago, 1967.

[126] T. STORER, Hurwitz on Hadamard designs. *Bull. Austral. Math. Soc.* 4 (1971), 109-112.

[127] J.J. SYLVESTER, Thoughts on inverse orthogonal matrices, simultaneous sign successions, and tesselated pavements in two or more colours, with applications to Newton's Rule, ornamental tile-work, and the theory of numbers. *Phil. Mag.* (4) 34, (1867), 461-475.

[128] E. and G. SZEKERES, On a problem of Schütte and Erdös. *Math. Gazette* 49 (1965), 290-293.

[129] G. SZEKERES, Tournaments and Hadamard matrices. *Enseignment Math.* 15 (1969), 269-278.

[130] G. SZEKERES, Cyclotomy and complementary difference sets. *Acta Arithmetica* 18 (1971), 349-353.

[131] K. TAKEUCHI, A table of difference sets generating balanced in complete block designs. *Review of Int. Stat. Inst.* 30 (1962), 361-366.

[132] K. TAKEUCHI, On the construction of a series of BIB designs. *Rep. Stat. Appl. Res. JUSE.* 10 (3) (1963), 48.

[133] OLGA TAUSSKY, An algebraic property of Laplace's differential equation. *Quart. J. Math. Oxford* 10 (1939), 99-103.

[134] OLGA TAUSSKY, Sums of squares. *Amer. Math. Monthly* 77 (1970), 805-830.

[135] OLGA TAUSSKY, (1,2,4,8)-sums of squares and Hadamard matrices. *Proc. Symp. in Pure Mathematics, Combinatorics,* Amer. Math. Soc. (1971), 229-234.

[136] R. TURYN, Optimum codes study. Final Report, Sylvania Electric Products Inc., 1960.

[137] R. TURYN, Finite binary sequences. Final Report, (Chapter VI), Sylvania Electric Products Inc., 1961.

[138] R.J. TURYN, Character sums and difference sets. *Pacific J. of Math.* 15 (1) (1965), 319-346.

[139] R.J. TURYN, The correlation function of a sequence of roots of 1. *IEEE Trans. on Information Theory* IT-13 (1967), 524-525.

[140] R.J. TURYN, Sequences with small correlation, in *Error Correcting Codes*, H.B. Mann (ed.), Wiley, New York, 195-228.

[141] R.J. TURYN, Complex Hadamard matrices. *Combinatorial Structures and their Applications*, Gordon and Breach, New York, 1970, 435-437.

[142] RICHARD J. TURYN, On C-matrices of arbitrary powers. *Can. J. Math.* 23 (1971), 531-535.

[143] RICHARD J. TURYN, Hadamard matrices, algebras, and composition theorems. *Notices Amer. Math. Soc.* 19 (1972), pA-388.

[144] R.J. TURYN, An infinite class of Williamson matrices. (to appear)

[145] S. VAJDA, *The Mathematics of Experimental Design, Incomplete Block Designs and Latin Squares*. No. 23, Griffins Statistical Monographs and Courses, Griffin & Co., London, 1967.

[146] J.H. VAN LINT and J.J. SEIDEL, Equilateral point sets in elliptic geometry. *Kon. Ned. Akad. Wetensch. Amst., Proceeding, Series A* 69 and *Indag Math.* 28 (1966), 335-348.

[147] I.M. VINOGRADOV, *Elements of Number Theory*. Reprinted by Dover, New York, 1954.

[148] JENNIFER WALLIS, *Hadamard Matrices*. M.Sc. Thesis, La Trobe University, 1969.

[149] JENNIFER WALLIS, A class of Hadamard matrices. *J. Combinatorial Th.* 6 (1969), 40-44.

[150] JENNIFER WALLIS, A note on a class of Hadamard matrices. *J. Combinatorial Th.* 6 (1969), 222-223.

[151] JENNIFER WALLIS, (v,k,λ)-configurations and Hadamard matrices. *J. Austral. Math. Soc.* 11 (1970), 297-309.

[152] JENNIFER WALLIS, Hadamard designs. *Bull. Aust. Math. Soc.* 2 (1970), 45-54.

[153] JENNIFER WALLIS, *Combinatorial Matrices.* Ph.D. Thesis, La Trobe University, 1971.

[154] JENNIFER WALLIS, Some (1,-1) matrices. *J. Combinatorial Th. Ser. B* 10 (1971), 1-11.

[155] JENNIFER WALLIS, Amicable Hadamard matrices. *J. Combinatorial Th. Ser. A* 11 (1971), 296-298.

[156] JENNIFER WALLIS, On integer matrices obeying certain matrix equations. *J. Combinatorial Th. Ser. A* 12 (1972), 112-118.

[157] JENNIFER WALLIS, Hadamard matrices of order 28m, 36m and 44m. *J. Combinatorial Th.*

[158] JENNIFER WALLIS, On supplementary difference sets. *Aeq. Math.*

[159] JENNIFER WALLIS, Complex Hadamard matrices. Department of Mathematics, University of Newcastle, Research Report No. 63, 1972.

[160] JENNIFER WALLIS and ALBERT LEON WHITEMAN, Some classes of Hadamard matrices with constant diagonal. Department of Mathematics, University of Newcastle, Research Report No. 69, 1972, and *Bull. Austral. Math. Soc.*

[161] W.D. WALLIS, Certain graphs arising from Hadamard matrices. *Bull. Aust. Math. Soc.* 1 (1969), 325-331.

[162] W.D. WALLIS, A non-existence theorem for $(v,k,\lambda)$-graphs. *J. Austral. Math. Soc.* 11 (1970), 381-383.

[163] W.D. WALLIS, Some non-isomorphic graphs. *J. Combinatorial Th.* 8 (1970), 448-449.

[164] W.D. WALLIS, A note on quasi-symmetric designs. *J. Combinatorial Th.* 8 (1970), 100-101.

[165] W.D. WALLIS, On extreme tournaments. *Ann. N.Y. Acad. Sci.* 175 (1970), 403-404.

[166] W.D. WALLIS, Some notes on integral equivalence of combinatorial matrices. *Israel J. Math.* 10 (1971), 457-464.

[167] W.D. WALLIS, Integral equivalence of Hadamard matrices. *Israel J. Math.* 10 (1971), 359-368.

[168] W.D. WALLIS, Hadamard matrices of order thirty-two. Research Report No. 49, Department of Mathematics, University of Newcastle, 1971.

[169] W.D. WALLIS, On a problem of K.A. Bush concerning Hadamard matrices. *Bull. Austral. Math. Soc.* 6 (1972), 321-326.

[170] W.D. WALLIS and JENNIFER WALLIS, Equivalence of Hadamard matrices. *Israel J. of Math.* 7 (1969), 122-128.

[171] A.L. WHITEMAN, A family of difference sets. *Illinois J. Math.* 6 (1962), 107-121.

[172] ALBERT LEON WHITEMAN, An infinite family of skew-Hadamard matrices. *Pacific J. Math.* 38 (1971), 817-822.

[173] ALBERT LEON WHITEMAN, An infinite family of Hadamard matrices of Williamson type. *J. Combinatorial Th.*

[174] ALBERT LEON WHITEMAN, Skew-Hadamard matrices of Goethals-Seidel type. (to appear)

[175] JOHN WILLIAMSON, Hadamard's determinant theorem and the sum of four squares. *Duke Math. J.* 11 (1944), 65-81.

[176] JOHN WILLIAMSON, Note on Hadamard's determinant theorem. *Bull. Amer. Math. Soc.* 53 (1947), 608-613.

[177] C.H. YANG, On designs of maximal (+1,-1)-matrices of order $n \equiv 2 \pmod 4$. *Math. of Computation* 22 (1968), 174-180.

[178] C.H. YANG, On designs of maximal (+1,-1)-matrices of order $n \equiv 2 \pmod 4$. II. *Math. of Computation* 23 (1969), 201-205.

[179] C.H. YANG, On Hadamard matrices constructible by circulant submatrices. *Math. of Computation* 25 (1971), 181-186.

[180] ALAN ZAME, Orthogonal sets of vectors over $Z_m$. *J. Combinatorial Th.* 9 (1970), 136-143.

[181] N. ZIERLER, Linear recurring sequences. *J. SIAM* 7 (1959), 31-48.

[182] H.O. LANCASTER, Pairwise Statistical Independence. *Ann. Math. Stat.* 36 (1965), 1313-1317.

[183] JOAN COOPER and JENNIFER WALLIS, A construction for Hadamard arrays. *Bull. Austral. Math. Soc.* (to appear)

# PART 5

## AFTERMATH

# CONTENTS

FINAL REMARKS 495

NOTATION 497

INDEX 499

# FINAL REMARKS

In the introduction we stated that the subject of this book is alive and research is currently proceeding. It is not surprising then that while the book was being typed, new results were found. We mention a few that have been brought to our attention, but doubtless other results have also been discovered.

1. H.P. Yap has written to say that he has part (but not all) of theorem 7.9 and that it is to appear in *J. Number Theory*; also he has found the maximal sum-free sets in $Z_{p^2} \oplus Z_p$ and it is to appear in *Nanta Math.*

2. P. Erdös and S. Shelah (1972) (to appear): Erdös wrote to point out they have additional results on the question of union-free sets. The paper is called, "On problems of Moser and Hanson".

3. Denis Hanson also pointed out further applications of the Erdös-Komlós method.

4. Dr. Richard J. Turyn [143] has announced an infinite class of Baumert-Hall arrays.

5. Professor A.L. Whiteman has written that he has a new infinite class of Hadamard matrices.

6. We understand that skew-Room squares of sides 15 and 21 have recently been discovered.

## NOTATION

The dimensions of all matrices, where not specifically stated, will be assumed to be compatible under binary operations and should be determined from the context. Groups are usually written additively.

The following symbols are used.

| | |
|---|---|
| $GF(p^n)$, $GF[p^n]$ | the Galois Field with $p^n$ elements |
| $I$ | the identity matrix |
| $J$ | the square matrix with every element one |
| $e$ | the 1 by n matrix of ones |
| $\times$ | the Kronecker product |
| $\Sigma' A\times B\times C$ | the sum obtained by circulating the letters formally. Thus $\Sigma' A\times B\times C = A\times B\times C + B\times C\times A + C\times A\times B$ |
| det $X$, $\|X\|$ | determinant of X, X a matrix |
| $\emptyset$ | the empty set |
| $S+T$ | $\{s+t \mid s \in S,\ t \in T\}$ |
| $S-T$ | $\{s-t \mid s \in S,\ t \in T\}$ |
| $S \setminus T$ | $\{ s \mid s \in S,\ s \notin T\}$ |
| $mS$ | $\{ms \mid s \in S\}$ |
| $\overline{S}$ | $\{g \in G,\ g \notin S\} = G \setminus S$ |
| k-set | set with k elements |
| $\#(S)$, $\|S\|$ | the number of elements in a set S |
| $\{x\colon Px\}$ | set of elements such that Px is true |
| $\#(x\colon Px)$ | the number of elements such that Px is true |
| $[x_1, x_2, \ldots, x_n]$ | a collection of elements |
| $[x\colon Px]$ | collection of elements such that Px is true |
| & | the join of two collections, e.g. [1,3,3,4] & [3,4] = [1,3,3,3,4,4] |
| $a \mid b$ | a divides b |
| $(x,y)$ | the greatest common divisor of x and y (but in context the open interval from x to y) |
| $[x]$ | the integer part of x |
| $\{x\}$ | the greatest integer less than x (= x-1 if x integral) |
| $\binom{n}{r}$ | the binomial coefficient |

| | |
|---|---|
| $\|\|x\|\|$ | distance of x from the nearest integer |
| $\{x,y\}$ | (in context) an edge from vertex a to vertex b |
| $K_n$ | complete graph on n vertices |
| $Z_n$ | cyclic group of order n = integers mod n |
| $R_n(k,2)$, $R(k_1,k_2,\ldots,k_n,r)$ | Ramsey numbers |
| $\phi(n)$ | the Euler function |
| $\langle x\rangle$ {or $\langle x_1,\ldots,x_n\rangle$} | meaning the group generated by x {or $x_1,\ldots,x_n$} |
| $\lambda(G)$ | cardinality of a maximal sum-free set in G |

# INDEX

A

Abbott, H.L., 137, 157, 167, 265
abelian, 24
abelian group
- sum-free sets in, see sum-free sets
- sum-free partitions of, see sum-free partitions
- transform of subsets of, see transform
adder, 42-49, 79-81, 112-113
- skew, 43-44, 48-49, 80-81
addition theorems in finite groups, 127, 186-204
adjacency, 21
admissible set
- of integers, 168
- bounds on cardinality of, 168-170, 178-185
- of integers modulo a prime, 170-171
Ahrens, R., 429, 478
Albert, A.A., 26, 478
amicable Hadamard matrix, see Hadamard matrix
antigroupoid, 265
aperiodic set, 192, 201
Archbold, J.W., 76, 78, 116, 118
arithmetic progression
- of integers in one block of partition, 141-147
- in abelian group, 187, 191, 202-204 (see also standard set)
Assmus, E.R. Jr., 478
automorphism
- of finite geometry, 76
- of one-factorization, 92, 94
- of Room square, 37-38, 97
avoidance, 212

B

back-circulant (see also incidence matrix), 14, 15
balanced incomplete block design, see BIBD
balanced orthogonal design, 433, 443
Barker, R.H., 434
Barker sequence, 434, 435
Baumert-Hall-(Welch) array, see Hadamard array
Baumert, Leonard D., 131, 137, 280, 354, 364, 386, 394, 408, 423, 448, 478, 486
BCH code, 440
Belevitch, V., 294, 306, 309, 316, 426, 437, 479
Berlekamp, E.R., 118, 157, 429, 479
Bhagawandas, 429, 478, 485
Bhaskar Rao, M., 431, 444, 478
Bhat, Vasanti N., 420, 479
BIBD, 12, 110-111, 342, 344, 444, 446
Blatt, David, 321, 323, 479
Bose, R.C., 429, 441, 442, 479
Brauer, A., 448, 479
Bridges, W.G., 479
bridge tournament, 109-110
Brown, E., 150, 484

B (cont.)

Bruck, R.H., 13, 26, 115, 116, 118
Bruck-Ryser-Chowla Theorem, 13
Burman, J.P., 484
Bush, K.A., 341, 430, 479
Bussemaker, F.C., 423, 429, 480
Butson, A.T., 401, 402, 404, 433, 435, 480, 483
Byleen, K., 81, 118

C

Camion, P., 480
Cauchy, A.L., 186, 204
Cayley loop, 363
Chandrasekharan, K., 185
character, 291, 331
Choi, S.L.G., 185
Chowla, 13
circulant matrix, see incidence matrix
C-matrix, see conference matrix and/or skew-Hadamard matrix
code, 440-442
Collens, R.J., 115, 118
colouring of graph, 247-248
commutative, 24
complement
- of block design, 12
- of graph, 21
- of Room square, 35-36, 60-62, 71-72, 74-75, 112
complete graph, 22-23, 127, 247-263
complex Hadamard matrix, see Hadamard matrix
conference matrix
- symmetric, 293, 295, 312, 323, 316, 318, 319, 349-353, 365, 368, 372-374, 444
- core of, 293, 312, 351, 375-378
configuration, (see also BIBD), 12
Cooper, Joan, 358, 359, 448
core, 292, 296, 309, 311, 312
covering, 212
critical pair, 187-191
cyclic code, 440
cyclic difference set, see difference set
cyclotomy, 10, 11, 322

D

Dade, E.C., 430, 480
Davenport, H., 204
Delsarte, P., 306, 317, 318, 404-406, 429, 435-437, 480
Dembowski, P., 480
design graphs, 429
Diananda, P.H., 204, 246, 266
Dickson, L.E., 131, 137
difference set
- cyclic, 19-20, 280, 374
- supplementary, 281, 283, 289-292, 298, 302, 334, 336, 378, 381, 392, 393, 395, 397, 398, 445
- group, 19-20, 369, 370, 372, 373, 376
- perfect, 19
distance (of code), 440
distribution of primes (see also large sieve inequality), 171-175

D (cont.)

division, see one-factorization
Doyle, P.H., 265

E

edge, 21
edge-colouring (see also colouring), 23
edge-disjoint, 22
Ehlich, H., 306, 309, 448, 480
Elliot, J.E.H., 480
equivalence
- Hadamard, H-equivalence, 16, 408-425
- integral, Z-equivalence, 408-425
- permutation, 313, 314
- Room squares, 35, 97-99, 106-108
- Seidel, S-equivalence, 313, 315, 408, 409, 422
Erdös, P., 130, 167, 185, 205, 246, 265
exponent, 24
exponential sums, 175-178

F

Federer, W.T., 26
Fermat number, 10, 85-87
Fermat's last theorem, 131, 137
finite projective geometry, 76-77
Fisher, R.A., 480
Folkman, Jon, 264, 361, 480
frame construction of Room squares, 66-70
Frasnay, C., 137, 167

G

Gallagher, P.X., 185
Galois field, 7
Gerwitz, A., 431
$G_2(\lambda)$ graph, 429
Giraud, G.R., 167
Gleason, A.M., 137, 167, 264
Goethals, J.M., 306, 309, 310, 313-319, 329, 330, 341, 342, 344, 355, 374, 392, 404-406, 429, 431, 436, 437, 448, 449, 453, 454
Goethals-Seidel matrix, type, array, 329, 333, 355, 356, 391-395
Golay code, 440
Goldberg, K., 310, 430, 448, 480, 481
Golomb, Solomon W., 137, 448, 478, 481
graph, 21
- complete, see complete graph
- colouring of, see colouring of graph
Greenwood, R.E., 137, 167, 264
Griffin, Harriet, 118, 294, 481
Groner, Alex, 118
group, 7, 20
- difference see (see also difference set), 19, 20
- sum-free partitions in, see sum-free partitions
- sum-free sets in, see sum-free sets
groupoid, 24
- sum-free sets in, see sum-free sets
G-string, 310-311

H

Hadamard array, 354-366, 444
- Baumert-Hall, 354, 364, 365, 394
- Baumert-Hall-Welch, 354, 356
- Williamson, 354
Hadamard group, 430
Hadamard matrix, 15, 367-398, 448-450
- amicable, 296, 300, 320, 337, 338, 348, 367, 371, 457, 470
- circulant, 435
- complex, 295, 297, 347-355, 368, 372, 444, 455, 456
- complex skew-Hadamard, 295, 348, 365
- generalized, 401-407
- modular, 433
- quaternion type, 364, 372, 382
- regular, 18
- regular symmetric, 280, 341-346
- skew-Hadamard, 292, 312, 318, 320-338, 355, 368-372, 378-381, 444, 445, 451, 453-454
- special, 296, 349, 350
- symmetric, 337, 339-346, 371, 376, 423, 444, 452
- Williamson type, 364, 381-391
Hadamard product, 307
Halberstam, H., 185
Hall, Marshall Jr., 5, 12, 14, 23, 26, 118, 130, 280, 354, 364, 386, 394, 419-421, 427, 431, 432, 448, 449, 478, 481
Hall polynomials, 399
Hanson, D., 167, 265-266
Harary, Frank, 23, 118
Hardy, G.H., 482
Harmuth, Henning F., 437, 482
Hedayat, A., 26
Higman, G., 431
homogeneous system of conditions, 139, 147, 149
Horton, J.D., 56, 115, 117-120
Howell rotation, 109-110, 429
Hunt, D.C., 325, 326, 451, 471
Hurwitz, Adolf, 364, 482
Hussain, Q.M., 418
Hwang, F.K., 118, 429, 479

I

idempotent quasigroup, 24
incidence matrix, 13, 20
- back-circulant, 284, 289, 325, 370, 392
- circulant, 14, 15, 284, 289, 315, 325, 329, 370, 392
- complex circulant, 296
- difference set, 20, 284
- multi-circulant, 314
- SBIBD, 279
- type 1, 284, 285, 287, 288, 289, 291, 303, 321, 325, 334, 352
- type 2, 284, 285, 287, 288, 289, 291, 303, 321, 325, 334, 353
intersection of graphs, 22
invariants of
- (0,1) matrix, 411
- Hadamard matrix, 414
- standard forms, 411-418
Iseki, K., 265
isomorphism
- of graphs, 23
- of Room squares, 35, 37-38, 97-99, 106-108, 112

J

Jacobthal's formula, 314
Johnsen, E.C., 426-428, 436, 482
Johnson, N.L., 76, 118

K

$(k,A_i)$-subset, 26
Kalbfleisch, J.G., 264, 429, 485
Kantor, W.M., 482
Kelly, J.B., 482
Kemperman, J.H.B., 204
Khinchin, A.Y., 157
Kim, J.B., 265
Kimberley, Marion E., 427, 482
Klarner, D.A., 185
Kleitman, D., 265
Kneser, M., 204
Komlós, J., 265
k-regular, 139, 147, 149, 150
Kronecker product, 18

L

Lancaster, H.O., 442, 483
large sieve inequality, 175-178
Latin squares, 25
- join, 66
- number of pairwise orthogonal symmetric, 100-105
- orthogonal, 25, 56, 61-62, 71-75
- orthogonal symmetric, 40-41, 100-105
Latin square graph, negative Latin square graph, pseudo-Latin square graph, 340, 341
Lawless, J.F., 116, 119
Legendre equation, 13
Legendre symbol, 9, 331
Lehmer, D.H., 482
Lehmer, E., 482
Levenshtein, V.I., 442, 483
Lindner, Charles C., 115, 119
line, 21
linear code, (n,k) linear code, 440
linear equations, simultaneous, see simultaneous linear equations
linear systems, regularity of, see regularity
line matrix, 313, 314
Liu, A.C., 157
loop, 24, 362, 363
Lukomskaya, M.A., 157

M

MacDuffee, C.C., 410, 483
MacWilliams, Mrs. F.J., 483
Majindar, K.N., 483
Mann, H.B., 204, 483
Marcus, M., 410, 411, 483
Marrero, O., 433, 483
Mathieu group, 431
Mattson, H.F. Jr., 478
maximal sum-free set, see sum-free set, maximal, 27

M (cont.)

McWorter, W.A., 204
Mendelsohn, N.S., 119
Menon, P. Kesava, 483
Mesner, Dale M., 483
Minc, H., 410, 411, 483
Mood, A.N., 120
Moore, E.H., 56, 119
Moore-type construction of Room squares, 56-60, 85-86, 88-89
- generalizations of, 60-65
Moser, L., 137, 157, 167, 246
Mullin, A.A., 265
Mullin, R.C., 42, 48, 115, 116, 118-121
multiplier, 19
mutants, 265

N

Nandi, H.K., 420, 483
negacyclic C-matrix, 318
Nemeth, E., 42, 48, 115, 120
Newman, Morris, 410, 483
(n,M,d)-code, see non-linear code
non-linear code, 442
Norton, D.A., 363, 483
nullgraph, 22

O

(0,1,-1) G-sequence, 436
Olsen, J., 204
one-factorization, 23, 39-41, 50, 91-95, 100
- and Room square, 39-41, 94-95
- and symmetric Latin square, 101
- divisions, 40, 51, 92-93
optimal code, 441
orthogonal
- array, 430
- Latin squares, see Latin square
- one-factorizations, see one-factorization
- starters, see starter
- Steiner triple systems, see Steiner triple system
O'Shaughnessy, C.D., 115, 120

P

pairwise balanced design, 81-84
pairwise statistical independence, 441
Paley matrix, 313
Paley, R.E.A.C., 306, 313, 320, 342, 426, 448, 484
Parker, E.T., 26, 120
Parseval relation, 390
partition
- of integers, 138
- induced, 138
- sum-free, see sum-free partition
perfect difference set, 19
periodic set, 192, 196
Pillai, S.S., 185

P (cont.)

Plackett, R.L., 484
Pless, Vera S., 440, 484
Plotkin, 441
point, 21
Posner, E.C., 484
primes, distribution of, see distribution of primes
primitive root, 7
product-free sets, 265
projective plane, 430

Q

quadratic character, 9
quadratic residue, 8-9
quasigroup, 24-25, 41
quasiperiodic set, 192-195, 197-200
quasi-symmetric block design, 150

R

Rado, R., 28, 157
Raghavarao, D., 294, 432, 484, 485
Ramsey, F.P., 130, 137
Ramsey number, 27-28, 127-129, 136-137, 160-161, 167, 247, 263
Ramsey's theorem, 129-130
Rao, C.R., 484
redundancy, 440, 441
regular
- equation, 28
- graph, 21
- Hadamard matrix, 18
regularity (of system of equations), 138-157
- law of, 139-141
Reid, K.B. Jr., 429, 484
residue, 7
- quadratic, 8
Rhemtulla, A.H., 246
Room, T.G., 33, 120
Room square, 33-121
- automorphisms, 37-38, 97
- complementary, 35-36, 60-62, 71-72, 74-75, 112
- embedded, 36, 62-63, 112
- equivalence, 35, 97-99, 106-108
- existence problem, 33-35, 85-90
- Hadamard, 35-36, 111-112
- isomorphism, 35, 37-38, 97-99, 106-108, 112
- skew, 35-38, 56-60, 82-83, 85-89, 97-99, 112
- standardized, 35, 37-38, 98, 106
Room t-design, 103-105
Rota, G.C., 5
Rumsey, H.C. Jr., 399, 400
Ryser, H.J., 12, 13, 14, 120, 130, 280, 344, 427, 479, 481, 484

S

Salié, H., 167
SBIBD, 279, 280, 426
Scarpis, V., 449, 485
Schellenberg, P.J., 116, 120

S (cont.)

Schur function, 28, 127, 131-136, 139-141, 159, 164-165, 167
- generalized, 158-161, 162-164, 164-167, 265-266
Schur, I., 127, 137
Schur's problem, 128, 131
segment of positive integers, 138, 141
Seidel-equivalent, see equivalence
Seidel, J.J., 294, 306, 309, 310, 313-318, 329, 330, 341, 342, 344, 355, 374, 392, 423, 426, 429, 431, 441, 448, 449, 453, 454, 479-481, 487
semigroup, 24
- sum-free sets in, see sum-free sets
sequences with small correlation, 434-437
(S)-free partition, 158-160
(S)-free set, 128, 158-160
Shah, K.R., 112, 120
Shannon, C., 440
shift, 19
Shrikhande, S.S., 343, 346, 420, 427, 429, 430, 435, 441, 442, 478, 479
sieve, large, see large sieve inequality
Sims, C., 431
simultaneous linear equations, 148-149, 150-157
Singer, J., 76
Singh, N.K., 341, 485
skew adder, see adder
skew-Hadamard matrix, see Hadamard matrix
skew Room square, see Room square
skew subsquare, 39, 56-60, 82-83, 85
skew-type matrix, 292, 325, 330
Skolem, Th., 130
Sloan, N., 441, 485
Smith, H.J.S., 485
Smith normal form, 410
spanning subgraph, 23
special Hadamard matrix, see Hadamard matrix
Spence, E., 361, 410, 485
Spencer, J., 361, 449, 485
Sprott, D., 448, 485, 486
standard form, 411
standard set, 187-191
standard pair, 197-200
Stanton, R.G., 42, 115, 119, 120, 121, 264, 429, 448, 485, 486
starter, 42-55, 76-77, 79-81, 101, 106-108, 112-113
- adder for, 42-44, 46, 79-81
- and one-factorization, 46-47
- from Steiner triple system, 50-53
- orthogonal, 45-48, 101
- patterned, 42, 45
- strong, 42-44, 47-55, 77, 101, 106-108, 112
Steiner difference blocks, 51-53
Steiner triple system, 50-53, 77
- used for starters, 50
- orthogonal, 50, 52
Stiffler, J.J., 408, 486
Storer, J., 435, 486
Storer, T., 10, 361, 364, 486
Straus, E.G., 185
Street, A.P., 246
strong graphs, 340, 453, 454
strongly regular graphs, 340, 429
subgraph, 23
- spanning, 23

S (cont.)

subsquare, 38-39, 56, 62-65, 82-84, 112
- skew, see skew-subsquare
sum-free partition
- of set of integers, 128, 131-135
- of group, 248-249, 259, 261, 263
sum-free sets, 27-28
- in abelian groups, 265
- in groupoids, 265
- in semigroups, 265
- in topological semigroups, 265
- maximal, 27
- maximal sum-free set in groups, 128, 205-246
- avoidance of subgroup by, 212
- bounds on cardinality of, 205, 207, 211-212, 245
- covering of subgroup by, 212
- in abelian groups
- with $q \| |G|$, $q \equiv 2 \pmod 3$, 208
- with $3 \| |G|$, 208-211
- with all divisors of $|G|$ congruent to 1 (mod 3), 211-242
- G cyclic, 211, 215-219, 227-242
- G elementary abelian, 212-214, 219-227
- in non-abelian groups, 242-245
- period of, 206-207
- of integers, 127
- strongly sum-free sets (of integers), 165
sum of Room squares, 35
sum-set, 127
- order of, 187, 195
- small, 201-202
switching, 423
Sylvester, J.J., 448, 486
symmetric conference matrix, see conference matrix
Szekeres, Esther, 429, 486
Szekeres, George, 130, 306, 320, 321, 323-325, 429, 451, 478, 479, 485, 486
Szekeres difference set, 302-304, 320-322, 324, 326, 377, 379, 444, 450, 453, 454

T

Takeuchi, K., 486
Taussky, Olga, 361, 486
Taylor, D.E., 318
T-character, 404, 405
Tharthare, S.K., 485
totality of differences, 282
transform of subsets of abelian group, 186-204
transversal, 26
Turán, P., 187
Turyn, Richard J., 295, 306, 310-312, 320, 330, 347, 366, 368, 389, 391, 434-436, 448, 449, 486, 487

U

union-free partitions, 265
union of graphs, 22

V

Vajda, S., 487
valency, 21, 429
Vandermonde matrix, 402
van der Waerden, B.L., 157
Vandiver, H.S., 137
van Lint, J.H., 294, 426, 479, 487
vertex, 21
Vinogradov, I., 8, 9, 487
(v,k,λ)-configuration, see SBIBD
(v,k,λ)-graphs, 429
Vosper, A.G., 204

W

Wallis, Jennifer, 300, 306, 310, 325, 333, 336, 337, 339, 344, 355, 356, 358, 359, 361, 372, 375, 376, 378, 393, 410, 433, 448, 449, 487-489
Wallis-Whiteman type, 334, 355, 393
Wallis, W.D., 115, 116, 120, 121, 341, 408, 410-417, 422-425, 429, 430, 488
Walsh functions, 437-439
Warne, R.J., 265
weighing designs, 431-432
well-spaced numbers, 176
Weisner, L., 121
Welch, L.R., 356, 448
Whitehead, E.G. Jr., 264
Whiteman, Albert Leon, 306, 320, 321, 323, 330, 331, 333, 334, 336, 355, 389, 391, 393, 398, 448, 488
Williamson, John, 306, 320, 330, 339, 354, 382-386, 448-450, 489
Williamson matrix, array, type, 325, 329, 391
Witt, 431
Wright, E.M., 482

Y

Yang, C.H., 395-398, 489
Yap, H.P., 246, 266
Yates, F., 480

Z

Zame, Alan, 489
Zierler, N., 489
Znám, S., 167

# Lecture Notes in Mathematics

**Comprehensive leaflet on request**

Vol. 111: K. H. Mayer, Relationen zwischen charakteristischen Zahlen. III, 99 Seiten. 1969. DM 16,–

Vol. 112: Colloquium on Methods of Optimization. Edited by N. N. Moiseev. IV, 293 pages. 1970. DM 18,–

Vol. 113: R. Wille, Kongruenzklassengeometrien. III, 99 Seiten. 1970. DM 16,–

Vol. 114: H. Jacquet and R. P. Langlands, Automorphic Forms on GL (2). VII, 548 pages. 1970. DM 24,–

Vol. 115: K. H. Roggenkamp and V. Huber-Dyson, Lattices over Orders I. XIX, 290 pages. 1970. DM 18,–

Vol. 116: Séminaire Pierre Lelong (Analyse) Année 1969. IV, 195 pages. 1970. DM 16,–

Vol. 117: Y. Meyer, Nombres de Pisot, Nombres de Salem et Analyse Harmonique. 63 pages. 1970. DM 16,–

Vol. 118: Proceedings of the 15th Scandinavian Congress, Oslo 1968. Edited by K. E. Aubert and W. Ljunggren. IV, 162 pages. 1970. DM 16,–

Vol. 119: M. Raynaud, Faisceaux amples sur les schémas en groupes et les espaces homogènes. III, 219 pages. 1970. DM 16,–

Vol. 120: D. Siefkes, Büchi's Monadic Second Order Successor Arithmetic. XII, 130 Seiten. 1970. DM 16,–

Vol. 121: H. S. Bear, Lectures on Gleason Parts. III, 47 pages. 1970. DM 16,–

Vol. 122: H. Zieschang, E. Vogt und H.-D. Coldewey, Flächen und ebene diskontinuierliche Gruppen. VIII, 203 Seiten. 1970. DM 16,–

Vol. 123: A. V. Jategaonkar, Left Principal Ideal Rings. VI, 145 pages. 1970. DM 16,–

Vol. 124: Séminare de Probabilités IV. Edited by P. A. Meyer. IV, 282 pages. 1970. DM 20,–

Vol. 125: Symposium on Automatic Demonstration. V, 310 pages. 1970. DM 20,–

Vol. 126: P. Schapira, Théorie des Hyperfonctions. XI, 157 pages. 1970. DM 16,–

Vol. 127: I. Stewart, Lie Algebras. IV, 97 pages. 1970. DM 16,–

Vol. 128: M. Takesaki, Tomita's Theory of Modular Hilbert Algebras and its Applications. II, 123 pages. 1970. DM 16,–

Vol. 129: K. H. Hofmann, The Duality of Compact Semigroups and C*-Bigebras. XII, 142 pages. 1970. DM 16,–

Vol. 130: F. Lorenz, Quadratische Formen über Körpern. II, 77 Seiten. 1970. DM 16,–

Vol. 131: A Borel et al., Seminar on Algebraic Groups and Related Finite Groups. VII, 321 pages. 1970. DM 22,–

Vol. 132: Symposium on Optimization. III, 348 pages. 1970. DM 22,–

Vol. 133: F. Topsøe, Topology and Measure. XIV, 79 pages. 1970. DM 16,–

Vol. 134: L. Smith, Lectures on the Eilenberg-Moore Spectral Sequence. VII, 142 pages. 1970. DM 16,–

Vol. 135: W. Stoll, Value Distribution of Holomorphic Maps into Compact Complex Manifolds. II, 267 pages. 1970. DM 18,–

Vol. 136: M. Karoubi et al., Séminaire Heidelberg-Saarbrücken-Strasbourg sur la K-Théorie. IV, 264 pages. 1970. DM 18,–

Vol. 137: Reports of the Midwest Category Seminar IV. Edited by S. MacLane. III, 139 pages. 1970. DM 16,–

Vol. 138: D. Foata et M. Schützenberger, Théorie Géométrique des Polynômes Eulériens. V, 94 pages. 1970. DM 16,–

Vol. 139: A. Badrikian, Séminaire sur les Fonctions Aléatoires Linéaires et les Mesures Cylindriques. VII, 221 pages. 1970. DM 18,–

Vol. 140: Lectures in Modern Analysis and Applications II. Edited by C. T. Taam. VI, 119 pages. 1970. DM 16,–

Vol. 141: G. Jameson, Ordered Linear Spaces. XV, 194 pages. 1970. DM 16,–

Vol. 142: K. W. Roggenkamp, Lattices over Orders II. V, 388 pages. 1970. DM 22,–

Vol. 143: K. W. Gruenberg, Cohomological Topics in Group Theory. XIV, 275 pages. 1970. DM 20,–

Vol. 144: Seminar on Differential Equations and Dynamical Systems, II. Edited by J. A. Yorke. VIII, 268 pages. 1970. DM 20,–

Vol. 145: E. J. Dubuc, Kan Extensions in Enriched Category Theory. XVI, 173 pages. 1970. DM 16,–

Vol. 146: A. B. Altman and S. Kleiman, Introduction to Grothendieck Duality Theory. II, 192 pages. 1970. DM 18,–

Vol. 147: D. E. Dobbs, Cech Cohomological Dimensions for Commutative Rings. VI, 176 pages. 1970. DM 16,–

Vol. 148: R. Azencott, Espaces de Poisson des Groupes Localement Compacts. IX, 141 pages. 1970. DM 16,–

Vol. 149: R. G. Swan and E. G. Evans, K-Theory of Finite Groups and Orders. IV, 237 pages. 1970. DM 20,–

Vol. 150: Heyer, Dualität lokalkompakter Gruppen. XIII, 372 Seiten. 1970. DM 20,–

Vol. 151: M. Demazure et A. Grothendieck, Schémas en Groupes I. (SGA 3). XV, 562 pages. 1970. DM 24,–

Vol. 152: M. Demazure et A. Grothendieck, Schémas en Groupes II. (SGA 3). IX, 654 pages. 1970. DM 24,–

Vol. 153: M. Demazure et A. Grothendieck, Schémas en Groupes III. (SGA 3). VIII, 529 pages. 1970. DM 24,–

Vol. 154: A. Lascoux et M. Berger, Variétés Kähleriennes Compactes. VII, 83 pages. 1970. DM 16,–

Vol. 155: Several Complex Variables I, Maryland 1970. Edited by J. Horváth. IV, 214 pages. 1970. DM 18,–

Vol. 156: R. Hartshorne, Ample Subvarieties of Algebraic Varieties. XIV, 256 pages. 1970. DM 20,–

Vol. 157: T. tom Dieck, K. H. Kamps und D. Puppe, Homotopietheorie. VI, 265 Seiten. 1970. DM 20,–

Vol. 158: T. G. Ostrom, Finite Translation Planes. IV. 112 pages. 1970. DM 16,–

Vol. 159: R. Ansorge und R. Hass. Konvergenz von Differenzenverfahren für lineare und nichtlineare Anfangswertaufgaben. VIII, 145 Seiten. 1970. DM 16,–

Vol. 160: L. Sucheston, Constributions to Ergodic Theory and Probability. VII, 277 pages. 1970. DM 20,–

Vol. 161: J. Stasheff, H-Spaces from a Homotopy Point of View. VI, 95 pages. 1970. DM 16,–

Vol. 162: Harish-Chandra and van Dijk, Harmonic Analysis on Reductive p-adic Groups. IV, 125 pages. 1970. DM 16,–

Vol. 163: P. Deligne, Equations Différentielles à Points Singuliers Reguliers. III, 133 pages. 1970. DM 16,–

Vol. 164: J. P. Ferrier, Seminaire sur les Algebres Complètes. II, 69 pages. 1970. DM 16,–

Vol. 165: J. M. Cohen, Stable Homotopy. V, 194 pages. 1970. DM 16,–

Vol. 166: A. J. Silberger, $PGL_2$ over the p-adics: its Representations, Spherical Functions, and Fourier Analysis. VII, 202 pages. 1970. DM 18,–

Vol. 167: Lavrentiev, Romanov and Vasiliev, Multidimensional Inverse Problems for Differential Equations. V, 59 pages. 1970. DM 16,–

Vol. 168: F. P. Peterson, The Steenrod Algebra and its Applications: A conference to Celebrate N. E. Steenrod's Sixtieth Birthday. VII, 317 pages. 1970. DM 22,–

Vol. 169: M. Raynaud, Anneaux Locaux Henséliens. V, 129 pages. 1970. DM 16,–

Vol. 170: Lectures in Modern Analysis and Applications III. Edited by C. T. Taam. VI, 213 pages. 1970. DM 18,–

Vol. 171: Set-Valued Mappings, Selections and Topological Properties of $2^X$. Edited by W. M. Fleischman. X, 110 pages. 1970. DM 16,–

Vol. 172: Y.-T. Siu and G. Trautmann, Gap-Sheaves and Extension of Coherent Analytic Subsheaves. V, 172 pages. 1971. DM 16,–

Vol. 173: J. N. Mordeson and B. Vinograde, Structure of Arbitrary Purely Inseparable Extension Fields. IV, 138 pages. 1970. DM 16,–

Vol. 174: B. Iversen, Linear Determinants with Applications to the Picard Scheme of a Family of Algebraic Curves. VI, 69 pages. 1970. DM 16,–

Vol. 175: M. Brelot, On Topologies and Boundaries in Potential Theory. VI, 176 pages. 1971. DM 18,–

Vol. 176: H. Popp, Fundamentalgruppen algebraischer Mannigfaltigkeiten. IV, 154 Seiten. 1970. DM 16,–

Vol. 177: J. Lambek, Torsion Theories, Additive Semantics and Rings of Quotients. VI, 94 pages. 1971. DM 16,–

Please turn over